Geophysical Monograph Series

Geophysical Monograph Series

234 **Pre-Earthquake Processes: A Multidisciplinary Approach to Earthquake Prediction Studies** *Dimitar Ouzounov, Sergey Pulinets, Katsumi Hattori, and Patrick Taylor (Eds.)*
235 **Electric Currents in Geospace and Beyond** *Andreas Keiling, Octav Marghitu, and Michael Wheatland (Eds.)*
236 **Quantifying Uncertainty in Subsurface Systems** *Celine Scheidt, Lewis Li, and Jef Caers (Eds.)*
237 **Petroleum Engineering** *Moshood Sanni (Ed.)*
238 **Geological Carbon Storage: Subsurface Seals and Caprock Integrity** *Stephanie Vialle, Jonathan Ajo-Franklin, and J. William Carey (Eds.)*
239 **Lithospheric Discontinuities** *Huaiyu Yuan and Barbara Romanowicz (Eds.)*
240 **Chemostratigraphy Across Major Chronological Eras** *Alcides N.Sial, Claudio Gaucher, Muthuvairavasamy Ramkumar, and Valderez Pinto Ferreira (Eds.)*
241 **Mathematical Geoenergy: Discovery, Depletion, and Renewal** *Paul Pukite, Dennis Coyne, and Daniel Challou (Eds.)*
242 **Ore Deposits: Origin, Exploration, and Exploitation** *Sophie Decree and Laurence Robb (Eds.)*
243 **Kuroshio Current: Physical, Biogeochemical and Ecosystem Dynamics** *Takeyoshi Nagai, Hiroaki Saito, Koji Suzuki, and Motomitsu Takahashi (Eds.)*
244 **Geomagnetically Induced Currents from the Sun to the Power Grid** *Jennifer L. Gannon, Andrei Swidinsky, and Zhonghua Xu (Eds.)*
245 **Shale: Subsurface Science and Engineering** *Thomas Dewers, Jason Heath, and Marcelo Sánchez (Eds.)*
246 **Submarine Landslides: Subaqueous Mass Transport Deposits From Outcrops to Seismic Profiles** *Kei Ogata, Andrea Festa, and Gian Andrea Pini (Eds.)*
247 **Iceland: Tectonics, Volcanics, and Glacial Features** *Tamie J. Jovanelly*
248 **Dayside Magnetosphere Interactions** *Qiugang Zong, Philippe Escoubet, David Sibeck, Guan Le, and Hui Zhang (Eds.)*
249 **Carbon in Earth's Interior** *Craig E. Manning, Jung-Fu Lin, and Wendy L. Mao (Eds.)*
250 **Nitrogen Overload: Environmental Degradation, Ramifications, and Economic Costs** *Brian G. Katz*
251 **Biogeochemical Cycles: Ecological Drivers and Environmental Impact** *Katerina Dontsova, Zsuzsanna Balogh-Brunstad, And Gaël Le Roux (Eds.)*
252 **Seismoelectric Exploration: Theory, Experiments, and Applications** *Niels Grobbe, André Revil, Zhenya Zhu, and Evert Slob (Eds.)*
253 **El Niño Southern Oscillation in a Changing Climate** *Michael J. McPhaden, Agus Santoso, and Wenju Cai (Eds.)*
254 **Dynamic Magma Evolution** *Francesco Vetere (Ed.)*
255 **Large Igneous Provinces: A Driver of Global Environmental and Biotic Changes** *Richard. E. Ernst, Alexander J. Dickson, and Andrey Bekker (Eds.)*
256 **Coastal Ecosystems in Transition: A Comparative Analysis of the Northern Adriatic and Chesapeake Bay** *Thomas C. Malone, Alenka Malej, and Jadran Faganeli (Eds.)*
257 **Hydrogeology, Chemical Weathering, and Soil Formation** *Allen Hunt, Markus Egli, and Boris Faybishenko (Eds.)*
258 **Solar Physics and Solar Wind** *Nour E. Raouafi and Angelos Vourlidas (Eds.)*
259 **Magnetospheres in the Solar System** *Romain Maggiolo, Nicolas André, Hiroshi Hasegawa, and Daniel T. Welling (Eds.)*
260 **Ionosphere Dynamics and Applications** *Chaosong Huang and Gang Lu (Eds.)*
261 **Upper Atmosphere Dynamics and Energetics** *Wenbin Wang and Yongliang Zhang (Eds.)*
262 **Space Weather Effects and Applications** *Anthea J. Coster, Philip J. Erickson, and Louis J. Lanzerotti (Eds.)*
263 **Mantle Convection and Surface Expressions** *Hauke Marquardt, Maxim Ballmer, Sanne Cottaar, and Jasper Konter (Eds.)*
264 **Crustal Magmatic System Evolution: Anatomy, Architecture, and Physico-Chemical Processes** *Matteo Masotta, Christoph Beier, and Silvio Mollo (Eds.)*
265 **Global Drought and Flood: Observation, Modeling, and Prediction** *Huan Wu, Dennis P. Lettenmaier, Qiuhong Tang, and Philip J. Ward (Eds.)*
266 **Magma Redox Geochemistry** *Roberto Moretti and Daniel R. Neuville (Eds.)*
267 **Wetland Carbon and Environmental Management** *Ken W. Krauss, Zhiliang Zhu, and Camille L. Stagg (Eds.)*
268 **Distributed Acoustic Sensing in Geophysics: Methods and Applications** *Yingping Li, Martin Karrenbach, and Jonathan B. Ajo-Franklin (Eds.)*
269 **Congo Basin Hydrology, Climate, and Biogeochemistry: A Foundation for the Future (English version)** *Raphael M. Tshimanga, Guy D. Moukandi N'kaya, and Douglas Alsdorf (Eds.)*
269 **Hydrologie, climat et biogéochimie du bassin du Congo: une base pour l'avenir (version française)** *Raphael M. Tshimanga, Guy D. Moukandi N'kaya, et Douglas Alsdorf (Éditeurs)*
270 **Muography: Exploring Earth's Subsurface with Elementary Particles** *László Oláh, Hiroyuki K. M. Tanaka, and Dezso˝ Varga (Eds.)*
271 **Remote Sensing of Water-Related Hazards** *Ke Zhang, Yang Hong, and Amir AghaKouchak (Eds.)*
272 **Geophysical Monitoring for Geologic Carbon Storage** *Lianjie Huang (Ed.)*
273 **Isotopic Constraints on Earth System Processes** *Kenneth W. W. Sims, Kate Maher, and Daniel P. Schrag (Eds.)*
274 **Earth Observation Applications and Global Policy Frameworks** *Argyro Kavvada, Douglas Cripe, and Lawrence Friedl (Eds.)*
275 **Threats to Springs in a Changing World: Science and Policies for Protection** *Matthew J. Currell and Brian G. Katz (Eds.)*
276 **Core-Mantle Co-Evolution: An Interdisciplinary Approach** *Takashi Nakagawa, Madhusoodhan Satish-Kumar, Taku Tsuchiya, and George Helffrich (Eds.)*
277 **Compressional Tectonics: Plate Convergence to Mountain Building (Tectonic Processes: A Global View, Volume 1)** *Elizabeth J. Catlos and Ibrahim Çemen (Eds.)*
278 **Extensional Tectonics: Continental Breakup to Formation of Oceanic Basins (Tectonic Processes: A Global View, Volume 2)** *Ibrahim Çemen and Elizabeth J. Catlos (Eds.)*
279 **Strike-Slip Tectonics: Oceanic Transform Faults to Continental Plate Boundaries (Tectonic Processes: A Global View, Volume 3)** *Ibrahim Çemen and Elizabeth J. Catlos (Eds.)*
280 **Landscape Fire, Smoke, and Health: Linking Biomass Burning Emissions to Human Well-Being** *Tatiana V. Loboda, Nancy H. F. French, and Robin C. Puett (Eds.)*
281 **Clouds and Their Climatic Impacts: Radiation, Circulation, and Precipitation** *Sylvia Sullivan and Corinna Hoose (Eds.)*
282 **Fast Processes in Large-Scale Atmospheric Models: Progress, Challenges, and Opportunities** *Yangang Liu and Pavlos Kollias (Eds.)*
283 **Helicities in Geophysics, Astrophysics, and Beyond** *Kirill Kuzanyan, Nobumitsu Yokoi, Manolis K. Georgoulis, and Rodion Stepanov (Eds.)*

Geophysical Monograph 283

Helicities in Geophysics, Astrophysics, and Beyond

Kirill Kuzanyan
Nobumitsu Yokoi
Manolis K. Georgoulis
Rodion Stepanov

Editors

This Work is a co-publication of the American Geophysical Union and John Wiley and Sons, Inc.

AGU
ADVANCING EARTH
AND SPACE SCIENCES

WILEY

This edition first published 2024

Published under the aegis of the AGU Publications Committee

Matthew Giampoala, Vice President, Publications
Carol Frost, Chair, Publications Committee
For details about the American Geophysical Union visit us at www.agu.org.

The right of Kirill Kuzanyan, Nobumitsu Yokoi, Manolis K. Georgoulis, and Rodion Stepanov to be identified as the editors of this work has been asserted in accordance with law.

Registered Office
John Wiley & Sons, Inc., 111 River Street, Hoboken, NJ 07030, USA

Editorial Office
111 River Street, Hoboken, NJ 07030, USA

For details of our global editorial offices, customer services, and more information about Wiley products visit us at www.wiley.com.

Wiley also publishes its books in a variety of electronic formats and by print-on-demand. Some content that appears in standard print versions of this book may not be available in other formats.

Library of Congress Cataloging-in-Publication Data

Names: Kuzanyan, Kirill, editor. | Yokoi, Nobumitsu, editor. | Georgoulis, Manolis K., editor. | Stepanov, Rodion, editor.
Title: Helicities in geophysics, astrophysics, and beyond / editors Kirill Kuzanyan, Nobumitsu Yokoi, Manolis K. Georgoulis, Rodion Stepanov.
Description: Hoboken, NJ : Wiley, 2024. | Includes bibliographical references and index.
Identifiers: LCCN 2023035109 (print) | LCCN 2023035110 (ebook) | ISBN 9781119841685 (hardback) | ISBN 9781119841692 (adobe pdf) | ISBN 9781119841708 (epub)
Subjects: LCSH: Particles (Nuclear physics)–Helicity. | Geophysics. | Astrophysics.
Classification: LCC QC793.3.H44 H45 2024 (print) | LCC QC793.3.H44 (ebook) | DDC 521–dc23/eng/20231017
LC record available at https://lccn.loc.gov/2023035109
LC ebook record available at https://lccn.loc.gov/2023035110

Cover Design: Wiley
Cover Image: © AceClipart_Etsy/Pixabay; Courtesy of Nobumitsu Yokoi

Set in 10/12pt TimesNewRomanMTStd by Straive, Chennai, India

CONTENTS

LIST OF CONTRIBUTORS

Arron N. Bale
Department of Mathematical Sciences
Durham University
Durham, UK

Mitchell A. Berger
Department of Mathematics
University of Exeter
Exeter, UK

Philippa K. Browning
Department of Physics and Astronomy
University of Manchester
Manchester, UK

Simon Candelaresi
School of Mathematics and Statistics
University of Glasgow
Glasgow, UK

Jie Chen
National Astronomical Observatories
Chinese Academy of Sciences
Beijing, China

Peng-Fei Chen
Key Lab of Modern Astronomy & Astrophysics
School of Astronomy and Space Science
Nanjing University
Nanjing, China

Otto Chkhetiani
A. M. Obukhov Institute of Atmospheric Physics
Russian Academy of Sciences
Moscow, Russia

Peter Davidson
Department of Engineering
University of Cambridge
Cambridge, UK

Fabio Del Sordo
Institute of Space Sciences
Barcelona, Spain

and

Catania Astrophysical Observatory
National Institute for Astrophysics
Catania, Italy

Maxim Dvornikov
Pushkov Institute of Terrestrial Magnetism, Ionosphere and Radiowave Propagation (IZMIRAN)
Russian Academy of Sciences
Moscow, Russia

Mykola Gordovskyy
Department of Physics and Astronomy
University of Manchester
Manchester, UK

Alan W. Hood
School of Mathematics and Statistics
University of St Andrews
St Andrews, UK

Nathan Kleeorin
Department of Mechanical Engineering
Ben-Gurion University of the Negev
Beer-Sheva, Israel

and

Institute of Continuous Media Mechanics
Russian Academy of Sciences
Perm, Russia

Michael Kurgansky
A. M. Obukhov Institute of Atmospheric Physics
Russian Academy of Sciences
Moscow, Russia

Kirill Kuzanyan
Pushkov Institute of Terrestrial Magnetism, Ionosphere and Radiowave Propagation (IZMIRAN)
Russian Academy of Sciences
Moscow, Russia

and

Institute of Continuous Media Mechanics
Russian Academy of Sciences
Perm, Russia

Jihong Liu
National Astronomical Observatories
Chinese Academy of Sciences
Beijing, China

and

School of Science
Shijiazhuang University
Shijiazhuang, China

David MacTaggart
School of Mathematics and Statistics
University of Glasgow
Glasgow, UK

Wolf-Christian Müller
Center for Astronomy and Astrophysics
Technical University of Berlin
Berlin, Germany

and

Max-Planck/Princeton Center for Plasma Physics
Princeton, NJ, USA

Yasuhito Narita
Institute for Theoretical Physics
Technical University of Braunschweig
Braunschweig, Germany

Sung-Hong Park
Korea Astronomy and Space Science Institute
Daejeon, Republic of Korea

and

Institute for Space-Earth Environmental Research
Nagoya University
Nagoya, Japan

and

W. W. Hansen Experimental Physics Laboratory
Stanford University
Stanford, CA, USA

Christopher Prior
Department of Mathematical Sciences
Durham University
Durham, UK

Avishek Ranjan
Department of Mechanical Engineering
Indian Institute of Technology Bombay
Mumbai, India

Igor Rogachevskii
Department of Mechanical Engineering
Ben-Gurion University of the Negev
Beer-Sheva, Israel

and

Nordic Institute for Theoretical Physics
KTH Royal Institute of Technology and Stockholm University
Stockholm, Sweden

Nikolai Safiullin
Department of Information Security
Ural Federal University
Ekaterinburg, Russia

and

Institute of Continuous Media Mechanics
Russian Academy of Sciences
Perm, Russia

Victor B. Semikoz
Pushkov Institute of Terrestrial Magnetism, Ionosphere and Radiowave Propagation (IZMIRAN)
Russian Academy of Sciences
Moscow, Russia

Dmitry D. Sokoloff
Pushkov Institute of Terrestrial Magnetism, Ionosphere and Radiowave Propagation (IZMIRAN)
Russian Academy of Sciences
Moscow, Russia

and

Department of Physics
Moscow State University
Moscow, Russia

Jean-Mathieu Teissier
Center for Astronomy and Astrophysics
Technical University of Berlin
Berlin, Germany

Shin Toriumi
Institute of Space and Astronautical Science
Japan Aerospace Exploration Agency
Sagamihara, Japan

Haiqing Xu
National Astronomical Observatories
Chinese Academy of Sciences
Beijing, China

Shangbin Yang
National Astronomical Observatories
Chinese Academy of Sciences
Beijing, China

and

School of Astronomy and Space Science
University of Chinese Academy of Sciences
Beijing, China

Xiao Yang
National Astronomical Observatories
Chinese Academy of Sciences
Beijing, China

Anthony R. Yeates
Department of Mathematical Sciences
Durham University
Durham, UK

Nobumitsu Yokoi
Institute of Industrial Science
University of Tokyo
Tokyo, Japan

Hongqi Zhang
National Astronomical Observatories
Chinese Academy of Sciences
Beijing, China

PREFACE

Helicities, defined by the volume integral of the inner product of a vector field and its curled counterpart in a generalized sense (including cross helicity and generalized helicity in Hall magnetohydrodynamics [MHD]), are known to play essential roles in several geophysical, astrophysical, and space plasma phenomena, from dynamo actions in geo/planetary magnetism and astrophysical objects to solar and stellar magnetic field evolution, mass condensation in star-forming regions, accreting jet formation near compact massive objects, amplification of the magnetic field in the Universe, magnetic confinement in fusion plasmas, etc. Aligning with these important roles in various research fields, helicities have been studied using various methods from different viewpoints.

This book is a synopsis of recent developments and achievements in helicity studies by leading scientists. It grew from a series of gatherings entitled "Online Advanced Study Program on Helicities in Astrophysics and Beyond" (https://helicity2020.izmiran.ru) in Fall 2020 and Spring 2021, which were held online as a substitute for in-person interactions within the research community during the Covid-19 pandemic.

These gatherings built on years of scientific studies and previous events. Magnetic helicity has been intensively studied in recent decades from observational, theoretical, and modeling viewpoints in fields such as general astrophysics, solar physics, plasma and fluid dynamics, and pure mathematics. A series of focused events on helicity have also taken place, such as the 1998 AGU Chapman Conference in Boulder, Colorado, USA; the 2009 and 2013 Helicity Thinkshops in China (https://sun10.bao.ac.cn/old/meetings/HT2009/ and https://sun10.bao.ac.cn/old/meetings/HT2013/); the 2017 Helicity Thinkshop in Japan (http://www.iis.u-tokyo.ac.jp/~nobyokoi/thinkshop/), part of the SEIKEN Symposium (https://science-media.org/conference/23); and the Program on Solar Helicities in Theory and Observations at NORDITA in 2019 (https://indico.fysik.su.se/event/6548/), among others.

The Online Advanced Study Program in 2020 and 2021 was hosted by IZMIRAN, the Russian Academy of Sciences, in Moscow. It featured seminars delivered by international experts in astrophysics, geophysics, and many fields of natural sciences involving observational and theoretical studies of magnetic, kinetic, and other helicities. We deliberately took an interdisciplinary and multidisciplinary approach to attract broader audiences. Some events exceeded 200 online attendees, and we thank all the participants for their energy and for directly and indirectly contributing to the contents of this book.

This book has two primary aims. The first is to provide a perspective on helicities relevant to geophysics, astrophysics, physical and space plasma sciences (including biological and quantum fluids). The second is to propose future directions for helicity studies in these fields using cohesive theoretical, observational, experimental, and numerical strategies for constructing models applicable to real-world phenomena. Compact, readable introductions in each chapter acquaint the reader with advanced topics and aspects of helicity studies.

We could not cover all topics in the natural sciences that involve helicities, but we have tried to give a reader the broadest possible representation of these fields. Certain important topics are lacking, such as the possible detection of magnetic helicity proxies in observable fast-rotating stars, a perspective on kinetic and magnetic helicities in the Earth's core, and laboratory fluid and plasma experiments. We hope such topics will be covered in future publications and books.

The book consists of 16 chapters divided into three parts. Part I discusses helicity basics and fundamental concepts. In Chapter 1, Anthony Yeates and Mitchell Berger propose that field-line helicity provides a finer local topological description of magnetic flux than the usual global magnetic helicity integral, with invariant properties preserved. They present a way to appropriately define field-line helicity in different volumes. They also discuss the time evolution of field-line helicity under both boundary motions and magnetic reconnection. In Chapter 2, David MacTaggart introduces the notion of magnetic winding from a theoretical perspective. Magnetic winding is a renormalization of magnetic helicity, directly measuring field-line topology. This new notion is complemented by an application to observations of solar active regions. In Chapter 3, Nobumitsu Yokoi constructs a turbulent transport model including helicity. The model reveals that the helical contribution may suppress momentum transport with a remarkable feature on the induction of a large-scale flow caused by an inhomogeneous coupling between helicity and rotation. This flow appears to have numerous applications to astro- and geophysical phenomena.

Part II contains several reviews of manifestations of helicities in various natural phenomena and their observations. Chapter 4, by Hongqi Zhang et al., reviews longstanding diagnostic tools available for inferring proxies of both the magnetic and current helicity in the solar atmosphere. They analyze solar atmospheric helicity at short and long timescales, in the latter showing analogs of the butterfly diagram for sunspots using the mean current helicity and twist, that have important implications for the solar dynamo. Chapter 5, by Peng-Fei Chen, casts doubt on how one can observationally infer the chirality of solar filaments by observing the skewness or bearing of the filament barbs. While this suggestion is gaining traction, it is far from unanimously accepted in the solar physics community but is eloquently presented to spur discussion and debate. Chapter 6, by Shin Toriumi and Sung-Hong Park, discusses various diagnostic tools for magnetic helicity in local (i.e., active-region) solar scales, aiming to understand and ultimately predict solar flares. Describing immense magnetic complexity, they make a twofold effort to distinguish populations of flaring active regions from the majority pool of non-flaring regions in order to understand the separator(s) of these populations and then use this knowledge for prediction purposes. Chapter 7, by Yasuhito Narita, reviews methods of evaluating magnetic helicity in the solar wind in terms of both single- and multipoint measurements. Methodologies invariably aim to resolve the transport problems of helicity, magnetic flux, and energy of the Sun into the heliosphere. On short (compared to magnetohydrodynamic turbulence) spatial scales in the ion-kinetic domain, the observations reveal nonzero helicity as a signature of linear-mode wave excitation, such as the kinetic Alfven waves and whistler waves. This differs in each solar hemisphere and varies with the solar cycle. In Chapter 8, Maxim Dvornikov reviews the role of magnetic helicity evolution in rotating neutron stars. He utilizes the conservation law for the sum of the chiral imbalance of charged particle densities and the density of magnetic helicity and explores the possibility of X-ray or gamma bursts observed in magnetars due to this mechanism. He argues that the quantum contribution dominates the classical contribution in the surface terms in standard MHD but only for neutron stars with rigid rotation. He shows that the characteristic time of the helicity change is in accord with the magnetic cycle period of certain pulsars. Chapter 9, by Christopher Prior and Arron Bale, deals with writhing and its prospects for wider interdisciplinary applications and interaction between biophysics, solar physics, and other disciplines. Writhing quantifies a structure's global self-entanglement (knotting) and plays a fundamental role in DNA compactification (supercoiling). Earlier results on magnetic helicity in solar physics can be used for biophysical applications such as understanding protein structures through their writhing measures. Chapter 10, by Otto Chkhetiani and Michael Kurgansky, explores kinetic helicity in the Earth's atmosphere and its role in atmospheric turbulence. The helicity balance and fluxes are used to analyze atmospheric vortices such as tropical cyclones, tornadoes, dust devils, and Ekman boundary layer dynamics. The helical properties of turbulence within the atmospheric boundary layer have been probed by direct pioneering measurements of turbulent helicity in natural atmospheric conditions.

Part III on theoretical and numerical modeling of helicities opens with Chapter 11 by Victor Semikoz and Dmitry Sokoloff, who review cosmological dynamos and explore the P-noninvariance of elementary particles. This presents a possibility for nonclassical dynamo generation in the early Universe based on the intrinsic mirror asymmetry of elementary particles. In Chapter 12, Simon Candelaresi and Fabio Del Sordo point out that magnetic helicity constrains the dynamics of plasmas. They discuss how magnetic helicity stabilizes the plasma and prevents its disruption, with reference to observations, numerical experiments, and analytical results. Several illustrative examples in the solar corona are presented, as well as fusion devices, galactic and extragalactic medium, and extragalactic bubbles. Chapter 13, by Philippa Browning et al., addresses the contentious topic of magnetic relaxation in the solar corona and its implications for magnetic helicity. They arrive logically at the concept of magnetic relaxation, likely in terms of the Taylor hypothesis, and show how ideal instabilities and merging magnetic flux ropes in the corona can lead to relaxation. Implications for laboratory plasmas and the elusive coronal heating mechanism(s) are also discussed. Chapter 14, by Jean-Mathieu Teissier and Wolf-Christian Müller, deals with the formation and sustainment of magnetic structures in supersonic isothermal magnetohydrodynamic turbulence. They review the first results obtained through direct numerical simulations of isothermal compressible MHD at Mach numbers ranging from subsonic to about 10, finding contributions of the local and nonlocal, direct, and inverse transfer of magnetic helicity in supersonic MHD regimes. Chapter 15, by Nathan Kleeorin et al., discusses nonlinear mean-field dynamos, with special reference to various mechanisms for sunspot formation and the prediction of solar activity. Based on nonlinear dynamo equations, including the model equation for magnetic helicity, they explain existing observations of magnetic helicity in the Sun and dynamical solar activity. The contributions of magnetic helicity, large-scale magnetic fields, and differential rotation to the mean tilt angle of sunspot bipolar regions are also discussed. Finally, Chapter 16, by Avishek Ranjan and Peter Davidson, is dedicated to the origin of the spatial segregation of kinetic helicity in dynamo simulations. They discuss various sources of kinetic helicity, including helical waves such as inertial waves. Strong spatial correlations of the segregation pattern of helicity and the source term due to buoyancy exhibited in numerical simulations are interpreted using helical wave propagation.

Together, these chapters present different aspects of helicities in diverse scientific contexts, from basic concepts and fundamental properties to manifestations in several natural phenomena, as well as theoretical and numerical models. The book also presents the possibility of tackling real-world problems via the many forms and flavors of helicity. We hope the breadth of information and evidence presented in this book will spur discussions and debates that will lead to an enhanced, broader scientific understanding of complexity in various natural systems and networks. We trust that this volume contributes to further developments in this fascinating, challenging, and ever-evolving subject.

We are grateful to the organizations that have hosted various in-person and online gatherings relating to helicity studies, such as the National Astronomical Observatories of China, University of Tokyo in Japan, NORDITA in Sweden, and IZMIRAN in Russia. We thank all the participants of these meetings for their contributions and discussions, especially the speakers who gave talks, posters, and online presentations. We would also like to acknowledge the support and hospitality of the Isaac Newton Institute for Mathematical Sciences in Cambridge, UK, during the Dynamo Theory Programme (DYT2) held in Fall 2022, where the contents of this book were finalized.

Our thanks go to the AGU Books Editorial Board and the reviewers of our proposal, who encouraged us and helped improve this book. We acknowledge the dedication of each of the chapter contributors for the time they spent preparing their manuscripts. We greatly appreciate the patience and professionalism of those who reviewed chapters to ensure that this book was of the highest standards as well as to increase its readability and appeal to the community. We also acknowledge support from AGU Publications and Wiley, including technical assistance and advice from their staff. Finally, we thank our colleagues, friends, and families for their patience and inspiration while we worked on this project.

Kirill Kuzanyan
Pushkov Institute of Terrestrial Magnetism, Ionosphere, and Radio Wave Propagation
Russian Academy of Sciences, Moscow, Russia
Nobumitsu Yokoi
Institute of Industrial Science
University of Tokyo, Tokyo, Japan
Manolis K. Georgoulis
Research Center for Astronomy and Applied Mathematics
Academy of Athens, Athens, Greece
Rodion Stepanov
Institute of Continuous Media Mechanics
Russian Academy of Sciences, Perm, Russia

Part I

Helicity Essentials: Basic and Fundamental Concepts

1

Introduction to Field Line Helicity

Anthony R. Yeates[1] **and Mitchell A. Berger**[2]

ABSTRACT

Field line helicity measures the net linking of magnetic flux with a single magnetic field line. It offers a finer topological description than the usual global magnetic helicity integral while still being invariant in an ideal evolution unless there is a flux of helicity through the domain boundary. In this chapter, we explore how to appropriately define field line helicity in different volumes in a way that preserves a meaningful topological interpretation. We also review the time evolution of field line helicity under both boundary motions and magnetic reconnection.

1.1. DEFINITIONS OF FIELD LINE HELICITY

We briefly review topological measures of magnetic field structure. Some of these refer to the structure of the total field within a volume (the *magnetic helicity*), and others to the relationship between individual pairs of field lines (linking and winding). Field line helicity is intermediate between these ideas, as it measures the net linking or winding of one field line with the total field. Helicity integrals depend on both the tangling of field lines with each other and the topology and geometry of the volume in which the field lines reside. A strong rationale for considering topological measures is that field line topology is conserved in any ideal evolution of the field (Moffatt, 1969).

1.1.1. Definitions of Field Line Helicity for Closed Volumes

The simplest situation occurs with two closed field lines residing within a volume $\mathcal{V}$ where the field lines do not cross the boundary of $\mathcal{V}$ (or within an infinite space). Historically, Gauss (1809) discovered a double line integral, which measures the linking of two closed curves L_1 and L_2. Let positions on the curves be given by $\vec{x}(\sigma)$ and $\vec{y}(\tau)$. Then

$$\mathcal{L}_{12} = -\frac{1}{4\pi}\oint_{L_2}\oint_{L_1}\frac{\mathrm{d}\vec{x}}{\mathrm{d}\sigma}\cdot\frac{\vec{r}}{r^3}\times\frac{\mathrm{d}\vec{y}}{\mathrm{d}\tau}\,d\sigma\,d\tau, \qquad (1.1)$$

where $\vec{r} = \vec{x} - \vec{y}$. This may be calculated by counting signed crossings (see Fig. 1.1).

For a magnetic field consisting of a finite collection of closed magnetic flux tubes (a very special case), we can define an overall invariant

$$H_N = \sum_{i=1}^{N}\sum_{j=1}^{N}\mathcal{L}_{ij}\Phi_i\Phi_j, \qquad (1.2)$$

where Φ_i represents the magnetic flux of each tube and $\mathcal{L}_{ij}$ is the linking number between the tubes. This invariant is called the *magnetic helicity*. For a more general magnetic field consisting entirely of closed field lines (still a special case), we can take $N \to \infty$ in equation (1.2) so that the sums become integrals. Accounting for the magnetic flux, the tangent vectors in equation (1.1) become magnetic

[1] *Department of Mathematical Sciences, Durham University, Durham, UK*

[2] *Department of Mathematics, University of Exeter, Exeter, UK*

Helicities in Geophysics, Astrophysics, and Beyond, Geophysical Monograph 283, First Edition.
Edited by Kirill Kuzanyan, Nobumitsu Yokoi, Manolis K. Georgoulis, and Rodion Stepanov.

DOI:10.1002/9781119841715.ch01

Figure 1.1 For this link (the Whitehead link), the Gauss linking number, $\mathcal{L}_{12}$, is zero. In general, the linking number equals half the difference between the number of positive crossings and the number of negative crossings, as seen in any plane projection.

field vectors, and the magnetic helicity may be written

$$H = -\frac{1}{4\pi}\int_{\mathcal{V}}\int_{\mathcal{V}} \vec{B}(\vec{x}) \cdot \frac{\vec{r}}{r^3} \times B(\vec{y})\, \mathrm{d}^3x\, \mathrm{d}^3y. \quad (1.3)$$

Here the integral is over all space or over a volume containing all field lines. In fact, in this equation, we do not need to assume that the field lines close upon themselves – some may ergodically fill a subvolume or twist around a toroidal surface with an irrational winding number. Such fields can be constructed as the limit as $N \to \infty$ of a sequence of fields consisting of N thin closed flux tubes (Arnol'd and Khesin, 1998). Thus, regardless of whether the field lines close upon themselves, we can approximate any magnetic field as a collection of such tubes.

Returning to the finite set of N flux tubes with fluxes Φ_i, notice that equation (1.2) can be written as

$$H_N = \sum_{i=1}^{N} \mathcal{A}_i \Phi_i; \qquad \mathcal{A}_i = \sum_{j=1}^{N} \mathcal{L}_{ij}\Phi_j. \quad (1.4)$$

The limit as $N \to \infty$ of $\mathcal{A}_i$ is another topological measure, this time defined for every individual magnetic field line – this is what we call the *field line helicity*.

Since the field lines do not cross the boundary of the volume, we can now use the Biot-Savart formula

$$\vec{A}(\vec{x}) = \mathrm{curl}^{-1}\vec{B}(\vec{x}) \equiv -\frac{1}{4\pi}\int_{\mathcal{V}} \frac{\vec{r}}{r^3} \times \vec{B}(\vec{y})\, \mathrm{d}^3y \quad (1.5)$$

to find

$$H = \int_{\mathcal{V}} \vec{A} \cdot \vec{B}\, \mathrm{d}^3x. \quad (1.6)$$

Historically, invariance of the expression in equation (1.6) was known before the interpretation in terms of Gauss linking number was identified (Moffatt, 1969; Moffatt and Ricca, 1992). For a closed line L within the field, the equivalent expression for the field line helicity will simply be

$$\mathcal{A}(L) = \oint_L \vec{A} \cdot \mathrm{d}l. \quad (1.7)$$

This is just the net magnetic flux encircled by L, and since L is a closed curve, it is manifestly gauge invariant. We remark that Yahalom (2013) interprets $\mathcal{A}(L)$ as a magnetohydrodynamic analog of the Aharonov-Bohm effect from quantum mechanics: if $\vec{B} = \nabla\chi \times \nabla\eta$, then $\vec{A} = \chi\nabla\eta + \nabla\zeta$, so that a nonzero H requires jumps in ζ around closed field lines, with such jumps giving $\mathcal{A}(L)$.

Suppose our volume $\mathcal{V}$ is simply connected (e.g., a sphere, not a torus). Note that because L is closed, we can gauge transform $\vec{A} \to \vec{A} + \nabla\phi$ for an arbitrary gauge function ϕ without affecting $\mathcal{A}(L)$. If we have a multiply connected volume, a difficulty arises: the wrong choice of $\vec{A}$ may imply the existence of magnetic flux in the external region that threads through a hole, which makes $\oint_L \vec{A} \cdot \mathrm{d}l$ dependent on the unknown external field. To remedy this, one can restrict the gauge of $\vec{A}$ so that $\oint_L \vec{A} \cdot \mathrm{d}l = 0$ for any closed curve on the boundary encircling a hole the long way around (see also MacTaggart and Valli, 2019).

1.1.2. Definitions of Field Line Helicity for Open Volumes

We now go to volumes with flux crossing the boundary. Suppose a field line forms a loop, with both ends on the same boundary (as, for example, in Fig. 1.2b). An early approach (Antiochos, 1987) involves drawing a geodesic between the two endpoints, thus forming a closed curve for which $\oint \vec{A} \cdot \mathrm{d}l$ can be measured. This scheme does not integrate to the full helicity, however. We wish to define field line helicity to be consistent with full helicity but also so that gauge transformations of $\vec{A}$ do not change our results. There are two principal approaches to this problem. One consists of defining topological quantities such as winding numbers and then summing to find the helicities. If the boundary flux does not move, ideal motions of the field lines will not change the windings, so field line helicity will be conserved. If the boundary flux does move, the topological structure can cross the boundary, leading to changes in helicity measures. This will be discussed in section 1.2.

The second approach is to measure helicity *relative* to a potential field: i.e., a field with zero current in the volume. As the potential field is the minimum energy state for given boundary conditions, its structure depends only on the shape of the boundary and the distribution of flux on the boundary (and for multiply connected volumes, specified new fluxes around each hole). For highly symmetric

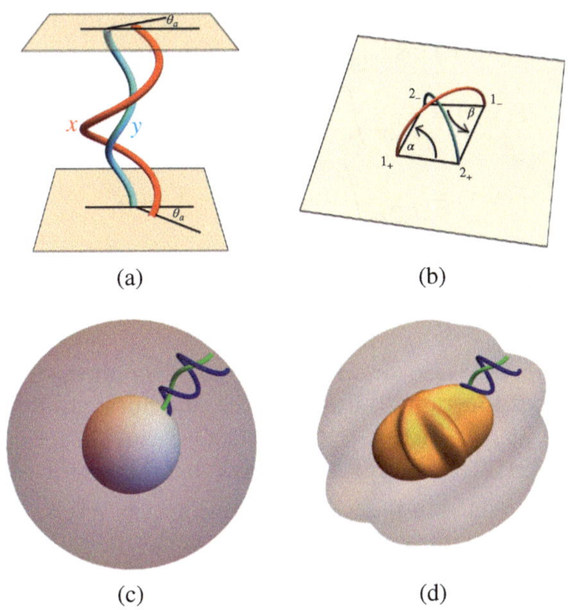

Figure 1.2 Examples of a pair of magnetic field lines in different domains, including (a) the volume between two parallel planes; (b) the volume above a plane; (c) a spherical shell (the volume between two concentric spheres); and (d) the volume between two more general surfaces.

volumes such as those bounded by planes or spheres, the two approaches give identical results.

For more complicated volumes, the shape of the boundary or boundaries can add extra terms that distinguish the two approaches, as discussed in section 1.1.2. A definition of field line helicity can also be based on relative helicity (Moraitis et al., 2019), to be discussed in section 1.1.3.

Volumes between Parallel Planes

Suppose the volume of interest $\mathcal{V}$ is the space between parallel planes, say $z = 0$ and $z = h$, as illustrated in Fig. 1.2a. We will first assume that all field lines start at $z = 0$ and end at $z = h$. Then no two field lines link in the Gauss sense; however, they will twist about each other through some angle $\delta\theta_{ij}$. The *winding number* makes this an integer for complete twists:

$$w_{ij} = \frac{1}{2\pi}\delta\theta_{ij}. \tag{1.8}$$

For the example in Fig. 1.2a, we have $w_{ij} = (\theta_b - \theta_a)/(2\pi) + 1$.

We can now adopt formulae for the helicity and field line helicity similar to the closed field case, with the winding number replacing the linking number. We denote horizontal positions by $\mathrm{x} = (x, y)$. Also, we distinguish positions in the bottom $x - y$ plane by $\tilde{\mathrm{x}} = (x, y, 0)$. Let L_i be the field line with foot point at $\tilde{\mathrm{x}}_i$. Also let $w(\tilde{\mathrm{x}}_i, \tilde{\mathrm{x}}_j)$ be the winding number between L_i and L_j. Then the field line helicity of L_i will be

$$\mathcal{A}(L_i) = \int w(\tilde{\mathrm{x}}_i, \tilde{\mathrm{x}}_j) B_z(\tilde{\mathrm{x}}_j)\, \mathrm{d}^2\tilde{\mathrm{x}}_j. \tag{1.9}$$

Let us build up $\mathcal{A}(L_i)$ one plane at a time – in other words, find expressions for $\mathrm{d}\mathcal{A}(L_i)/\mathrm{d}z$. For now, we make the simplifying assumption that $B_z > 0$ everywhere. There are two contributions to the z derivative: first (I), horizontal flux in the plane can wrap around the field line L_i. Second (II), L_i can move horizontally about other field lines. Now $\hat{\theta}_{ij} = \hat{z} \times \hat{r}_{ij}$ is the angular direction at x_j about x_i, with $\vec{r}_{ij} = \mathrm{x}_j - \mathrm{x}_i$. Also, a field line traveling from (r, θ, z) to $(r + \mathrm{d}r, \theta + \mathrm{d}\theta, z + \mathrm{d}z)$ satisfies the equation

$$\frac{\mathrm{d}r}{B_r} = \frac{r\mathrm{d}\theta}{B_\theta} = \frac{\mathrm{d}z}{B_z}. \tag{1.10}$$

Thus for a field line at $\mathrm{x}_j = (x, y, z)$ (writing $\vec{B}_j = B(\mathrm{x}_j)$, etc.),

$$\left(\frac{\mathrm{d}\theta_{ij}}{\mathrm{d}z}\right)_I = \frac{B_{\theta j}}{r_{ij} B_{zj}} = \frac{\vec{B}_j \cdot \hat{z} \times \vec{r}_{ij}}{r_{ij}^2 B_{zj}} = \hat{z} \cdot \frac{\vec{r}_{ij} \times \vec{B}_j}{r_{ij}^2 B_{zj}}. \tag{1.11}$$

Next, the field line L_i can move horizontally around other field lines. In this case,

$$\left(\frac{\mathrm{d}\theta_{ij}}{\mathrm{d}z}\right)_{II} = -\hat{z} \cdot \frac{\vec{r}_{ij} \times \vec{B}_i}{r_{ij}^2 B_{zi}} = \frac{\vec{B}_i}{B_{zi}} \cdot \frac{\vec{r}_{ij} \times B_{zj}\hat{z}}{r_{ij}^2 B_{zj}}. \tag{1.12}$$

Summing the last two equations, the total change in winding between lines L_i and L_j is

$$\frac{\mathrm{d}\theta_{ij}}{\mathrm{d}z} = \frac{\vec{B}_i}{B_{zi}} \cdot \frac{\vec{r}_{ij} \times \vec{B}_j}{r_{ij}^2 B_{zj}}. \tag{1.13}$$

Let S_z be the horizontal plane at height z. Suppose we add up contributions from all over the plane: we multiply the previous expression by $B_{zj}\mathrm{d}^2\mathrm{x}_j$ and integrate. Dividing by 2π to make complete turns into integers,

$$\frac{\mathrm{d}\mathcal{A}(L_i)}{\mathrm{d}z} = \frac{1}{2\pi}\frac{\vec{B}_i}{B_{zi}} \cdot \int_{S_z} \frac{\vec{r}_{ij} \times \vec{B}_j}{r_{ij}^2}\, \mathrm{d}^2\mathrm{x}_j. \tag{1.14}$$

This expression for the field line helicity can be expressed very simply using a special vector potential – the *winding gauge* (Prior and Yeates, 2014): for a point on field line L_i at (x, y, z),

$$\vec{A}^W(x, y, z) = \frac{1}{2\pi}\int_{S_z} \frac{\vec{r}_{ij} \times \vec{B}_j}{r_{ij}^2}\, \mathrm{d}^2\mathrm{x}_j. \tag{1.15}$$

We have

$$\mathcal{A}(L_i) = \int \vec{A}^W \cdot \mathrm{d}\mathbf{l} = \int \vec{A}^W \cdot \frac{\mathrm{d}\mathbf{l}}{\mathrm{d}z}\, \mathrm{d}z, \tag{1.16}$$

so

$$\frac{dA(L_i)}{dz} = \vec{A}^W \cdot \frac{\mathbf{dl}}{dz} = \vec{A}^W \cdot \frac{\vec{B}_i(x,y,z)}{B_{zi}}. \quad (1.17)$$

Volume above a Plane

In the previous section, we assumed that $B_z > 0$ everywhere. Thus the field lines were braided about each other, and simple winding numbers sufficed to measure pairwise entanglements. If we remove the restriction on B_z, field lines can go up and down. We will also remove the restriction of an upper plane in the following discussion, although that is not strictly necessary. Such a domain is illustrated in Fig. 1.2b.

If field lines can form loops that go up and down, we can measure topological structure in several ways. One method is to cut the field lines at any local minima or maxima where $B_z = 0$ and then sum the winding numbers as before (Berger and Prior, 2006). The winding gauge will still work in this situation, as the formulae for calculating $\vec{A}^W$ do not require $B_z > 0$.

A second method, when considering the winding between two loops, consists of examining the angles of the quadrilateral formed by the four foot points of the loops (Berger, 1986; Demoulin et al., 2006). Consider the upper half space $\{z > 0\}$, shown in Fig. 1.2b. The foot points of loops 1 and 2 are labeled 1^+, 1^-, 2^+, and 2^-, where $B_z > 0$ at 1^+ and 2^+. If the loops cross as seen from above, we assume that loop 1 is the upper loop. Consider the quadrilateral $1_+2_+1_-2_-$. Let α and β be the angles at vertices 1_+ and 1_-, respectively. Then (from considering the net change of helicity found when bringing the two loops in from infinity) Berger (1984) showed that

$$w_{12} = \frac{1}{2\pi}(\alpha + \beta). \quad (1.18)$$

Poloidal-Toroidal Representation

The winding gauge employed in the previous sections is equivalent to the vector potential found when employing a poloidal-toroidal representation for the field (Moffatt, 1978; Berger and Hornig, 2018). Because the magnetic field $\vec{B}$ is a three-vector subject to one condition ($\nabla \cdot \vec{B} = 0$), we can often express it in terms of two scalar functions. Here we let P and T be the *poloidal* and *toroidal* functions. Let the horizontal surface Laplacian be $\Delta_\| = \partial^2/\partial_x^2 + \partial^2/\partial_y^2$. In any plane $z =$ const., P is determined by the vertical flux,

$$\Delta_\| P = -B_z, \quad (1.19)$$

while T is determined by the vertical current,

$$\Delta_\| T = -J_z = -(\nabla \times \vec{B})_z. \quad (1.20)$$

We assume T, P, B_z, and J_z all vanish at infinity. Then the surface Laplacians have unique solutions,

$$P(x,y) = -\frac{1}{2\pi}\int_{S_z} B_z(\vec{x}')\ln|\vec{x} - \vec{x}'|\, d^2x'; \quad (1.21)$$

$$T(x,y) = -\frac{1}{2\pi}\int_{S_z} J_z(\vec{x}')\ln|\vec{x} - \vec{x}'|\, d^2x'. \quad (1.22)$$

We can now create a vector potential

$$\vec{A} = \nabla \times P\hat{z} + T\hat{z}. \quad (1.23)$$

Note that the horizontal components of $\vec{A}$ only involve the poloidal function P. Furthermore, the horizontal divergence of $\vec{A}$ vanishes: $\nabla_\| \cdot \vec{A}_\| = 0$. One can then employ equations (1.21) and (1.22) to show that this vector potential is identical to the winding gauge, $\vec{A}^W = \nabla \times P\hat{z} + T\hat{z}$, provided that the volume considered is infinite in the x and y directions. We can again employ this vector potential to calculate field line helicities. The total helicity can also be regarded as the net linking of toroidal and poloidal fields (Berger and Hornig, 2018),

$$H = 2\int_V \nabla \times (T\hat{z}) \cdot \nabla \times (P\hat{z})\, d^3x. \quad (1.24)$$

Volumes Bounded by a Sphere

Spherical boundaries – such as in Fig. 1.2c – can, for the most part, be treated the same way as planar boundaries, with the radial unit vector $\hat{r}$ replacing the vertical vector $\hat{z}$. In essence, winding angles become azimuthal angles. Suppose we consider three points A, B, and C on a sphere. Rotate the spherical coordinates so that B is at the North pole and A is at the azimuthal coordinate $\phi = 0$. Then the angle $\angle ABC$ equals the azimuthal coordinate of C.

The poloidal and toroidal flux functions are (Kimura and Okamoto, 1987)

$$P(\theta,\phi) = \frac{1}{4\pi}\int_S B_r(\vec{x}')\ln(1 - \cos\xi)\, d^2x'; \quad (1.25)$$

$$T(\theta,\phi) = \frac{1}{4\pi}\int_S J_r(\vec{x}')\ln(1 - \cos\xi)\, d^2x', \quad (1.26)$$

where

$$\cos\xi = \cos\theta\cos\theta' + \sin\theta\sin\theta'\cos(\phi - \phi') \quad (1.27)$$

is the spherical distance between (θ,ϕ) and (θ',ϕ'). With these functions,

$$\vec{A} = \nabla P \times \hat{r} + T\hat{r}; \qquad \vec{B} = \nabla \times (\nabla P \times \hat{r}) + \nabla T \times \hat{r}. \quad (1.28)$$

Some care must be taken in spherical geometries. One must ensure that no magnetic monopoles are hiding inside the sphere – i.e., the net flux through the sphere must vanish (similarly for the net current).

Also, in the special case of a spherical shell geometry where flux enters at the inner sphere and exits at the outer sphere (as in Fig. 1.2c), the separation of helicity into *self helicity* and *mutual helicity* can be ambiguous (Campbell and Berger, 2014). Self helicity measures the twist and writhe of individual tubes, while mutual helicity measures linking and intertwining between tubes. This will not affect the field line helicity, however.

More General Volumes

For volumes bounded by planes or spheres, there are natural definitions of *angle* and *winding angle* depending only on the Euclidean metric. The winding gauge, or the poloidal-toroidal vector potential, captures this natural definition. We could also consider a volume with boundaries consisting of a bottom plane and side boundaries, where flux only crosses the bottom plane. Such a volume could represent a closed active region. We could choose winding angles without regard to the side boundaries, using the same winding gauge. An alternative will be described later when using relative measures of helicity.

Next, consider a volume bounded by one or two simply connected surfaces that lack the symmetry of planes or spheres. An example is shown in Fig. 1.2d. Here, definitions of angle are less obvious. However, we can still employ a generalization of the poloidal-toroidal representation (Berger and Hornig, 2018) to define the winding gauge.

Let our volume $\mathcal{V}$ be sliced into nested surfaces. Employ coordinates (u, v, w), where $w =$ const. labels one of the nested surfaces. The toroidal field (which is responsible for the current perpendicular to surfaces of constant w) can be written in terms of a toroidal function T as before:

$$\vec{B}_T = \nabla \times \vec{A}_T; \qquad \vec{A}_T = T\nabla w. \tag{1.29}$$

However, the poloidal field is more complicated. Let the normal component of the curl be given by the operator $\mathcal{D}$,

$$\mathcal{D}\vec{V} = \hat{n} \cdot \nabla \times \vec{V}. \tag{1.30}$$

This operator has an inverse $\mathcal{D}^{-1}$. To make the inverse unique, we require that the inverse normal curl gives a divergence-free vector field parallel to the surface. Thus

$$\mathcal{D}^{-1}f \cdot \hat{n} = 0; \qquad \nabla \cdot \mathcal{D}^{-1}f = 0. \tag{1.31}$$

Let B_n be the normal magnetic field. Ordinarily, we would write the poloidal field as

$$\vec{A}'_P = \mathcal{D}^{-1}B_n; \qquad \vec{B}_P = \nabla \times \vec{A}'_P. \tag{1.32}$$

However, without spherical or planar symmetry, this magnetic field may have nonzero normal current J_n – but the toroidal field has already taken care of J_n. Thus (Berger and Hornig, 2018) we must add a *shape* term

$$\vec{A}_S = T_S\nabla w, \tag{1.33}$$

where

$$\nabla \times \nabla \times A_s \cdot \hat{n} = -\nabla \times \nabla \times \vec{A}'_P \cdot \hat{n}. \tag{1.34}$$

Putting it all together, the generalized winding gauge becomes

$$\vec{A}^W = \mathcal{D}^{-1}B_n + (T + T_S)\nabla w. \tag{1.35}$$

1.1.3. Relative Field Line Helicity

In the previous sections, helicity and field line helicity have been defined in terms of geometrical quantities such as winding and linking. The helicity of open fields in a volume $\mathcal{V}$ can also be defined in terms of how much the magnetic structure within $\mathcal{V}$ contributes to the helicity of all space (Berger and Field, 1984). Let the field inside $\mathcal{V}$ be called $\vec{B}_\mathcal{V}$. Also consider a potential (zero current) field $\vec{P}_\mathcal{V}$ inside $\mathcal{V}$ with the same boundary conditions $\vec{B}_\mathcal{V} \cdot \hat{n}|_S = \vec{P}_\mathcal{V} \cdot \hat{n}|_S$. A potential field minimizes the magnetic energy in the volume subject to the constraint of the given boundary flux distribution. Hence potential fields can be said to have the minimum structure (more accurately, their structure only depends on the boundary conditions, so they add zero *additional* structure to the field).

Here we compare the total helicity of all space with the helicity that would be obtained if, inside $\mathcal{V}$, $\vec{B}_\mathcal{V}$ were replaced by $\vec{P}_\mathcal{V}$. All integrals are gauge invariant, so the difference between the two helicities will also be gauge invariant. Let space external to $\mathcal{V}$ be $\mathcal{V}'$. Then the relative helicity $H_R(\mathcal{V})$ is

$$H_R(\mathcal{V}) = H(B_\mathcal{V}, B_{\mathcal{V}'}) - H(P_\mathcal{V}, B_{\mathcal{V}'}). \tag{1.36}$$

It is important to note that the calculation of $H_R(\mathcal{V})$ does not require knowledge of the external field $B_{\mathcal{V}'}$ (Berger and Field, 1984; Finn and Antonsen, 1985).

For symmetric volumes (i.e., boundaries consisting of a plane or parallel planes or one or two concentric spheres), the topological and relative definitions give the same answers (Berger and Field, 1984; Berger and Hornig, 2018). For more complicated geometries, they may differ. For example, suppose the volume $\mathcal{V}$ consists of a helical tube rising between parallel planes. Suppose we place a potential field both inside and outside $\mathcal{V}$. Then the helicity of all space will be nonzero because of the helical structure. Now add some axial current inside $\mathcal{V}$ so the field has an extra twist. Then the relative helicity inside $\mathcal{V}$ will only measure this extra twist because the helicity due to the axis writhe is subtracted with the potential term.

The relative helicity of a potential field is, by definition, zero. Suppose again that the boundary is planar or spherical. Then the topological helicity (summing winding numbers as in previous sections) will also be zero: the potential field is purely poloidal, so there is no linkage between toroidal and poloidal components. However, for a potential field, the field line helicities of individual field lines may not be zero, even if they sum to zero (Yeates, 2020). These potential field line helicities reflect irregularities and lack of mirror symmetry in the distribution of boundary flux (Bourdin and Brandenburg, 2018). Thus the field line helicity of the potential field can help characterize magnetic distributions, for example, in the solar photosphere.

This raises a question, however: if there are currents inside a volume, how do they change the field line helicities from their minimal (potential field) values? We need to subtract the potential field line helicities to remove the contributions from the boundary flux distribution. Thus, we can adopt the relative helicity viewpoint for field line helicity as well (Yeates and Page, 2018; Moraitis et al., 2019; Moraitis et al., 2021). First, to avoid gauge ambiguity, we employ the same gauge conditions as apply in the poloidal-toroidal formulation. Let the potential field be $\vec{P}$, with vector potential $\vec{A}_P$. In particular, for the components of the vector potentials parallel to the boundary, we have

$$\nabla \cdot \vec{A}_{\|} = \nabla \cdot \vec{A}_{P\|} = 0; \tag{1.37}$$

$$\vec{A}_{\|} = \vec{A}_{P\|}. \tag{1.38}$$

It can be shown (Yeates and Page, 2018) that this choice of gauge for $\vec{A}_P$ minimizes the boundary integral $\oint_S |\vec{A}_P \times \hat{n}|^2 \, \mathrm{d}^2x$. Next, let α_+ be the positive endpoint of a field line. The field lines through α_+ for $\vec{B}$ and $\vec{P}$ will generally have different trajectories and connect to different exit points α_- and α_{P-}. We can now take the difference in field line helicities

$$\mathcal{A}^+(\alpha_+) = \int_{\alpha_+}^{\alpha_-} \vec{A} \cdot \mathrm{d}\vec{l} - \int_{\alpha_+}^{\alpha_{P-}} \vec{A} \cdot \mathrm{d}\vec{l}. \tag{1.39}$$

We could also do this for the negative endpoints. Starting at the same negative endpoint α_-, we could track the field lines backward to either α_+ or α_{P+}. In this case, we could write

$$\mathcal{A}^-(\alpha_-) = \int_{\alpha_+}^{\alpha_-} \vec{A} \cdot \mathrm{d}\vec{l} - \int_{\alpha_{P+}}^{\alpha_-} \vec{A} \cdot \mathrm{d}\vec{l}. \tag{1.40}$$

Finally, we can take the average of the two (for the endpoints α_+ and α_- of the true field line $\vec{B}$) to find

$$\mathcal{A}^0 = \frac{1}{2}(\mathcal{A}^+(\alpha_+) + \mathcal{A}^-(\alpha_-)). \tag{1.41}$$

This is the measure of *relative field line helicity* adopted by Moraitis et al. (2019), although those authors use a slightly different gauge in their computations.

1.2. IDEAL EVOLUTION

Just like the total magnetic helicity, the field line helicities within a given domain $\mathcal{V}$ may change in two ways: by (i) boundary motions or (ii) non-ideal processes in the interior. In this section, we show how to derive an evolution equation for case (i), while case (ii) is deferred to section 1.3. We will give some examples where these have been applied, although much remains to be explored.

As we have seen, $\mathcal{A}$ is determined entirely by the choice of tangential components $\vec{A} \times \hat{n}$ on the boundary $\partial\mathcal{V}$, along with an extra condition when the boundary has two disconnected components. Typically, we think of setting $\vec{A} \times \hat{n} = \vec{A}_P \times \hat{n}$ for some reference vector potential $\vec{A}_P$. Thus when $\vec{B} \cdot \hat{n}$ on $\partial\mathcal{V}$ is evolving, we cannot keep a fixed reference but must evolve $\vec{A}_P$ in some suitable way over time that preserves the physical interpretation. For example, Yeates (2020) chooses the so-called minimal gauge.

If $\vec{B}$ evolves ideally in some volume $\mathcal{V}$ so that

$$\frac{\partial \vec{B}}{\partial t} = \nabla \times \left(\vec{u} \times \vec{B}\right), \tag{1.42}$$

then it is well known that magnetic field lines retain their identity: if two fluid elements lie on the same magnetic field line at some initial time, they will continue to do so at all later times. Moreover, by Alfvén's theorem, field lines cannot change their mutual linkage. It follows immediately that $\mathcal{A}(L)$ is preserved for every *closed* field line L in $\mathcal{V}$. Moreover, any functional of $\mathcal{A}$ will also be preserved, including not only the total helicity

$$H = \int_{\mathcal{V}} \vec{A} \cdot \vec{B} \, \mathrm{d}^3x = \int_L \mathcal{A}(L) \mathrm{d}\Phi \tag{1.43}$$

but also the helicity of any subregion composed of closed field lines, as well as the unsigned helicity defined by

$$\overline{H} = \int_L |\mathcal{A}(L)| \mathrm{d}\Phi. \tag{1.44}$$

To inject or remove field line helicity by continuous ideal evolution therefore requires *open* field lines where $B_n \equiv \vec{B} \cdot \hat{n}\big|_S \neq 0$ along with $\vec{u} \neq \vec{0}$ on the boundary S, as illustrated in Fig. 1.3(b). In this case, the fundamental issue of gauge choice is important. A physically meaningful approach is to fix something like the winding or poloidal-toroidal gauge for $\vec{A}$ at all times. Consider an open field line L_t, where the subscript t denotes that this line moves with the fluid. To derive the evolution equation for $\mathcal{A}(L_t)$, note first that uncurling equation (1.42) gives

$$\frac{\partial \vec{A}}{\partial t} = \vec{u} \times \vec{B} + \nabla\phi \tag{1.45}$$

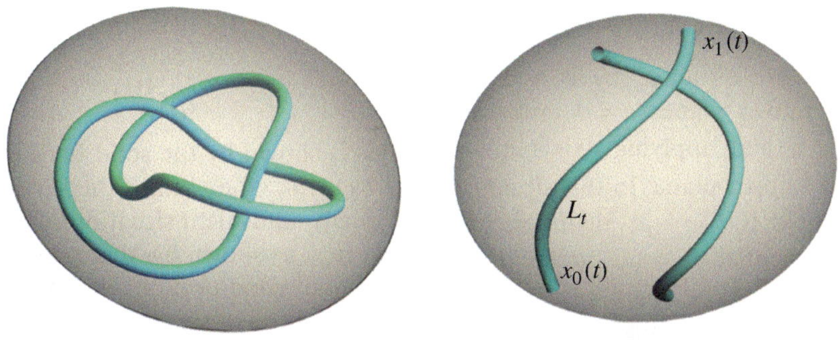

Figure 1.3 Examples of closed (left) and open (right) field lines in a domain V with connected boundary $\partial\mathcal{V}$. Foot points of the open field line L_t are denoted $x_0(t)$, where $\vec{B}\cdot\hat{n}\big|_S < 0$, and $x_1(t)$, where $\vec{B}\cdot\hat{n}\big|_S > 0$.

for some scalar function $\phi(\vec{x}, t)$ that depends on the chosen gauge of $\vec{A}$ and will generally be nonzero for the poloidal-toroidal gauge.

Now let $\mathcal{V}_{t,\epsilon}$ be an infinitesimal flux tube surrounding L_t and moving with the fluid. Let $\Phi(\mathcal{V}_{t,\epsilon})$ and $h(\mathcal{V}_{t,\epsilon}) = \int_{\mathcal{V}_{t,\epsilon}} \vec{A}\cdot\vec{B}\,\mathrm{d}^3x$ be the magnetic flux and helicity of this flux tube. Then the rate of change of field line helicity is

$$\frac{\mathrm{d}}{\mathrm{d}t}\mathcal{A}(L_t) = \frac{\partial}{\partial t}\lim_{\epsilon\to 0}\frac{h(\mathcal{V}_{t,\epsilon})}{\Phi(\mathcal{V}_{t,\epsilon})} \tag{1.46}$$

$$= \lim_{\epsilon\to 0}\frac{1}{\Phi(\mathcal{V}_{t,\epsilon})}\frac{\mathrm{d}h(\mathcal{V}_{t,\epsilon})}{\mathrm{d}t}. \tag{1.47}$$

By the Reynolds transport theorem, since $\mathcal{V}_{t,\epsilon}$ is a material volume, we have

$$\frac{\mathrm{d}}{\mathrm{d}t}h(\mathcal{V}_{t,\epsilon}) = \int_{\mathcal{V}_{t,\epsilon}}\frac{\partial}{\partial t}(\vec{A}\cdot\vec{B})\mathrm{d}^3x + \oint_{S_{t,\epsilon}}(\vec{A}\cdot\vec{B})\vec{u}\cdot\hat{n}\,d^2x, \tag{1.48}$$

Which, using equation (1.45), reduces to

$$\frac{\mathrm{d}}{\mathrm{d}t}h(\mathcal{V}_{t,\epsilon}) = \oint_{S_{t,\epsilon}}\left(\phi + \vec{A}\cdot\vec{u}\right)\vec{B}\cdot\hat{n}\,d^2x. \tag{1.49}$$

This is just the standard expression for the evolution of magnetic helicity in a material volume (e.g., Moffatt and Dormy (2019) p. 61). Since $\mathcal{V}_{t,\epsilon}$ is a magnetic flux tube, $\vec{B}\cdot\hat{n}$ is nonzero only on the two portions of the tube boundary $S_{t,\epsilon}$ that coincide with the boundary of $\mathcal{V}$ – in other words, the ends of the tube. So substituting into equation (1.47) and taking the limit, noting that $\vec{B}\cdot\hat{n}\to\Phi(\mathcal{V}_{t,\epsilon})$ at both ends of the tube, gives

$$\frac{\mathrm{d}}{\mathrm{d}t}\mathcal{A}(L_t) = \left[\phi + \vec{A}\cdot\vec{u}\right]_{x_0(t)}^{x_1(t)}, \tag{1.50}$$

where $x_0(t)$ and $x_1(t)$ denote the endpoints of L_t on the boundary of $\mathcal{V}$.

The expression in equation (1.50) shows that even though field lines L_t can be uniquely tracked in time during an ideal evolution, their field line helicities $\mathcal{A}(L_t)$ depend on the chosen gauge of $\vec{A}$ on the boundary over time, since this changes both $\vec{A}\cdot\vec{u}$ and ϕ, the latter being defined for a given choice of $\vec{A}$ through equation (1.45). For meaningful results, this gauge should be chosen objectively – such as the poloidal-toroidal gauge – rather than for computational convenience. Indeed, it is always possible to "cancel" the effect of an ideal evolution by choosing $\phi = -\vec{A}\cdot\vec{u}$ in equation (1.45) so that $\mathcal{A}$ remains invariant for all field lines, but this is a highly artificial choice; in effect, it corresponds to measuring winding with respect to a frame moving with the boundary motions. Of course, if there are no boundary motions, $\mathcal{A}$ will be invariant in time for all field lines provided that ϕ on the boundary is not varied. In particular, if the boundary is a connected surface, it is sufficient to fix $\vec{A}\times\hat{n}$. In cases like the volume between two spheres, one must also fix the integral of $\vec{A}$ along a line between the two boundaries.

When there are boundary motions, the winding gauge makes it possible to express the evolution of $\mathcal{A}$ in terms of the mean angular motions of field line endpoints around one another (Berger, 1988; MacTaggart and Prior, 2021). In the restricted case where there are boundary motions but they preserve the boundary distribution $B\cdot\hat{n}|_S$ and, in addition, the magnetic field has a "simple topology" (meaning the mapping from positive to negative endpoints is continuous), Aly (2018) derived the alternative formula

$$\frac{\mathrm{d}}{\mathrm{d}t}\mathcal{A}(L_t) = \left[U\frac{\partial\zeta}{\partial U} - \zeta\right]_{x_0(t)}^{x_1(t)}. \tag{1.51}$$

Here the boundary motions have been written as $\vec{u} = (\hat{n}\times\nabla\zeta)/\vec{B}\cdot\hat{n}$, and the gauge of $\vec{A}$ has been chosen in a particular way. Specifically, the parallel components of $\vec{A}$ on the boundary are set by equation (1.38), but with $\vec{A}_P = U\nabla V$, where U and V are Euler potentials for the reference field $\vec{P} = \nabla U\times\nabla V$. Formula (1.51) may be shown to be equivalent to equation (1.50) for this gauge choice.

1.2.1. Simple Examples

The simplest examples are obtained by considering a magnetic field between two planes $z = 0$ and $z = 1$ on which $\vec{B} \cdot \hat{n}\big|_S = B_z = 1$ and applying axisymmetric boundary motions on the upper plane. In this situation, the potential reference field is $\vec{P} = \hat{z}$ at all times, with poloidal-toroidal vector potential $\vec{A}_P = (r/2)\hat{\theta}$. We fix the gauge of $\mathcal{A}$ by setting $\vec{A} \times \hat{z} = \vec{A}_P \times \hat{z}$ on $z = 0$ and $z = 1$ and, by the additional condition (because the boundary has two disconnected components $z = 0$ and $z = 1$), that $\int_L \vec{A} \cdot \mathrm{d}l = \int_L \vec{A}_P \cdot \mathrm{d}l = 0$ for the vertical line L at $r = 0$. This gives us $\mathcal{A}$ in the winding, or equivalently poloidal-toroidal gauge (the two are equivalent in this volume).

Suppose first that we apply a rigid rotation of the upper boundary with respect to the lower one, given by $\vec{u} = rz\hat{\theta}$. In this case, $\vec{u} \times \vec{B} = u_\theta B_z \hat{r} = rz\hat{r}$, so maintaining the gauge condition $\vec{A} \times \hat{z} = \vec{A}_P \times \hat{z}$ requires, according to equation (1.45), that

$$\frac{\partial \phi}{\partial r} = -rz \quad \Longrightarrow \quad \phi = -\frac{r^2 z}{2}. \tag{1.52}$$

Thus for a field line at radius r, equation (1.50) gives

$$\frac{\mathrm{d}}{\mathrm{d}t}\mathcal{A}(r) = \left[-\frac{r^2 z}{2} + \left(\frac{r}{2}\right) rz\right]_{z=0}^{z=1} = 0. \tag{1.53}$$

This shows that a rigid rotation generates no field line helicity (as noted by Yeates et al. (2021)). In effect, all the field lines rotate together so that none of them acquires any twist with respect to the others. This example is shown by the dashed lines in the right-hand panels of Fig. 1.4.

For an example with nontrivial injection of field line helicity, we can apply a localized rotation $\vec{u} = 2r\mathrm{e}^{-r^2}\hat{\phi}$ for which a similar computation leads to a localized patch of field line helicity:

$$\frac{\mathrm{d}}{\mathrm{d}t}\mathcal{A}(r) = (r^2 + 1)\mathrm{e}^{-r^2}. \tag{1.54}$$

This is shown by the solid lines in the right-hand panels of Fig. 1.4. One obtains the same $\mathcal{A}$ pattern in the topologically equivalent situation of a toroidal flux ring localized in z, as shown by Yeates and Hornig (2014). Indeed, the initial condition in Fig. 1.8 (later) is produced by the superposition of six such flux rings, offset so the field lines follow a tangled pattern.

A final analytical example demonstrates how field line helicity may be injected into a magnetic field defined on the half-space $z > 0$ through shearing motions on the lower boundary $z = 0$ instead of twisting. This models the shearing of an arcade of coronal magnetic loops in the Sun's atmosphere. We suppose that the arcade is initially current-free and (for simplicity) invariant in x, with $\vec{B} = \vec{P} = z\hat{y} - y\hat{z}$, the solar surface being the plane $z = 0$ and the volume V with $z > 0$ representing the corona. A suitable poloidal-toroidal reference vector potential is $\vec{A}_P = (y^2/2 + z^2/2)\hat{x}$. Suppose we apply the constant shearing velocity $\vec{u} = y\mathrm{e}^{-y^2}\hat{x}$, localized near the polarity inversion line $y = 0$. Here $\vec{u} \times \vec{B} = -u_x B_z \hat{y} + u_x B_y \hat{z}$, so requiring that $\hat{n} \times \vec{A} = \hat{n} \times \vec{A}_P$ on the boundary $z = 0$ implies that

$$\frac{\partial \phi}{\partial y} = u_x B_z = -y^2 \mathrm{e}^{-y^2} \quad \Longrightarrow$$
$$\phi = -\frac{1}{4}\left(\sqrt{\pi}\,\mathrm{erf}(y) - 2y\mathrm{e}^{-y^2}\right). \tag{1.55}$$

Substituting into equation (1.50) and evaluating on the field line rooted at $y > 0$ then gives

$$\frac{\mathrm{d}}{\mathrm{d}t}\mathcal{A}(y) = (y + y^3)\mathrm{e}^{-y^2} - \frac{\sqrt{\pi}}{2}\mathrm{erf}(y). \tag{1.56}$$

This profile is shown in Fig. 1.5. The first term vanishes on the unsheared field lines at large y. But the second

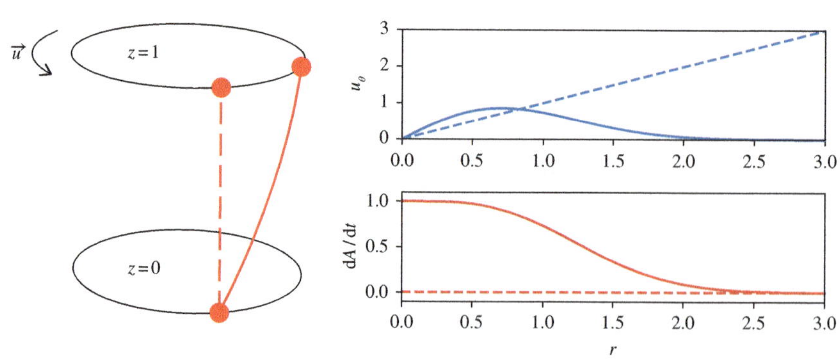

Figure 1.4 Simple example where the boundary has two parallel components, one of which is rotated with respect to the other. The graphs show the azimuthal velocity (top) and rate of generation of field line helicity (bottom) for the two examples in the text (dashed and solid lines).

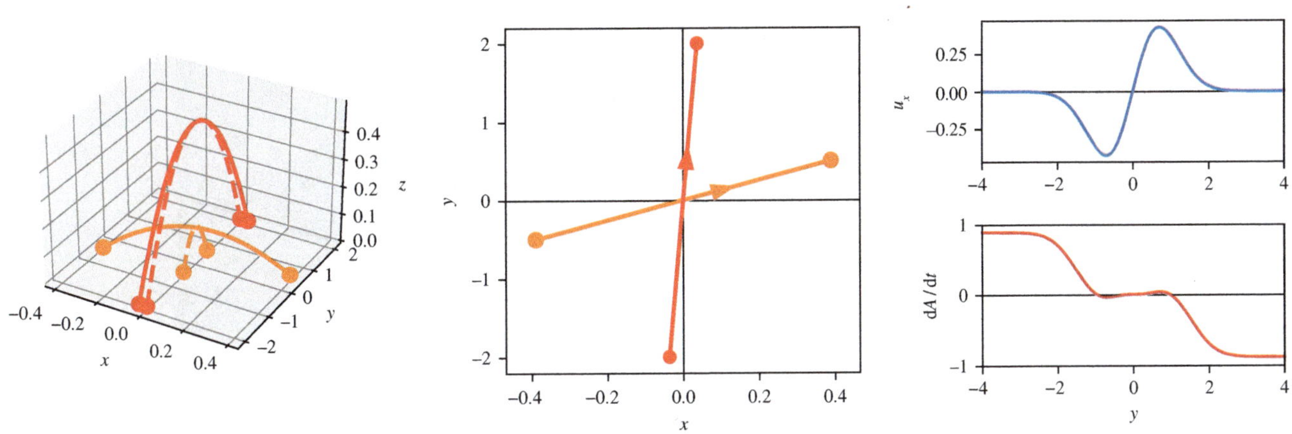

Figure 1.5 Simple example of an initial potential "arcade" being sheared by a flow on the lower boundary. Although they are largely unsheared, the overlying field lines acquire negative field line helicity, as may be seen from their negative crossings with the sheared field lines beneath.

term does not. This demonstrates the nonlocal nature of helicity: the field lines in the overlying arcade gain a field line helicity because the core of the arcade is sheared. Effectively, a magnetic flux is passing through them. Indeed, this was precisely the way Antiochos (1987) proposed to define field line helicity, as mentioned earlier. Notice here that the sign of the erf(y) term is negative, consistent with Stokes's theorem and the direction of the sheared field relative to the orientation of the overlying arcade. One could also infer this sign by looking at the arcade from above and observing a negative crossing (Fig. 1.5).

1.2.2. Application to the Global Solar Corona

To a first approximation, the magnetic field in the Sun's atmosphere evolves ideally in response to the emergence of new magnetic active regions from inside the Sun, decay of these strong magnetic fields due to convective shredding, and transport of the resulting magnetic flux by large-scale motions such as the Sun's differential rotation. During these processes, magnetic helicity builds up in the corona and is ejected into the heliosphere. Field line helicity offers the exciting prospect of a localized measure for studying where this helicity is located within the corona.

Yeates and Hornig (2016) used a global magneto-frictional model in a spherical shell to study how field line helicity evolves in the corona in response to the evolution of the solar surface magnetic field (Fig. 1.6). In the simulation shown, active region emergence is neglected so that the dominant injection of helicity is shearing by differential rotation of field line foot points on the solar surface. The behavior of field line helicity is found to be different on open and closed magnetic field lines. On open field lines (meaning in this context that one field line endpoint is on the outer boundary), injected field line helicity is lost at a steady rate through the outer boundary by relaxation of the field line. By contrast, closed field lines, with both endpoints on the solar surface, can store field line helicity – as in the simple arcade example in Fig. 1.5.

This storage of helicity on closed field lines is important because eventually, these form twisted flux ropes that lose equilibrium and erupt, leading to bursts of helicity flux output and believed to explain the origin of coronal mass ejections. Lowder and Yeates (2017) studied these eruptions in the magneto-frictional model and used field line helicity as a diagnostic tool to define flux ropes in the first place. The overall magnetic flux and helicity content of these structures was found to be comparable to that estimated in observations of interplanetary magnetic clouds. Recently, Bhowmik and Yeates (2021) have used field line helicity to show that episodic losses of helicity in the magneto-frictional model come not only from the eruption of flux ropes formed along polarity inversion lines in the low corona but also from a second type of eruption generated in the overlying streamers. Similar eruptions are known in magnetohydrodynamic (MHD) simulations (Linker and Mikic, 1995) and have been suggested as a possible explanation for so-called stealth coronal mass ejections (CMEs) that lack an obvious low-coronal source (Lynch et al., 2016). Many unanswered questions remain, not least the role of active regions in the global helicity balance or the possibility of long-term storage of their helicity in the corona. Field line helicity will greatly facilitate these investigations.

1.2.3. Application to Solar Active Regions

Field line helicity also offers the possibility to identify locations of helicity storage on a smaller scale, within

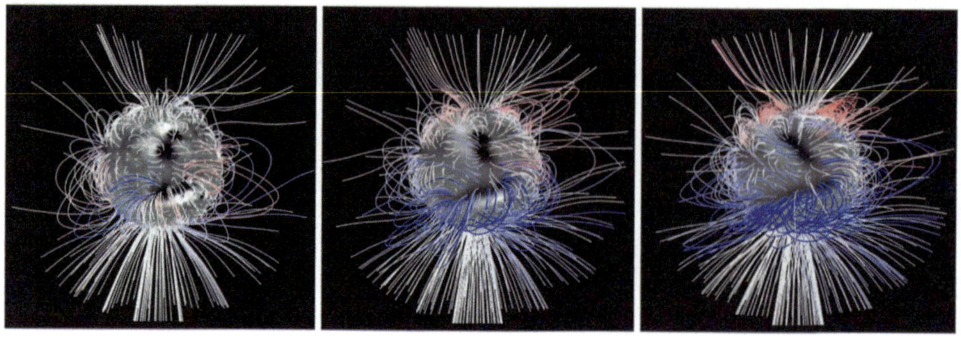

(a) (b) (c)

Figure 1.6 Evolution of field line helicity in a magneto-frictional model of the solar corona as an initial potential field (a) is sheared by the Sun's differential rotation for (b) 25 days and (c) 50 days. Field lines are colored by their field line helicity in poloidal-toroidal gauge, with red for positive and blue for negative (for details of the simulation, see (Yeates and Hornig, 2016)).

individual active regions, provided three-dimensional magnetic field models are available. Even in a potential field model, a simple bipolar active region can have nonzero field line helicity if its field lines are linked with the overlying background field (Yeates, 2020). But much larger values of field line helicity are expected in more realistic current-carrying models of active regions. This has been confirmed by Moraitis et al. (2019) both in idealized MHD models and in extrapolations. For example, Moraitis et al. (2021) have computed field line helicity in nonlinear force-free extrapolations of a real active region NOAA 11158 (Fig. 1.7).

Figure 1.7 shows a highly sheared magnetic field in the core of the active region, which has positive field line helicity. The authors found that during a flare, the region lost 25% of its relative magnetic helicity. Comparing the field line helicity between the two extrapolations shown at 01:11 UT and 01:59 UT reveals that the decrease in helicity took place within the same region (the green box) where emission from a large X-class solar flare was observed in extreme ultraviolet (EUV). This supports the idea that a decrease in helicity was indeed associated with the flare. It is worth noting that Moraitis et al. (2021) uses relative field line helicity (section 1.1.3). However, they show that using the ordinary field line helicity (in poloidal-toroidal gauge) leads to the same qualitative conclusions in this example, albeit with lower values.

1.3. NON-IDEAL EVOLUTION

Suppose equation (1.42) is generalized to

$$\frac{\partial \vec{B}}{\partial t} = \nabla \times \left(\vec{u} \times \vec{B}\right) - \nabla \times \vec{N}, \qquad (1.57)$$

where $\vec{N} = \vec{E} + \vec{u} \times \vec{B}$ represents some non-ideal term in Ohm's law. A common example would be $\vec{N} = \eta \vec{J}$ corresponding to resistive MHD. It is well-known that the total helicity is no longer conserved but can be dissipated within the volume when $\vec{N} \cdot \vec{B} \neq 0$. In this section, we show how – for magnetic fields of simple topology without null points – equations (1.50) and (1.51) have been generalized to this non-ideal case. For magnetic fields of more complex topology – for example, the solar coronal examples in sections 1.2.2 and 1.2.3 – evolution equations have not yet been derived explicitly. Indeed, differential equations may not be appropriate since the distribution of $\mathcal{A}$ is discontinuous across magnetic separatrices between different connectivity domains. Reconnection can transport $\mathcal{A}$ across these separatrices, but its evolution in such a situation remains to be studied in detail.

There is an important caveat to what follows: field line helicity can hold physical significance in a non-ideal evolution only if the magnetic field lines themselves retain sufficient identity over time for their topology to play a physical role. In practice, this means the magnetic Reynolds number (Rm) must be sufficiently large or, equivalently, the (effective) resistivity must be sufficiently small. There is no precise threshold for this, but it is already accessible for the parameters achievable in numerical simulations, as will be illustrated in section 1.3.2.

1.3.1. Evolution Equation for Non-Null Magnetic Fields

The trick for generalizing equation (1.50) or equation (1.51) to non-ideal evolution in the non-null case is to decompose $\vec{N}$ into parallel and perpendicular parts (Yeates and Hornig, 2011), writing

$$\vec{N} = -\vec{v} \times \vec{B} + \nabla \psi. \qquad (1.58)$$

When the magnetic field has a simple topology, this decomposition exists globally, with $\vec{v}$ and ψ continuous

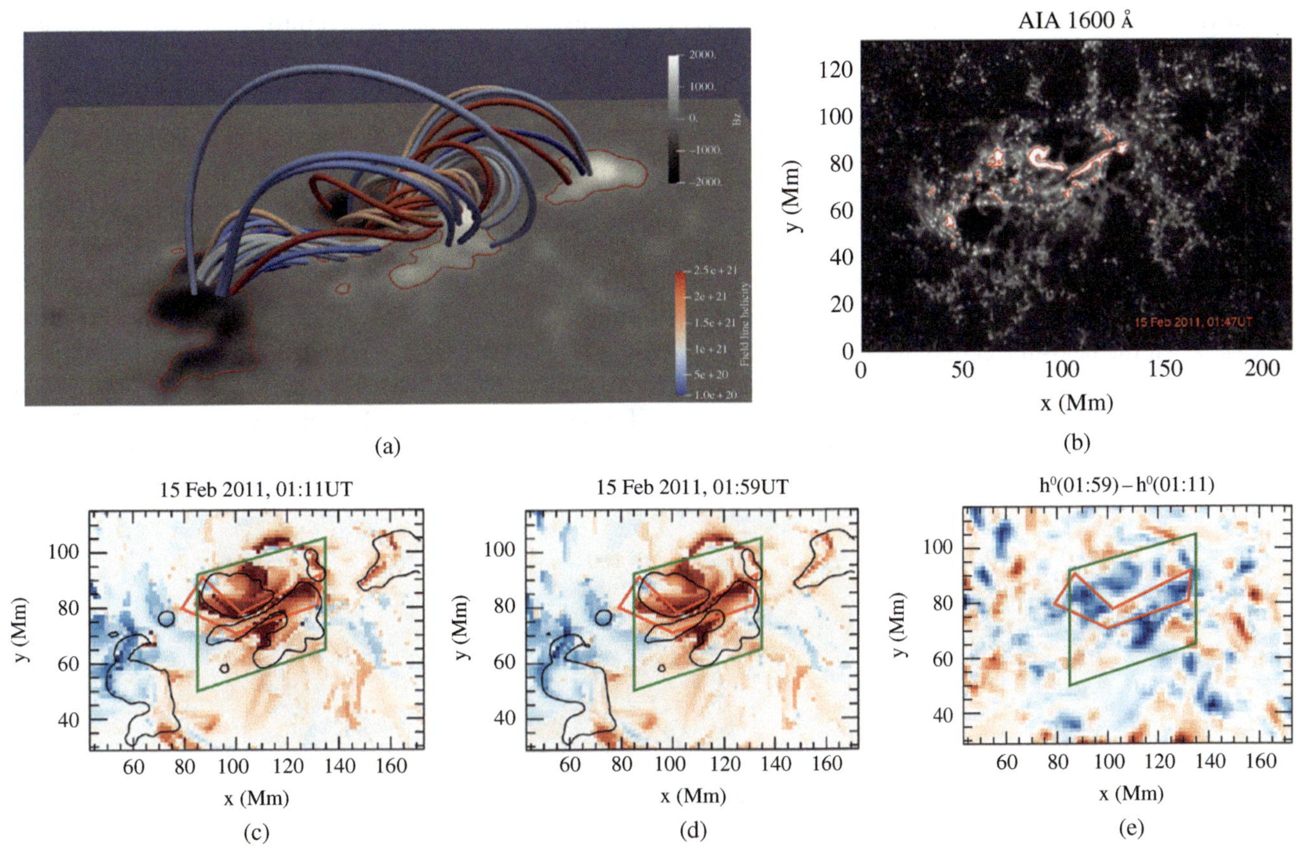

Figure 1.7 Field line helicity in a nonlinear force-free model of active region NOAA 11158, from Moraitis et al. (2021). Panel (a) shows the field lines colored by (relative) field line helicity in the extrapolation at 01:11 UT, while panel (b) shows an image of EUV emission from SDO/AIA during the X-class flare at 01:47 UT. Panels (c) and (d) show (relative) field line helicity before and after the flare (blue/red), while panel (e) shows their difference. Credit: Moraitis et al., 2021/The European Southern Observatory (ESO).

throughout V, although they are not unique (as we shall discuss shortly). Substituting this decomposition into equation (1.57) shows that in a non-null magnetic field, we can write

$$\frac{\partial \vec{B}}{\partial t} = \nabla \times \left(\vec{w} \times \vec{B}\right), \quad \text{where} \quad \vec{w} = \vec{u} + \vec{v}, \tag{1.59}$$

which shows that the magnetic field is still frozen, but into the *field line (transport) velocity* $\vec{w}$ rather than the plasma velocity $\vec{u}$. Thus, in a non-ideal evolution where $\vec{v} \neq \vec{0}$, the field lines slip at some velocity $\vec{v}$ through the plasma (Newcomb, 1958; Priest and Forbes, 1992; Aulanier et al., 2006; Schindler, 2010). Uncurling, we have

$$\frac{\partial \vec{A}}{\partial t} = \vec{w} \times \vec{B} + \nabla(\phi - \psi). \tag{1.60}$$

Thus, if we identify the same field line L_t at different times by the fact that it is frozen into the flow of $\vec{w}$, an argument similar to section 1.2 shows that

$$\frac{\mathrm{d}}{\mathrm{d}t}\mathcal{A}(L_t) = \left[\phi + \vec{A} \cdot \vec{w} - \psi\right]_{x_0(t)}^{x_1(t)}. \tag{1.61}$$

Similarly, Aly (2018) showed that equation (1.51) becomes

$$\frac{\mathrm{d}}{\mathrm{d}t}\mathcal{A}(L_t) = \left[U\frac{\partial(\zeta + \psi)}{\partial U} - (\zeta + \psi)\right]_{x_0(t)}^{x_1(t)}. \tag{1.62}$$

Notice that $[\psi]_{x_0(t)}^{x_1(t)} = \int_{L_t} \vec{E} \cdot \mathrm{d}l$, which is precisely the parallel electric field used to define the reconnection rate in the theory of general magnetic reconnection (Schindler et al., 1988). It represents the change in $\mathcal{A}(L_t)$ due to a (nonlocal) change in the magnetic flux linked with L_t. The term $\vec{A} \cdot \vec{w}$ is harder to interpret in general because it depends on the choices of $\vec{A}$ and $\vec{w}$. With the gauge condition $\hat{n} \times \vec{A} = \hat{n} \times \vec{A}_P$ on the boundary, the $[\vec{A}_P \cdot \vec{w}]_{x_0(t)}^{x_1(t)}$ term effectively represents "work done" by motion of the field line L_t with respect to the reference field (Russell et al., 2015).

For a given non-ideal evolution of $\vec{B}$, the field line velocity $\vec{w}$ is not uniquely defined, and consequently, the identification of the field line L_t over time is not unique. While the component of $\vec{w}$ parallel to $\vec{B}$ is arbitrary – as is clear from equation (1.58) – it is the non-uniqueness of the perpendicular component $\vec{w}_\perp$ that changes the identification of L_t. To see that this component is non-unique, note that equation (1.58) implies

$$\vec{w}_\perp = \vec{v}_\perp = \frac{\vec{B} \times (\nabla\psi - \vec{E})}{B^2}. \tag{1.63}$$

On each field line, we can specify an initial value of ψ. This does not change $[\psi]_{x_0(t)}^{x_1(t)}$ but does change $\vec{w}_\perp$.

One situation where a natural choice of $\vec{w}_\perp$ arises is the case when the field-line endpoints are line-tied on the boundary, meaning $\vec{u} = \vec{N} = \vec{E} = \vec{0}$ there (Russell et al., 2015). In that case, we can identify L_t over time by fixing one endpoint – say, $\vec{w}_\perp(x_0) = \vec{0}$ – for all field lines. This is achieved by choosing $\psi = 0$ throughout the region of the boundary where $\vec{B} \cdot \hat{n}|_S < 0$. At the opposite endpoints $x_1(t)$, the values of ψ will then be fixed by $[\psi]_{x_0(t)}^{x_1(t)}$, leading to $\vec{w}_\perp(x_1) \neq \vec{0}$ in general so that these endpoints will move in time in a non-ideal evolution. We could equally well fix $\psi = 0$ on the region with $\vec{B} \cdot \hat{n}|_S > 0$, thus defining the field lines by fixed x_1 positions so that $x_0(t)$ varies over time.

If $\vec{u}$ is nonzero on the boundary but $\vec{N}$ remains zero there (or can be neglected), then it is possible to subtract the ideal term $[\vec{A} \cdot \vec{u}]_{x_0(t)}^{x_1(t)}$ from equation (1.61) and isolate the change in field line helicity coming from non-ideal evolution. An analogous calculation was implemented for the magnetic winding measure by Gekelman et al. (2020) in

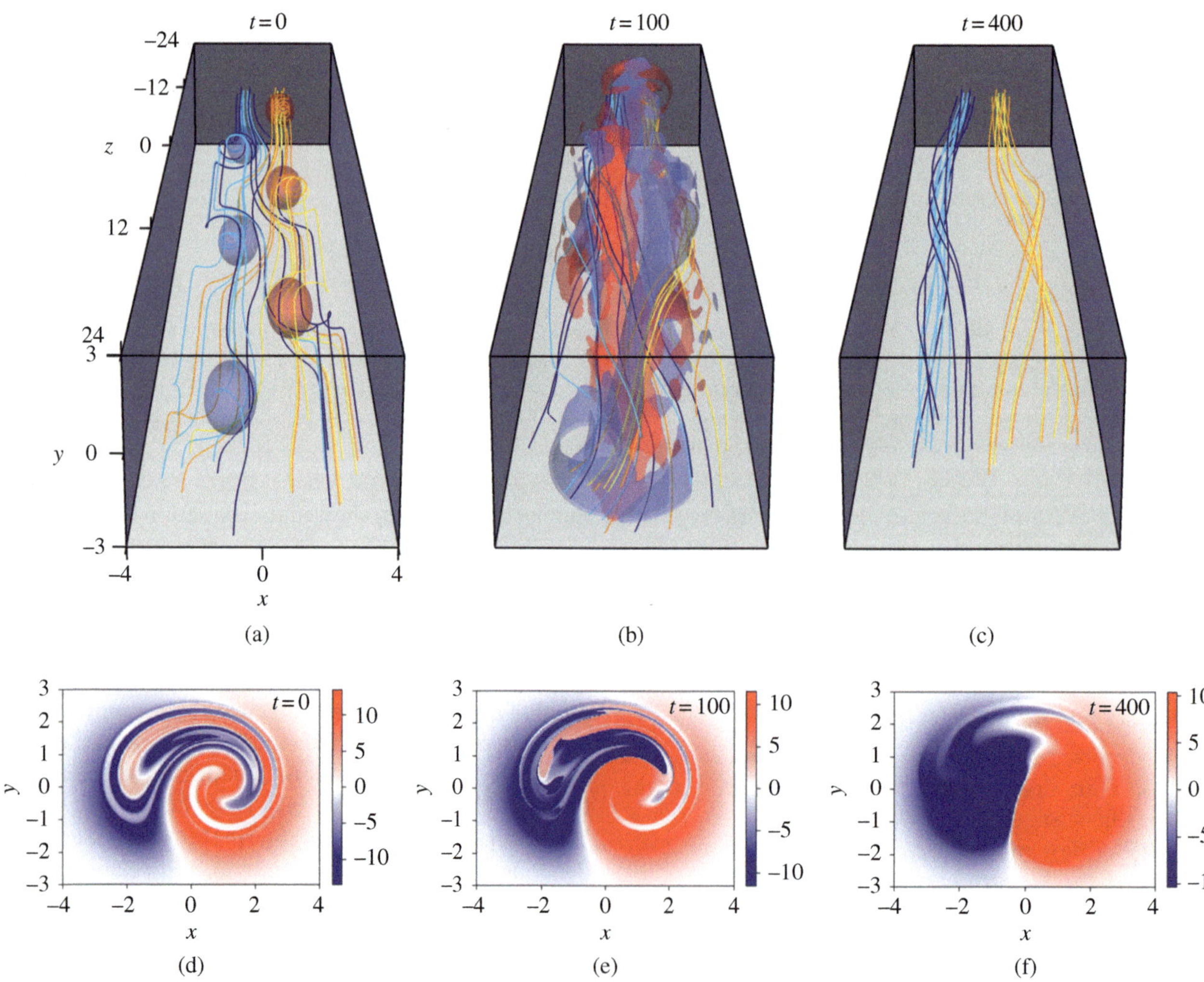

Figure 1.8 Evolution of field line helicity during line-tied resistive relaxation of a braided magnetic field (see Yeates et al. (2021) for details of the simulation). Panels a–c show magnetic field lines and isosurfaces of current density (a) in the initial condition, (b) during the turbulent relaxation, and (c) in the relaxed state. Panels d–f show cross sections of field line helicity on the $z = -24$ boundary at the same time. Chen et al. 2021/Cambridge University Press/Licensed under CC BY 4.0.

data from a laboratory experiment of interacting magnetic flux ropes.

1.3.2. Application to Turbulent Magnetic Relaxation

The non-ideal evolution of $\mathcal{A}$ has been explored only in the context of braided magnetic fields, where all field lines connect between two planar boundaries $z = 0$ and $z = 1$ at which $\vec{u} = \vec{N} = \vec{0}$. The most significant finding to date is that when the magnetic field line mapping from $z = 0$ to $z = 1$ is complex with sharp gradients, the evolution of $\mathcal{A}$ for high Rm is dominated by redistribution between field lines, rather than dissipation (Russell et al., 2015; Yeates et al., 2021).

To see this, consider the evolution of $\mathcal{A}$ on the field line traced from a fixed point x_0 on the lower boundary $z = 0$. Russell et al. (2015) made the natural choice $\vec{w} = \vec{0}$ on $z = 0$, in which case equation (1.61) reduces to

$$\frac{\partial \mathcal{A}(x_0)}{\partial t} = \vec{A}(x_1(t)) \cdot \vec{w}(x_1(t)) - \psi(x_1(t)), \qquad (1.64)$$

where L_t is the field line defined by tracing from the fixed point x_0 at subsequent times. They argued that the $\vec{A} \cdot \vec{w}$ dominates the ψ term in a magnetic field with complex mapping, because $|\vec{w}|$ scales like $|\nabla\psi|/B$ from equation (1.63). Since $\psi(x_1(t))$ is a field-line integrated quantity, and since the field line mapping has sharp gradients, we would expect $|\vec{A} \cdot \vec{w}| \gg |\psi|$ for high Rm. This was demonstrated originally in kinematic examples (see also Yeates (2019)) and has recently been confirmed by direct calculation of the terms in equation (1.64) in resistive MHD simulations (Yeates et al., 2021). One such simulation is illustrated in Fig. 1.8.

Watching the full evolution of $\mathcal{A}(x_0)$ in the Fig. 1.8 simulation suggests that the $\mathcal{A}$ pattern evolves predominantly by rearrangement with some fictitious flow. To see this from the evolution equation, we can employ the seemingly unhelpful trick of labeling field lines by fixing their endpoint x_1 on $z = 1$ so that $\psi = 0$ and $\vec{w}_\perp = \vec{0}$ on $z = 1$. Then instead of equation (1.64), equation (1.50) becomes

$$\frac{\mathrm{d}}{\mathrm{d}t}\mathcal{A}(L_t) = -\vec{A}(x_0(t)) \cdot \vec{w}(x_0(t)) + \psi(x_0(t)). \qquad (1.65)$$

To compute the evolution of $\mathcal{A}$ at a fixed position x_0, we must now account for the fact that the field lines are moving past this point with the local field line velocity $\vec{w}(x_0)$. Thus the time derivative in equation (1.65) is a Lagrangian one, so at the fixed position x_0, we have

$$\frac{\partial \mathcal{A}(x_0)}{\partial t} = -\vec{w} \cdot \nabla\mathcal{A}(x_0) - \vec{A}(x_0) \cdot \vec{w}(x_0) + \psi(x_0). \quad (1.66)$$

Notice that all the terms here are "local," although ψ and $\mathcal{A}$ are still field-line integrated quantities. The advection term $\vec{w} \cdot \nabla\mathcal{A}$ is effectively a product of two gradients of field-line integrated quantities, so one might expect this to be the largest. It is therefore no surprise that the dominant behavior observed is a rearrangement of the $\mathcal{A}$ pattern by advection.

Chen et al. (2021) took this idea and developed a variational model for turbulent magnetic relaxation that predicts the relaxed state to have the "simplest" pattern of field line helicity achievable by pure advection. This already predicts the two oppositely twisted magnetic "flux tubes" in Fig. 1.8(f). However, closer inspection of the numerical simulations by Yeates et al. (2021) shows that the other terms in equation (1.66) play a role in establishing the substructure of the final state.

REFERENCES

Aly, J. J. (2018). New formulae for magnetic relative helicity and field line helicity. *Fluid Dynamics Research*, *50*(1), 011408. https://doi.org/10.1088/1873-7005/aa737a

Antiochos, S. K. (1987). The topology of force-free magnetic fields and its implications for coronal activity. *The Astrophysical Journal*, *312*, 886.

Arnold, V. I., & Khesin, B. A. (1998). *Topological magnetohydrodynamics*. Springer.

Aulanier, G., Pariat, E., Démoulin, P., & DeVore, C. R. (2006). Slip-running reconnection in quasi-separatrix layers. *Solar Physics*, *238*(2), 347–376. https://doi.org/10.1007/s11207-006-0230-2

Berger, M. A. (1984). Rigorous new limits on magnetic helicity dissipation in the solar corona. *Geophysical & Astrophysical Fluid Dynamics*, *30*(1–2), 79–104. https://doi.org/10.1080/03091928408210078

Berger, M. A. (1986). Topological invariants of field lines rooted to planes. *Journal of Physics A: Mathematical and General*, *34*, 265–281.

Berger, M. A. (1988). An energy formula for nonlinear force-free magnetic fields. *Astronomy and Astrophysics*, *201*, 355–361.

Berger, M. A., & Field, G. B. (1984). The topological properties of magnetic helicity. *Journal of Fluid Mechanics*, *147*, 133–148.

Berger, M. A., & Hornig, G. (2018). A generalized poloidal–toroidal decomposition and an absolute measure of helicity. *Journal of Physics A: Mathematical and Theoretical*, *51*(49), 495501.

Berger, M. A., & Prior, C. (2006). The writhe of open and closed curves. *Journal of Physics A: Mathematical and General*, *39*, 8321–8348.

Bhowmik, P., & Yeates, A. R. (2021). Two classes of eruptive events during solar minimum. *Solar Physics*, *296*(7), 109. https://doi.org/10.1007/s11207-021-01845-x

Bourdin, P. A., & Brandenburg, A. (2018). Magnetic helicity from multipolar regions on the solar surface. *The Astrophysical Journal*, *869*, 3.

Campbell, J., & Berger, M. A. (2014). Helicity, linking, and writhe in a spherical geometry. *Journal of Physics: Conference Series*, *544*, 012001. https://doi.org/10.1088/1742-6596/544/1/012001

Chen, L., Yeates, A. R., & Russell, A. J. B. (2021). Optimal unstirred state of a passive scalar. *Journal of Fluid Mechanics*, *911*, A30. https://doi.org/10.1017/jfm.2020.1154

Demoulin, P., Pariat, E., & Berger, M. A. (2006). Basic properties of mutual magnetic helicity. *Solar Physics*, *233*, 3–27.

Finn, J. M., & Antonsen, T. M. Jr. (1985). Magnetic helicity: What is it and what is it good for? *Comments on Plasma Physics and Controlled Fusion*, *9*, 11.

Gekelman, W., DeHaas, T., Prior, C. B., & Yeates, A. R. (2020). Using topology to locate the position where fully three-dimensional reconnection occurs. *SN Applied Sciences*, *2*(12), 2187. https://doi.org/10.1007/s42452-020-03896-4

Kimura, Y., & Okamoto, H. (1987). Vortex motion on a sphere. *Journal of the Physical Society of Japan*, *56*, 2024.

Linker, J. A., & Mikic, Z. (1995). Disruption of a helmet streamer by photospheric shear. *The Astrophysical Journal Letters*, *438*, L45. https://doi.org/10.1086/187711

Lowder, C., & Yeates, A. (2017). Magnetic flux rope identification and characterization from observationally driven solar coronal models. *The Astrophysical Journal*, *846*, 106. https://doi.org/10.3847/1538-4357/aa86b1

Lynch, B. J., Masson, S., Li, Y., DeVore, C. R., Luhmann, J. G., Antiochos, S. K., & Fisher, G. H. (2016). A model for stealth coronal mass ejections. *Journal of Geophysical Research: Space Physics*, *121*(11), 10,677–10,697. https://doi.org/10.1002/2016JA023432

MacTaggart, D., & Prior, C. (2021). Helicity and winding fluxes as indicators of twisted flux emergence. *Geophysical and Astrophysical Fluid Dynamics*, *115*(1), 85–124. https://doi.org/10.1080/03091929.2020.1740925

MacTaggart, D., & Valli, A. (2019). Magnetic helicity in multiply connected domains. *Journal of Plasma Physics*, *85*(5), 775850501. https://doi.org/10.1017/S0022377819000576

Moffatt, H. K. (1969). The degree of knottedness of tangled vortex lines. *Journal of Fluid Mechanics*, *35*, 117–129.

Moffatt, H. K., & Ricca, R. L. (1992). Helicity and the Calugareanu invariant. *Proceedings of the Royal Society of London Series A: Mathematical, Physical and Engineering Sciences*, *439*(1906), 411–429. https://doi.org/10.1098/rspa.1992.0159

Moffatt, H. K. (1978). *Magnetic field generation in electrically conducting fluids*. Cambridge University Press.

Moffatt, H. K., & Dormy, E. (2019). *Self-exciting fluid dynamos*. Cambridge Texts in Applied Mathematics. Cambridge University Press. https://doi.org/10.1017/9781108766508

Moraitis, K., Pariat, E., Valori, G., & Dalmasse, K. (2019). Relative magnetic field line helicity. *Astronomy and Astrophysics*, *624*, A51. https://doi.org/10.1051/0004-6361/201834668

Moraitis, K., Patsourakos, S., & Nindos, A. (2021). Relative field line helicity of a large eruptive solar active region. *Astronomy and Astrophysics*, *649*, A107.

Newcomb, W. A. (1958). Motion of magnetic lines of force. *Annals of Physics*, *3*(4), 347–385. https://doi.org/10.1016/0003-4916(58)90024-1

Priest, E. R., & Forbes, T. G. (1992). Magnetic flipping: reconnection in three dimensions without null points. *Journal of Geophysical Research*, *97*(A2), 1521–1531. https://doi.org/10.1029/91JA02435

Prior, C., & Yeates, A. R. (2014). On the helicity of open magnetic fields. *The Astrophysical Journal*, *787*, 100.

Russell, A. J.B., Yeates, A. R., Hornig, G., & Wilmot-Smith, A. L. (2015). Evolution of field line helicity during magnetic reconnection. *Physics of Plasmas*, *22*(3), 032106. https://doi.org/10.1063/1.4913489

Schindler, K., Hesse, M., & Birn, J. (1988). General magnetic reconnection, parallel electric fields, and helicity. *Journal of Geophysical Research*, *93*(A6), 5547–5557. https://doi.org/10.1029/JA093iA06p05547

Schindler, K. (2010). *Physics of space plasma activity*. Cambridge University Press.

Yahalom, A. (2013). Aharonov-Bohm effects in magnetohydrodynamics. *Physics Letters A*, *377*(31–33), 1898–1904. https://doi.org/10.1016/j.physleta.2013.05.037

Yeates, A. R. (2019). *Magnetohydrodynamic relaxation theory*. CISM International Centre for Mechanical Sciences. Springer International Publishing. https://doi.org/10.1007/978-3-030-16344-0

Yeates, A. R. (2020). The minimal helicity of solar coronal magnetic fields. *The Astrophysical Journal Letters*, *898*, L49.

Yeates, A. R., & Hornig, G. (2011). A generalized flux function for three-dimensional magnetic reconnection. *Physics of Plasmas*, *18*(10), 102118. https://doi.org/10.1063/1.3657424

Yeates, A. R., & Hornig, G. (2014). A complete topological invariant for braided magnetic fields. *Journal of Physics Conference Series*, *544*, 012002. https://doi.org/10.1088/1742-6596/544/1/012002

Yeates, A. R., & Hornig, G. (2016). The global distribution of magnetic helicity in the solar corona. *Astronomy and Astrophysics*, *594*, A98. https://doi.org/10.1051/0004-6361/201629122

Yeates, A. R., & Page, M. H. (2018). The minimal helicity of solar coronal magnetic fields. *Journal of Plasma Physics*, *84*(775840602).

Yeates, A. R., Russell, A. J. B., & Hornig, G. (2021). Evolution of field line helicity in magnetic relaxation. https://arxiv.org/abs/2108.01346

2

Magnetic Winding: Theory and Applications

David MacTaggart

ABSTRACT

Magnetic winding is a renormalization of magnetic helicity that provides a direct measure of field line topology. Despite its close connection to magnetic helicity, magnetic winding can provide different and important information about field line topology that is not clear from an analysis of magnetic helicity alone. In this chapter, we introduce magnetic winding from a theoretical perspective, with a keen eye on its role in understanding magnetic helicity and field line topology. We also provide practical applications of magnetic winding, highlighting that it can be used as an important measure for understanding the evolution of active regions in solar observations.

2.1. INTRODUCTION

Magnetic helicity is a conserved quantity of ideal magnetohydrodynamics (MHD). What it conserves relates to the topology of magnetic field lines. To understand this concept, we will consider magnetic helicity written in a useful gauge. In a bounded domain $\Omega \subset \mathbb{R}^3$ with genus g that is magnetically closed ($\boldsymbol{B} \cdot \boldsymbol{n} = 0$ on $\partial\Omega$, for surface unit normal vector $\boldsymbol{n}$), the gauge invariant form of magnetic helicity can be written as

$$H = \int_\Omega \boldsymbol{A} \cdot \boldsymbol{B} \, \mathrm{d}^3 x - \sum_{i=1}^{g} \left(\int_{\gamma_i} \boldsymbol{A} \cdot \boldsymbol{t}_i \, \mathrm{d}x \right) \left(\int_{\Sigma_i} \boldsymbol{B} \cdot \boldsymbol{n}_{\Sigma_i} \, \mathrm{d}^2 x \right), \tag{2.1}$$

where $\boldsymbol{B}$ is the magnetic field, $\boldsymbol{A}$ is a magnetic vector potential, γ_i are closed paths around "holes" in the domain (e.g., the hole of a torus) and have unit tangent vectors $\boldsymbol{t}_i$, and the Σ_i are the surfaces of cuts through the domain with unit normal vectors $\boldsymbol{n}_{\Sigma_i}$. Equation (2.1) is derived in MacTaggart and Valli (2019), which describes more details of the geometrical setup (see also Faraco et al. (2022) for an application to a rigorous proof of Taylor's conjecture).

Although equation (2.1) holds for any vector potential $\boldsymbol{A}$ in domains of arbitrary topological complexity, it is too general to yield useful information about magnetic field line topology, which, as mentioned at the start, is related to what H conserves. If, again with $\boldsymbol{B} \cdot \boldsymbol{n} = 0$ on $\partial\Omega$, we make the gauge choice $\boldsymbol{A} = \mathrm{BS}(\boldsymbol{B})$, where $\mathrm{BS}(\boldsymbol{B})$ is the Biot-Savart operator

$$\mathrm{BS}(\boldsymbol{B}) = \frac{1}{4\pi} \int_\Omega \boldsymbol{B}(\boldsymbol{y}) \times \frac{\boldsymbol{x} - \boldsymbol{y}}{|\boldsymbol{x} - \boldsymbol{y}|^3} \, \mathrm{d}^3 y, \tag{2.2}$$

where $\boldsymbol{x}$ and $\boldsymbol{y}$ are position vectors in $\mathbb{R}^3$, then equation (2.1) reduces to the classical form of magnetic helicity,

$$\begin{aligned} H &= \int_\Omega \mathrm{BS}(\boldsymbol{B}) \cdot \boldsymbol{B} \, \mathrm{d}^3 x, \\ &= \frac{1}{4\pi} \int_\Omega \int_\Omega \boldsymbol{B}(\boldsymbol{x}) \cdot \boldsymbol{B}(\boldsymbol{y}) \times \frac{\boldsymbol{x} - \boldsymbol{y}}{|\boldsymbol{x} - \boldsymbol{y}|^3} \, \mathrm{d}^3 x \, \mathrm{d}^3 y. \end{aligned} \tag{2.3}$$

Equation (2.3) holds the key to understanding how helicity includes field line topology. Indeed, it was this form of helicity that Moffatt considered in his seminal work on the subject (Moffatt, 1969).

Let us now take a side step and consider a fundamental topological description of linked loops. Let C_1 and C_2 be two closed and distinct loops with assigned directions, as shown in Fig. 2.1.

School of Mathematics and Statistics, University of Glasgow, Glasgow, UK

Helicities in Geophysics, Astrophysics, and Beyond, Geophysical Monograph 283, First Edition.
Edited by Kirill Kuzanyan, Nobumitsu Yokoi, Manolis K. Georgoulis, and Rodion Stepanov.

DOI:10.1002/9781119841715.ch02

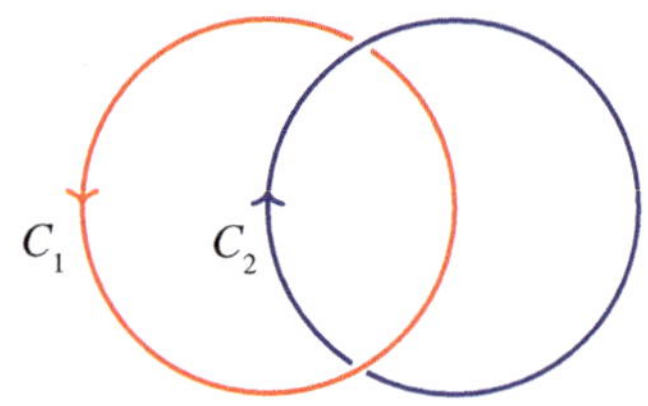

Figure 2.1 Two closed and linked loops, C_1 and C_2, with specified directions.

Any deformation that does not break the loops preserves the linkage. Therefore, linkage, in this scenario, is a topological quantity. If position vectors of points on C_1 are denoted by $\boldsymbol{x}$ and those on C_2 by $\boldsymbol{y}$, the Gauss linking number, which measures the net algebraic linkage, is given by

$$Lk = \frac{1}{4\pi} \int_{x(s_1)} \int_{y(s_2)} \frac{\mathrm{d}\boldsymbol{x}}{\mathrm{d}s_1} \cdot \frac{\mathrm{d}\boldsymbol{y}}{\mathrm{d}s_2} \times \frac{\boldsymbol{x}-\boldsymbol{y}}{|\boldsymbol{x}-\boldsymbol{y}|^3} \, \mathrm{d}s_1 \, \mathrm{d}s_2, \quad (2.4)$$

where s_1 and s_2 are parameterizations of C_1 and C_2, respectively. The linkage Lk is an integer if the loops in the link are closed. The closure of the links is also necessary for Lk to be a topological invariant: i.e., indifferent to deformations that do not break closed loops. For a historical account of the development of equation (2.4), the reader is directed to Ricca and Nipoti (2011).

For the example in Fig. 2.1, the two loops are clearly linked once. Due to the directions of the arrows, $Lk = 1$. We can also perform the calculation in equation (2.4) in reverse: i.e., assign $\boldsymbol{x}$ to points on C_2 and $\boldsymbol{y}$ to points on C_1 and reach the same result. Another way to write this is $Lk_{12} = Lk_{21} = 1$.

In Fig. 2.1, we now transform the closed loops into magnetic flux tubes. That is, each loop becomes the axis of a thin flux tube with flux Φ_i, $i = 1, 2$. All field lines are parallel to the tube axis: i.e., for the moment, we do not consider internal twist. With this construction of the flux tubes, we have that

$$\Phi_i \boldsymbol{t}_i \, \mathrm{d}x = \boldsymbol{B} \, \mathrm{d}^3x,$$

which connects flux and line elements to magnetic field and volume elements. This means by rescaling $\mathrm{d}\boldsymbol{x}$ of C_1 and $\mathrm{d}\boldsymbol{y}$ of C_2 in equation (2.4) by the magnetic flux Φ_i associated with each loop, we find the helicity in equation (2.3). In other words, the magnetic helicity in equation (2.3) of closed magnetic fields measures the (pairwise) Gauss linking number of field lines weighted by magnetic flux. For the example in Fig. 2.1, the value is $H = Lk_{12}\Phi_1\Phi_2 + Lk_{21}\Phi_2\Phi_1 = 2\Phi_1\Phi_2$. Although magnetic helicity can be extended to incorporate more complex magnetic fields that have internal twist (Moffatt and Ricca, 1992) and ergodic field lines (Laurence and Avellaneda, 1993), the basic picture we have presented here holds. That is, magnetic helicity is a measure of field line topology weighted by magnetic flux. This conceptual definition of helicity will be useful for the rest of this chapter.

2.2. RELATIVE HELICITY

So far, we have considered magnetic helicity in a magnetically closed domain. However, there are many applications, particularly in solar physics, for which, at some boundary, nonzero normal components of the magnetic field must be included. In this situation, equation (2.1) no longer represents a gauge invariant quantity. Instead, a relative measure of magnetic helicity, which compares two magnetic fields with the same normal boundary conditions, is required. From this moment onward in this chapter, although not strictly necessary for everything we introduce, our domain Ω will be simply connected. This choice not only simplifies many descriptions but also relates to practical applications – the domains of most solar simulations and observations are simply connected.

Following Berger and Field (1984) and Finn and Antonsen (1985), relative magnetic helicity has the form

$$H_R(\boldsymbol{B}, \boldsymbol{B}') = \int_\Omega (\boldsymbol{A} + \boldsymbol{A}') \cdot (\boldsymbol{B} - \boldsymbol{B}') \, \mathrm{d}^3x, \quad (2.5)$$

Where $\boldsymbol{B}'$ is a reference magnetic field with $\boldsymbol{B}' \cdot \boldsymbol{n} = \boldsymbol{B} \cdot \boldsymbol{n}$ on $\partial\Omega$ and $\boldsymbol{A}'$ is a reference vector potential. Similar to equation (2.1), equation (2.5) is too general to reveal clear information about the field line topology. We cannot just substitute in equation (2.2) as before, since field lines are not necessarily closed, so the Gauss linking number is no longer a topological invariant. Instead, we must find another description of field line topology, this time starting from equation (2.5). To achieve this, we will further restrict our attention to "tubular" domains, as illustrated in Fig. 2.2.

By "tubular," we mean the following: the domain Ω is bounded above and below by horizontal planes. The magnetic field in Ω must be tangent to side boundaries. Field lines can connect to both, one, or neither of the upper and lower boundaries. Although this restricts what kind

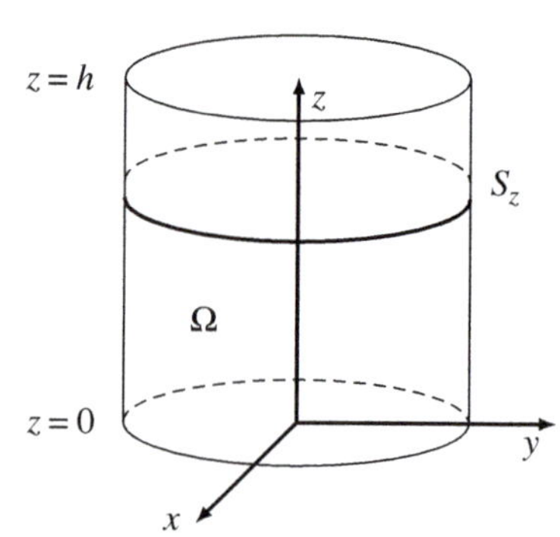

Figure 2.2 A "tubular" domain Ω between two horizontal planes at $z = 0$ and $z = h$. At each height in Ω, the horizontal slice is denoted S_z.

of magnetic field we consider, these conditions include a huge variety of configurations that are of interest to us in applications of the theory (to MHD simulations and solar observations).

Now that we have a fixed but general set of conditions on the magnetic field, we can consider a particular gauge that will reveal the nature of the field line topology, just as the Biot-Savart operator does for classical (closed-field) helicity. The gauge of choice is called the *winding gauge*, for reasons that will be made clear shortly, and is given by

$$A^W(B)(x_1, x_2, z) = \frac{1}{2\pi}\int_{S_z} B(y_1, y_2, z) \times \frac{x - y}{|x - y|^2}\, \mathrm{d}^2 y, \tag{2.6}$$

where S_z is a horizontal slice of Ω at height z and x and y are horizontal position vectors, on S_z, pointing to the intersection of field lines with the plane S_z.

With this choice of gauge, the relative helicity in equation (2.5) can be written as

$$H_R(B, B') = H^W(B) - H^W(B'), \tag{2.7}$$

where H^W is the *winding helicity*, given by

$$H^W(B) = \int_{\Omega} A^W \cdot B\, \mathrm{d}^3 x. \tag{2.8}$$

The effect of the winding gauge in equation (2.7) is to remove the cross term, thus reducing the relative helicity to the difference between two winding helicities. If the reference field B' is chosen such that its winding helicity is zero, a choice that can always be made (Prior and Yeates, 2014), the winding helicity of B is equivalent to its relative helicity. For the rest of this chapter, the winding and relative helicities will be considered interchangeable.

2.3. MAGNETIC WINDING

2.3.1. Winding Helicity

We are now in a position to consider the field line topology of (relative or winding) helicity in Ω. Inserting the expression for A^W in equation (2.8) leads to

$$H^W(B) = \frac{1}{2\pi}\int_0^h \int_{S_z}\int_{S_z} B(x) \cdot B(y) \times \frac{x - y}{|x - y|^2}\, \mathrm{d}^2 x\, \mathrm{d}^2 y\, \mathrm{d}z. \tag{2.9}$$

Note the similarity between equations (2.9) and (2.3), suggesting that they share common topological properties. This is indeed the case; and to show this, we need to do some more work, as the Gauss linking number is no longer viable as a topological measure of open magnetic fields. Prior and Yeates (2014) showed that the winding helicity in Ω can be written as

$$H^W(B) = \frac{1}{2\pi}\int_0^h \int_{S_z}\int_{S_z} \frac{\mathrm{d}}{\mathrm{d}z}\theta(x, y) B_z(x) B_z(y)\, \mathrm{d}^2 x\, \mathrm{d}^2 y\, \mathrm{d}z, \tag{2.10}$$

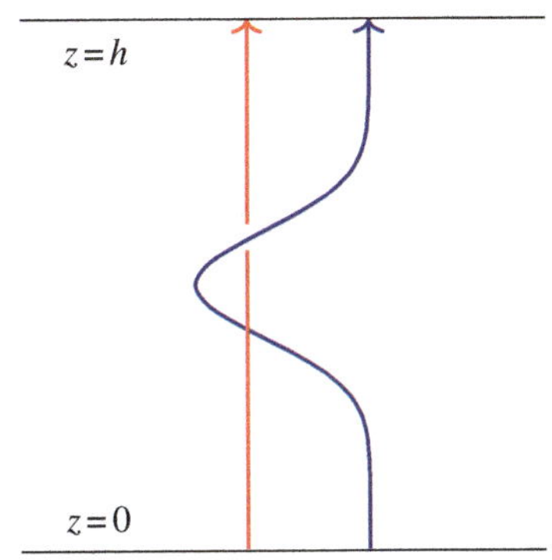

Figure 2.3 Two discrete and braided flux tubes between $z = 0$ and $z = h$. Both tubes have the same magnetic flux Φ.

where $\theta(x, y)$ is the polar angle of $x - y$, given by

$$\theta(x, y) = \arctan\left[\frac{(x - y)\cdot e_y}{(x - y)\cdot e_x}\right], \tag{2.11}$$

with the standard Cartesian basis $\{e_x, e_y, e_z\}$. From equation (2.10), the underlying field line topology of winding helicity becomes clear. The winding of the polar angle $\theta(x, y)$ from the base to the top of Ω provides a measure of the field line topology. This quantity is weighted by the vertical magnetic flux at locations x and y. Thus, the winding helicity represents the net pairwise winding of field lines weighted by magnetic flux. As an example, consider the magnetic field shown in Fig. 2.3, which is a periodic braid formed by two discrete, thin, and untwisted flux tubes with the same magnetic flux Φ.

In this simple example, the winding and flux parts of the winding helicity integral can be performed separately, giving

$$H^W = \frac{\theta(x, y)|_{z=0}^{z=h}}{2\pi}\Phi^2 + \frac{\theta(y, x)|_{z=0}^{z=h}}{2\pi}\Phi^2, \tag{2.12}$$

where the two parts on the right-hand side of equation (2.12) correspond to the winding of one tube about the other and vice versa. From Fig. 2.3, the flux tubes wind about each other for a full rotation, so the difference in θ between the top and bottom of Ω is 2π. Thus, the winding helicity for the field in Fig. 2.3 is $H^W = 2\Phi^2$. Note that this is the same formula as for the link example from Fig. 2.1. If we were to connect the top and bottom planes of Ω without introducing any twist, we would re-create the field in Fig. 2.1. In other words, the pairwise winding number for periodic strings is equivalent numerically to the Gauss linking number of the link formed by closing the braid (Berger and Prior, 2006). In this sense, the pairwise winding of field lines can be considered a generalization of the pairwise linkage of field lines.

2.3.2. Separating Out the Topology

Building on the discussion in the last section, we can create a measure of the field line topology independent of the magnetic field strength. To do this, we renormalize the winding helicity to create the *magnetic winding*.

$$L(\boldsymbol{B}) = \frac{1}{2\pi}\int_{S_z}\int_{S_z}\frac{\mathrm{d}}{\mathrm{d}z}\theta(\boldsymbol{x},\boldsymbol{y})\sigma(\boldsymbol{x})\sigma(\boldsymbol{y})\,\mathrm{d}^2x\,\mathrm{d}^2y\,\mathrm{d}z, \quad (2.13)$$

where σ is an indicator function given by

$$\sigma(\boldsymbol{x}) = \begin{cases} 1 & \text{if } B_z(\boldsymbol{x}) > 0, \\ -1 & \text{if } B_z(\boldsymbol{x}) < 0, \\ 0 & \text{if } B_z(\boldsymbol{x}) = 0. \end{cases} \quad (2.14)$$

Equation (2.13) represents the net pairwise winding of all field lines integrated over slices in Ω. Berger and Prior (2006) defined a generalized form of winding for which the winding of monotonically increasing field lines is a special case. Let two field lines $\boldsymbol{x}$ and $\boldsymbol{y}$ have n and m distinct turning points, respectively – that is, points where $\mathrm{d}x_z/\mathrm{d}z = 0$ with $\boldsymbol{x}\cdot\boldsymbol{e}_z = x_z$ or $\mathrm{d}y_z/\mathrm{d}z = 0$ with $\boldsymbol{y}\cdot\boldsymbol{e}_z = y_z$. Let $\boldsymbol{x}$ be partitioned into $n+1$ regions and $\boldsymbol{y}$ into $m+1$ regions so that curve sections $\boldsymbol{x}_i$ and $\boldsymbol{y}_j$ share a mutual z-range $[z_{ij}^{\min}, z_{ij}^{\max}]$ in each region. The total winding is defined as the sum of the weighted pairwise winding in each region,

$$L(\boldsymbol{x},\boldsymbol{y}) = \sum_{i=1}^{n+1}\sum_{j=1}^{m+1}\frac{\sigma(\boldsymbol{x}_i)\sigma(\boldsymbol{y}_j)}{2\pi}\int_{z_{ij}^{\min}}^{z_{ij}^{\max}}\frac{\mathrm{d}}{\mathrm{d}z}\theta(\boldsymbol{x}_i(z),\boldsymbol{y}_j(z))\,\mathrm{d}z, \quad (2.15)$$

where $\sigma(\boldsymbol{x}_i)$ is the indicator function in equation (2.14). This generalized winding of two field lines is represented in equation (2.13), although now it is interpreted as the net winding between all pairs of field lines over an area.

2.3.3. Magnetic Winding Flux

In solar physics applications, the flux of magnetic helicity through the photosphere can be estimated directly from observations rather than the full three-dimensional field calculation. It has long been known (Berger, 1988; Démoulin and Berger, 2003) that the time-integrated relative helicity flux through a plane P (here taken to represent the photosphere) can be written as

$$H_R = -\frac{1}{2\pi}\int_0^T\int_P\int_P\frac{\mathrm{d}}{\mathrm{d}t}\theta(\boldsymbol{x},\boldsymbol{y})B_z(\boldsymbol{x})B_z(\boldsymbol{y})\,\mathrm{d}^2x\,\mathrm{d}^2y\,\mathrm{d}t, \quad (2.16)$$

where $t\in[0,T]$ is the time in a fixed period and all other symbols have their standard meanings. Notice that equation (2.16) is equivalent mathematically to the winding helicity in equation (2.10), with the transformation $z\in[0,h]\leftrightarrow t\in[T,0]$. Thus, the time-integrated relative helicity inherits winding as its underlying topological description. Instead of slices S_z stacked in height z, we now have different realizations of the photosphere P stacked in time t. For more discussion of the connections between winding helicity and the time-integrated helicity flux, the reader is directed to Prior and MacTaggart (2020) and Candelaresi et al. (2021).

Similar to the renormalization of the winding helicity, the time-integrated magnetic winding flux is given by

$$L_R = -\frac{1}{2\pi}\int_0^T\int_P\int_P\frac{\mathrm{d}}{\mathrm{d}t}\theta(\boldsymbol{x},\boldsymbol{y})\sigma(\boldsymbol{x})\sigma(\boldsymbol{y})\,\mathrm{d}^2x\,\mathrm{d}^2y\,\mathrm{d}t. \quad (2.17)$$

This formula allows for a direct measure of field line topology to be estimated from solar observations without the need for additional models, such as three-dimensional extrapolations of the magnetic field.

2.4. APPLICATIONS

2.4.1. Theoretical Examples

Until now, we have shown that magnetic helicity can be thought of as a combination of "field line topology" (winding) and "magnetic flux." We may ask, if magnetic helicity already contains magnetic winding, why is there any need to separate the two? The answer is that despite the intimate relationship between both quantities, they can produce different information about the magnetic field.

Consider a magnetic field that is weakly entangled but has a very strong field strength. The magnetic helicity of such a field could be large (due to the field strength weighting) but the magnetic winding small (as the effect of the field strength has been removed, leaving a direct measure of field line topology). Similarly, a field with a strong net entanglement but weak field strength may yield the opposite result.

Quasistatic Decay

One simple analytical example of how magnetic helicity and magnetic winding differ is the behavior of a linear force-free field decaying subject to constant magnetic diffusion. Assuming no flow and a magnetic diffusion η, the induction equation becomes the canonical diffusion equation

$$\frac{\partial\boldsymbol{B}}{\partial t} = \eta\nabla^2\boldsymbol{B}. \quad (2.18)$$

Jette (1970) proved that the only force-free fields that remain force-free while subject to equation (2.18) are linear force-free fields, $\nabla\times\boldsymbol{B} = \alpha\boldsymbol{B}$ with constant α. It was shown in Prior and MacTaggart (2020) that the winding helicity (although the same can also be shown for classical helicity) satisfies

$$H^W = H_0^W \exp(-2\alpha^2 \eta t), \tag{2.19}$$

where H_0^W is the winding helicity at $t = 0$. In equation (2.19), only the magnitude of the helicity is decaying in time; the field line topology encoded in H_0^W remains unchanged. Thus $L = L_0$ for $t \geq 0$, and the field line topology and magnetic flux are behaving separately in different ways. This example reveals that even subject to simple conditions, the magnetic helicity and magnetic winding describe different characteristics of the evolving magnetic field.

Discrete Flux Tubes

Explicit formulae have been developed for the relative mutual helicities of discrete, thin flux tubes (Berger, 1986). Here, we present a simple demonstration to show that winding provides the underlying topology of these formulae.

Figure 2.4 shows two discrete and crossing flux tubes that connect to the same plane. Figure 2.4(a) shows the two flux tubes α (with flux Φ_α) and β (with flux Φ_β), with arrows indicating the field directions. Each tube is split in two, on either side of the crossing point, with the section labels $\alpha = (\alpha_1, \alpha_2)$ and $\beta = (\beta_1, \beta_2)$. Figure 2.4(b) shows the orthogonal projection of the tubes onto the lower plane and indicates the key winding angles (which should be viewed from this projection). The winding formula in equation (2.13) reduces to a sum when the flux tubes are treated as strings. Angles rotating in the counterclockwise direction are positive. Using equation (2.14) to account for the directions of the flux tubes, the pairwise winding values of various segments, starting from α, are

$$L(\alpha_1, \beta_1) = \frac{\theta_1}{2\pi}, \quad L(\alpha_1, \beta_2) = \frac{\theta_2}{2\pi}, \quad L(\alpha_2, \beta_1) = \frac{\theta_3}{2\pi},$$
$$L(\alpha_2, \beta_2) = \frac{\theta_4}{2\pi}.$$

The total winding is the sum of these component parts, so the mutual helicity of α relative to β is given by

$$H_{\alpha\beta}^W = \frac{1}{2\pi}\left(\sum_{i=1}^{4} \theta_i\right)\Phi_\alpha \Phi_\beta. \tag{2.20}$$

Since $H_{\alpha\beta}^W = H_{\beta\alpha}^W$ (Berger, 1986; Candelaresi et al., 2021), the total mutual helicity is

$$H_{\alpha\beta}^W = \frac{1}{\pi}\left(\sum_{i=1}^{4} \theta_i\right)\Phi_\alpha \Phi_\beta. \tag{2.21}$$

This expression for the mutual helicity of two crossing flux tubes matches that reported in other works (e.g., Berger, 1986; Wright, 2019). This example clearly shows how the pairwise winding of the flux tubes is a fundamental constituent of the mutual helicity.

Flux Emergence

We now consider a more complex example of a magnetic flux tube with a nontrivial internal field line topology emerging and submerging through a fixed horizontal plane, as shown in Fig. 2.5. Although this example represents a highly idealized realization of flux emergence, it clearly shows how magnetic helicity and magnetic winding can provide distinct information about a magnetic field.

Although all the field lines in the tube are twisted, there is a strong concentration of twist near the apex of the

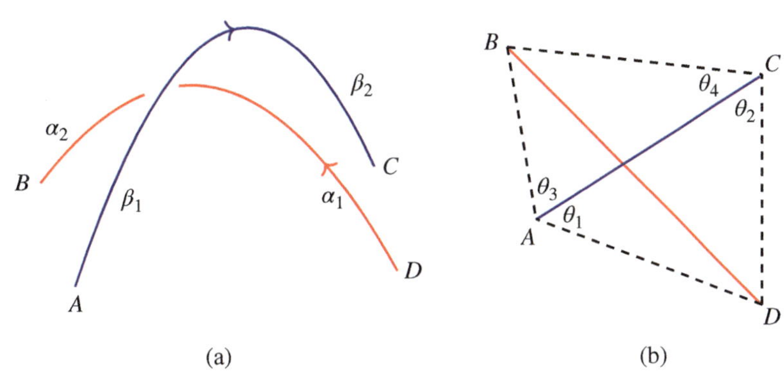

Figure 2.4 A diagram of two crossing flux tubes connected to the same lower boundary. (a) A three-dimensional representation of the tubes, with different parts of the tubes given different labels. (b) A projection orthogonal to the lower boundary, indicating the winding angles.

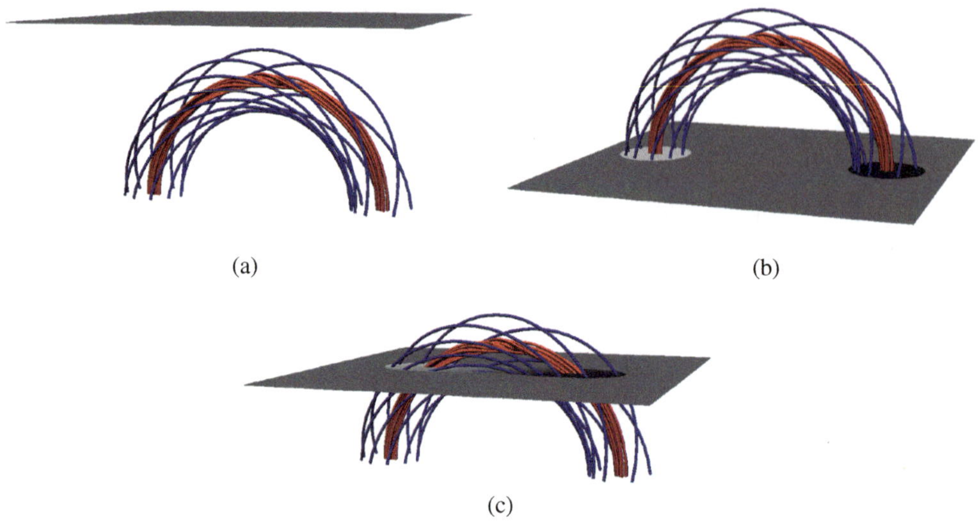

Figure 2.5 The emergence and submergence of a twisted flux tube. (a) The tube before emergence. (b) The tube when it has fully emerged. (c) The tube partially submerged, with the small region of strong twist still above the surface. MacTaggart and Prior, 2021a/IOP Publishing/Licensed under CC BY 3.0.

tube, displayed in Fig. 2.5 with red field lines (outer field lines are shown in blue). Because this localized region has much greater twist than the rest of the tube, the winding associated with this region will be larger than that of the rest of the tube. Performing the emergence and submergence motion shown in Fig. 2.5 reveals differences in what the magnetic winding and magnetic helicity measure. The accumulations of the magnetic helicity and winding fluxes are displayed in Fig. 2.6.

The accumulation of magnetic winding shows a rise and plateau that corresponds to the detection of the localized region of strong twist passing through the photosphere. After the localized region passes through the photosphere, no more topologically significant field passes through this plane (the twist in the rest of the field is much weaker than that in the localized region). Thus

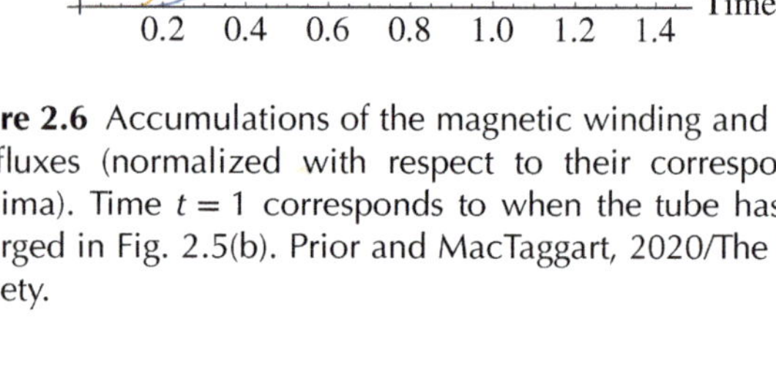

Figure 2.6 Accumulations of the magnetic winding and helicity fluxes (normalized with respect to their corresponding maxima). Time $t = 1$ corresponds to when the tube has fully emerged in Fig. 2.5(b). Prior and MacTaggart, 2020/The Royal Society.

the change in the magnetic winding flux is minimal after the emergence of the localized region.

The magnetic helicity flux, on the other hand, tells a different story. Although magnetic helicity can be said to contain magnetic winding, it combines magnetic winding with the strength of the magnetic field, which affects the profile of the helicity accumulation. The significant removal of magnetic flux during the submergence phase causes the helicity accumulation to diminish significantly. Thus the identification of the localized region is not as clear from the helicity flux as from the winding flux.

2.4.2. Solar Observations

As has been shown so far, magnetic winding and magnetic helicity provide powerful tools for analyzing magnetic fields. MacTaggart and Prior (2021b) investigated the behavior of both magnetic winding and magnetic helicity fluxes in solar flux emergence simulations. They found that a "rise and plateau" signature in the winding accumulation is characteristic of an emerging twisted flux tube. This result has also been found in observations, confirming that solar active regions can be created by twisted flux tube emergence (MacTaggart et al., 2021). As a brief example of this, Fig. 2.7 displays the winding accumulation for the active region AR11318 during its initial emergence. The total winding accumulation L is decomposed into contributions due to the emergence of field lines, L_{emerge}, and the braiding of field lines at the photosphere, L_{braid}. More details about these decompositions can be found in MacTaggart et al. (2021).

The "rise and plateau" signature can be seen in Fig. 2.7. Note that there is a data gap between 20 and 30 hours,

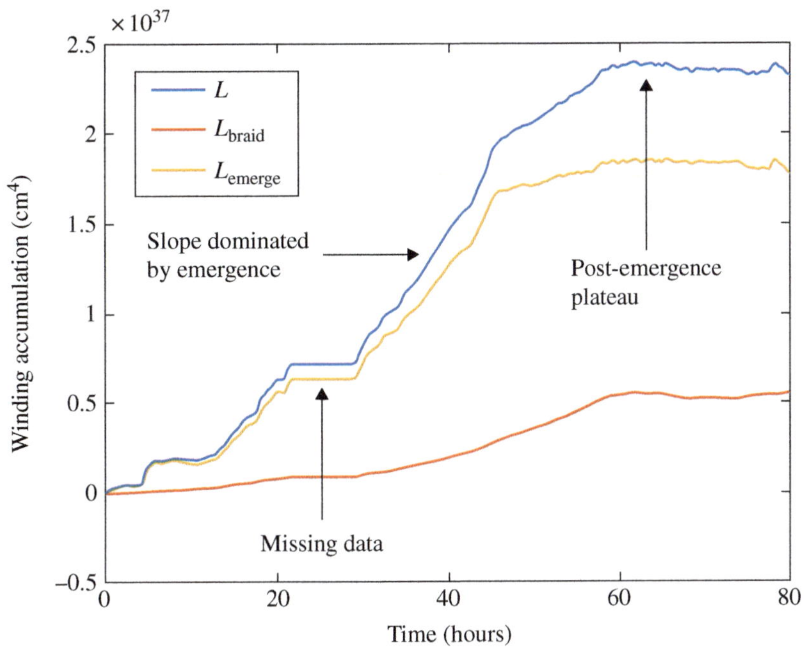

Figure 2.7 The magnetic winding accumulation for the first 80 hours of the emergence of AR11318. The accumulations of L_{emerge} and L_{braid} are shown, with $L = L_{\text{emerge}} + L_{\text{braid}}$. MacTaggart et al., 2021/Springer Nature/Licensed under CC BY 4.0.

which creates a false mini-plateau. Despite this, the global rise and plateau signature is still very clear. Further, L is dominated by L_{emerge}, meaning the winding signature is dominated by a pretwisted magnetic field emerging into the atmosphere. The interested reader can find more details in MacTaggart et al. (2021).

2.5. BRIEF HISTORICAL NOTES

The use of winding as a topological description of magnetic fields and magnetic helicity can be traced back to the work of Berger (1986). In this work, Berger derived formulae for the mutual helicity of discrete, thin flux tubes connected to the photosphere. This line of research has been developed in several other works (Wright and Berger, 1989, 1990; Wright, 2019; Démoulin et al., 2006) and has been applied to observational-based extrapolations (Tziotziou et al., 2012, 2013). Berger's original proof of the mutual helicity formulae was based on linkage rather than winding per se, and Candelaresi et al. (2021) showed that Berger's formulae can readily be derived based on winding alone (see also the discussion in this chapter). Further, Candelaresi et al. (2021) extended previous results to find, based on magnetic winding, a self-mutual helicity decomposition for open magnetic fields, thus generalizing the calculations from thin to thick flux tubes.

Berger (1993) developed the idea of winding in terms of the average crossing number in a magnetic braid. This was used to provide an energy bound for braided flux tubes, with further extensions and practical calculations given in Yeates et al. (2014). The winding gauge was developed in Prior and Yeates (2014), with the connection to winding elaborated in Prior and MacTaggart (2020). Magnetic winding flux was developed in Prior and MacTaggart (2019) and MacTaggart and Prior (2021b) in the context of flux emergence simulations. The calculation of magnetic winding flux from observational data has recently been achieved in MacTaggart et al. (2021), confirming that twisted flux tubes can create solar active regions. Almost all solar observational studies of magnetic helicity flux (using the formulations of Berger (1988), Démoulin and Berger (2003), and Pariat et al. (2005)) are based on winding being the underlying description of the field line topology. Theoretical interpretations of winding underlying magnetic helicity flux can be found in Prior and MacTaggart (2020) and MacTaggart and Prior (2021a).

REFERENCES

Berger, M. A. (1986). Topological invariants of field lines rooted to planes. *Geophysical and Astrophysical Fluid Dynamics*, *34*, 265–281.

Berger, M. A., & Field, G. B. (1984). The topological properties of magnetic helicity. *Journal of Fluid Mechanics*, *147*, 133–148.

Berger, M. A., & Prior, C. (2006). The writhe of open and closed curves. *Journal of Physics A: Mathematical and General*, *39*(26), 8321.

Berger, M. A. (1988). An energy formula for nonlinear force-free magnetic fields. *Astronomy and Astrophysics*, *201*, 355–361.

Berger, M. A. (1993). Energy-crossing number relations for braided magnetic fields. *Physical Review Letters*, *70*, 705–708.

Candelaresi, S., Hornig, G., MacTaggart, D., & Simitev, R. D. (2021). On self and mutual winding helicity. *Communications in Nonlinear Science and Numerical Simulation*, *103*, 106015.

Démoulin, P., & Berger, M. A. (2003). Magnetic energy and helicity fluxes at the photospheric level. *Solar Physics*, *215*(2), 203–215.

Démoulin, P., Pariat, E., & Berger, M. A. (2006). Basic properties of mutual magnetic helicity. *Solar Physics*, *233*, 3–27.

Faraco, D., Lindberg, S., MacTaggart, D., & Valli, A. (2022). On the proof of Taylor's conjecture in multiply connected domains. *Applied Mathematics Letters*, *124*, 107654.

Finn, J. M., & Antonsen, T. M. (1985). Magnetic helicity: What is it and what is it good for? *Comments on Plasma Physics and Controlled Fusion*, *9*(3), 111–126.

Jette, A. D. (1970). Force-free magnetic fields in resistive magnetohydrostatics. *Journal of Mathematical Analysis and Applications*, *29*(1), 109–122.

Laurence, P., & Avellaneda, M. (1993). A Moffatt-Arnold formula for the mutual helicity of linked flux tubes. *Geophysical & Astrophysical Fluid Dynamics*, *69*, 243–256.

MacTaggart, D., & Prior, C. (2021a). Magnetic winding – a key to unlocking topological complexity in flux emergence. *Journal of Physics: Conference Series*, *1730*, 012013.

MacTaggart, D., & Prior, C. (2021b). Helicity and winding fluxes as indicators of twisted flux emergence. *Geophysical & Astrophysical Fluid Dynamics*, *115*, 85–124.

MacTaggart, D., Prior, C., Raphaldini, B., Romano, P., & Guglielmino, S. L. (2021). Direct evidence that twisted flux tube emergence creates solar active regions. *Nature Communications*, *12*, 6621.

MacTaggart, D., & Valli, A. (2019). Magnetic helicity in multiply connected domains. *Journal of Plasma Physics*, *85*(5), 775850501.

Moffatt, H. K. (1969). The degree of knottedness of tangled vortex lines. *Journal of Fluid Mechanics*, *35*, 117–129.

Moffatt, H. K., & Ricca, R. L. (1992). Helicity and the Călugăreanu invariant. *Proceedings of the Royal Society of London. Series A: Mathematical and Physical Sciences*, *439*, 411–429.

Pariat, E., Démoulin, P., & Berger, M. A. (2005). Photospheric flux density of magnetic helicity. *Astronomy and Astrophysics*, *439*, 1191–1203.

Prior, C., & MacTaggart, D. (2019). Interpreting magnetic helicity flux in solar flux emergence. *Journal of Plasma Physics*, *85*(2).

Prior, C., & MacTaggart, D. (2020). Magnetic winding: what is it and what is it good for? *Proceedings of the Royal Society A: Mathematical, Physical and Engineering Sciences*, *476*(2242), 20200483.

Prior, C., & Yeates, A. R. (2014). On the helicity of open magnetic fields. *The Astrophysical Journal*, *787*(2), 100.

Ricca, R., & Nipoti, B. (2011). Gauss' linking number revisited. *Journal of Knot Theory and Its Ramifications*, *20*, 1325–1343.

Tziotziou, K., Georgoulis, M. K., & Liu, Y. (2013). Interpreting eruptive behaviour in NOAA AR 11158 via the region's magnetic energy and relative-helicity budgets. *Astrophysical Journal*, *772*, 115.

Tziotziou, K., Georgoulis, M. K., & Raouafi, N.-E. (2012). The magnetic energy-helicity diagram of solar active regions. *Astrophysical Journal Letters*, *759*, L4.

Wright, A. N. (2019). Partitioning of magnetic helicity in reconnected flux tubes. *Astrophysical Journal*, *878*, 102.

Wright, A. N., & Berger, M. A. (1989). The effect of reconnection upon the linkage and interior structure of magnetic flux tubes. *Journal of Geophysical Research*, *94*, 1295–1302.

Wright, A. N., & Berger, M. A. (1990). The interior structure of reconnected flux tubes in a sheared plasma flow. *Journal of Geophysical Research*, *95*, 8029–8036.

Yeates, A. R., Bianchi, F., Welsh, B. T., & Bushby, P. J. (2014). The coronal energy input from magnetic braiding. *Astronomy and Astrophysics*, *564*, A131.

3

Transport in Helical Fluid Turbulence

Nobumitsu Yokoi

ABSTRACT

Kinetic helicity (hereafter helicity) is defined as the correlation between velocity and flow-aligned vorticity. Helicity, like energy, is an inviscid invariant of hydrodynamic equations. In contrast to energy, which is a measure of turbulent intensity, turbulent helicity, representing the right-handed and left-handed twist associated with a fluctuating motion, provides a measure of the structural or topological property of the fluctuation. The helicity effect on turbulent transport can be obtained analytically in the framework of multiple-scale renormalized perturbation expansion theory by including the non-reflectionally symmetric part for the lowest-order (homogeneous and isotropic) velocity correlation. The physical significance of the helicity-related contribution to momentum transport is explained in this chapter. By utilizing the analytical expression of Reynolds stress, a turbulence model is constructed, incorporating the helicity effect (helicity model). This helicity model is applied to a swirling flow to show its validity in describing the flow's significant properties. In addition to transport suppression, inhomogeneous helicity coupled with rotation can induce a large-scale flow. The results of direct numerical simulations (DNSs) confirming global flow generation by helicity will also be reviewed, followed by several possible applications in geophysical and astrophysical flow phenomena.

3.1. INTRODUCTION

Kinetic helicity measures broken mirror symmetry and represents the geometrical and topological properties of fluid turbulence. The presence of helicity alters the dynamic and statistical properties of turbulence. Beyond fluid dynamics, helicity plays a role in areas of research such as DNA and biochemistry, ventral nodal flow in the generation of left-right asymmetry, vortex entanglement in quantum fluids, chirality in neutrino left-right skewness asymmetry in the universe, zonal flow generation in fusion plasma devices, etc.

The first discussions of the importance of helicity were in the context of magnetic-field generation: dynamos. If fluid motion is helical, the magnetic field frozen into the fluid can also be twisted. Then the mean electric-current density configuration parallel or antiparallel to the original mean magnetic flux tube can be generated. Assuming cyclonic fluid motion – which is related to cyclones and anticyclones in the atmosphere – in small scales, Parker (1955) proposed a mechanism in which an azimuthal or toroidal magnetic field is deformed to become a dipole or poloidal magnetic field and vice versa. This is the α dynamo, considered one of the most relevant processes in the stellar migratory dynamo cycle. Helicity in turbulence is an essential ingredient of this dynamo. An inhomogeneous velocity fluctuation $\mathbf{u}'$ along the large-scale magnetic field $\mathbf{B}$ induces a magnetic-field fluctuation $\mathbf{b}'$. In the presence of kinetic helicity in turbulence, $\langle \mathbf{u}' \cdot (\nabla \times \mathbf{u}') \rangle$, an electromotive force component, $\langle \mathbf{u}' \times \mathbf{b}' \rangle$, parallel or antiparallel to the large-scale magnetic field can be induced, depending

Institute of Industrial Science, University of Tokyo, Tokyo Japan

Helicities in Geophysics, Astrophysics, and Beyond, Geophysical Monograph 283, First Edition.
Edited by Kirill Kuzanyan, Nobumitsu Yokoi, Manolis K. Georgoulis, and Rodion Stepanov.

DOI:10.1002/9781119841715.ch03

on the sign of the turbulent helicity ($\langle\cdots\rangle$: ensemble average). This contrasts with turbulent energy, which always induces an electromotive force antiparallel to large-scale electric-current density, enhancing magnetic diffusivity due to turbulence. In the presence of current helicity in turbulence, $\langle \mathbf{b}' \cdot (\nabla \times \mathbf{b}') \rangle$, mediated by part of the fluctuating Lorentz force $(\nabla \times \mathbf{b}') \times \mathbf{B}$, an electromotive force can be induced in a direction parallel or antiparallel to the large-scale magnetic field $\mathbf{B}$. For the kinetic-helicity effect in a dynamo as well as other helicities effects, the reader is referred to textbooks and review papers on dynamos (Moffatt, 1978; Parker, 1979; Krause & Rädler, 1980; Brandenburg & Subramanian, 2005; Yokoi, 2013; Moffatt & Dormy, 2019; Tobias, 2021).

Since helicity is a measure of broken mirror symmetry, a prerequisite for nonzero helicity is an element that breaks mirror symmetry. Rotation is one such element that causes the breakage of mirror symmetry. As we see later in the transport equation for turbulent helicity (section 3.4.2), rotation coupled with inhomogeneity provides turbulence with a finite helicity through the flux across the boundary. In geophysical and astrophysical flows, rotation and density stratification are basic ingredients of the system. Strong stratification (low Froude number $Fr = U/(N_{\rm BV} h) \lesssim 1$, where U is the horizontal velocity scale, h is the vertical length scale, and $N_{\rm BV}$ is the Brunt–Väisälä frequency) causes the fluid to be deflected almost purely horizontally, and the flow becomes two-dimensional. Even with this two-dimensionality in the large-scale flow structure, small-scale turbulent motions are usually more complex and three-dimensional. For example, the generation of helicity due to rotation and stratification is a concept dating back to Steenbeck, Krause & Rädler (1966) and Hide (1989), who demonstrated how it could occur. In the presence of inhomogeneity along the rotation axis, turbulence in a stratified fluid can be helical (Marino, Rosenberg & Pouquet, 2013; Marino, et al., 2013).

Another interesting point related to the dimensionality of turbulence is the helicity effect in two dimensions with a three-component configuration (2D3C). In the presence of strong rotation, depending on the boundary conditions, the flow becomes a quasi-2D3C state, where the helicity invariants are still present. From the viewpoint of the helicity effect, which we present in section 3.3.3, helicity inhomogeneity in space is essential. Inhomogeneity of helicity cannot be present in the third direction perpendicular to the 2D or horizontal plane in the 2D3C configuration. On the other hand, helicity inhomogeneity is possible in the horizontal plane. This horizontally inhomogeneous helicity coupled with perpendicular mean absolute vorticity (associated with rotation and non-uniform horizontal flows) may contribute to horizontal momentum transfer through Reynolds stress. These features of the 2D3C configuration would be an interesting subject to explore in the future as well as the cascade direction and scalings of energy and helicity associated with constraints involving invariants other than energy.

In fundamental theoretical studies of turbulence, how and how much helicity influences energy cascade and dissipation has often been investigated using homogeneous isotropic turbulence (HIT). By constructing statistical mechanics for turbulence in a truncated system of equations, Kraichnan (1973) postulated a cascade of helicity toward small scales. Since the classical work by André & Lesieure (1977), the influence of helicity on the evolution of turbulent spectra has been examined. In addition, the relationship between the high-helicity region and low-dissipation region in turbulence has been studied by Rogers & Moin (1987) in several turbulent flow geometries. Recent progress in understanding helical processes in fluid turbulence has been made in various ways. Introducing the helical decomposition of velocity, Waleffe (1993) argued that energy and helicity cascade with two classes of triad interaction of turbulence, depending on whether the longest legs of the triad are helical modes of the same or opposite sign. Using direct numerical simulation (DNS) of HIT 3D turbulence, it was shown in Chen, Chen, & Eyink (2003) that strong fluxes of positive and negative helical modes cancel with each other at high wave numbers, leading to a return to full isotropy at small scale. By examining the subcategories of nonlinear interaction, it has been found that different subsets of interaction conserve different combinations of turbulent energy and helicity evolutions, leading to the possibility of a partial inverse cascade of the energy of helical turbulence (Biferale, Musacchio & Toschi, 2012; Alexaxis & Biferale, 2018). By using DNSs of HIT, Taylor–Green (TG) flow with laboratory experiments of mirror-symmetric HIT and chiral von Kármán flows, the relationship between the linking number of tracer trajectories and helicity has been established. This paves the way for experimentally measuring helicity on a firm mathematical basis (Angriman, et al., 2021). Developments in homogeneous turbulence studies are briefly reviewed in Pouquet & Yokoi (2022).

In this chapter, we will not deal with the problem of cascade and dissipation in homogeneous helical turbulence; rather, we will focus on transport in inhomogeneous helical turbulence. What is the role of turbulent kinetic helicity in the dynamics of large-scale or mean velocity and vorticity? We present how to tackle this problem theoretically in inhomogeneous turbulence.

From the viewpoint of transport, the primary effect of turbulence is to drastically enhance effective transport. Eddy viscosity, eddy diffusivity, turbulent resistivity, etc. are transport enhancement effects determined by the intensity and timescale of turbulence. In some cases where symmetry is broken in turbulence, effective transport may also be suppressed. In this suppression process, the quantity related to the breakage of symmetry plays a key role in counterbalancing transport enhancement. As will be shown in later sections, helicity is crucial in transport suppression in non-mirror-symmetric fluid turbulence.

In contrast to homogeneous turbulence, where energy injection by external forcing is indispensable for sustaining turbulence, in inhomogeneous turbulence, turbulent energy and helicity are naturally provided by the system itself, arising from large-scale inhomogeneous fields such as mean velocity shear, vorticity, temperature gradient, etc., and fluxes from the boundaries. Mean fields are determined by turbulence through turbulent transport coefficients, and turbulence fields are determined by the production and transport rates directly linked to large-scale inhomogeneities. To treat this nonlinear mean–turbulence interaction consistently, we must simultaneously consider and solve the mean and turbulence fields. DNSs of real inhomogeneous turbulence at realistic parameters for geophysical and astrophysical phenomena with complex geometry and boundaries are impossible. In this situation, the turbulence modeling approach provides a powerful tool for analyzing real turbulent flows. In this chapter, we deal with incorporating such suppression effects of helicity into turbulence modeling.

The organization of this chapter is as follows. First, in section 3.2, we present the definitions and basic properties of kinetic helicity. In section 3.3, we show how to theoretically treat the helicity effect in inhomogeneous turbulence at a very high Reynolds number. A key procedure in the theoretical formulation of helical turbulence is adopting the non-mirror-symmetric part and energy-related mirror-symmetric part for the homogeneous isotropic two-mode, two-time correlation function of a fluctuating field. In section 3.4, utilizing the analytical results for relevant turbulent fluxes obtained by the theoretical formulation, we construct a turbulence model for turbulent helical flows. In the model, in addition to the eddy viscosity, which contributes to the destruction of large-scale structures, the helicity-related transport coefficient appears. This helicity effect contributes to sustaining and generating large-scale flow structures. In section 3.5, the helicity turbulence model is applied to a physically interesting and practically important flow configuration: turbulent swirling flow. It will be shown that the helicity model successfully reproduces prominent features of turbulent swirling flow that cannot be reproduced by the standard eddy-viscosity turbulence model. In section 3.6, we discuss global flow generation due to inhomogeneous turbulent helicity with theoretical and numerical analyses. In section 3.7, we summarize our arguments.

3.2. HELICITY: DEFINITION AND PROPERTIES

Helicity possesses several distinctive and interesting mathematical and physical properties. We will define helicity and then present these characteristic properties, which include helicity as a pseudoscalar and conserved quantity, geometrical and topological interpretations of helicity, and the role of helicity in cascade suppression.

(A) Definition Helicity is defined as the volume integral of the inner product of the velocity and its curl (vorticity):

$$\mathcal{H} = \int_V \mathbf{u} \cdot (\nabla \times \mathbf{u})\, dV. \tag{3.1}$$

Helicity density (hereafter, simply denoted as helicity) is defined by the inner product of velocity $\mathbf{u}$ and vorticity $\boldsymbol{\omega}(= \nabla \times \mathbf{u})$, $\mathbf{u} \cdot \boldsymbol{\omega}$. A positive helicity ($\mathbf{u} \cdot \boldsymbol{\omega} > 0$) represents a right-handed twist or screw of a fluid element along its motion $\mathbf{u}$. On the other hand, a negative helicity ($\mathbf{u} \cdot \boldsymbol{\omega} < 0$) represents a left-handed twist or screw (Fig. 3.1).

It is noteworthy that fully helical (Beltrami) situations such as Fig. 3.1 are unstable. They are fully helical but eventually become unstable and return to isotropy at small scales. This means the constraint of helicity should be less stringent at a small scale. Related to this point, the scale dependence of the relative importance of the helicity effect in Reynolds stress will be discussed at the end of section 3.3.

(B) Pseudoscalars Based on the parity property, vector quantities are divided into two categories. One is the polar or pure vector, which changes its sign under the inversion of the coordinate system. The other is the axial vector or pseudovector, which does not change its sign under inversion. For the sake of simplicity, avoiding a complex argument about transformation between left- and right-handed systems, here we adopt the argument in terms of inversion rather than reflection with respect to each symmetric surface. Under this categorization, the velocity $\mathbf{u}$ is a polar vector, while the vorticity $\boldsymbol{\omega}(= \nabla \times \mathbf{u})$ is an axial vector. It follows that the helicity $\mathbf{u} \cdot \boldsymbol{\omega}$, defined by the inner product of the polar and axial vectors, is a scalar that changes its sign under inversion of the

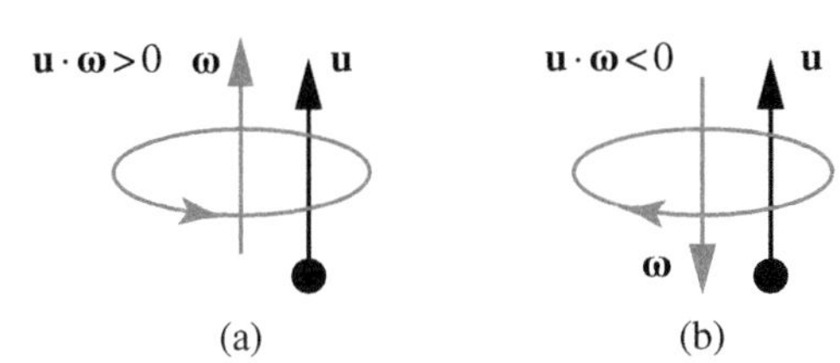

Figure 3.1 Helicity as the velocity–vorticity correlation. (a) Positive helicity; (b) negative helicity.

coordinate system. Such a scalar is called a pseudoscalar, whereas a scalar that does not change its sign under inversion is called a pure scalar.

[Note: Transformation of a vector $\mathbf{f} = \{f_i\}$ is written $f'_i = |\mathcal{A}|\mathcal{A}_{ij}f_j$, where $\mathcal{A}_{ij}$ is the transformation matrix and $|\mathcal{A}|$ is its determinant. For pseudovectors $|\mathcal{A}| = -1$. A reflection can be written in terms of a combination of inversion $\mathcal{I}$ and rotation C. For instance, a reflection or mirror in the xy plane represented by $\Sigma = \begin{pmatrix} 1 & 0 & 0 \\ 0 & 1 & 0 \\ 0 & 0 & -1 \end{pmatrix}$ can be constructed by combining the inversion $\mathcal{I} = \begin{pmatrix} -1 & 0 & 0 \\ 0 & -1 & 0 \\ 0 & 0 & -1 \end{pmatrix}$ and the π-angle rotation about the z axis, $C = \begin{pmatrix} -1 & 0 & 0 \\ 0 & -1 & 0 \\ 0 & 0 & 1 \end{pmatrix}$ as $\Sigma = \mathcal{I}C$. Since for all rotation, the determinant $|C|$ is always +1 ($|C| = +1$), the determinant of the reflection, $|\Sigma|$, is represented by that of the inversion, $|\mathcal{I}|$. In this sense, we can argue for nonmirror symmetry in terms of inversion.]

The statistical average of a pseudoscalar vanishes in a mirror-symmetric system. This can be shown as follows. In a mirror-symmetric system, by the definition of mirror symmetry, any statistical quantity $f(\mathbf{x})$ satisfies $f(\mathbf{x}) = f(-\mathbf{x})$ under inversion of the coordinate system: $\mathbf{x} \to -\mathbf{x}$. At the same time, by definition, a pseudoscalar changes its sign $f(\mathbf{x}) = -f(-\mathbf{x})$ under inversion. As a consequence, a pseudoscalar statistical quantity $f(\mathbf{x})$ satisfies $f(\mathbf{x}) = -f(-\mathbf{x}) = -f(\mathbf{x})$, leading to $f(\mathbf{x}) = 0$: the pseudoscalar statistical quantity vanishes in a mirror-symmetric system. Conversely, a finite or nonzero pseudoscalar implies that the mirror or reflectional symmetry of the system is broken. Since helicity is a pseudoscalar, it serves as a measure of broken mirror symmetry.

(C) Conserved Quantities A prominent characteristic of helicity is that helicity, like energy, is an inviscid invariant of the system of fluid dynamics. It is a conserved quantity at the limit of vanishing viscosity.

The equation that governs velocity $\mathbf{u}$ in an incompressible fluid is

$$\frac{\partial \mathbf{u}}{\partial t} + (\mathbf{u} \cdot \nabla)\,\mathbf{u} = -\nabla p + \nu\nabla^2\mathbf{u} \tag{3.2}$$

or, equivalently,

$$\frac{\partial \mathbf{u}}{\partial t} = \mathbf{u} \times \boldsymbol{\omega} - \nabla\left(p + \frac{\mathbf{u}^2}{2}\right) + \nu\nabla^2\mathbf{u}, \tag{3.3}$$

and the solenoidal condition of $\mathbf{u}$ is

$$\nabla \cdot \mathbf{u} = 0, \tag{3.4}$$

where p is the pressure normalized by a constant density ρ, and ν is the kinematic viscosity. Associated with these equations, the vorticity $\boldsymbol{\omega}(= \nabla \times \mathbf{u})$ obeys

$$\frac{\partial \boldsymbol{\omega}}{\partial t} + (\mathbf{u} \cdot \nabla)\,\boldsymbol{\omega} = (\boldsymbol{\omega} \cdot \nabla)\,\mathbf{u} + \nu\nabla^2\boldsymbol{\omega} \tag{3.5}$$

or, equivalently in the rotational form,

$$\frac{\partial \boldsymbol{\omega}}{\partial t} = \nabla \times (\mathbf{u} \times \boldsymbol{\omega}) + \nu\nabla^2\boldsymbol{\omega}, \tag{3.6}$$

and the solenoidal condition of $\boldsymbol{\omega}$ is

$$\nabla \cdot \boldsymbol{\omega} = 0. \tag{3.7}$$

It follows from equations (3.2)–(3.7) that the governing equation of helicity is given by

$$\begin{aligned} \frac{d}{dt}\mathbf{u} \cdot \boldsymbol{\omega} &\equiv \left(\frac{\partial}{\partial t} + \mathbf{u} \cdot \nabla\right) \mathbf{u} \cdot \boldsymbol{\omega} \\ &= \nabla \cdot \left[\left(\frac{1}{2}\mathbf{u}^2 - p\right)\boldsymbol{\omega} + \nu\nabla(\mathbf{u} \cdot \boldsymbol{\omega})\right] \\ &\quad -2\nu\frac{\partial u_j}{\partial x_i}\frac{\partial \omega_j}{\partial x_i}. \end{aligned} \tag{3.8}$$

Integrating equation (3.8) over the volume V and putting $\nu = 0$, we obtain

$$\frac{d}{dt}\int_V \mathbf{u} \cdot \boldsymbol{\omega}\, dV = \int_S \left[\left(\frac{1}{2}\mathbf{u}^2 - p\right)\boldsymbol{\omega}\right] \cdot \mathbf{n}\, dS. \tag{3.9}$$

This means in the absence of influx or outflux through the integral surface S, the (total) helicity $\int_V \mathbf{u} \cdot \boldsymbol{\omega}\, dV$ does not vary with the fluid motion $\mathbf{u}$. Helicity is an inviscid invariant of fluid motion, as is the energy (per unit mass) $\int_V \mathbf{u}^2/2\, dV$. This fact strongly suggests that helicity and energy are of fundamental importance in fluid dynamics.

[Note: Of course, putting $\nu = 0$ is not permissible in treating turbulent flows since viscosity plays an essential role in turbulent dissipation. For example, kinetic helicity decays on a turbulent time scale regardless of the nature of the forcing and the value of the Reynolds number (Brandenburg & Petrosyan, 2012).]

(D) Geometrical and Topological Interpretations Helicity and its properties can be interpreted and explained in terms of topology (Moffatt, 1978, 1992; Moffatt & Dormy, 2019). Helicity corresponds to the topological properties of flows through knots, links, twists, and writhes of vortex tubes. As the simplest case for links, we consider the Hopf link of two vortex tubes, Φ_1 and Φ_2, which are along closed loops C_1 and C_2, respectively. They are linked as in Fig. 3.2(a). The surface areas spanned by closed loops C_1 and C_2 are given as $S_1(C_1)$ and $S_2(C_2)$, respectively. The volumes of the vortex-tube regions are denoted as V_1 and V_2.

When vorticity is localized only in the regions of V_1 and V_2, the integral of the helicity over the whole space volume V can be expressed by the sum of the volume integrals over V_1 and V_2:

$$\int_V \mathbf{u} \cdot \boldsymbol{\omega}\, dV = \int_{V_1} \mathbf{u} \cdot \boldsymbol{\omega}\, dV_1 + \int_{V_2} \mathbf{u} \cdot \boldsymbol{\omega}\, dV_2. \tag{3.10}$$

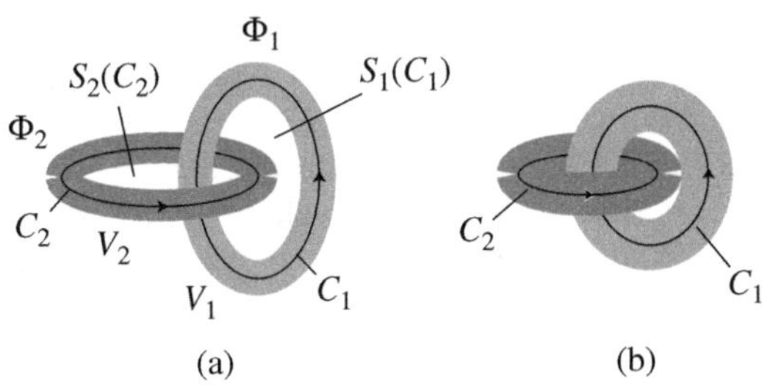

Figure 3.2 Topological interpretation of helicity. (a) Linked vortex tubes (Hopf link); (b) constraint on relaxation due to helicity invariance.

Assuming that the magnitude of the vorticity $\boldsymbol{\omega}$ is uniform across the cross-section of the vortex-tube region V_1, we have

$$\int_{V_1} \mathbf{u}\cdot\boldsymbol{\omega}\, dV_1 = \Phi_1 \oint_{C_1} \mathbf{u}\cdot d\mathbf{s} = \Phi_1 \int_{S_1(C_1)} (\nabla\times\mathbf{u})\cdot\mathbf{n}_1\, dS_1$$
$$= \Phi_1 \int_{S_1(C_1)} \boldsymbol{\omega}_2\cdot\mathbf{n}_1\, dS_1 = \Phi_1\Phi_2. \qquad (3.11)$$

Here we have made use of the assumption that the vorticity is localized, so the vorticity threading its way through the surface $S_1(C_1)$ is solely $\boldsymbol{\omega}_2$. With a similar calculation on region V_2, total helicity is

$$\int_V \mathbf{u}\cdot\boldsymbol{\omega}\, dV = 2\Phi_1\Phi_2. \qquad (3.12)$$

In the general case where the number of links is N (N: winding number), total helicity is expressed as

$$\int_V \mathbf{u}\cdot\boldsymbol{\omega}\, dV = 2N\Phi_1\Phi_2. \qquad (3.13)$$

This shows that helicity represents a link between vortex tubes.

Helicity conservation in the inviscid case corresponds to the fact that vortex tubes are immortal (neither arising nor ceasing) and the linking number of the vortex tubes is invariant in the case of vanishing viscosity ($\nu = 0$).

Due to the $(\boldsymbol{\omega}\cdot\nabla)\mathbf{u}$ term in the vorticity equation (3.5), vorticity is enhanced (or reduced) by stretching (or shrinking) the vortex tube. As we saw in equation (3.13), helicity conservation is equivalent to the invariance of the linking number. Because of the invariance of the linking number, vortex shrinking is restricted as in Fig. 3.2(b). This topological constraint is linked to the suppression of the energy cascade due to helicity conservation.

(E) Cascade Suppression Between the velocity and vorticity vectors, $\mathbf{u}$ and $\boldsymbol{\omega}$, we have the Pythagorean identity:

$$\frac{(\mathbf{u}\cdot\boldsymbol{\omega})^2}{|\mathbf{u}|^2|\boldsymbol{\omega}|^2} + \frac{(\mathbf{u}\times\boldsymbol{\omega})^2}{|\mathbf{u}|^2|\boldsymbol{\omega}|^2} = 1. \qquad (3.14)$$

This relationship implies that in a flow with greater $|\mathbf{u}\cdot\boldsymbol{\omega}|$, $|\mathbf{u}\times\boldsymbol{\omega}|$ is relatively smaller. Since $\mathbf{u}\times\boldsymbol{\omega}$ represents part of the nonlinear interaction in equations (3.3) and (3.6), in a flow with a large $|\mathbf{u}\cdot\boldsymbol{\omega}|$, the energy cascade due to nonlinear interaction may be suppressed. With this expectation, the role of helicity in turbulence cascades has been studied for many years. These classical studies showed that energy dissipation can be locally suppressed in a flow region where helicity is relatively strong, but with several reservations (Rogers & Moin, 1987). It was also suggested that the energy cascade can be suppressed in helical turbulence. However, such a helicity effect on energy-cascade suppression cannot be retained. The effect ceases once helicity starts cascading to smaller scales (André & Lesieure, 1977). For the helicity effect to be retained in turbulence, helicity must be sustained by mechanisms that provide turbulence with helicity. When considering such helicity generation mechanisms, we understand the importance of treating the helicity effect in inhomogeneous turbulence such as rotating turbulence, helicity externally injected into turbulence, etc.

3.3. THEORETICAL ANALYSIS OF HELICITY EFFECTS IN INHOMOGENEOUS TURBULENCE

In the previous section, we briefly reviewed how helicity is defined and its properties. In this section, we address the problem of helicity effects in inhomogeneous turbulence and discuss how turbulent transport may be altered by helicity. We do so with an analysis with a statistical theory of inhomogeneous turbulence.

3.3.1. Equations for Mean Fields and Fluctuation Fields

We consider a system rotating with a constant angular velocity $\boldsymbol{\omega}_F$. We adopt the Reynolds decomposition of a field f as

$$f = F + f', \quad F = \langle f \rangle \qquad (3.15)$$

with

$$f = (\mathbf{u}, \boldsymbol{\omega}, p), \qquad (3.16a)$$
$$F = (\mathbf{U}, \boldsymbol{\Omega}, P), \qquad (3.16b)$$
$$f = (\mathbf{u}', \boldsymbol{\omega}', p'), \qquad (3.16c)$$

where $\mathbf{u}$ is the velocity, $\boldsymbol{\omega}(= \nabla\times\mathbf{u})$ is the vorticity, p is the pressure (scaled by a reference density), and $\langle\cdots\rangle$ denotes ensemble averaging. Under this decomposition, the equations governing the mean velocity $\mathbf{U}$ of an incompressible fluid are

$$\left(\frac{\partial}{\partial t} + \mathbf{U}\cdot\nabla\right)\mathbf{U} = \mathbf{U}\times 2\boldsymbol{\omega}_F - \nabla P - \nabla\cdot\mathcal{R} + \nu\nabla^2\mathbf{U} \qquad (3.17)$$

and the solenoidal condition

$$\nabla \cdot \mathbf{U} = 0, \tag{3.18}$$

where P is the mean pressure normalized by the density. The mean vorticity $\mathbf{\Omega}(= \nabla \times \mathbf{U})$ is subject to

$$\frac{\partial \mathbf{\Omega}}{\partial t} = \nabla \times \left[\mathbf{U} \times (\mathbf{\Omega} + 2\boldsymbol{\omega}_{\mathrm{F}}) - \nu \nabla \times \mathbf{\Omega} \right] + \nabla \times \mathbf{V}_{\mathrm{M}}, \tag{3.19}$$

$$\nabla \cdot \mathbf{\Omega} = 0. \tag{3.20}$$

In equations (3.17) and (3.19), the Reynolds stress $\boldsymbol{\mathcal{R}} = \{\mathcal{R}_{ij}\}$ and the ponderomotive or vortex-motive force $\mathbf{V}_{\mathrm{M}}$ are defined by

$$\mathcal{R}_{ij} = \langle u_i' u_j' \rangle, \tag{3.21}$$

$$\mathbf{V}_{\mathrm{M}} = \langle \mathbf{u}' \times \boldsymbol{\omega}' \rangle, \tag{3.22}$$

respectively. They represent the sole direct effects of fluctuations in the mean fields.

The velocity associated with rotation with angular velocity $\boldsymbol{\omega}_{\mathrm{F}}$ is $\mathbf{u} = \boldsymbol{\omega}_{\mathrm{F}} \times \mathbf{x}$. Taking a curl operation, we have

$$\begin{aligned} \boldsymbol{\omega} &= \nabla \times \mathbf{u} \\ &= \nabla \times (\boldsymbol{\omega}_{\mathrm{F}} \times \mathbf{x}) = \boldsymbol{\omega}_{\mathrm{F}}(\nabla \cdot \mathbf{x}) - \boldsymbol{\omega}_{\mathrm{F}} = 2\boldsymbol{\omega}_{\mathrm{F}}. \end{aligned} \tag{3.23}$$

This shows that system rotation with angular velocity $\boldsymbol{\omega}_{\mathrm{F}}$ is locally equivalent to vorticity of $2\boldsymbol{\omega}_{\mathrm{F}}$. We utilize this relationship as follows: thanks to this local equivalence between rotation and vorticity, by considering the helicity effect in the turbulence correlations in a rotating frame, we can selectively extract the effect of mean vorticity $\mathbf{\Omega}(= \nabla \times \mathbf{U})$ coupled with helicity using a lower-order calculation.

3.3.2. Statistical Analytical Theory for Inhomogeneous Turbulence

In turbulent transport, a primary aim of the statistical analytical theory of turbulence is obtaining expressions for the turbulent correlations in the mean-field equations. Turbulent correlations such as the Reynolds stress (equation (3.21)) and the vortex-motive force (equation (3.22)) determine the effective transport due to fluctuating motions. Two-scale direct-interaction approximation (TSDIA) or multiple-scale DIA (MSDIA) – a renormalized perturbation expansion theory for strongly nonlinear turbulence combined with multiple-scale analysis – is a theoretical framework that enables us to treat inhomogeneous turbulence through derivative expansion. DIA is one of the oldest modern turbulence theories, consisting of renormalized perturbation theory for homogeneous isotropic turbulence at a very large Reynolds number (Kraichnan, 1959). For details of MSDIA, the reader is referred to Yoshizawa (1984) and Yokoi (2020). In this section, we note only some points that should be considered when treating helical turbulence with MSDIA analysis.

[Note: In DIA formulation, introducing the response or Green's function as well as the velocity correlation completes the statistical closure of the system of equations for the propagators (correlation and response functions). With the aid of renormalization procedures for the propagators, summations of the expansion terms are performed partially but up to an infinite order of expansion. Regarding field theory, DIA is a line (propagator) renormalization theory with truncation of the vertex at the lowest order. It is not necessary to invoke this type of elaborate closure scheme for weak nonlinear turbulence, but in numerical simulations, the general behavior of DIA on the energy spectrum, energy transfer spectrum, microscale length scale, velocity derivative skewness, etc. at low Reynolds numbers was found to be quite good (Kraichnan, 1964). For detailed justifications of DIA, the reader is referred to classical textbooks (Leslie, 1973; McComb, 1990).]

(A) Multiple-Scale Analysis Two-scale DIA or MSDIA combines DIA and multiple-scale analysis. We introduce the slow and fast variables as

$$\mathbf{X} = \delta_x \mathbf{x}, \quad \boldsymbol{\xi} = \mathbf{x}; \quad T = \delta_t t, \quad \tau = t, \tag{3.24}$$

where δ_x and δ_t are scale parameters of the space and time variables, respectively. If δ_x and δ_t are small, $\mathbf{X}$ and T vary significantly only when the original variables $\mathbf{x}$ and t change considerably. In this sense, $\mathbf{X}$ and T are suitable for describing large-scale and slow evolution and are called slow variables. If the scale parameters δ_x and δ_t are small, we have significant scale separations between the slow and fast variables.

With these two scale variables $\mathbf{X}$, $\boldsymbol{\xi}$, T, and τ, a field quantity $f(\mathbf{x}; t)$ is divided into slowly and quickly (fast) varying components as follows:

$$f(\mathbf{x}; t) = F(\mathbf{X}; T) + f'(\mathbf{X}, \boldsymbol{\xi}; T, \tau). \tag{3.25}$$

In the two-scale formulation, the spatial and temporal derivatives are expressed as

$$\frac{\partial}{\partial x_i} = \frac{\partial}{\partial \xi_i} + \delta_x \frac{\partial}{\partial X_i}, \quad \frac{\partial}{\partial t} = \frac{\partial}{\partial \tau} + \delta_t \frac{\partial}{\partial T}. \tag{3.26}$$

This means large-scale inhomogeneities appear with scale parameters δ_x and δ_t. In this sense, the two-scale analysis is a derivative expansion with respect to the slow-variable inhomogeneities. At this stage, scale separation affects the validity of the derivative expansion. For instance, in a gradient-diffusion-type expression, such as eddy viscosity, it is not appropriate for the starting-order approximation; taking only lower-order expansion terms in the derivative expansion does not give the best approximation.

We apply this two-scale formulation to the fundamental equations of hydrodynamics. For simplicity of the analysis, hereafter we assume that the space and time scale parameters are the same and put $\delta_x = \delta_t = \delta$. In the case of an incompressible fluid, the governing equations for the velocity fluctuation are written as

$$\frac{\partial u'_i}{\partial \tau} + U_j \frac{\partial u'_i}{\partial \xi_j} + \frac{\partial}{\partial \xi_j} u'_j u'_i + \frac{\partial p'}{\partial \xi_i} - \nu \nabla_\xi^2 u'_i - 2\epsilon_{ij\ell}\omega_{\mathrm{F}j} u'_\ell$$
$$= \delta \left(-u'_j \frac{\partial U_i}{\partial X_j} - \frac{D u'_i}{DT} - \frac{\partial p'}{\partial X_i} - \frac{\partial}{\partial X_j}\left(u'_j u'_i - \langle u'_j u'_i \rangle \right) \right.$$
$$\left. -2\nu \frac{\partial^2 u'_i}{\partial X_j \partial \xi_j} \right) + \delta^2 \left(\nu \nabla_X^2 u'_i \right), \tag{3.27}$$

and the solenoidal condition as

$$\frac{\partial u'_j}{\partial \xi_j} + \delta \frac{\partial u'_j}{\partial X_j} = 0, \tag{3.28}$$

where the mean-flow advective derivative D/DT is defined as

$$\frac{D}{DT} = \frac{\partial}{\partial T} + \mathbf{U} \cdot \nabla_X. \tag{3.29}$$

(B) Fourier Representation with Respect to Fast Variables We assume that the fluctuation field is homogeneous with respect to the fast space variable $\boldsymbol{\xi}$. Then a fluctuation field is Fourier transformed as follows:

$$f'(\boldsymbol{\xi}, \mathbf{X}; \tau, T) = \int f'(\mathbf{k}, \mathbf{X}; \tau, T) \exp[-i\mathbf{k} \cdot (\boldsymbol{\xi} - \mathbf{U}\tau)]\, d\mathbf{k}. \tag{3.30}$$

Here, the factor $\mathbf{k} \cdot (\xi - \mathbf{U}\tau)$ denotes that the Fourier transform is performed in the frame traveling with the mean flow $\mathbf{U}$.

Using this Fourier representation, the equations for velocity fluctuation in the wave-number space are written as

$$\frac{\partial u'_i(\mathbf{k};\tau)}{\partial \tau} + \nu k^2 u'_i(\mathbf{k};\tau) + 2\epsilon_{ij\ell}\omega_{\mathrm{F}j} u'_\ell(\mathbf{k};\tau)$$
$$- ik_j \iint \delta(\mathbf{k}-\mathbf{p}-\mathbf{q}) u'_i(\mathbf{p};\tau) u'_j(\mathbf{q};\tau)\, d\mathbf{p} d\mathbf{q}$$
$$= \delta \left(-u'_j(\mathbf{k};\tau) \frac{\partial U_i}{\partial X_j} - \frac{D u'_i(\mathbf{k};\tau)}{\partial T_{\mathrm{I}}} - \frac{\partial p'}{\partial X_{\mathrm{I}i}} \right.$$
$$- \iint \delta(\mathbf{k}-\mathbf{p}-\mathbf{q}) \frac{\partial}{\partial X_{\mathrm{I}j}} u'_i(\mathbf{p};\tau) u'_j(\mathbf{q};\tau)\, d\mathbf{p} d\mathbf{q}$$
$$\left. + \delta(\mathbf{k}) \frac{\partial R_{ji}}{\partial X_j} \right), \tag{3.31}$$

and the solenoidal condition as

$$\mathbf{k} \cdot \mathbf{u}'(\mathbf{k};\tau) = \delta \left(-i \frac{\partial u'_j(\mathbf{k};\tau)}{\partial X_{\mathrm{I}j}} \right), \tag{3.32}$$

where

$$\left(\frac{\partial}{\partial T_{\mathrm{I}}}, \nabla_{X\mathrm{I}} \right) = \exp(-i\mathbf{k} \cdot \mathbf{U}\tau) \left(\frac{\partial}{\partial T}, \nabla_X \right) \exp(i\mathbf{k} \cdot \mathbf{U}\tau). \tag{3.33}$$

The fluctuation fields depend on slow variables $\mathbf{X}$ and T as well as on fast variables $\boldsymbol{\xi}$ and τ. However, for simplicity of notation, we suppress denoting $\mathbf{X}$ and T and just denote $\mathbf{u}'(\mathbf{k};\tau)$ in wave-number space.

(C) Scale-Parameter Expansion We expand a field quantity f with scale parameter δ as follows:

$$f' = f'_0 + \delta f'_1 + \delta^2 f'_2 + \cdots = \sum_n \delta^n f'_n. \tag{3.34}$$

Substituting equation (3.34) into the velocity fluctuation equation (3.31) and eliminating the pressure p', the lowest- or zeroth-order velocity equation is

$$\frac{\partial u'_{0i}(\mathbf{k};\tau)}{\partial \tau} + \nu k^2 u'_{0i}(\mathbf{k};\tau) + 2\epsilon_{ij\ell}\omega_{\mathrm{F}j} u'_{0\ell}(\mathbf{k};\tau)$$
$$- iM_{ij\ell}(\mathbf{k}) \iint \delta(\mathbf{k}-\mathbf{p}-\mathbf{q}) u'_{0j}(\mathbf{p};\tau) u'_{0\ell}(\mathbf{q};\tau)\, d\mathbf{p} d\mathbf{q} = 0. \tag{3.35}$$

(D) External-Field Expansion As discussed in section 3.2, helicity is a measure of reflectional symmetry. To delve into the effects of helicity, here we consider hydrodynamic turbulence in a system with rotation, which breaks reflectional symmetry. The rotation axis gives a particular direction, so a system with rotation is no longer isotropic. To utilize some properties of isotropic turbulence, we expand a turbulence field with respect to the angular velocity of rotation $\omega_{\mathrm{F}} = |\boldsymbol{\omega}_{\mathrm{F}}|$. The turbulent velocity field of the nth order in scale parameter δ expansion, $\mathbf{u}'_n$ is expanded with ω_{F} as follows:

$$\mathbf{u}'_n = \sum_{m=0}^{\infty} \omega_{\mathrm{F}}^m \mathbf{u}'_{nm}. \tag{3.36}$$

Specifically, the lowest-order field $\mathbf{u}'_{00}$ is called the basic field and denoted as $\mathbf{u}'_{\mathrm{B}}$. The equation for $\mathbf{u}'_{\mathrm{B}}$ is written as

$$\frac{\partial u'_{\mathrm{B}i}(\mathbf{k};\tau)}{\partial \tau} + \nu k^2 u'_{\mathrm{B}i}(\mathbf{k};\tau)$$
$$- iM_{ij\ell}(\mathbf{k}) \iint \delta(\mathbf{k}-\mathbf{p}-\mathbf{q}) u'_{\mathrm{B}j}(\mathbf{p};\tau) u'_{\mathrm{B}\ell}(\mathbf{q};\tau)\, d\mathbf{p} d\mathbf{q} = 0, \tag{3.37}$$

which is the same as for homogeneous isotropic turbulence.

We introduce the Green's function associated with the basic-field equation: $G'_{ij}(\mathbf{k};\tau,\tau')$. The equation for $G'_{ij}(\mathbf{k};\tau,\tau')$ is defined by

$$\frac{\partial G'_{ij}(\mathbf{k};\tau,\tau')}{\partial \tau} + \nu k^2 G'_{ij}(\mathbf{k};\tau,\tau')$$
$$- 2iM_{i\ell m}(\mathbf{k}) \iint \delta(\mathbf{k}-\mathbf{p}-\mathbf{q}) u'_{\mathrm{B}\ell}(\mathbf{p};\tau) G'_{mj}(\mathbf{q};\tau,\tau')\, d\mathbf{p} d\mathbf{q}$$
$$= D_{ij}(\mathbf{k}) \delta(\tau - \tau'). \tag{3.38}$$

Reflecting the properties of the basic-field equation (3.37), the Green's function (equation (3.38)) is the same as for homogeneous isotropic turbulence.

The first-order equation is given as

$$\begin{aligned}
&\frac{\partial u'_{1i}(\mathbf{k};\tau)}{\partial \tau} + \nu k^2 u'_{1i}(\mathbf{k};\tau) \\
&\quad - 2iM_{ij\ell}(\mathbf{k}) \iint \delta(\mathbf{k}-\mathbf{p}-\mathbf{q}) u'_{\mathrm{B}j}(\mathbf{p};\tau) u'_{\mathrm{S1}\ell}(\mathbf{q};\tau)\, d\mathbf{p}d\mathbf{q} \\
&\quad + 2\epsilon_{ij\ell}\omega_{\mathrm{F}j} u'_{\mathrm{B}\ell}(\mathbf{k};\tau) \\
&= -D_{im}(\mathbf{k}) u'_{\mathrm{B}\ell} \frac{\partial U_m}{\partial X_\ell} - D_{i\ell}(\mathbf{k}) \frac{D u'_{\mathrm{B}\ell}(\mathbf{k};\tau)}{D T_{\mathrm{I}}} \\
&\quad + 2M_{i\ell m}(\mathbf{k}) \iint \delta(\mathbf{k}-\mathbf{p}-\mathbf{q}) \frac{q_m}{q^2} u'_{\mathrm{B}j}(\mathbf{p};\tau) \\
&\quad \times \frac{\partial u'_{\mathrm{B}n}}{\partial X_{\mathrm{I}n}}(\mathbf{q};\tau)\, d\mathbf{p}d\mathbf{q} \\
&\quad + D_{ip}(\mathbf{k}) M_{\ell mnp}(\mathbf{k}) \iint \delta(\mathbf{k}-\mathbf{p}-\mathbf{q}) \frac{\partial}{\partial X_{\mathrm{I}n}} \\
&\quad \times u'_{\mathrm{B}\ell}(\mathbf{p};\tau) u'_{\mathrm{B}m}(\mathbf{q};\tau)\, d\mathbf{p}d\mathbf{q}
\end{aligned} \tag{3.39}$$

with the solenoidal condition based on equation (3.28)

$$\mathbf{u}'_1(\mathbf{k};\tau) = \mathbf{u}'_{\mathrm{S1}}(\mathbf{k};\tau) - i\frac{\mathbf{k}}{k^2}\frac{\partial u'_{\mathrm{B}\ell}(\mathbf{k};\tau)}{\partial X_{\mathrm{I}\ell}}, \tag{3.40}$$

where $\mathbf{u}'_{\mathrm{S1}}$ is the first-order solenoidal velocity satisfying

$$\nabla \cdot \mathbf{u}'_{\mathrm{S1}}(\mathbf{k};\tau) = 0. \tag{3.41}$$

We see from equation (3.39) that all large-scale inhomogeneity effects enter the right-hand side (r.h.s.) with a scale parameter δ originating from the r.h.s. of equation (3.31). On the other hand, the left-hand side (l.h.s.) of equation (3.39) is in the same form as the equation for $\mathbf{u}'_{\mathrm{B}}$ (equation (3.37)). By treating the r.h.s. of equation (3.39) as force terms, we formally solve $\mathbf{u}'_{\mathrm{S1}}$ in terms of Green's function G defined by equation (3.38). Utilizing the expressions for $\mathbf{u}'_{\mathrm{B}}$, $\mathbf{u}'_{\mathrm{S1}}$, etc., we calculate the turbulent correlations with the aid of the DIA.

(E) Basic Field with Broken Mirror Symmetry For helicity effects to work, helicity must be supplied to turbulence. As discussed in section 3.4, system rotation coupled with inhomogeneity is one such supply mechanism. Since helicity is a measure of broken mirror symmetry, we must properly capture this symmetry breakage in the theoretical formulation.

In the standard two-scale analysis, the basic field $\mathbf{u}'_{\mathrm{B}}$ is assumed to be homogeneous and isotropic. To take helicity effects into account, we assume in this work that the basic field is homogeneous and isotropic but non-mirror-symmetric.

First, we assume a generic expression for the correlation of the basic velocity field in the configuration space. In homogeneous turbulence, statistical properties do not depend on where the origin is located but are determined only by the relative position $\mathbf{r} = \mathbf{x}' - \mathbf{x}$. Then the two-point, two-time velocity correlation is written as follows:

$$\tilde{R}_{ij}(\mathbf{r};t,t') = \langle u'_i(\mathbf{x};t) u'_j(\mathbf{x}';t')\rangle = \langle u'_i(\mathbf{0};t) u'_j(\mathbf{r};t')\rangle. \tag{3.42}$$

The general forms of the correlation tensors for various geometries have been derived in Robertson (1940) and Chandrasekhar (1950). The generic expression for isotropic and homogeneous $\tilde{R}_{ij}$ can be obtained as follows (Lesieur, 2008). From the two-point, two-time correlation $\tilde{R}_{ij}$ and two arbitrary fixed vector $\mathbf{a}$ and $\mathbf{b}$, we construct a quantity $a_i \tilde{R}_{ij}(\mathbf{r};t,t') b_j$ that is a scalar and consequently isotropic. This scalar quantity should be invariant under rotation of $\mathbf{r}$, $\mathbf{a}$, and $\mathbf{b}$. So, it depends only on the lengths, relative angles, and orientations of this set of vectors: $\mathbf{r}\cdot\mathbf{r}$, $\mathbf{a}\cdot\mathbf{a}$, $\mathbf{b}\cdot\mathbf{b}$, $\mathbf{r}\cdot\mathbf{a}$, $\mathbf{r}\cdot\mathbf{b}$, $\mathbf{a}\cdot\mathbf{b}$, and $\mathbf{r}\cdot(\mathbf{a}\times\mathbf{b})$. Hence, we have

$$\begin{aligned}
a_i \tilde{R}_{ij}(\mathbf{r};t,t') b_j &= A(\mathbf{r};t,t')\mathbf{a}\cdot\mathbf{b} + B(\mathbf{r};t,t')(\mathbf{r}\cdot\mathbf{a})(\mathbf{r}\cdot\mathbf{b}) \\
&\quad + C(\mathbf{r};t,t')\mathbf{r}\cdot(\mathbf{a}\times\mathbf{b}).
\end{aligned} \tag{3.43}$$

If we choose $\mathbf{a}$ and $\mathbf{b}$ as the unit vectors in the i and j directions, $\mathbf{e}_i$ and $\mathbf{e}_j$, we obtain

$$\tilde{R}_{ij}(\mathbf{r};t,t') = A(\mathbf{r};t,t')\delta_{ij} + B(\mathbf{r};t,t') r_i r_j + C(\mathbf{r};t,t')\epsilon_{ij\ell} r_\ell, \tag{3.44}$$

or equivalently,

$$\tilde{R}_{ij} = g\left(\delta_{ij} - \frac{r_i r_j}{r^2}\right) + f\frac{r_i r_j}{r^2} + h\epsilon_{ij\ell}\frac{r_\ell}{r}, \tag{3.45}$$

with $g = A$, $f = A + r^2 B$, and $h = rC$. Here g represents the transverse velocity correlation, f is the longitudinal velocity correlation, and h is the cross-velocity correlation (Fig. 3.3). The cross-velocity correlation divided by r, limiting $r \to 0$, is equivalent to the local helicity $\langle \mathbf{u}'\cdot\boldsymbol{\omega}'\rangle$.

Corresponding to equation (3.45) in the configuration space, we have an expression in the wave-number space.

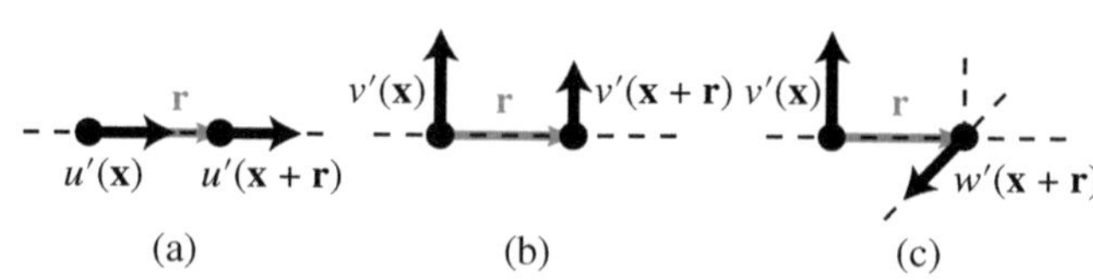

Figure 3.3 Velocity correlations. (a) Longitudinal, (b) transverse, and (c) cross-correlations. Here, $\mathbf{r}$ is the displacement vector, u' is the longitudinal component of the velocity fluctuation along $\mathbf{r}$, and v' and w' are the transverse components perpendicular to each other.

The statistical property of the basic field is assumed to be

$$\frac{\langle u'_{Bi}(\mathbf{k},\mathbf{X};\tau,T)u'_{Bj}(\mathbf{k}',\mathbf{X};\tau',T)\rangle}{\delta(\mathbf{k}+\mathbf{k}')} = D_{ij}(\mathbf{k})Q(k,\mathbf{X};\tau,\tau',T) + \Pi_{ij}(\mathbf{k})Q_C(k,\mathbf{X};\tau,\tau',T) + \frac{i}{2}\frac{k_\ell}{k^2}\epsilon_{ij\ell}H(k,\mathbf{X};\tau,\tau',T), \tag{3.46}$$

where Q is the energy spectral density of the basic field, Q_C is the compressible energy spectral density, and H is the helicity spectral density, $D_{ij}(\mathbf{k})(=\delta_{ij}-k_ik_j/k^2)$ is the solenoidal projection operator, $\Pi_{ij}(\mathbf{k})(=k_ik_j/k^2)$ is the compressible projection operator, and $\epsilon_{ij\ell}$ is the alternate tensor. In the case of an incompressible fluid, equation (3.46) is reduced to

$$\frac{\langle u'_{Bi}(\mathbf{k},\mathbf{X};\tau,T)u'_{Bj}(\mathbf{k}',\mathbf{X};\tau',T)\rangle}{\delta(\mathbf{k}+\mathbf{k}')} = D_{ij}(\mathbf{k})Q(k,\mathbf{X};\tau,\tau',T) + \frac{i}{2}\frac{k_\ell}{k^2}\epsilon_{ij\ell}H(k,\mathbf{X};\tau,\tau',T). \tag{3.47}$$

This is the most generic expression of homogeneous isotropic non-mirror-symmetric incompressible turbulence (Batchelor, 1953). At the same time, Green's function $G'_{ij}(\mathbf{k};\tau,\tau')$ is assumed to be isotropic:

$$G'_{ij}(\mathbf{k};\tau,\tau') = D_{ij}(\mathbf{k})G(k;\tau,\tau'). \tag{3.48}$$

As we see in equation (3.51) in section 3.3.3, the correlation function in the configuration space can be calculated using the spectral integral of the spectral function. It follows from equation (3.47) that turbulent energy is calculated as

$$\langle \mathbf{u}'^2_B\rangle/2 = \langle u'_{B\ell}u'_{B\ell}\rangle/2 = \int Q(k;\tau,\tau)\,d\mathbf{k}, \tag{3.49}$$

which shows that the H-related part of equation (3.47) does not contribute to turbulent energy. On the other hand, turbulent helicity is calculated as follows:

$$\langle \mathbf{u}'_B\cdot\boldsymbol{\omega}'_B\rangle = \left\langle u'_{B\ell}\epsilon_{\ell mn}\frac{\partial u'_{Bn}}{\partial x_m}\right\rangle = \int H(k;\tau,\tau)\,d\mathbf{k}. \tag{3.50}$$

Only the non-mirror-symmetric part of equation (3.47) contributes to turbulent helicity.

3.3.3. Analytical Expressions for Reynolds Stress and Vortex-Motive Force

Reynolds stress in the configuration space can be expressed using its wave-number space counterpart as

$$\mathcal{R}_{ij} = \langle u'_i(\boldsymbol{\xi},\mathbf{X};\tau,T)u'_j(\boldsymbol{\xi},\mathbf{X};\tau,T)\rangle = \int \mathcal{R}_{ij}(\mathbf{k},\mathbf{X};\tau,\tau,T)\,d\mathbf{k}, \tag{3.51}$$

where the integrand is abbreviated $\mathcal{R}_{ij}(\mathbf{k})$ and expressed as follows

$$\begin{aligned}\mathcal{R}_{ij}(\mathbf{k}) &= \frac{\langle u'_i(\mathbf{k};\tau)u'_j(\mathbf{k}';\tau)\rangle}{\delta(\mathbf{k}+\mathbf{k}')} = \frac{\langle u'_{0i}(\mathbf{k};\tau)u'_{0j}(\mathbf{k}';\tau)\rangle}{\delta(\mathbf{k}+\mathbf{k}')} \\ &+ \delta\left(\frac{\langle u'_{0i}(\mathbf{k};\tau)u'_{1j}(\mathbf{k}';\tau)\rangle}{\delta(\mathbf{k}+\mathbf{k}')} + \frac{\langle u'_{1i}(\mathbf{k};\tau)u'_{0j}(\mathbf{k}';\tau)\rangle}{\delta(\mathbf{k}+\mathbf{k}')}\right) + O(\delta^2) \\ &= \frac{\langle u'_{Bi}(\mathbf{k};\tau)u'_{Bj}(\mathbf{k}';\tau)\rangle}{\delta(\mathbf{k}+\mathbf{k}')} \\ &+ \frac{\langle u'_{Bi}(\mathbf{k};\tau)u'_{01j}(\mathbf{k}';\tau)\rangle}{\delta(\mathbf{k}+\mathbf{k}')} + \frac{\langle u'_{01i}(\mathbf{k};\tau)u'_{Bj}(\mathbf{k}';\tau)\rangle}{\delta(\mathbf{k}+\mathbf{k}')} \\ &+ \delta\left(\frac{\langle u'_{Bi}(\mathbf{k};\tau)u'_{10j}(\mathbf{k}';\tau)\rangle}{\delta(\mathbf{k}+\mathbf{k}')} + \frac{\langle u'_{10i}(\mathbf{k};\tau)u'_{Bj}(\mathbf{k}';\tau)\rangle}{\delta(\mathbf{k}+\mathbf{k}')} + \cdots\right) \\ &+ O(\delta^2).\end{aligned} \tag{3.52}$$

Substituting $\mathbf{u}'_B(\mathbf{k};\tau)$ with the statistical properties from equation (3.47) and $\mathbf{u}'_{01}(\mathbf{k};\tau)$ and $\mathbf{u}'_{10}(\mathbf{k};\tau)$ into equation (3.52), we obtain the Reynolds stress expression

$$\mathcal{R}_{ij} = \frac{2}{3}K\delta_{ij} - \nu_T S_{ij} + \left[\Gamma_i\Omega_{*j} + \Gamma_j\Omega_{*i}\right]_D, \tag{3.53}$$

where D denotes the deviatoric or traceless part of a tensor $\mathcal{A}_{ijD} = \mathcal{A}_{ij} - (1/3)\mathcal{A}_{\ell\ell}\delta_{ij}$, $\mathcal{S} = \{S_{ij}\}$ is the mean velocity strain rate defined by

$$S_{ij} = \frac{\partial U_i}{\partial x_j} + \frac{\partial U_j}{\partial x_i} - \frac{2}{3}\nabla\cdot\mathbf{U}\delta_{ij}, \tag{3.54}$$

and $\boldsymbol{\Omega}_*$ is the mean absolute vorticity $\boldsymbol{\Omega}_* = \boldsymbol{\Omega} + 2\boldsymbol{\omega}_F$. In equation (3.53), the transport coefficients ν_T and $\boldsymbol{\Gamma}$ are given as

$$\nu_T = \frac{7}{15}\int d\mathbf{k}\int_{-\infty}^{t} d\tau_1 G(k;\tau,\tau_1)Q(k;\tau,\tau_1), \tag{3.55}$$

$$\boldsymbol{\Gamma} = \frac{1}{30}\int k^{-2}d\mathbf{k}\int_{-\infty}^{t} d\tau_1 G(k;\tau,\tau_1)\nabla H(k;\tau,\tau_1). \tag{3.56}$$

Here, ν_T (equation (3.55)) is the transport coefficient coupled with the mean velocity strain $\mathcal{S}$ in equation (3.53) and is called the eddy or turbulent viscosity; it is determined by Green's function G and energy spectral function Q. In the simplest case, where the time integral of the Green's function can be evaluated independently of that of the energy spectral function, $\int_{-\infty}^{t} G(k;\tau,\tau_1)d\tau_1$ gives the turbulence timescale τ, and the spectral integral of energy spectrum $\int d\mathbf{k}Q(k;\tau,\tau_1)$ gives the turbulent energy $K \sim u'^2$ (u' is the magnitude of the turbulence velocity). In this case, equation (3.55) is reduced to

$$\nu_T \sim \tau K \sim u'\ell, \tag{3.57}$$

where $\ell \sim u'\tau$ is the characteristic length of turbulence. Equation (3.57) corresponds to the mixing-length expression for the turbulent viscosity.

On the other hand, $\mathbf{\Gamma}$ (equation (3.56)), is the coupling coefficient for the mean absolute vorticity $\mathbf{\Omega}_*$ in equation (3.53). This $\mathbf{\Gamma}$ is determined by G and the spatial gradient of the turbulent helicity spectral function, ∇H. Equation (3.56) implies that inhomogeneity of turbulent helicity is essential for the helicity effect in momentum transport. In a manner similar to equation (3.57), the simplest expression for $\mathbf{\Gamma}$ can be written as follows:

$$\mathbf{\Gamma} \sim \ell^2 \tau \nabla H. \tag{3.58}$$

This form of $\mathbf{\Gamma}$, proportional to ∇H instead of H, can be naturally understood by the symmetry argument of the Reynolds stress. The Reynolds stress $\mathcal{R} = \{\langle u_i' u_j' \rangle\}$ is symmetric with respect to inversion of the coordinate system $\mathbf{x} \to \hat{\mathbf{x}} = -\mathbf{x}$ and has even parity: $\mathcal{R}_{ij} \to \hat{\mathcal{R}}_{ij} = \langle \hat{u}_i' \hat{u}_j' \rangle = \langle (-u_i')(-u_j') \rangle = \langle u_i' u_j' \rangle = \mathcal{R}_{ij}$. The mean (absolute) vorticity is an axial vector or pseudovector, which does not change its sign under inversion. On the other hand, turbulent helicity is a pseudoscalar that changes its sign $H \to \hat{H} = -H$ under inversion. Consequently, the helicity H (odd parity) cannot enter the Reynolds stress expression as the proportional coefficient coupled with $\mathbf{\Omega}_*$, but the gradient of helicity, ∇H, which has even parity $\nabla H \to \hat{\nabla} \hat{H} = (-\nabla)(-H) = \nabla H$, can. This ∇H dependence was obtained in equation (3.56).

Note that $\mathbf{\Gamma}$ is a vector and that the direction of $\mathbf{\Gamma}$ is determined by the spatial distribution of turbulent helicity H through ∇H. Depending on the spatial distribution of H, the sign of each component of $\Gamma_i \Omega_{*j} + \Gamma_j \Omega_{*i}$ in equation (3.53) can be positive or negative. If its sign is opposite that of the eddy-viscosity term $-\nu_{\mathrm{T}} \mathcal{S}_{ij}$, the $\mathbf{\Gamma}$- or helicity-related term suppresses the eddy-viscosity effect. Otherwise, it enhances the eddy-viscosity effect. However, the sign of the $\mathbf{\Gamma}$- or helicity-related term is not arbitrarily distributed. Since the generation of turbulent helicity depends on the mean vorticity and rotation, the helicity inhomogeneity coupled with the mean vorticity and rotation, $\mathbf{\Gamma \Omega}_* = \{\Gamma_i \Omega_{*j}\}$, is often expected to have a certain sign. Detailed discussions based on the transport equation of turbulent helicity from equation (3.69) with equations (3.73)–(3.75) and the model equation (3.78) are in section 3.4.2.

The vortex-motive force $\mathbf{V}_{\mathrm{M}} = \{V_{\mathrm{M}i}\}$ in the mean vorticity (equation (3.19)) can be obtained from the Reynolds stress $\mathcal{R} = \{\mathcal{R}_{ij}\}$ through the exact relationship

$$\mathbf{V}_{\mathrm{M}} = -\nabla \cdot \mathcal{R} + \nabla K, \tag{3.59}$$

where $K(= \langle \mathbf{u}'^2 \rangle /2)$ is the turbulent energy. The second or K-related term does not contribute to the evolution of mean vorticity $\mathbf{\Omega}$ since $\nabla \times (\nabla K) = 0$.

Substituting the Reynolds-stress expression from equation (3.53) into equation (3.59), we have the expression of the vortex-motive force

$$\mathbf{V}_{\mathrm{M}} = -D_{\Gamma}(\mathbf{\Omega} + 2\boldsymbol{\omega}_{\mathrm{F}}) - [(\mathbf{\Omega} + 2\boldsymbol{\omega}_{\mathrm{F}}) \cdot \nabla]\mathbf{\Gamma} - \nu_{\mathrm{T}} \nabla \times \mathbf{\Omega}, \tag{3.60}$$

with

$$D_{\Gamma} = \nabla \cdot \mathbf{\Gamma}. \tag{3.61}$$

In equation (3.60), the third or ν_{T}-related term represents the destruction of the mean vorticity $\mathbf{\Omega}$ due to eddy viscosity. On the other hand, the first and second terms related to $\mathbf{\Gamma}$ represent the possible generation or sustainment of $\mathbf{\Omega}$ against the eddy-viscosity effect. We see from equation (3.60) that inhomogeneous turbulent helicity coupled with mean absolute vorticity (mean vorticity and rotation) may contribute to mean-vorticity generation, whereas turbulent energy contributes to mean-vorticity destruction.

From the spectral expression of the Reynolds stress in equation (3.53) with equations (3.55) and (3.56), the relative magnitude of the helicity effect to the eddy-viscosity effect may be written as

$$\frac{\text{(helicity effect)}}{\text{(eddy-viscosity effect)}} = \frac{|\mathbf{\Gamma \Omega}_*|}{|\nu_{\mathrm{T}} \mathcal{S}|} \sim \frac{\Omega_*}{\mathcal{S}} \frac{|H(k)|}{kE(k)}, \tag{3.62}$$

where $\Omega_* = \sqrt{\Omega_{*ij} \Omega_{*ij}}$ and $\mathcal{S} = \sqrt{\mathcal{S}_{ij} \mathcal{S}_{ij}}$ are the magnitudes of mean absolute vorticity and mean velocity strain, respectively. If the scaling of the helicity spectrum $H(k)$ is the same as that of the energy $E(k)$, which is supported by DNSs, the relative helicity $|H(k)|/[kE(k)]$ behaves like $\propto k^{-1}$. This implies that the relative importance of the helicity effect decreases as we go to the smaller scale. This point should be considered when we apply the helicity effect to subgrid-scale (SGS) modeling of large-eddy simulations (LESs) with a small filter width Δ.

3.4. TURBULENCE MODELING WITH HELICITY

In the previous sections, we showed that the mean velocity and vorticity fields are subjected to the effects of turbulent helicity through the Reynolds stress $\mathcal{R} = \{\mathcal{R}_{ij}\}$ and turbulent vortex-motive force $\mathbf{V}_{\mathrm{M}} = \{V_{\mathrm{M}i}\}$. From a practical viewpoint, it is important to construct a turbulence model incorporating helicity effects. Helical turbulence, in which velocity and vorticity fluctuations are (at least partly) aligned, is essentially three-dimensional. At the same time, the helical flow structure often shows a strong anisotropy at large scales. Consequently, we must consider a huge three-dimensional simulation domain throughout all scales, large and small. This is very demanding for DNSs. With a turbulence model incorporating the helicity effect, we can consistently perform numerical analyses of the helical turbulent flow configuration relevant to astrophysical, geophysical, plasma physics, and engineering phenomena. Here, we present an example of such a helicity turbulence model with an emphasis on the basic notion of modeling.

3.4.1. Choice of One-Point Statistical Quantities

As pointed out in section 3.2, the total amount of helicity and the total amount of energy are inviscid invariants of a system of fluid equations. In this sense, the local density of helicity, as well as the local density of energy, is a fundamental quantity that represents the properties of fluid dynamics.

The turbulent energy $\langle \mathbf{u}'^2 \rangle /2 (\equiv K)$ represents the local intensity of fluctuating motion. This point is reflected by the fact that turbulent energy determines turbulent transport, such as eddy viscosity. In contrast to turbulent energy, turbulent helicity $\langle \mathbf{u}' \cdot \boldsymbol{\omega}' \rangle (\equiv H)$ represents left- and right-handed twist. In this sense, turbulent helicity can serve as a measure of the structure of fluctuating motion.

The $K - \varepsilon$ model is one of the most-used turbulence models in engineering and scientific fields. In the model, to solve both the mean and turbulent fields in a self-consistent manner, the equations of the turbulent statistical quantities turbulent energy $K = \langle \mathbf{u}'^2 \rangle /2$ and its dissipation rate ε are solved simultaneously with mean-field equations (Launder & Spalding, 1972; Rodi, 1993). The turbulent transport coefficients, such as eddy viscosity, eddy diffusivity, etc., are expressed in terms of K and ε. Since the transport equations of K and ε are solved with the dynamics of the mean fields, the $K - \varepsilon$ model holds the direct link between the fluctuation velocity and the mean velocity gradients. In this sense, the $K - \varepsilon$ model is much more suitable than the mixing-length theory in self-consistently treating nonlinear interaction between the mean and fluctuation dynamics.

In addition to turbulent energy per mass K and its dissipation rate ε, we adopt turbulent helicity H as a turbulent statistical quantity. These three turbulent statistical quantities are defined as follows:

$$K = \langle \mathbf{u}'^2 \rangle /2, \tag{3.63}$$

$$\varepsilon = \nu \left\langle \frac{\partial u'_j}{\partial x_i} \frac{\partial u'_j}{\partial x_i} \right\rangle, \tag{3.64}$$

$$H = \langle \mathbf{u}' \cdot \boldsymbol{\omega}' \rangle. \tag{3.65}$$

In addition, we can adopt the dissipation rate of turbulent helicity, ε_H, as a turbulent statistical quantity:

$$\varepsilon_H = 2\nu \left\langle \frac{\partial u'_j}{\partial x_i} \frac{\partial \omega'_j}{\partial x_i} \right\rangle. \tag{3.66}$$

However, compared with the model equation of ε, that of ε_H is not well formulated. So, we do not adopt ε_H as a turbulent statistical quantity. For more on modeling the helicity dissipation-rate equation, see Yokoi (2016).

The expressions for eddy viscosity $\nu_{\rm T}$ in equation (3.55) and the helicity-related transport coefficient $\boldsymbol{\Gamma}$ in equation (3.56) with the integrals of the spectral and response functions in the wave-number space are too heavy for practical use. Using the one-point statistical quantities K, ε, and H, the eddy viscosity $\nu_{\rm T}$ and helicity-related coefficient $\boldsymbol{\Gamma}$ in the Reynolds stress $\mathcal{R}$ (equation (3.53)) and vortex-motive force $\mathbf{V}_{\rm M}$ (equation (3.60)) are simplified and modeled as

$$\nu_{\rm T} = C_\mu \frac{K^2}{\varepsilon}, \tag{3.67}$$

$$\boldsymbol{\Gamma} = C_\gamma \frac{K^4}{\varepsilon^3} \nabla H, \tag{3.68}$$

where C_μ and C_γ are model constants. Note that these expressions are written in terms of the turbulent statistical quantities K, ε, and H, with $\tau \sim K/\varepsilon$ and $\ell \sim K^{3/2}/\varepsilon$, while equations (3.57) and (3.58) are expressed in terms of the mixing length ℓ and the characteristic timescale τ of the turbulence.

3.4.2. Helicity Turbulence Model

The helicity turbulence model consists of the mean velocity (equation (3.17)), the Reynolds stress (equation (3.53)), transport coefficients $\nu_{\rm T}$ (equation (3.55)) and $\boldsymbol{\Gamma}$ (equation (3.56)), and the model equations for the three turbulent statistical quantities K, ε, and H.

From the fundamental equations, the exact equations for turbulent energy and helicity are

$$\left(\frac{\partial}{\partial t} + \mathbf{U} \cdot \nabla \right) F = P_F - \varepsilon_F + \nabla \cdot \mathbf{T}_F, \tag{3.69}$$

where F denotes the turbulent energy K or turbulent helicity H as $F = (K, H)$. In equation (3.69), P_F, ε_F, and $\mathbf{T}_F$ are the production rate, dissipation rate, and transport rate flux of F, respectively. They are defined by

$$P_K = - \left\langle u'_\ell u'_m \right\rangle \frac{\partial U_m}{\partial x_\ell}, \tag{3.70}$$

$$\varepsilon_K = \nu \left\langle \frac{\partial u'_m}{\partial x_\ell} \frac{\partial u'_m}{\partial x_\ell} \right\rangle \equiv \varepsilon, \tag{3.71}$$

$$\mathbf{T}_K = - \left\langle p' \mathbf{u}' \right\rangle - \left\langle \frac{1}{2} \mathbf{u}'^2 \mathbf{u}' \right\rangle + \nu \nabla \left(\frac{1}{2} \left\langle \mathbf{u}'^2 \right\rangle \right), \tag{3.72}$$

$$P_H = - \left\langle u'_m u'_\ell \right\rangle \frac{\partial \Omega_m}{\partial x_\ell} + \Omega_\ell \frac{\partial}{\partial x_m} \left\langle u'_m u'_\ell \right\rangle, \tag{3.73}$$

$$\varepsilon_H = 2\nu \left\langle \frac{\partial u'_m}{\partial x_\ell} \frac{\partial \omega'_m}{\partial x_\ell} \right\rangle, \tag{3.74}$$

$$\mathbf{T}_H = - \left\langle p' \boldsymbol{\omega}' \right\rangle + \frac{1}{2} \left\langle \mathbf{u}'^2 \boldsymbol{\omega}' \right\rangle - \left\langle (\mathbf{u}' \cdot \boldsymbol{\omega}') \mathbf{u}' \right\rangle + 2\boldsymbol{\omega}_{\rm F} \cdot \langle \mathbf{u}'\mathbf{u}' \rangle + \nu \nabla \left\langle \mathbf{u}' \cdot \boldsymbol{\omega}' \right\rangle, \tag{3.75}$$

where $[2\boldsymbol{\omega}_{\rm F} \cdot \langle \mathbf{u}'\mathbf{u}' \rangle]_\ell = 2\omega_{{\rm F}m} \left\langle u'_m u'_\ell \right\rangle$.

From equation (3.69) with equations (3.70)–(3.75), the model equations of K, ε, and H may be given as

$$\left(\frac{\partial}{\partial t} + \mathbf{U}\cdot\nabla\right) K = P_K - \varepsilon + \nabla\cdot\left(\frac{\nu_\mathrm{T}}{\sigma_K}\nabla K\right), \tag{3.76}$$

$$\left(\frac{\partial}{\partial t} + \mathbf{U}\cdot\nabla\right) \varepsilon = C_{\varepsilon 1}\frac{\varepsilon}{K}P_K - C_{\varepsilon 2}\frac{\varepsilon}{K}\varepsilon + \nabla\cdot\left(\frac{\nu_\mathrm{T}}{\sigma_\varepsilon}\nabla \varepsilon\right), \tag{3.77}$$

$$\left(\frac{\partial}{\partial t} + \mathbf{U}\cdot\nabla\right) H = P_H - \varepsilon_H + \nabla\cdot\left(2\boldsymbol{\omega}_\mathrm{F}K + \frac{\nu_\mathrm{T}}{\sigma_H}\nabla H\right). \tag{3.78}$$

Here, σ_K, σ_ε, and σ_H are the turbulent Prandtl numbers for the energy, dissipation rate, and helicity, and $C_{\varepsilon 1}$, C_{ε_2} and C_H are the model constants related to the energy and helicity dissipation rates.

For ε_H in the r.h.s. of equation (3.78), we adopt an algebraic model:

$$\varepsilon_H = C_H \frac{\varepsilon}{K} H. \tag{3.79}$$

Alternatively, we can construct the transport equation of the helicity dissipation rate ε_H and solve it as well as equations (3.76)–(3.78) without resorting to the algebraic model in equation (3.79). We assume that, like a passive scalar, helicity is determined by the scale k and the energy and helicity transfer rates ε and ε_H. The spectra in the inertial range are assumed to be

$$\frac{\sigma_K(k,\mathbf{x};t)}{\varepsilon} = \sigma_{K0}\varepsilon^{-1/3}k^{-11/3}, \tag{3.80}$$

$$\frac{\sigma_H(k,\mathbf{x};t)}{\varepsilon_H} = \sigma_{H0}\varepsilon^{-1/3}k^{-11/3}, \tag{3.81}$$

where σ_{K0} is the Kolmogorov constant and σ_{H0} is the counterpart of the Kolmogorov constant for helicity. With the aid of TS-DIA, up to the lowest-order analysis, we obtain the algebraic relation among H, ε, ε_H, and the size of the largest energy-containing eddies ℓ_C as equation (3.79). This corresponds to the estimate of the turbulent helicity dissipation rate ε_H in HIT. If we proceed to first-order analysis, the equation of the helicity dissipation rate ε_H is obtained as

$$\frac{D\varepsilon_H}{Dt} = C_{\varepsilon H1}\frac{\varepsilon_H}{K}P_K - C_{\varepsilon H2}\frac{\varepsilon_H}{K}\varepsilon + C_{\varepsilon H3}\frac{\varepsilon_H}{H}P_H - C_{\varepsilon H4}\frac{\varepsilon_H}{H}\varepsilon_H, \tag{3.82}$$

where the model constants are theoretically estimated as

$$C_{\varepsilon H1} = 0.36,\quad C_{\varepsilon H2} = 0.49,\quad C_{\varepsilon H3} = C_{\varepsilon H4} = 1.1. \tag{3.83}$$

Reflecting the assumption that the helicity spectrum depends on both ε and ε_H as in equation (3.81), the ε_H equation depends not only on P_H and ε_H but also on P_K and ε. For details of the derivation of the model, see Yokoi (2016).

Since the validity of the ε_H equation has not yet been numerically examined, hereafter we adopt equation (3.78) with the algebraic model in equation (3.79).

[Note: However, it should be noted that in some flow configurations, the algebraic model results in discrepancies. For instance, in DNSs of the Ekman boundary layer, the spatial distribution of the turbulent helicity H shows a sign reversal at some heights, while the dissipation rate of turbulent helicity ε_H does not vanish there (Deusebio & Lindborg, 2014).]

In the equations for mean-flow energy and helicity, $\mathbf{U}^2/2$ and $\mathbf{U}\cdot\boldsymbol{\Omega}$, we have the same terms but opposite signs as P_K (equation (3.70)) and P_H (equation (3.73)). This means the sink (or source) of the mean-field energy and helicity work as the source (or sink) of their turbulent counterparts. In this sense, the production terms P_K and P_H represent cascades from the mean-field energy and helicity to their turbulent counterparts. We see from equations (3.70) and (3.73) that inhomogeneous mean fields coupled with Reynolds stress cause such cascades and contribute to the production of turbulent energy and helicity.

As we see in the third or σ-related terms of model equations (3.76)–(3.78), transport rate terms such as $\mathbf{T}_K$, $\mathbf{T}_H$, etc. are modeled with gradient diffusion approximations. On the other hand, the $2\boldsymbol{\omega}_\mathrm{F}K$ term in equation (3.78) represents turbulent helicity generation arising from inhomogeneity along the rotation:

$$\nabla\cdot\left(2\boldsymbol{\omega}_\mathrm{F}K\right) = (2\boldsymbol{\omega}_\mathrm{F}\cdot\nabla)K. \tag{3.84}$$

This term is expected to be crucial in turbulent helicity generation in geophysical and astrophysical flows, where rotation and stratification are key ingredients. For instance, in typhoons and hurricanes, rotation $\boldsymbol{\omega}_\mathrm{F}$ coupled with vertical inhomogeneity of flow may contribute to turbulent helicity generation. On the other hand, in a tornado, because of the small dimension of the motion, the system rotation $\boldsymbol{\omega}_\mathrm{F}$ is negligibly small compared to the inhomogeneous mean velocity effect. But the mean relative vorticity coupled with inhomogeneous Reynolds stress in the second term of P_H (equation (3.73)) may contribute to the generation of turbulent helicity. Possible applications to some geophysical and astrophysical flows will be discussed in section 3.6.4.

In the absence of turbulent helicity ($H = 0$), this three-equation ($K - \varepsilon - H$) model should be reduced to the standard two-equation ($K - \varepsilon$) model. So, the model constant related to the $K - \varepsilon$ model should be the same as in the standard $K - \varepsilon$ model:

$$C_\mu = 0.09,\ \sigma_K = 1.0,\ C_{\varepsilon 1} = 1.4,\ C_{\varepsilon 2} = 1.9,\ \sigma_\varepsilon = 1.3. \tag{3.85}$$

Note that the $K - \varepsilon$ model with fixed constants has been successfully applied to many two- and three-dimensional wall boundary layers, duct flows, free shear flows, recirculating flows, confined flows, and jets. A useful account of the model and model constants can be seen in Launder & Spalding (1972), Rodi (1993). In this sense, the model constants (equation (3.85)) in the $K - \varepsilon$ model have been well optimized and should be fixed as far as possible. It is known that a 5% change in $C_{\varepsilon 1}$ or $C_{\varepsilon 2}$ results in a 20% change in the spreading rate of a jet (Rodi, 1993). As will be discussed in section 3.5.3, without the helicity effect, a considerable reduction of $C_\mu = 0.09 \to 0.001$ is required for a practical application of the standard $K - \varepsilon$ model to a swirling pipe flow. However, there is no firm basis for this ad hoc treatment.

The rest of the model constants, which are intrinsic to the $K - \varepsilon - H$ model, should be determined through practical applications of the present helicity model to turbulent helical flows such as swirling flows.

Utilizing the helicity effect, we can construct a set of SGS turbulence models for LESs. In LESs, we apply a filter to a field quantity f to divide it into grid-scale (GS) and subgrid-scale (SGS) components, $\overline{f}$ and f''. In the simulation, large-scale or GS motions (large eddies) are solved, whereas small-scale or SGS motions are modeled (SGS modeling). The Smagorinsky model is the most commonly used SGS model. In this model, the SGS viscosity $\nu_{\rm S}$ in the GS equation is expressed in terms of the GS velocity strain $\overline{S}(= \sqrt{\overline{S}_{ij}^2/2})$ and the filter width Δ as $\nu_{\rm S} = (C_{\rm S}\Delta)^2\overline{S}$ ($\overline{S}_{ij}$ is the GS rate-of-strain tensor, and $C_{\rm S}$ is the Smagorinsky constant). As this expression implies, the Smagorinsky model is the SGS counterpart of the mixing-length model. (Note that Δ is the largest scale of unresolved or SGS motions.) This simplicity has made the Smagorinsky model commonly used in various turbulent flows. However, this model has several deficiencies. One is the need for constant adjustment. The Smagorinsky constant $C_{\rm S}$ must be adjusted from flow to flow: $C_{\rm S} = 0.18$ for isotropic turbulence, 0.15 for mixing layer turbulence, and 0.10 for wall turbulence. This means the Smagorinsky model with model constant $C_{\rm S} = 1.8$, optimized for homogeneous isotropic turbulence, is more than three times $((1.8/1.0)^2 = 3.24)$ too dissipative if it is applied to near-wall turbulence where streamwise vorticity structures are ubiquitously observed.

Since streamwise vorticity strongly implies the local presence of turbulent helicity, it is expected that turbulent helicity plays an important role in reducing effective viscosity in shear turbulence, such as the mixing layer and wall turbulence. The effects of turbulence structures represented by streamwise vorticity are incorporated into the SGS model through SGS helicity. The problem of overestimating SGS viscosity in the presence of SGS shear structures, including streamwise vorticity, is alleviated by implementing the SGS helicity effect even in the framework of the classical Smagorinsky model (Yokoi & Yoshizawa, 2017).

3.5. APPLICATION TO SWIRLING FLOW

3.5.1. Swirling Flows

The helicity turbulence model consists of the mean-velocity equation and turbulent-field equations, where the Reynolds stress is given by equation (3.53), with the turbulent viscosity $\nu_{\rm T}$ and helicity-related coefficient $\boldsymbol{\Gamma}$ expressed as equations (3.67) and (3.68), respectively. The model equations of the turbulent statistical quantities are equation (3.76) for the turbulent energy K, equation (3.77) for energy dissipation rate ε, and equation (3.78) for turbulent helicity H, with an algebraic model for the helicity dissipation rate ε_H (equation (3.79)).

We apply the present helicity model to turbulent swirling flows. First we summarize the characteristic properties of swirling flow. Swirling flow consists of the main or axial flow in the mainstream or axial direction and swirl or circumferential flow around the axis. This configuration is so simple that swirling flow is ubiquitously observed in the scientific and engineering communities. In engineering, it is utilized in devices including swirling jet nozzles, combustion fuel gas injection chambers, etc. In the scientific field, swirling flows are typically observed in tornadoes and cyclones. In these geophysical flows, rotating flows are often accompanied by ascending and/or descending flows (Fig. 3.4). These helical flow structures are considered relevant to the genesis and sustainment of large-scale coherent flow structures in tornadoes and cyclones (Lilly, 1986; Levina and Montgomery, 2014). In astrophysical flow phenomena, bipolar jets ejected from the central region of an accretion disk are considered helical. These outflows vertical to the disk are typically accompanied by circumferential flows associated with azimuthal rotation in the accretion disk.

One of the prominent features of swirling jets is a dent in the jet-velocity profile near the central axis. In several types of swirling flows, it is ubiquitously observed that axial velocity decelerates in the central region

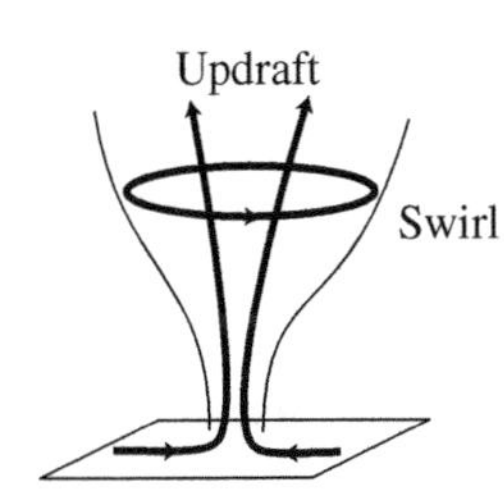

Figure 3.4 Cyclone swirl accompanied by updraft.

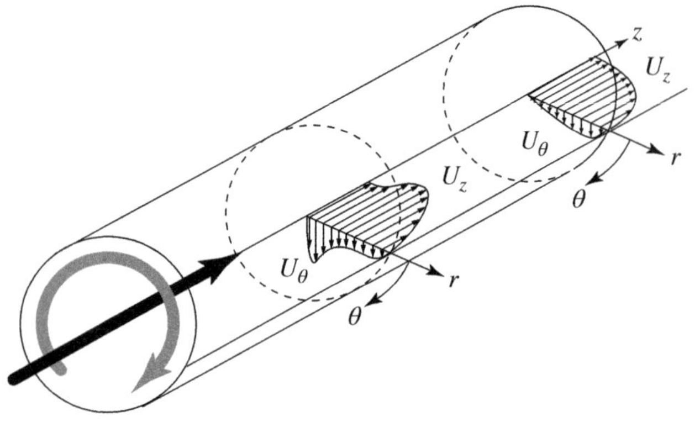

Figure 3.5 Configuration of swirling flow in a straight pipe.

compared to the outer core region. The mean axial and circumferential velocity profiles in turbulent swirling flow in a straight pipe are schematically depicted in Fig. 3.5. In the upstream region near the inlet, where the swirling motion is injected near the central region, the mean axial velocity shows a dent profile near the central axis. This dent profile decays toward the downstream region and finally becomes a flat profile observed in the usual turbulent pipe flow without swirl.

The deceleration of axial velocity in swirling flow can be easily understood in laminar flow. We consider a cylindrical coordinate system (r, θ, z) with the swirling axis in the z direction. For a stationary axisymmetric swirling flow, we assume that the radial flow u_r is negligibly smaller than the circumferential and axial velocities, u_θ and u_z. In this case, the radial momentum balance is achieved by the balance between the centripetal acceleration and radial pressure gradient

$$-\frac{(u_\theta)^2}{r} = -\frac{1}{\rho}\frac{\partial p}{\partial r}, \tag{3.86}$$

where ρ is the fluid density. Integrating equation (3.86) with respect to r from the axis center $r = 0$ to a radial position $r = r_*$ in the outer periphery region, and differentiating with respect to z, we have

$$\left.\frac{\partial p}{\partial z}\right|_{r=0} - \left.\frac{\partial p}{\partial z}\right|_{r=r_*} = -\rho \int_0^{r_*} \frac{\partial}{\partial z}\frac{(u_\theta)^2}{r} dr. \tag{3.87}$$

This shows that the axial gradients of pressure differ between the center $(r = 0)$ and outer peripheral point $(r = r_*)$, depending on the magnitude of the circumferential velocity u_θ there. On the other hand, the axial balance of momentum is given by

$$u_z \frac{\partial u_z}{\partial z} = -\frac{1}{\rho}\frac{\partial p}{\partial z}. \tag{3.88}$$

With equation (3.87), this reads

$$\begin{aligned}\left.\frac{\partial}{\partial z}\frac{1}{2}(u_z)^2\right|_{r=0} &= -\left.\frac{1}{\rho}\frac{\partial p}{\partial z}\right|_{r=0} \\ &= -\left.\frac{1}{\rho}\frac{\partial p}{\partial z}\right|_{r=r_*} + \int_0^{r_*} \frac{\partial}{\partial z}\frac{(u_\theta)^2}{r} dr \\ &= \left.\frac{\partial}{\partial z}\frac{1}{2}(u_z)^2\right|_{r=r_*} + \int_0^{r_*} \frac{\partial}{\partial z}\frac{(u_\theta)^2}{r} dr. \end{aligned} \tag{3.89}$$

This means there is a difference between the central and periphery regions of the evolution of axial velocity. The difference depends on the axial evolution of the circumferential velocity profile through the second term in the final r.h.s. of equation (3.89). If the swirling motion is driven in the central and inner core regions of the pipe at the inlet and decays in the core region, the second or $(u_\theta)^2$-related term in equation (3.89) is negative:

$$\int_0^{r_*} \frac{\partial}{\partial z}\frac{(u_\theta)^2}{r} dr < 0. \tag{3.90}$$

In this case, we have

$$\left.\frac{\partial}{\partial z}\frac{1}{2}(u_z)^2\right|_{r=0} < \left.\frac{\partial}{\partial z}\frac{1}{2}(u_z)^2\right|_{r=r_*}, \tag{3.91}$$

which means in the downstream region, the axial velocity at the center $(r = 0)$ is relatively small compared to the outer region $(r = r_*)$, resulting in a dent profile in the central region. On the other hand, if the swirling motion is driven at the outer periphery, in the main portion of the flow except the near wall, the swirling velocity will increase along the axis. Then the second term in the final line of equation (3.89) is positive. In this case, the axial velocity at the center should be greater than in the outer region, leading to acceleration of the axial velocity in the central region. This occurs in the case of axially rotating pipe flows, where the rotation of the outer boundary pipe wall is the source of the fluid's swirling motion. This argument clearly shows that decay of swirling velocity along the axial direction is directly connected to deceleration of axial velocity in the central region through pressure interactions.

The situation in turbulent swirling flow is much more complicated because of turbulent fluxes. In such a case, proper evaluation of the Reynolds stresses is crucial for accurately describing and correctly predicting swirling flow.

3.5.2. Characteristics of Turbulent Swirling Flow

Because of its simple geometry, swirling flow in a straight pipe has been intensively investigated using experiments. These experimental studies provided elaborated data on the mean axial and circumferential velocities and turbulence statistics in swirls (Kitoh, 1991; Steenbergen, 1995; Hoekstra, Derksen, & Van Den Akker, 1999; Gupta & Kumar, 2007).

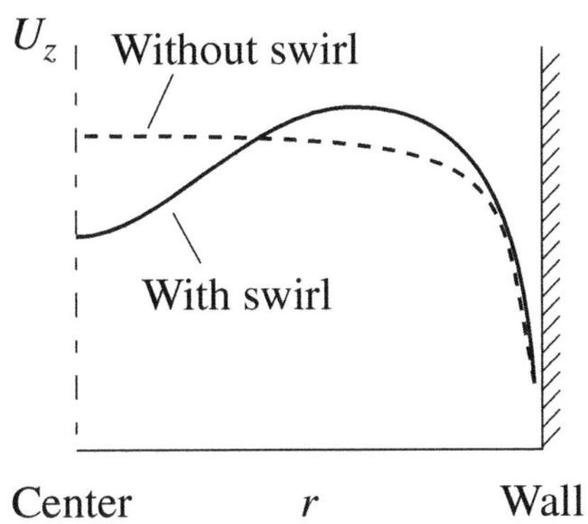

Figure 3.6 Mean axial velocity profile with and without swirl.

The prominent features of turbulent swirling flow in a straight pipe are as follows:

(i) The mean axial flow is decelerated in the central axis region compared to the core region (dent at the center). The degree of deceleration increases with swirl intensity (scaled axial flux of angular momentum).
(ii) Swirl intensity decays exponentially with the downstream or axial distance from a reference point.

Here, swirl intensity S_W is defined by the axial flux of the angular momentum scaled by the axial flux of the linear momentum as

$$S_W = \frac{\int_0^a 2\pi r^2 U_\theta U_z dr}{\pi a^3 U_m^2}, \qquad (3.92)$$

where a is the radius of the pipe and U_m is the bulk velocity defined by $U_m = \int_0^a 2\pi U_z r dr / \pi a^2$.

These two features are related. The first, a dent in axial velocity at the center, is depicted schematically in Fig. 3.6.

The axial mean flow reverses in the central axis region when the swirl intensity is strong. This reversal is very important in engineering applications since the flow direction associated with combustion, fuel supply, etc. changes at some points. Axial flow reversal was investigated in variational analyses with helicity as the constraint of the variational calculus (Yoshizawa, et al., 2001; Yokoi, et al., 2004). It was shown theoretically that if the mean-flow helicity is greater than critical helicity, which is proportional to the square of the mean axial momentum flux, the mean axial velocity at the central axis is reversed.

3.5.3. Conventional Turbulence Model for Swirling Flow in a Straight Pipe

It is well known that the standard eddy-viscosity-type $K - \varepsilon$ model poorly reproduces features (i) and (ii) described in the previous section. The standard $K - \varepsilon$ model is recovered from equation (3.53) with the third or helicity-related term dropped. The standard $K - \varepsilon$ model applied to turbulent swirling flow in a pipe gives a flat mean axial velocity profile, which is observed in the axial velocity profile of a turbulent pipe flow without swirling motion (Fig. 3.6). This is because in the presence of fully developed turbulence, the eddy-viscosity effect is so strong that no inhomogeneous mean velocity structures can exist except the boundary layer near the wall, where viscosity dominates nonlinearity. To alleviate this deficiency of the standard $K - \varepsilon$ model and reproduce the dent axial velocity profile in the framework of the model, some modifications have been proposed in the engineering field. In this modified $K - \varepsilon$ model, the model constant associated with the r-θ component of the Reynolds stress, $\mathcal{R}_{r\theta}$, was artificially made extremely small: for example, $C_\mu = 0.09 \rightarrow 0.001$. Then the dent profile of the mean axial velocity was retained in the model calculation (Kobayashi & Yoda, 1987). However, given the requirement of the universality of model constants, which any good turbulence model should satisfy, this treatment is ad hoc and not preferable.

3.5.4. Analysis of Turbulent Swirling Flow with the Helicity Turbulence Model

We apply the present helicity turbulence model ($K - \varepsilon - H$ model) to turbulent swirling flow in a straight pipe. In the model, in addition to equation (3.17) (mean velocity) with transport coefficients ν_T (equation (3.67)) and $\mathbf{\Gamma}$ (equation (3.68)), the three transport equations of the turbulent statistical quantities K (equation (3.76)), ε (equation (3.77)), and H (equation (3.78)) are simultaneously solved. Then the mean fields and turbulence fields are self-consistently determined.

In Fig. 3.7, we plot the profile of the mean axial velocity U_z and the decay of the swirl intensity S_W calculated by the $K - \varepsilon - H$ model compared to those from experiments and the standard $K - \varepsilon$ model. As mentioned in

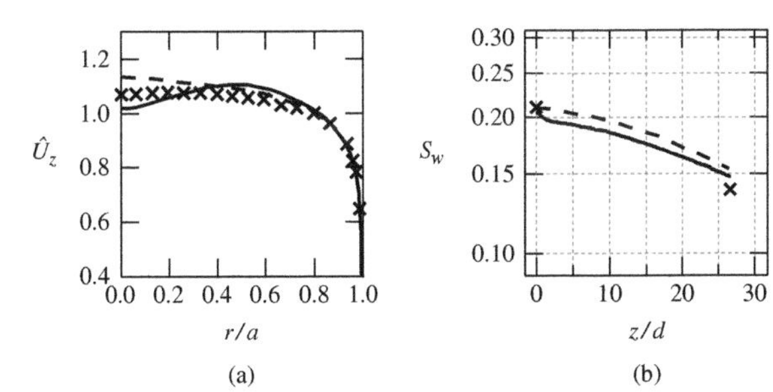

Figure 3.7 The radial profile of the mean axial velocity and axial decay of swirl intensity. (a) Axial velocity normalized by bulk velocity $\hat{U}_z = U_z/U_M$; (b) swirl intensity S_W. × denotes the experiment, – – – the standard $K - \varepsilon$ model, and –– the helicity ($K - \varepsilon - H$) model. Adapted from Yokoi & Yoshizawa (1993).

section 3.5.3, the dent profile of the mean axial velocity in the central region cannot be reproduced with the standard $K-\varepsilon$ model. In marked contrast, such a dent profile is naturally reproduced with the $K-\varepsilon-H$ model without any artificial adjustment of the model constants. The exponential decay of the swirl intensity S_W is also reproduced better by the helicity ($K-\varepsilon-H$) model than the standard $K-\varepsilon$ model.

These results clearly show the relevance of the helicity effect for treating the flow evolution of turbulent swirling flow. Inhomogeneous turbulent helicity coupled with rotation and/or large-scale vorticity (the antisymmetric component of the mean velocity shear) contributes to balancing the enhanced transport due to turbulent energy through eddy viscosity, coupled with large-scale velocity strain (the symmetric component of mean velocity shear).

3.6. LARGE-SCALE FLOW GENERATION BY INHOMOGENEOUS HELICITY

The previous section discussed the importance of the helicity effect on transport suppression. Turbulent helicity contributes to the dynamic balance between transport enhancement and transport suppression. It contributes to transport suppression by counterbalancing transport enhancement due to turbulent energy through eddy viscosity. At the same time, turbulent helicity may contribute to global flow structure. In this section, we discuss the role of turbulent helicity in generating a large-scale flow.

3.6.1. Vortex Dynamos

Replacing vorticity $\boldsymbol{\omega}$ with a magnetic field $\mathbf{b}$ and viscosity ν with magnetic diffusivity η in the equation (3.6) (fluid vorticity), we obtain the magnetic induction equation:

$$\frac{\partial \mathbf{b}}{\partial t} = \nabla \times (\mathbf{u} \times \mathbf{b}) + \eta \nabla^2 \mathbf{b}. \tag{3.93}$$

Equation (3.93) leads to the possibility of a dynamo. Corresponding to the mean vorticity induction in equation (3.19), the induction equation for the mean magnetic field $\mathbf{B}$ is written as

$$\frac{\partial \mathbf{B}}{\partial t} = \nabla \times (\mathbf{U} \times \mathbf{B} - \eta \nabla \times \mathbf{B}) + \nabla \times \mathbf{E}_\mathrm{M}, \tag{3.94}$$

with the solenoidal condition for $\mathbf{B}$

$$\nabla \cdot \mathbf{B} = 0, \tag{3.95}$$

where $\mathbf{E}_\mathrm{M}$ is the turbulent electromotive force defined by

$$\mathbf{E}_\mathrm{M} = \langle \mathbf{u}' \times \mathbf{b}' \rangle. \tag{3.96}$$

The mean magnetic induction equation (3.94) leads to the generation and sustainment of a large-scale magnetic field through the turbulent electromotive force $\mathbf{E}_\mathrm{M}$. Despite several fundamental differences between equation (3.6) and equation (3.93) (Moffatt & Dormy, 2019), the similarity suggests the possibility of a vortex dynamo: the generation and sustainment of large-scale vorticity through the turbulent vortex-motive force $\mathbf{V}_\mathrm{M} = \langle \mathbf{u}' \times \boldsymbol{\omega}' \rangle$ in the mean vorticity equation. Several studies have been undertaken regarding this vortex dynamo problem (Krause & Rüdiger, 1974; Moiseev, et al., 1983; Chkhetiani & Gvaramadze, 1988; Khomenko, Moiseev & Tur, 1991; Kitchatinov, Rüdiger & Khomenko, 1994). Through these studies, it has been recognized that to generate large-scale flow, we need breakage of symmetry in the Reynolds stress, such as anisotropy, compressibility, mean flow, etc. For instance, Elperin and co-workers studied the effect of mean velocity shear on homogeneous nonhelical turbulence (Elperin, Kleeorin & Rogachevskii, 2003) and on a large-scale instability for the formation of vortical structure (Elperin, et al., 2007). Spontaneous formation of large-scale flow patterns in shear-flow turbulence was numerically confirmed (Käpylä, Mitra & Brandenburg, 2009). These shear-flow effects in nonhelical turbulence correspond to the anisotropic state of turbulence related to the second-order nonlinear model of the Reynolds stress. The velocity shear $\nabla \mathbf{U} = \{\partial U_j / \partial x_i\}$ is divided into the rate-of-strain tensor $\mathcal{S}_{ij}$ and vorticity tensor Ω_{ij}. The nonlinear effects concerning $\mathcal{S}_{ij}$ and Ω_{ij} are expressed by the quadratic strain-tensor and vorticity-tensor terms, such as $\mathcal{S}_{i\ell}\mathcal{S}_{\ell j}$, $\mathcal{S}_{i\ell}\Omega_{\ell j}$, and $\Omega_{i\ell}\Omega_{\ell j}$. These nonlinear models lead to secondary flows in a duct flow (Speziale, 1991). Note that these nonlinear expressions of the Reynolds stress can be derived from the higher-order $O(\delta^2)$ calculation in the TS-DIA formulation without resorting to non-mirror symmetry (Yoshizawa, 1984, 1993).

In the following, we treat the vortex dynamo problem from the viewpoint of the inhomogeneous helicity effect (Yokoi & Yoshizawa, 1993; Yokoi & Brandenburg, 2016; Kleeorin & Rogachevskii, 2018).

Two mechanisms are similar to the present helicity effects: the so-called anisotropic kinetic alpha (AKA) effect (Frisch, She, & Sulem, 1987) and other Λ effect (Rüdiger, 1980, 1989). The AKA effect is a large-scale instability for global flow generation in non-mirror-symmetric turbulence. In the AKA formulation, the turbulent Reynolds number is assumed to be small, and the Reynolds stress can be expressed using the Taylor expansion with respect to a small and uniform mean velocity. Under these assumptions, it is shown that the evolution of mean velocity is subject to an anisotropic transport coefficient coupled with mean velocity shear. As these assumptions suggest, the AKA effect operates only at low Reynolds numbers when eddy viscosity is negligible in the momentum equation. This contrasts with the helicity effect, which operates in strongly nonlinear turbulence. On the other hand, the Λ effect contributes to Reynolds stress arising from the anisotropy of turbulence. In this formulation, the expression of the Reynolds stress is assumed to be a linear combination of the functionals of the angular velocity $\boldsymbol{\omega}_{\rm F}$. The Λ effect is similar to the present helicity effect in that both operate in the strong turbulence regime. This is in strong contrast to the AKA effect. However, the Reynolds stress expression in the Λ effect is given as an Ansatz. In this sense, the physics that determines the transport coefficients and their validity should be examined for each application. For further descriptions of the AKA and Λ effects, the reader is referred to the appendix of Yokoi & Brandenburg (2016).

Substituting the expression for vortex-motive force $\mathbf{V}_{\rm M}$ (equation (3.60)) into the mean vorticity equation (3.19), we have

$$\frac{\partial \boldsymbol{\Omega}}{\partial t} = \nabla \times \left[\mathbf{U} \times (\boldsymbol{\Omega} + 2\boldsymbol{\omega}_{\rm F}) - \nu \nabla \times \boldsymbol{\Omega}\right] + \nabla \times \mathbf{V}_{\rm M}$$
$$= \nabla \times \left[\mathbf{U} \times (\boldsymbol{\Omega} + 2\boldsymbol{\omega}_{\rm F})\right] - \nabla \times \left[(\nu + \nu_{\rm T}) \nabla \times \boldsymbol{\Omega}\right] + \mathbf{I}_{\rm V}. \qquad (3.97)$$

Here, the second term represents enhanced transport due to turbulence. The effective viscosity is enhanced from ν to $\nu + \nu_{\rm T}$ due to the eddy viscosity $\nu_{\rm T}$. The third term $\mathbf{I}_{\rm V}$ is defined by

$$\mathbf{I}_{\rm V} = \nabla \times \left\{-D_\Gamma(\boldsymbol{\Omega} + 2\boldsymbol{\omega}_{\rm F}) - \left[(\boldsymbol{\Omega} + 2\boldsymbol{\omega}_{\rm F}) \cdot \nabla\right] \boldsymbol{\Gamma}\right\}, \qquad (3.98)$$

which denotes the possible vorticity induction due to the inhomogeneous helicity effect [$\boldsymbol{\Gamma} \propto \nabla H$ (equation (3.56))]. The sign of $\mathbf{I}_{\rm V}$ can be positive or negative. The sign depends on the spatial distribution of the turbulent helicity through the signs of $D_\Gamma = \nabla^2 H$ and $(\boldsymbol{\Omega}_* \cdot \nabla)\boldsymbol{\Gamma}$, where $\boldsymbol{\Omega}_*$ is the mean absolute vorticity defined by $\boldsymbol{\Omega}_* = \boldsymbol{\Omega} + 2\boldsymbol{\omega}_{\rm F}$. In the case of negative $\mathbf{I}_{\rm V}$, the turbulent helicity effect enhances the eddy-viscosity effect.

During the development of large-scale vorticity, where $|\boldsymbol{\Omega}| \ll |2\boldsymbol{\omega}_{\rm F}|$, induction due to inhomogeneous helicity can be approximated as

$$\mathbf{I}_{\rm V} = \nabla \times \left[-2D_\Gamma \boldsymbol{\omega}_{\rm F} - \left(2\boldsymbol{\omega}_{\rm F} \cdot \nabla\right) \boldsymbol{\Gamma}\right]$$
$$= -2(\nabla D_\Gamma) \times \boldsymbol{\omega}_{\rm F} - 2\nabla \times \left[(\boldsymbol{\omega}_{\rm F} \cdot \nabla)\boldsymbol{\Gamma}\right]. \qquad (3.99)$$

3.6.2. Simple Argument on the Origin of Flow Generation Due to the Helicity Effect

Uncurling the mean vorticity equation (3.97), we obtain

$$\frac{\partial \mathbf{U}}{\partial t} = \mathbf{U} \times (\boldsymbol{\Omega} + 2\boldsymbol{\omega}_{\rm F}) - (\nu + \nu_{\rm T}) \nabla \times \boldsymbol{\Omega} + \mathbf{V}_{\rm M} + \nabla \varphi, \qquad (3.100)$$

where φ is a potential field determined by boundary conditions whose detail is not argued here. With the turbulent vortex-motive force $\mathbf{V}_{\rm M}$ (equation (3.60)), the mean

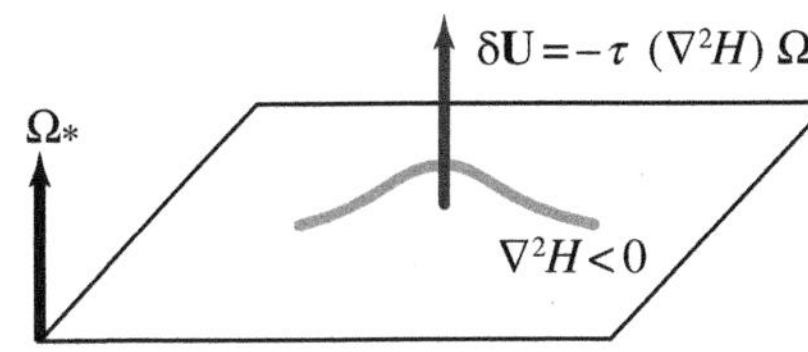

Figure 3.8 The possible physical origin of global flow induction is due to the inhomogeneous helicity effect.

velocity induced by turbulence, $\delta\mathbf{U}$, may be approximated by time integration of equation (3.100):

$$\delta \mathbf{U} \sim -\tau D_\Gamma \boldsymbol{\Omega}_* \sim -\tau(\nabla^2 H)\boldsymbol{\Omega}_*. \qquad (3.101)$$

This implies that a global flow $\delta\mathbf{U}$ is induced in the presence of inhomogeneous turbulent helicity represented by the Laplacian of turbulent helicity $\nabla^2 H = D_\Gamma$. The flow is parallel to the mean absolute vorticity $\boldsymbol{\Omega}_*(= \boldsymbol{\Omega} + 2\boldsymbol{\omega}_{\rm F})$ for $\nabla^2 H < 0$ and antiparallel for $\nabla^2 H > 0$. The Laplacian of a quantity represents how much the quantity is locally distributed more or less than the surroundings. It follows from equation (3.101) that a mean flow $\delta\mathbf{U}$ is generated in a direction parallel to the mean absolute vorticity $\boldsymbol{\Omega}_*$ where the turbulent helicity is locally more distributed ($\nabla^2 H < 0$), and antiparallel to $\boldsymbol{\Omega}_*$ where turbulent helicity is less distributed than the surroundings ($\nabla^2 H > 0$). In this sense, this global flow is induced by the mean absolute vorticity (relative vorticity and rotation), consuming turbulent helicity redundant over that of the surroundings (Fig. 3.8). Since the induced global velocity is parallel (or antiparallel) to the original mean absolute vorticity $\boldsymbol{\Omega}_*$ for $\nabla^2 H < 0$ (or $\nabla^2 H > 0$), a positive (or negative) mean-field helicity $\mathbf{U} \cdot \boldsymbol{\Omega}$ is generated. In this sense, backscattering of helicity from smaller to larger scales occurs. Is this an inverse cascade of helicity with constant fluxes? What are the values of the parameter in what regime for this phenomenon? These points should be examined further in the future.

3.6.3. Numerical Validation of Global Flow Generation Due to the Helicity Effect

Using the multiple-scale renormalized perturbation expansion method, the expression for the Reynolds stress was obtained as equation (3.53). The essential ingredients of the helicity effect in equation (3.53) are inhomogeneous turbulent helicity represented by $\boldsymbol{\Gamma}(\propto \nabla H)$ and the antisymmetric part of the mean velocity shear represented by $\boldsymbol{\Omega}_*(= \nabla \times \mathbf{U} + 2\boldsymbol{\omega}_{\rm F})$. We check the validity of the Reynolds-stress expression in equation (3.53) with the aid of DNSs of a turbulent flow with inhomogeneous helicity [Yokoi & Brandenburg, 2016]. For this purpose, we adopt a numerical setup of helical turbulence in a box

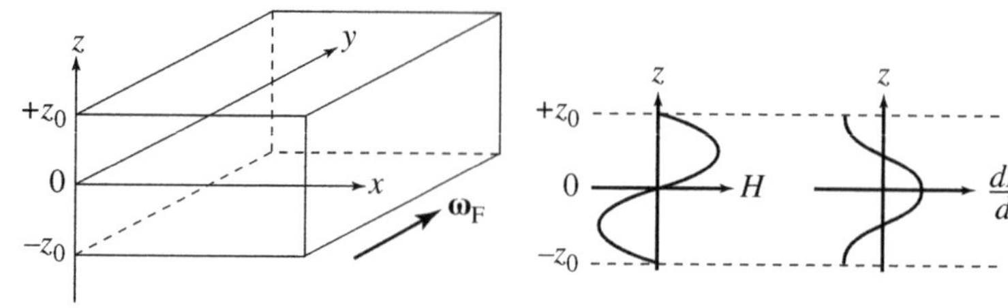

Figure 3.9 Setup of DNSs for validating the helicity effect. A periodic box configuration with imposed rotation $\omega_F = (0, \omega_F, 0)$ (left), and the spatial distributions of imposed turbulent helicity H and its gradient dH/dz (right).

in a Cartesian coordinate system (x, y, z) with imposed rotation $\boldsymbol{\omega}_F$, whose rotation axis in the y direction is

$$\boldsymbol{\omega}_F = (\omega_{Fx}, \omega_{Fy}, \omega_{Fz}) = (0, \omega_F, 0) \tag{3.102}$$

as in Fig. 3.9. The turbulent helicity $H = \langle \mathbf{u}' \cdot \boldsymbol{\omega}' \rangle$ is externally imposed by sinusoidal forcing.

In real geophysical and astrophysical flow phenomena, helicity and its inhomogeneity are supplied by helicity arising from mean-field inhomogeneities through the field configuration (see equation (3.78)) and/or boundary conditions (related to the diffusion terms in equations (3.76) and (3.78)). For example, rotation coupled with inhomogeneity, such as energy inhomogeneity, density stratification, asymmetry of inertial wave propagation, etc., may produce turbulent helicity. Here, for the sake of simplicity of the numerical setup, we adopt external forcing as the mechanism supplying helicity to the turbulence.

At the initial state, our setup does not have mean or large-scale velocity ($\mathbf{U} = 0$). It follows from equation (3.53) that in the early stage of development, the y-z component of the Reynolds stress, $\mathcal{R}_{yz}$, is given by

$$\mathcal{R}_{yz} = \langle u'_y u'_z \rangle = \eta 2\omega_F \frac{\partial H}{\partial z}, \tag{3.103}$$

where

$$\eta = C_\eta \tau \ell^2 = C_\eta \frac{K}{\varepsilon} \frac{K^3}{\varepsilon^2} \tag{3.104}$$

(C_η: model constant).

Once the helicity effect begins to generate mean flow, the Reynolds stress $\mathcal{R}_{yz}$ is expressed as

$$\mathcal{R}_{yz} = \langle u'_y u'_z \rangle = -\nu_T \frac{\partial U_y}{\partial z} + \eta 2\omega_F \frac{\partial H}{\partial z}. \tag{3.105}$$

Here we have assumed that the generated mean relative vorticity $\boldsymbol{\Omega} = \nabla \times \mathbf{U}$ is much smaller than the imposed rotation ($|\boldsymbol{\Omega}| \ll 2|\boldsymbol{\omega}_F|$) to be neglected in the generation term due to helicity. The first or ν_T-related term in equation (3.105) represents the eddy-viscosity effect. The second or helicity-gradient-related term in equation (3.105) represents the generation of large-scale velocity due to inhomogeneous helicity. If the eddy-viscosity term ν_T and the helicity term are balanced with each other in the statistically stationary state, we have $\mathcal{R}_{yz} = \langle u'_y u'_z \rangle \simeq 0$. In this case, we have

$$U_y \simeq \frac{\eta}{\nu_T} 2\omega_F H. \tag{3.106}$$

This implies that in such a balanced state, a mean flow $\mathbf{U}$ is induced in the direction of the imposed rotation $\boldsymbol{\omega}_F$. The induced flow $\mathbf{U}$ is parallel to the rotation $\boldsymbol{\omega}_F$ for the positive-helicity region ($H > 0$) and antiparallel for the negative-helicity region ($H < 0$). Note that inhomogeneity of turbulent helicity is indispensable for generating large-scale flow. No large-scale flow generation is observed if we impose uniform turbulent helicity.

In Fig. 3.10, we present the temporal evolution of large-scale flow ($\mathbf{U}$) and turbulent helicity H. The spatial (z) distribution of the contours of U_y and H are plotted against time. Since turbulent helicity is externally imposed by sinusoidal forcing throughout the period of evolution, a stationary sinusoidal pattern (positive H for $z > 0$ and negative H for $z < 0$) is observed (top). On the other hand, we have no large-scale flow ($\mathbf{U} = 0$) at the initial time (bottom). Due to the inhomogeneous helicity effect,

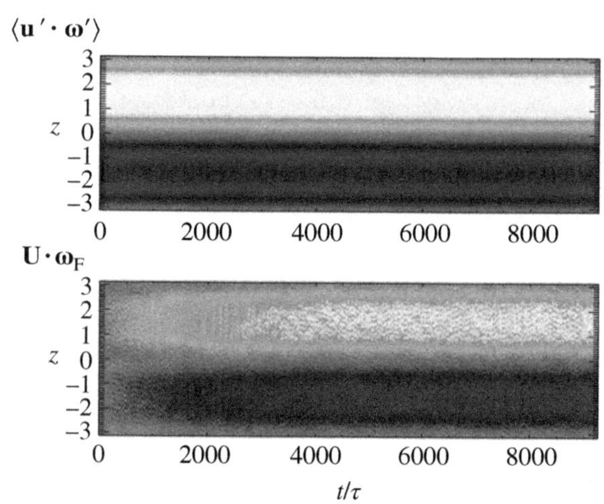

Figure 3.10 Temporal evolution of turbulent helicity (top) and mean flow (bottom) with z dependence. Adapted from Yokoi & Brandenburg (2016).

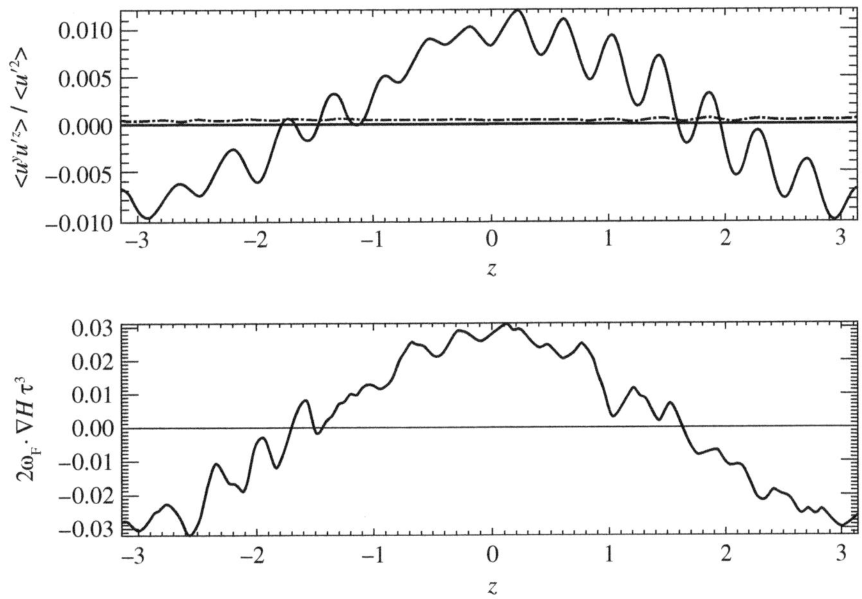

Figure 3.11 Reynolds stress and helicity gradient at the early stage of evolution. The y-z component of the Reynolds stress (top) and the turbulent helicity gradient term $\omega_F(\nabla H)_z$ (bottom). Averaged over time from $t/\tau = 40$ to 200. Yokoi & Brandenburg, 2016 / American Physical Society.

a large-scale flow is generated as time proceeds (early stage) and reaches a stationary state (developed stage).

In Fig. 3.11, we plot the y-z component of the Reynolds stress, $\mathcal{R}_{yz}$, and the gradient of turbulent helicity multiplied by the rotation, $2\omega_F(\partial H/\partial z)$, at the early stage of evolution (averaged over time from $t/\tau = 40$ to 200). The spatial profile of $\mathcal{R}_{yz} = \langle u'_y u'_z \rangle$ is in good agreement with $2\omega_F(\partial H/\partial z)$. This suggests that inhomogeneous helicity coupled with rotation works for large-scale flow generation.

In Fig. 3.12, we plot the mean axial velocity U_y and turbulent helicity multiplied by rotation, $2\omega_F H$, at the developed equilibrium stage (averaged over time from $t/\tau = 0$ to 2000) against z. The spatial profile of the induced mean axial velocity U_y is in fairly good agreement with that of turbulent helicity $2\omega_F H$. This implies that at the developed equilibrium stage, the balance between the eddy-viscosity effect $\nu_T \mathcal{S} = \{\nu_T \mathcal{S}_{ij}\}$ and the inhomogeneous helicity effect $\mathbf{\Gamma\Omega}_* = \{\Gamma_i \Omega_{*j}\}$ (equation (3.106)) is well achieved and maintained.

We see from these results that the eddy-viscosity effect is not sufficient, and the inhomogeneous helicity effect should be considered in the analysis of turbulent transport in a system with rotating motion.

Inhomogeneous helicity effects in flow generation and transport suppression were also confirmed by LESs of rotating turbulence with externally imposed localized helicity. In the work, turbulent helicity was locally imposed by forcing with and without rotation (Inagaki, Yokoi & Hamba, 2017). Global flow generation was observed only when both turbulent helicity and rotation were simultaneously imposed. Examination of each component of the Reynolds stress revealed that counterbalancers to eddy viscosity were the Coriolis-force term and the pressure-diffusion term in the Reynolds stress equations. This is natural since the effect of rotation $\boldsymbol{\omega}_F$ can enter turbulence dynamics only through the Coriolis-force and fluctuating-pressure terms. This result suggests that improving equation (3.78) for the turbulent helicity model may be possible by elaborate modeling of the fluctuating pressure-related terms.

3.6.4. Possible Applications in Spherical Geometry

Spherical Surfaces

To see the physical consequences of induction due to inhomogeneous helicity, we consider local Cartesian coordinates (x, y, z) on a tangential spherical surface, as in Fig. 3.13. Here, x denotes the azimuthal direction, y the latitudinal polar direction toward the north pole, and z the radial direction outward normal to the local tangential surface. We consider two limiting cases: first, length scales in the horizontal directions (in the x and y directions) that are much larger than the normal length scale (in the z direction); and second, a normal length scale (z) much larger than the horizontal length scales (x and y).

(A) Case with Large Horizontal Scales If the length scales of the horizontal directions are much larger than the normal length scale, the gradient operator in the local

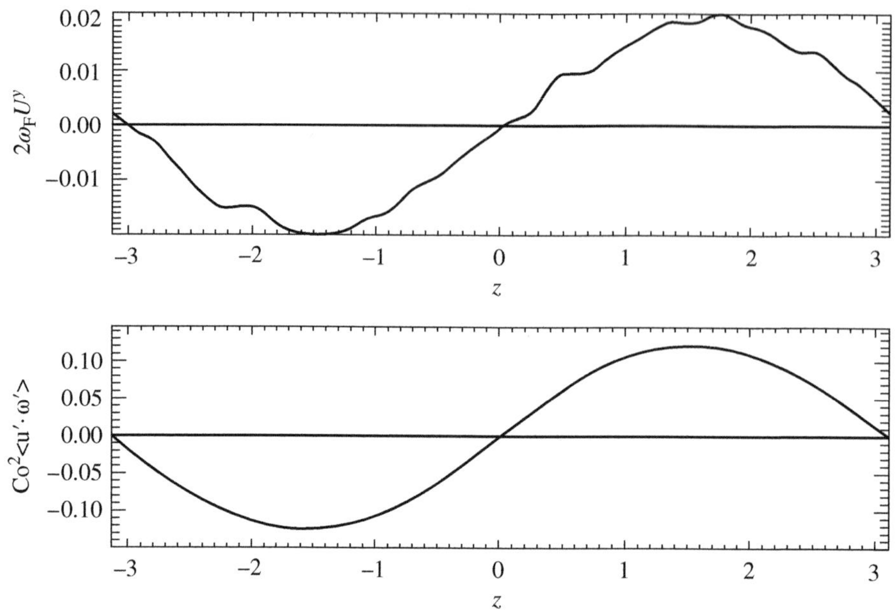

Figure 3.12 Induced mean velocity and turbulent helicity at the developed stage of evolution. The induced mean velocity U_y (top) and the turbulent helicity H (bottom) (Co: Corliolis number defined by $\mathrm{Co} = \omega_F \tau$) . Averaged over time from $t/\tau = 0$ to 2000. Yokoi & Brandenburg, 2016 / American Physical Society.

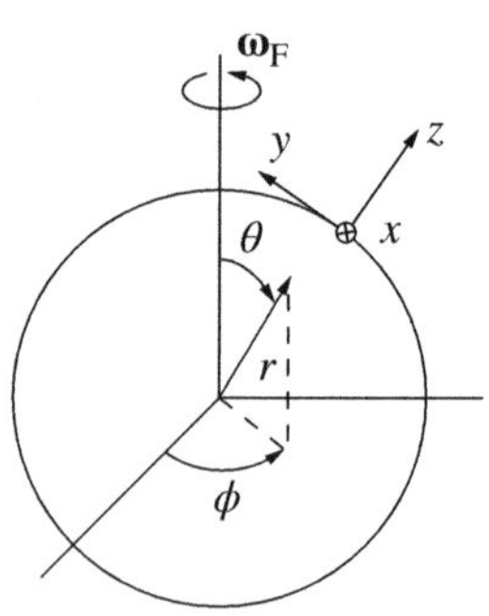

Figure 3.13 Local Cartesian coordinate (x, y, z).

Cartesian coordinates (x, y, z) can be approximated as

$$\nabla = \left(\frac{\partial}{\partial x}, \frac{\partial}{\partial y}, \frac{\partial}{\partial z} \right) \simeq \left(0, 0, \frac{\partial}{\partial z} \right). \tag{3.107}$$

In this case, the second term in equation (3.99) vanishes. Then the vorticity induction due to inhomogeneous helicity is given by

$$\mathbf{I}_V = -2(\nabla D_\Gamma) \times \boldsymbol{\omega}_F. \tag{3.108}$$

Since the angular velocity vector $\boldsymbol{\omega}_F$ in the local Cartesian coordinate is given in terms of the colatitudinal polar angle θ,

$$\boldsymbol{\omega}_F = \left(\omega_{Fx}, \omega_{Fy}, \omega_{Fz} \right) = \left(0, \omega_F \sin\theta, \omega_F \cos\theta \right), \tag{3.109}$$

equation (3.108) is expressed as

$$\mathbf{I}_V = -2\nabla(\nabla \cdot \boldsymbol{\Gamma}) \times \boldsymbol{\omega}_F = \left(2\frac{\partial D_\gamma}{\partial z} \omega_F \sin\theta, 0, 0 \right). \tag{3.110}$$

This implies that the rotation $\boldsymbol{\omega}_F$ coupled with inhomogeneous turbulent helicity may induce large-scale vorticity in the x or azimuthal direction, especially in the low-latitude region ($0 \ll \theta < \pi/2$). Once large-scale vorticity is generated in the azimuthal direction, Ω_x, the first term in equation (3.98) starts working and gives

$$I_{Vy} \simeq \left\{ \nabla \times \left[-D_\Gamma(\boldsymbol{\Omega} + 2\boldsymbol{\omega}_F) \right] \right\}_y = -\frac{\partial}{\partial z}(D_\Gamma \Omega_x), \tag{3.111}$$

inducing the latitudinal component of large-scale vorticity, Ω_y. This mechanism causes a helical large-scale vortical structure in the low-latitude region at the surface of a rotating stellar and/or planetary object.

(B) Case with Large Normal Scale If the length scale in the direction normal to the tangential surface is much larger than the length scales in the horizontal directions, the gradient operator is approximated as

$$\nabla = \left(\frac{\partial}{\partial x}, \frac{\partial}{\partial y}, \frac{\partial}{\partial z} \right) \simeq \left(\frac{\partial}{\partial x}, \frac{\partial}{\partial y}, 0 \right). \tag{3.112}$$

In this case, the vector $\boldsymbol{\Gamma}$ related to the helicity gradient in equation (3.98) is written as

$$\boldsymbol{\Gamma} = \left(\Gamma_x, \Gamma_y, 0 \right) \simeq \left(\frac{\partial H}{\partial x}, \frac{\partial H}{\partial y}, 0 \right). \tag{3.113}$$

At the early stage of vorticity generation ($|\boldsymbol{\Omega}| \ll |2\boldsymbol{\omega}_F|$), we see from equation (3.99) that the helicity contribution to vorticity induction is given by

$$\mathbf{I}_V = \nabla \times \left[-2D_\Gamma \boldsymbol{\omega}_F - \left(2\boldsymbol{\omega}_F \cdot \nabla \right) \boldsymbol{\Gamma} \right]$$

$$= \left(-2\frac{\partial D_\Gamma}{\partial y}\omega_F \cos\theta, 2\frac{\partial D_\Gamma}{\partial x}\omega_F \cos\theta, -2\frac{\partial D_\Gamma}{\partial x}\omega_F \sin\theta\right) + \left(0, 0, 2\omega_F \sin\theta \left(\frac{\partial^2 \Gamma_x}{\partial y^2} - \frac{\partial^2 \Gamma_y}{\partial x \partial y}\right)\right). \quad (3.114)$$

For the lower-latitude region, this is reduced to

$$\mathbf{I}_V = \left(0, 0, 2\omega_F \sin\theta \left(-\frac{\partial D_\Gamma}{\partial x} + \frac{\partial^2 \Gamma_x}{\partial y^2} - \frac{\partial^2 \Gamma_y}{\partial x \partial y}\right)\right). \quad (3.115)$$

This implies that the rotation vector coupled with inhomogeneous turbulence helicity induces large-scale vorticity in the direction normal to the tangential surface in the lower-latitude region. This may enhance the genesis of cyclones in the near-equatorial region (Yokoi & Yoshizawa, 1993).

Angular Momentum Transport

Equation (3.101) implies that inhomogeneous spatial distribution of turbulent helicity coupled with mean vortical motion and/or rotation can induce a large-scale flow in the direction of the mean vorticity/rotation. In other words, angular momentum is transported to the domain of flow acceleration and/or generation due to the inhomogeneous turbulent helicity effect. Two interesting cases are suggested next.

(A) Stellar Angular Momentum Transport Stellar and planetary interior convective motions play key roles in forming and sustaining stellar and planetary magnetic fields through dynamo actions. Recent developments in helioseismology unraveled flow configurations inside the Sun: the latitudinal and radial distributions of solar angular velocity and basic patterns of meridional circulation. It has been established that solar azimuthal rotation represented by solar angular velocity is faster near the equator and the surface (Miesch, 2005).

Although the detailed pattern of meridional circulation is still an open problem, the basic pattern may be considered, as shown in Fig. 3.14(a). In the surface region, the meridional circulation $\mathbf{U}_{MC}$ flows poleward, while at some depth, it returns equatorward. Then the mean vorticity associated with the meridional circulation $\mathbf{\Omega}_{MC}(= \nabla \times \mathbf{U}_{MC})$ is in the westward azimuthal direction.

On the other hand, the spatial distribution of turbulent kinetic helicity has been obtained using DNSs of the spherical shell domain mimicking stellar and planetary convective zones (Duarte, et al., 2016). Following the numerical results, the radial distribution of turbulent helicity is schematically depicted in Fig. 3.14(b). At the bottom boundary of the convective zone, helicity is null; and near the bottom, it is positive. Positive helicity decreases with radius and becomes negative in the shallower domain of the convective zone. The magnitude of the negative helicity decreases as it approaches the surface and finally becomes null at the surface boundary. This spatial distribution of helicity is naturally understood from the helicity-generation mechanisms due to inhomogeneity along the rotation $(\boldsymbol{\omega}_F \cdot \nabla)K$ (equation (3.84)) discussed in section 3.4.2. If a lump of fluid with angular momentum approaches the outer boundary, it expands horizontally due to impinging on the boundary. Due to the conservation of local angular momentum, the lump's rotation decreases, leading to a negative swirl motion compared to ambient fluids. This is alternatively expressed by inhomogeneity along the rotation term.

We apply the inhomogeneous helicity effect on the mean velocity (equation (3.53) with equation (3.58)) to the problem of angular-momentum transport inside the Sun. Let us consider angular-momentum transport in the spherical polar coordinate system (r, θ, ϕ). The angular momentum

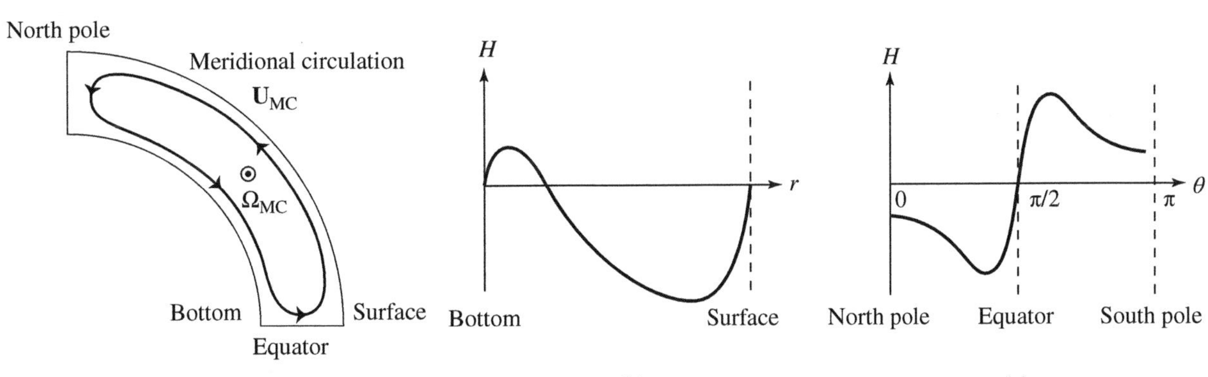

Figure 3.14 Angular momentum transport in the Sun based on the inhomogeneous helicity effect. (a) Basic pattern of meridional circulation, where $\mathbf{U}_{MC}$ is the meridional circulation velocity and $\mathbf{\Omega}_{MC}(= \nabla \times \mathbf{U}_{MC})$ is the vorticity associated with meridional circulation. (b) Radial variation of turbulent helicity. (c) Colatitudinal variation of turbulent helicity. All figures are schematically depicted.

around the rotation axis, L, is defined by

$$L = \Lambda r^2 \omega_{\mathrm{F}} + \Lambda r U_\phi, \tag{3.116}$$

with $\Lambda = \sin\theta$. The transport of L is subject to

$$\frac{\partial}{\partial t}\rho L + \nabla \cdot \left(\rho \mathbf{F}_L\right) = 0. \tag{3.117}$$

Here, ρ is the density, and $\mathbf{F}_L$ is the vector flux of the angular momentum, expressed as

$$F_{Lr} = LU_r + r\Lambda\mathcal{R}_{r\phi}, \tag{3.118}$$

$$F_{L\theta} = LU_\theta + r\Lambda\mathcal{R}_{\theta\phi}, \tag{3.119}$$

where $\mathcal{R}(=\{\mathcal{R}_{ij}\} = \{\langle u_i' u_j'\rangle\})$ is the Reynolds stress.

In the axisymmetric case $(\partial/\partial\phi = 0)$, the inhomogeneous helicity contribution to the Reynolds stress can be written

$$\mathcal{R}_{\theta\phi}^{(H)} = +\frac{1}{r}\frac{\partial H}{\partial\theta}\left(\frac{1}{r}\frac{\partial}{\partial r} rU_\theta - \frac{1}{r}\frac{\partial U_r}{\partial\theta}\right), \tag{3.120}$$

$$\mathcal{R}_{r\phi}^{(H)} = +\frac{\partial H}{\partial r}\left(\frac{1}{r}\frac{\partial}{\partial r} rU_\theta - \frac{1}{r}\frac{\partial U_r}{\partial\theta}\right). \tag{3.121}$$

In the presence of meridional circulation in the poleward direction at the solar surface, the toroidal mean vorticity is negative in the northern hemisphere:

$$\frac{1}{r}\frac{\partial}{\partial r} rU_\theta \begin{cases} < 0 & \text{(Northern hemisphere)} \\ > 0 & \text{(Southern hemisphere)} \end{cases} \tag{3.122}$$

On the other hand, turbulent helicity in the northern hemisphere is expected to be more negative in shallow regions (except for the vicinity of the surface) than in deeper regions:

$$\frac{\partial H}{\partial r} \begin{cases} < 0 & \text{(Northern hemisphere)} \\ > 0 & \text{(Southern hemisphere)} \end{cases} \tag{3.123}$$

At the same time, the colatitudinal gradient of turbulent helicity is negative in both hemispheres, except in the vicinity of the equator, where the sign of the helicity changes from negative to positive (Fig. 3.14(c)). Then we expect

$$\frac{1}{r}\frac{\partial H}{\partial\theta} \begin{cases} < 0 & \text{(Northern hemisphere)} \\ < 0 & \text{(Southern hemisphere)} \end{cases} \tag{3.124}$$

It follows from equations (3.122)–(3.124) that the helicity contributions to the angular-momentum fluxes are written as follows:

$$r\Lambda\mathcal{R}_{\theta\phi}^{(H)} \simeq +\Lambda\frac{\partial H}{\partial\theta}\frac{1}{r}\frac{\partial}{\partial r} rU_\theta > 0, \tag{3.125}$$

$$r\Lambda\mathcal{R}_{r\phi}^{(H)} \simeq +r\Lambda\frac{\partial H}{\partial r}\frac{1}{r}\frac{\partial}{\partial r} rU_\theta > 0. \tag{3.126}$$

Equations (3.125) and (3.126) show that the helicity contributions to the θ-ϕ and r-ϕ components are positive. This means the helicity effect accelerates azimuthal velocity by transporting angular momentum toward shallower and lower-latitude (near the equator) regions. This is in agreement with the results of helioseismic observations. Preliminary studies utilizing DNSs of a spherical shell mimicking the solar convective zone show that the magnitudes of the helicity terms, $|(\nabla H)\boldsymbol{\Omega}| = \{|(\partial H/\partial x_i)\Omega_j|\}$, with the parameters for the solar interior are comparable to those of $\mathcal{R}_{r\phi}$ and $\mathcal{R}_{\theta\phi}$ (M. Miesch, private communication, 2017). This suggests that this inhomogeneous helicity effect should be further investigated in the solar interior.

More intuitively, we can argue the role of the inhomogeneous helicity effect on angular momentum transport in the Sun as follows. We see from equation (3.101) that the mean velocity variation due to the helicity effect is given as

$$\delta\mathbf{U}_{\mathrm{MC}} \sim -\tau(\nabla^2 H)\boldsymbol{\Omega}_{\mathrm{MC}}. \tag{3.127}$$

Since $\nabla^2 H$ is positive in the shallow region (Fig. 3.14b), the variation of the mean velocity $\delta\mathbf{U}_{\mathrm{MC}}$ (equation (3.127)) is expected to be antiparallel to the mean vorticity due to meridional circulation, $\boldsymbol{\Omega}_{\mathrm{MC}}$. We see that this inhomogeneous helicity effect works for the acceleration of azimuthal angular velocity in the shallow convective zone, which again matches helioseismic observations.

(B) Wind Acceleration near Cyclone Eyewalls Another possibly interesting application of the inhomogeneous helicity effect is wind acceleration in the formation of eyewalls in tropical cyclones. It is observationally known that a double-eyewall structure is formed in the core of a cyclone (Fig. 3.15), and a distinct maximum wind speed (tangential velocity of the cyclone) is located at the radial position of each eyewall. Such double-eyewall formation and maximum wind speed at eyewalls have been previously discussed in terms of radial propagation of the vortex Rossby wave and its stagnation at eyewall formation regions (Montgomery & Kallenbach, 1997; Huang, Montgomery & Wu, 2012; House, 2010).

Since eyewall regions are related to inhomogeneous helicity generation, it may be possible to explain such wind acceleration by the inhomogeneous helicity effect. The common configurations of cyclones are upward drafts at the inner and outer eyewalls and dry forced descent at the eye (Fig. 3.15). The updraft and its circulated downflow constitute the azimuthal component of the mean vorticity. At the same time, the presence of vertical flows in the cyclone eye region coupled with vertical mean vorticity associated with cyclone wind velocity indicates a mean-flow helicity $\mathbf{U}\cdot\boldsymbol{\Omega}$. This suggests that turbulent helicity must be highly inhomogeneous near the eyewall. Such strong inhomogeneous mean helicity naturally

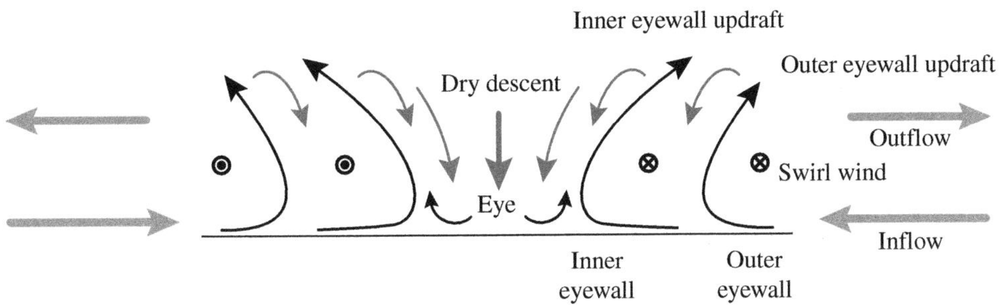

Figure 3.15 Swirling wind configuration of a tropical cyclone.

results in the inhomogeneous spatial distribution of turbulent helicity $H = \langle \mathbf{u}' \cdot \boldsymbol{\omega}' \rangle$. The combination of mean azimuthal vorticity and inhomogeneous helicity near the eyewall may induce a large-scale flow in the azimuthal direction, following equation (3.101).

3.7. SUMMARY AND CONCLUSION

Helicity contributes to the alteration of flow dynamics and fundamental statistical properties in turbulence. In contrast to energy, which represents the intensity properties of fluctuations, helicity represents the structural properties of fluctuations. Incorporating helicity effects into the statistical theory of turbulence is achieved by including cross-flow or cross-component correlation and longitudinal and transverse velocity correlations in equation (3.45). In the configuration space, longitudinal and transverse velocity correlations are linked to turbulent energy, while cross-correlation is linked to turbulent helicity. Since helicity is a measure of the breakage of mirror symmetry, its effects are incorporated into theoretical analysis by taking into account the non-mirror-symmetric part of the expression of fluctuation correlations in the wave-number space (equations (3.46) and (3.47)).

Turbulent transport is expressed by turbulent fluxes such as the Reynolds stress in the mean momentum equation and vortex-motive force in the mean vorticity equation. In the mirror-symmetric case, turbulent transport is determined solely by the intensity of fluctuation. The more intense the fluctuation is, the more effective the transport. From the viewpoint of the turbulence model, eddy viscosity represents such enhanced transport due to turbulence. Turbulent viscosity is expressed in terms of turbulent energy and its timescale. In the non-mirror-symmetric case, helicity and energy enter the expressions of turbulent fluxes as the descriptor of turbulent transport.

With the aid of the multiple-scale renormalized perturbation expansion theory, we obtained analytical expressions for the Reynolds stress and turbulent vortex-motive force. In these theoretical results, the transport coefficients are expressed in terms of the spectral correlation functions and response function (Green's function). The eddy viscosity, the transport coefficient coupled with the symmetric part of the mean velocity shear (mean velocity strain), is expressed by the spectral and time integral of the energy spectral function and the Green's functions in equation (3.55). In a system lacking reflectional symmetry, in addition to eddy viscosity, another transport coefficient coupled with the antisymmetric part of the mean velocity shear (mean vorticity and rotation) appears. This transport coefficient is expressed by the inhomogeneity of the helicity spectral function and Green's function (equation (3.56)).

Based on these analytical results, we constructed a turbulence model with the structural effect incorporated through turbulent helicity. To construct a self-consistent system of model equations, the mean and turbulence fields should be solved simultaneously. For this purpose, the transport equations of turbulent energy K, its dissipation rate ε, and turbulent helicity H were proposed ($K - \varepsilon - H$ model). This model was applied to swirling flow in a straight pipe. This helicity turbulence model successfully reproduced the main characteristics of swirling flow – deceleration of mean axial velocity and exponential decay of swirl intensity – which could not be reproduced with the standard $K - \varepsilon$ model with eddy-viscosity representation. This numerical result shows that the eddy viscosity, which enhances effective transport too much for swirling flow, can be successfully counterbalanced by the inhomogeneous helicity effect.

The helicity effect was also validated using direct numerical simulations (DNSs) of a rotating triple-periodic symmetrical box with inhomogeneous helicity externally imposed by force. Starting with no initial mean flow, it was shown that inhomogeneous helicity coupled with rotation contributes to the induction of global flow in the direction of the rotation vector. It was shown that the Reynolds stress shows a high correlation with inhomogeneous helicity at the early stage of flow evolution, where the eddy-viscosity effect is weak because of the absence of mean velocity strain. Also, high correlation between induced mean flow and local turbulent helicity at the developed stage confirms the balance between eddy viscosity and inhomogeneous helicity effects. These numerical validations show that in the non-mirror-symmetric case, the helicity effect coupled with mean absolute vorticity (mean vorticity and rotation) should be considered in addition to the usual eddy-viscosity effect coupled with mean velocity strain.

From these theoretical and numerical results, we can summarize the roles of helicity in turbulent fluid transport. In a non-mirror-symmetric system, the Reynolds stress is schematically expressed as

$$\langle \mathbf{u}'\mathbf{u}' \rangle := \overbrace{-\nu_\mathrm{T} \mathcal{S}}^{\text{Eddy viscosity}} \quad \overbrace{+\boldsymbol{\Gamma}\boldsymbol{\Omega}_*}^{\text{Helicity effect}} . \tag{3.128}$$

Transport: Enhancement Suppression

Flow/Structure: Destruction Generation

Here, for the sake of brevity, the notations for the deviatoric and transposed components of a tensor are suppressed, $:=$ denotes "schematically represents," $\mathcal{S}$ is the mean velocity strain tensor, $\boldsymbol{\Gamma} \propto \nabla H$ is the gradient of helicity, and $\boldsymbol{\Omega}_* (= \boldsymbol{\Omega} + 2\boldsymbol{\omega}_\mathrm{F})$ is the mean absolute vorticity. The eddy viscosity ν_T is determined by the turbulent energy and timescale of turbulence, while the helicity-related transport coefficient $\boldsymbol{\Gamma}$ is determined by the gradient of helicity and the timescale and length scale of turbulence. Notably, the turbulence timescale and length scale are altered by strong helicity, rotation, velocity shear, non-equilibrium effects, etc. (Yokoi, Masada & Takiwaki, 2022; Yokoi, 2023). The implications of the strain- and vorticity-tensor effect on the turbulence timescale and length scale have been explored in conventional nonlinear turbulence modeling studies (Speziale, 1991).

In treating turbulent transport in non-mirror-symmetric turbulence, helicity effects, representing transport suppression and structure generation, must be incorporated into turbulent fluxes, their model expressions, and the turbulent energy effect. In the present helicity effects, the direction of change in turbulent transport and global flow generation/destruction depends on the spatial distribution of turbulent helicity. In this sense, the spatiotemporal evolution of turbulent helicity and its production, dissipation, and transport mechanisms play crucial roles. In real-world turbulence, such mechanisms are determined by nonlinear dynamics between the mean and turbulence fields. A self-consistent approach based on turbulence modeling described in section 3.4 is of fundamental importance in exploring global flow-generation mechanisms in nature.

Much more attention should be paid to helicity. In particular, detailed laboratory experiments and in situ and/or remote observations, numerical calculations of spatiotemporal distributions of helicity (kinetic helicity, magnetic helicity, current helicity, residual helicity, cross helicity, etc.) are required to develop our insights in turbulence and nonlinear flow phenomena.

ACKNOWLEDGMENTS

My thanks to the anonymous reviewers for their thoughtful comments based on their knowledge and insight, which contributed greatly to improving the manuscript. I cordially dedicate this work to the memory of my parents, Hisashi Yokoi (14 March 1928–18 July 1999) and Taduko Yokoi (13 October 1929–22 April 2020), who provided agapeic encouragement and led me into a scholarly life.

REFERENCES

Alexakis, A., & Biferale, L. (2018). Cascades and transitions in turbulent flows. *Physics Reports*, *524*, 1–139. https://doi.org/10.1016/j.physrep.2018.08.001

André, J. C., & Lesieur, M. (1977). Influence of helicity on the evolution of isotropic turbulence at high Reynolds number. *Journal of Fluid Mechanics*, *81*, 187–207.

Angriman, S., Cobelli, P. J., Bourgoin, M., Huisman, S. G., Volk, R., & Mininni, P. D. (2021). Broken mirror symmetry of tracer's trajectories in turbulence. *Physical Review Letters*, *127*, 254502.

Batchelor, G. K. (1953). *The theory of homogeneous turbulence*. Cambridge University Press.

Biferale, L., Musacchio, S., & Toschi, F. (2012). Inverse energy cascade in three-dimensional isotropic turbulence. *Physical Review Letters*, *108*, 164501. https://doi.org/110.1103/PhysRevLett.108.164501

Brandenburg, A., & Subramanian, K. (2005). Astrophysical magnetic fields and nonlinear dynamo theory. *Physics Reports*, *417*, 1–209.

Brandenburg, A., & Petrosyan, A. (2012). Kinetic helicity decay in linearly forced turbulence. *Astronomische Nachrichten*, *333*, 195–201.

Chandrasekhar, S. (1950). The theory of axisymmetric turbulence. *Philosophical Transactions of the Royal Society A*, *242*, 557–577.

Chen, Q., Chen, S., & Eyink, G. L. (2003). The joint cascade of energy and helicity in three-dimensional turbulence. *Physics of Fluids*, *15*, 361–374.

Chkhetiani, O. G., & Gvaramadze, V. V. (1988). Generation of large-scale vortices in compressible helical turbulence. In *Proceedings of the Joint Varenna-Abastumani International School & Workshop on Plasma Astrophysics*, Varenna, Italy, 24 Aug. to 3 Sept. 1988 (ESA SP-285, 1988), *1*, 363–369.

Deusebio, E., & Lindborg, E. (2014). Helicity in the Ekman boundary layer. *Journal of Fluid Mechanics*, *755*, 654–671.

Duarte, L. D. V., Wicht, J., Browning, M. K., & Gastine, T. (2016). Helicity inversion in spherical convection as a means for equatorward dynamo wave propagation. *Monthly Notices of the Royal Astronomical Society*, *456*, 1708–1722.

Elperin, T., Kleeorin, N., & Rogachevskii, I. (2003). Generation of large-scale vorticity in a homogeneous turbulence with a mean velocity shear. *Physical Review E*, *68*, 016311-1–8.

Elperin, T., Golubev, I., Kleeorin, N., & Rogachevskii, I. (2007). Large-scale instability in a sheared nonhelical turbulence: Formation of vortical structures. *Physical Review E*, *76*, 066310-1–12.

Frisch, U., She, Z. S., & Sulem, P. L. (1987). Large scale flow driven by the anisotropic kinetic alpha effect. *Physica D*, *28*, 382–392.

Gupta, A., & Kumar, R. (2007). Three-dimensional turbulent swirling flow in a cylinder: Experiments and computations. *International Journal of Heat and Fluid Flow*, *28*, 249–261.

Hide, R. (1989). Superhelicity, helicity and potential vorticity. *Geophysical & Astrophysical Fluid Dynamics*, *48*, 69–79.

Hoekstra, A. J., Derksen, J. J., & Van Den Akker, H. E. A. (1999). An experimental and numerical study of turbulent swirling flow in gas cyclones. *Chemical Engineering Science*, *54*, 2055–2065.

House Jr., R. A. (2010). Clouds in tropical cyclones. *Monthly Weather Review*, *138*, 293–344.

Huang, Y.-H., Montgomery, M. T., & Wu, C.-C. (2012). Concentric eyewall formation in typhoon Sinlaku (2008). Part II: Axisymmetric dynamical processes. *Journal of the Atmospheric Sciences*, *69*, 662–674. https://doi.org/10.1175/JAS-D-11-0114.1

Inagaki, K., Yokoi, N., & Hamba, F. (2017). Mechanism of mean flow generation in rotating turbulence through inhomogeneous helicity. *Physical Review Fluids*, *2*, 114605. https://doi.org/10.1103/PhysRevFluids.2.114605

Käpylä, P. J., Mitra, D., & Brandenburg, A. (2009). Numerical study of large-scale vorticity generation in shear-flow turbulence. *Physical Review E*, *79*, 016302-1–8.

Khomenko, G. A., Moiseev, S. S., & Tur, A. V. (1991). The hydrodynamical alpha-effect in a compressible medium. *Journal of Fluid Mechanics*, *225*, 355.

Kitchatinov, L. L., Rüdiger, G., & Khomenko, G. (1994). Large-scale vortices in rotating stratified disks. *Astronomy & Astrophysics*, *287*, 320.

Kitoh, O. (1991). Experimental study of turbulent swirling flow in a straight pipe. *Journal of Fluid Mechanics*, *225*, 445–479.

Kleeorin, N., & Rogachevskii, I. (2018). Generation of large-scale vorticity in rotating stratified turbulence with inhomogeneous helicity: Mean-field theory. *Journal of Plasma Physics*, *84*, 735840303.

Kobayashi, T., & Yoda, M. (1987). Modified $k-\epsilon$ model for turbulent swirling flow in a straight pipe. *JSME International Journal*, *30*, 66–71.

Kraichnan, R. H. (1959). The structure of isotropic turbulence at very high Reynolds numbers. *Journal of Fluid Mechanics*, *5*, 497–543.

Kraichnan, R. H. (1964). Decay of isotropic turbulence in the direct-interaction approximation. *Physics of Fluids*, *7*, 1030–1048.

Kraichnan, R. H. (1973). Helical turbulence and absolute equilibrium. *Journal of Fluid Mechanics*, *59*, 745–752.

Krause, F., & Rädler, K.-H. (1980). *Mean-field magnetohydrodynamics and dynamo theory*. Pergamon Press.

Krause, F., & Rüdiger, G. (1974). On the Reynolds stresses in mean field hydrodynamics I. Incompressible homogeneous isotropic turbulence. *Astronomische Nachrichten*, *295*, 93.

Launder, B. E., & Spalding, D. B. (1972). *Lectures in mathematical models of turbulence*. Academic Press.

Lesieur, M. (2008). *Turbulence in fluids* (4th ed.). Springer Verlag.

Leslie, D. C. (1973). *Developments in the theory of turbulence*. Clarendon Press.

Levina, G. V., & Montgomery, M. T. (2014). Tropical cyclogenesis: a numerical diagnosis based on helical flow organization. *Journal of Physics: Conference Series*, *544*, 012013. https://doi.org/10.1088/1742-6596/544/1/012013

Lilly, D. K. (1986). The structure, energetics and propagation of rotating convective storms. Part II: Helicity and storm stabilization. *Journal of the Atmospheric Sciences*, *43*, 126–140.

Marino, R., Rosenberg, D., & Pouquet, A. (2013). Inverse cascades in rotating stratified turbulence: Fast growth of large scales. *Europhysics Letters*, *102*, 44006.

Marino, R., Mininni, P. D., Rosenberg, D., & Pouquet, A. (2013). Emergence of helicity in rotating stratified turbulence. *Physical Review E*, *87*, 033016.

McComb, W. D. (1990). *The physics of fluid turbulence*. Clarendon Press.

Miesch, M. (2005). Large-scale dynamics of the convection zone and tachocline. *Living Reviews in Solar Physics*, *2*, 1–139. http://www.livingreviews.org/lrsp-2005-1

Moffatt, H. K. (1978). *Magnetic field generation in electrically conducting fluids*. Cambridge University Press.

Moffatt, H. K. (1992). Relaxation under topological constraints. In H. K. Moffatt, G. M. Zaslavsky, P. Comte, & M. Tabor (Eds.), *Topological aspects of dynamics of fluids and plasmas* (pp. 3–28). Kluwer Academic Publishers.

Moffatt, H. K., & Dormy, E. (2019). *Self-exciting fluid dynamos*. Cambridge University Press.

Moiseev, S. S., Sagdeev, R. Z., Tur, A. V., Khomenko, G. A., & Yanovsky, V. V. (1983). A theory of large-scale structure origination in hydrodynamic turbulence. *Soviet Physics Journal of Experimental and Theoretical Physics*, *53*, 1149.

Montgomery, M. T., & Kallenbach, R. J. (1997). A theory for vortex Rossby-waves and its application to spiral bands and intensity changes in hurricanes. *Quarterly Journal of the Royal Meteorological Society*, *123*, 435–465.

Parker, E. N. (1955). Hydromagnetic dynamo models. *The Astrophysical Journal*, *122*, 293–314.

Parker, E. N. (1979). *Cosmical magnetic field*. Oxford University Press.

Pouquet, A., & Yokoi, N. (2022). Helical fluid and (Hall)-MHD turbulence: A brief review. *Philosophical Transactions of the Royal Society A: Mathematical, Physical and Engineering Sciences*, *380*, 20210087. doi:10.1098/rsta.2021.0087

Robertson, H. P. (1940). The invariant theory of isotropic turbulence. *Proceedings of the Cambridge Philosophical Society*, *36*, 209–223.

Rodi, W. (1993). *Turbulence models and their application in hydraulics: a state-of-the-art review*. Taylor and Francis.

Rogers, M. M., & Moin, P. (1987). Helicity fluctuations in incompressible turbulent flows. *Physics of Fluids*, *30*, 2662–2671.

Rüdiger, G. (1980). Reynolds stresses and differential rotation I. On recent calculations of zonal fluxes in slowly rotating stars. *Geophysical & Astrophysical Fluid Dynamics*, *16*, 239–261.

Rüdiger, G. (1989). *Differential rotation and stellar convection: Sun and solar-type stars*. Gordon and Breach.

Speziale, C. G. (1991). Analytical methods for the development of Reynolds-stress closures in turbulence. *Annual Review of Fluid Mechanics*, *23*, 107–157.

Steenbergen, W. (1995). *Turbulent pipe flow with swirl* (dissertation). Eindhoven University of Technology.

Steenbeck, M., Krause, F., & Rädler, K.-H. (1966). Berechnung der mittleren Lorentz-Feldstärke $\overline{\mathbf{v} \times \mathbf{B}}$ für ein elektrisch leitendes Medium in turbulenter, durch Coriolis-Kräfte beeinflußter Bewegung. *Zeitschrift für Naturforschung Teil A*, *21*, 369–376.

Tobias, S. M. (2021). The turbulent dynamo. *Journal of Fluid Mechanics: Perspectives*, *912*, 1–76. https://doi.org/10.1017/jfm.2020.1055

Waleffe, F. (1993). Inertial transfers in helical decomposition. *Physics of Fluids A: Fluid Dynamics*, *5*, 677–685.

Yokoi, N. (2013). Cross helicity and related dynamos. *Geophysical & Astrophysical Fluid Dynamics*, *104*, 114–184.

Yokoi, N. (2016). Modeling helicity dissipation-rate equation. In J. Peinke et al. (Eds.), *Progress in turbulence VI, Springer proceedings in physics* (Vol. *165*, pp. 93–96). Springer Verlag. https://doi.org/10.1007/978-3-319-29130-7_17

Yokoi, N. (2020). Turbulence, transport and reconnection. In D. MacTaggart & A. Hillier (Eds.), *Topics in magnetohydrodynamic topology, reconnection and stability theory* (Vol. 591, pp. 177–265). CISM International Centre for Mechanical Sciences. Springer Verlag. https://doi.org/10.1007/978-3-030-16343-3_6

Yokoi, N. (2023). Non-equilibrium turbulent transport in convective plumes obtained from closure theory. *Atmosphere*, *14*, 1013.

Yokoi, N., & Yoshizawa, A. (1993). Statistical analysis of the effects of helicity in inhomogeneous turbulence. *Physics of Fluids A: Fluid Dynamics*, *5*, 464–477.

Yokoi, N., & Brandenburg, A. (2016). Large-scale flow generation by inhomogeneous helicity. *Physical Review E*, *93*, 033125.

Yokoi, N., & Yoshizawa, A. (2017). Subgrid-scale model with structural effects incorporated through the helicity. In R. Örlü et al. (Eds.), *Progress in turbulence VII, Springer proceedings in physics* (vol. *196*, pp. 115–121). Springer Verlag. https://doi.org/10.1007/978-3-319-57934-4_17

Yokoi, N., Yoshizawa, A., Itoh, K., & Itoh, S.-I. (2004). A variational analysis of flow-reversal condition in a turbulent swirling pipe flow using the bulk-helicity concept. *Physics of Fluids*, *16*, 1186–1202.

Yokoi, N., Masada, Y., & Takiwaki, T. (2022). Modelling stellar convective transport with plumes: I. Non-equilibrium turbulence effect in double-averaging formulation. *Monthly Notices of the Royal Astronomical Society*, *16*, 1186–1202. https://doi.org/10.1093/mnras/stac1181

Yoshizawa, A. (1984). Statistical analysis of the deviation of the Reynolds stress from its eddy-viscosity representation. *Physics of Fluids*, *27*, 1377–1387.

Yoshizawa, A. (1993). Derivation of a model Reynolds-stress transport equation using the renormalization of the eddy-viscosity-type representation. *Physics of Fluids A: Fluid Dynamics*, *5*, 707–715.

Yoshizawa, A., Yokoi, N., Nisizima, S., Itoh, S.-I., & Itoh, K. (2001). Variational approach to a turbulent swirling pipe flow with the aid of helicity. *Physics of Fluids*, *13*, 2309–2319.

Part II

Helicity Manifestations in Nature and Their Observations

4

Observations of Magnetic Helicity Proxies in the Solar Photosphere: Helicity With Solar Cycles

Hongqi Zhang[1], Shangbin Yang[1,2], Haiqing Xu[1], Xiao Yang[1], Jie Chen[1], and Jihong Liu[1,3]

ABSTRACT

Observations of magnetic helicity transportation through the solar photosphere reflect the interaction of turbulent plasma movements and magnetic fields in the solar dynamo process. In this chapter, we review the research on magnetic helicity inferred from observed solar magnetic fields in the photosphere and solar morphological configurations with solar cycles. After introducing some achievements in the study of magnetic helicity, we summarize several key points. The magnetic (current) helicity in the solar surface layer presents a statistical distribution similar to that of a sunspot butterfly diagram, but its maximum value is delayed compared to the extreme value of the sunspot butterfly diagram and corresponds to the phase with the statistical eruption of solar flares. During the spatial transport of magnetic (current) helicity from the interior of the Sun into interplanetary space at the time-space scale of the solar cycle, the helicity shows a statistical distribution and fluctuation with the hemispheric sign rule. These show that the current helicity and magnetic helicity transport calculation methods complement each other. We also see that the study of the inherent relationship between magnetic helicity and the solar cycle depends on the observed accuracy of the solar magnetic field.

4.1. INTRODUCTION

The study of solar magnetic helicity with solar cycles is an interesting topic. The solar activity cycle was first found from observations of sunspots, which have 11-year cycles (Schwabe, 1843). Observation of solar magnetic fields with magnetographs has brought human understanding of solar activity to a new stage (Hale, 1908).

A large number of observations show that magnetic field energy on the solar surface is most concentrated in solar active regions (cf. Fisher et al., 2000; Schmieder et al., 2014; Toriumi et al., 2020). Accompanied by flares or coronal mass ejections, the energy of the magnetic field is transported from the solar subatmosphere into interplanetary space in strong solar storms (cf. Wang & Xu, 2002; Aschwanden, 2005). It has been found that intense solar eruptive phenomena are often connected with strong magnetic helicity (cf. Low, 1996; Bothmer & Schwenn, 1998; Yurchyshyn et al., 2001; Nindos & Zhang, 2002; Lynch et al., 2005; Török & Kliem, 2005).

Due to the opacity of the solar atmosphere, despite the use of helioseismology to diagnose the Sun's structure, little is known about magnetic fields inside the Sun (Parker, 1979; Fan, 2009). The magnetic field distribution on the Sun's surface provides important information about the formation and transport of magnetic field energy with helicity in the subatmosphere.

The theory of solar dynamos has been used to explore the formation of the magnetic field inside the Sun and explain the statistical phenomenon that solar activity

[1] *National Astronomical Observatories, Chinese Academy of Sciences, Beijing, China*
[2] *School of Astronomy and Space Science, University of Chinese Academy of Sciences, Beijing, China*
[3] *School of Science, Shijiazhuang University, Shijiazhuang, China*

Helicities in Geophysics, Astrophysics, and Beyond, Geophysical Monograph 283, First Edition.
Edited by Kirill Kuzanyan, Nobumitsu Yokoi, Manolis K. Georgoulis, and Rodion Stepanov.

DOI:10.1002/9781119841715.ch04

presents an 11-year cycle (cf. Parker, 1955; Moffatt, 1978; Krause & Rädler, 1980; Zeldovich et al., 1983). The usual theories state that the chirality (helicity) of this magnetic field is important evidence about its formation and that helicity tends to have opposite signs in both solar hemispheres during the solar dynamo process (cf. Seehafer, 1994). It is generally believed that the observed magnetic helicity on the solar surface demonstrates the α-effect in the deep solar convection zone from the viewpoint of the mean-field $\alpha\omega$ dynamo model (cf. Brandenburg & Subramanian, 2005) and the flux-transport model (cf. Choudhuri et al., 2004).

Statistical analysis of the observed helical magnetic field in the solar atmosphere, such as in solar active regions, provides a window for studying the possible mechanism of forming active cycles with solar dynamos.

In this chapter, we present the current helicity in active regions, the injection of magnetic helicity with solar cycles inferred from observed photospheric magnetic fields, and the relationship with solar eruptive activities. We also discuss questions arising from observations of solar magnetic helicity.

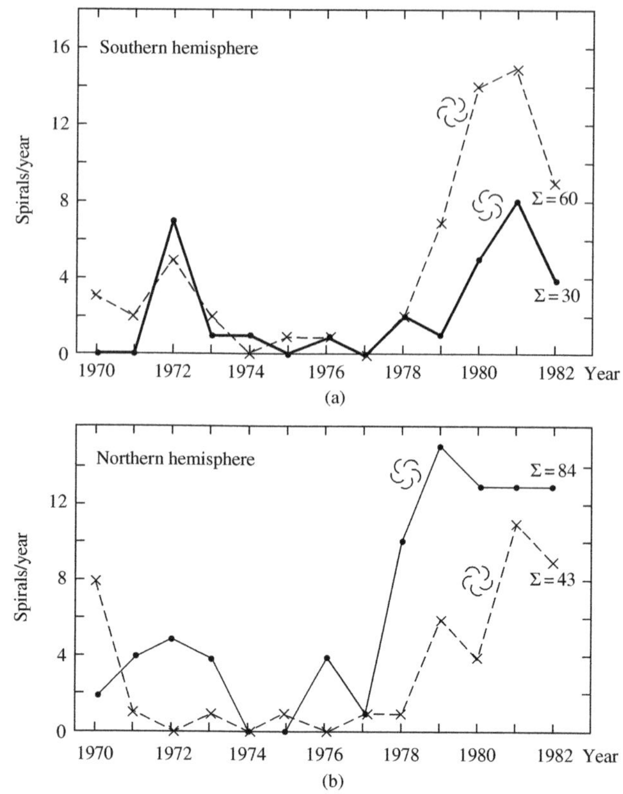

Figure 4.1 Distribution of spiral patterns of active regions in the solar (a) southern and (b) northern hemispheres. The ordinates (spirals/year) are the statistical number of twisted sunspots per year. The solid (dashed) lines indicate the twist of sunspots in right (left) handedness. Ding et al., 1987/Springer Nature.

4.2. MAGNETIC HELICITY FROM OBSERVATIONS

4.2.1. Early Morphological Observations of Helical Patterns in the Solar Atmosphere

The chiral patterns of active regions on the Sun were first found by Hale (1908), Richardson (1941), whose observations showed that most sunspots (about 80%) with discernible whirls had phenomenological counter-clockwise (clockwise) rotations of the fibril patterns in the northern (southern) hemisphere. This was confirmed and statistically demonstrated for the distribution of helical sunspots in both hemispheres in 1970–1982 (approximately a solar cycle) by Ding et al. (1987), as shown in Fig. 4.1. Interestingly, only about 66% of sunspots follow the helical sign rule.

Observing magnetic fields in the solar photosphere using solar magnetographs or Stokes polarimeters is a general method to diagnose the twisted pattern of the field in sunspot regions. For example, Fig. 4.2 shows the photospheric vector and chromospheric longitudinal magnetic field in a delta active region (Zhang, 2019, 2020). The horizontal twisting component of the photospheric vector magnetic field of the active region is consistent with features of the Hβ chromospheric magnetic field and chromospheric morphological features. These confirm that the morphological pattern of spiral spot features as observed previously by Hale (1908), Richardson (1941), Ding et al. (1987) reflects the configuration of the magnetic fields in the atmosphere (such as those marked with an *f* in Fig. 4.2).

Evidence of the hemispheric sign rule for helical features in the chromosphere was also indicated by Martin et al. (1994). They found morphologically that most quiescent filaments were dextral/sinistral in the northern/southern hemisphere. Rust (1994) pointed out the relationship of the chirality patterns in sunspots, active regions, filaments, and interplanetary magnetic clouds. These imply that helical patterns on the Sun relate to different scales of magnetic fields due to the dominant controls of the magnetic fields in active regions, filaments, magnetic clouds, and so on. This means the helical features extend from the photosphere into the high atmosphere and interplanetary space. It has been noted that the low sensitivity of chromospheric spectral lines to magnetic field observations, nonlocal thermal dynamic equilibrium, and disturbance of photospheric blended spectral lines (such as in the wing of the Hβ line) affect the measurement of the solar chromospheric magnetic field and make it difficul to obtain the chromospheric vector magnetic field (Zhang, 2019, 2020).

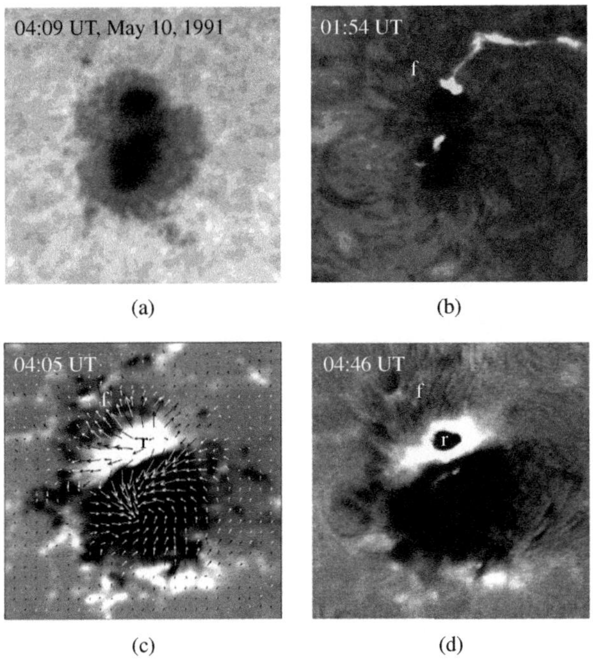

Figure 4.2 Active region NOAA 6619 observed on 10 May 1991. (a) Photospheric filtergram. (b) Hβ filtergram. (c) Photospheric vector magnetogram. (d) longitudinal magnetogram in the chromosphere from Hβ. Black (white) indicates negative (positive) polarity in the magnetograms. Zhang [2019], Springer Nature.

The reversal patch r in the umbra of the chromospheric Hβ magnetogram in Fig. 4.2 is caused by disturbance of photospheric blended spectral lines in the wing of the Hβ line.

4.2.2. Magnetic Helicity from Solar Vector Magnetic Fields

The magnetic helicity is the volume integral (Woltjer, 1958a,b)

$$H_m = \int_V \mathbf{A} \cdot \nabla \times \mathbf{A} d^3x, \tag{4.1}$$

where $\mathbf{A}$ is the magnetic vector potential, is not an immediately observed quantity, and does not satisfy the requirement of gauge invariance. To simplify calculations, we can employ the Coulomb gauge (Moffatt, 1969; Arnold, 1974; Berger & Field, 1984)

$$\mathbf{A}(\mathbf{x}) = -\frac{1}{4\pi} \int d^3x' \frac{\mathbf{r}}{r^3} \times \mathbf{B}(\mathbf{x}'), \tag{4.2}$$

which gives

$$H_m = -\frac{1}{4\pi} \int d^3x \int d^3x' \mathbf{B}(\mathbf{x}) \cdot \left[\frac{\mathbf{r}}{r^3} \times \mathbf{B}(\mathbf{x}')\right], \tag{4.3}$$

with the relationship $\mathbf{B} = \nabla \times \mathbf{A}$ (Low, 2015).

The current helicity H_c is defined as

$$H_c = \int_V \mathbf{B} \cdot \nabla \times \mathbf{B} d^3x, \tag{4.4}$$

where $\mathbf{B}$ is the magnetic field and the current helicity density can be defined as

$$h_c = \mathbf{B} \cdot \nabla \times \mathbf{B} = (\mathbf{B} \cdot \nabla \times \mathbf{B})_z + (\mathbf{B} \cdot \nabla \times \mathbf{B})_t. \tag{4.5}$$

The subscripts z and t indicate the longitudinal and transverse components of current helicity, respectively. The current helicity density $h_{cz} = (\mathbf{B} \cdot \nabla \times \mathbf{B})_z$ is observable in the lower solar atmosphere as shown in the photospheric vector magnetogram in Fig. 4.3, while $(\mathbf{B} \cdot \nabla \times \mathbf{B})_t$

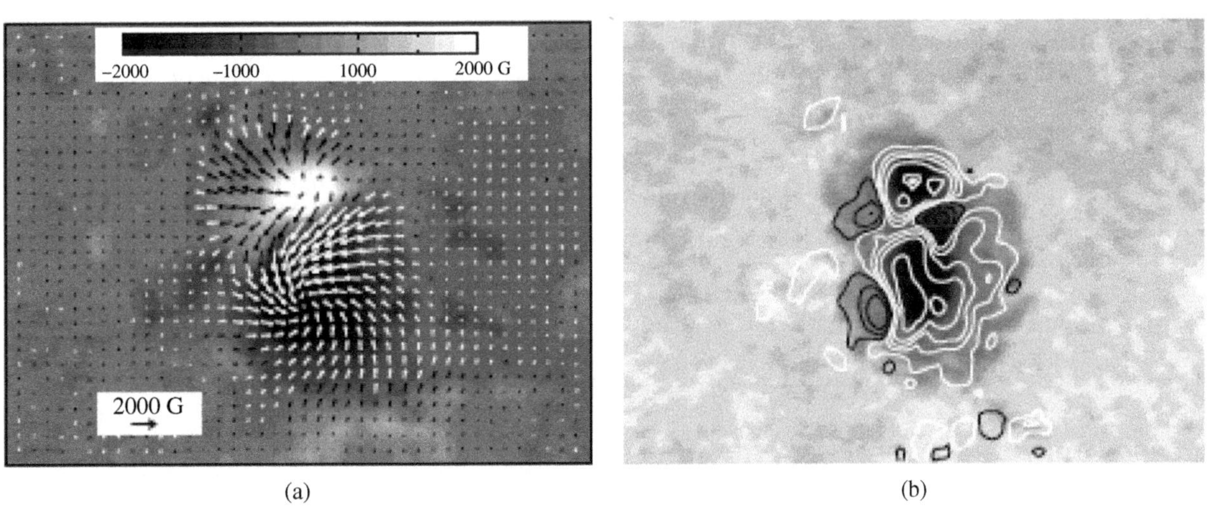

Figure 4.3 The active region NOAA 6619, taken at Huairou Solar Observing Station on 11 May 1991, 03:26 UT. (a) Photospheric vector magnetogram. (b) Electric current helicity density $h_{cz} = (\mathbf{B} \cdot \nabla \times \mathbf{B})_z$ (contours) overlapped by a filtergram of this active region; black (white) contours correspond to positive (negative) values of 0.01, 0.05, 0.1, 0.2, and 0.5 G^2m^{-1}.

(a)

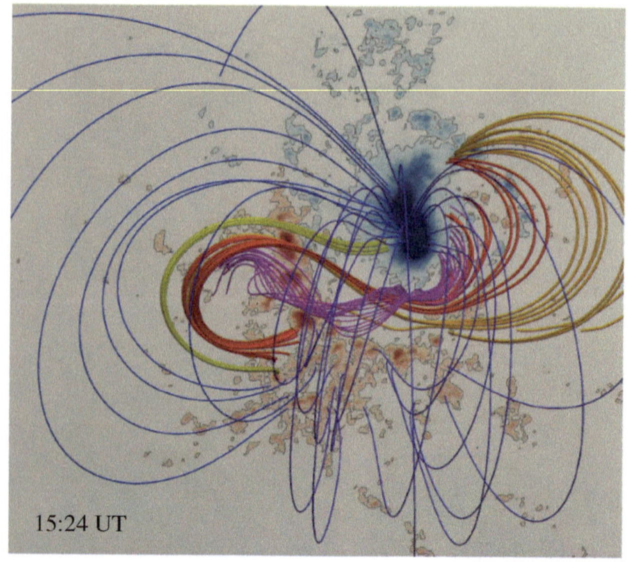

(b)

Figure 4.4 Sigmoidal field lines wrapping around a flux rope in active region 12158, which has a negative helicity (reverse S, left-hand twist). Taken 10 September 2014, two hours before a flare. (a) Observation in AIA 94 Å; (b) Sigmoidal magnetic field lines using a nonlinear force-free extrapolation based on the Grad-Rubin method (Gilchrist & Wheatland, 2014), overlying a longitudinal component of the photospheric magnetic field saturated at 2000 gauss. Schmieder & Aulanier, 2018/John Wiley & Sons.

is difficult to determine because it is impossible to construct the real transverse component of electric current through observations of solar magnetic fields.

The chirality of magnetic structures can be analyzed morphologically due to the magnetic freezing effect of the plasma in the solar atmosphere. It reflects the spatial characteristics of the magnetic field helicity. This also expands the distribution and range of magnetic helicity in the solar atmosphere that we can analyze through observation. Figure 4.4 shows the extrapolated nonlinear force-free (NLFF) magnetic field calculated from the photospheric magnetic field in an active region to fit AIA 94Å flare loops (Zhao, 2016; Schmieder & Aulanier, 2018), which has a negative helicity.

From this discussion, it can be seen that the distribution of the magnetic (current) helicity in the solar atmosphere and its evolution can be analyzed from various perspectives (the magnetic vector potential, magnetic field, and morphology). The key parameter is the magnetic field, especially from observations of the photospheric magnetic field.

4.3. ELECTRIC CURRENT HELICITY FROM OBSERVATIONS

4.3.1. Helicity from Solar Vector Magnetic Fields

Early statistical research on magnetic (current) helicity from observations of solar magnetic fields was performed by Seehafer (1990), followed by a series of studies by Pevtsov, Canfield, and Metcalf (1995), Abramenko et al. (1996), Bao & Zhang (1998), Hagino & Sakurai (2005), Xu et al. (2007) and others. The observable current helicity density averaged over an active region is

$$\overline{h_{cz}} = \overline{(\mathbf{B} \cdot \nabla \times \mathbf{B})_z}, \tag{4.6}$$

which can be derived from photospheric vector magnetograms (see Fig. 4.3). In the approximation of local homogeneity and isotropy, the value of h_{cz} can be used to analyze the total current helicity h_c (Xu et al., 2015). Note that $h_{cz} = 1/3h_c$ is an estimated value due to incompleteness for inversion of the solar magnetic field through solar polarized spectral light (Zhang, 2019, 2020).

Another proxy of the magnetic field is twist: i.e., the force-free factor $\alpha = \mathbf{B} \cdot (\nabla \times \mathbf{B})/B^2$. An equivalent of the quantity averaged over an active region is the ratio

$$\alpha_{av} = \overline{(\nabla \times \mathbf{B})_z / B_z} \tag{4.7}$$

calculated from observed vector magnetograms.

Pevtsov et al. (2005) reported the result of a study of magnetic helicity in solar active regions in 1980–2000 (cycles 21–23) using α_{av} as the proxy for current helicity. They did not see consistency between instruments operated at different observatories in regard to years of disobeying the hemispheric helicity rule. They found that present data sets do not allow for making statistically significant inferences about possible cyclic

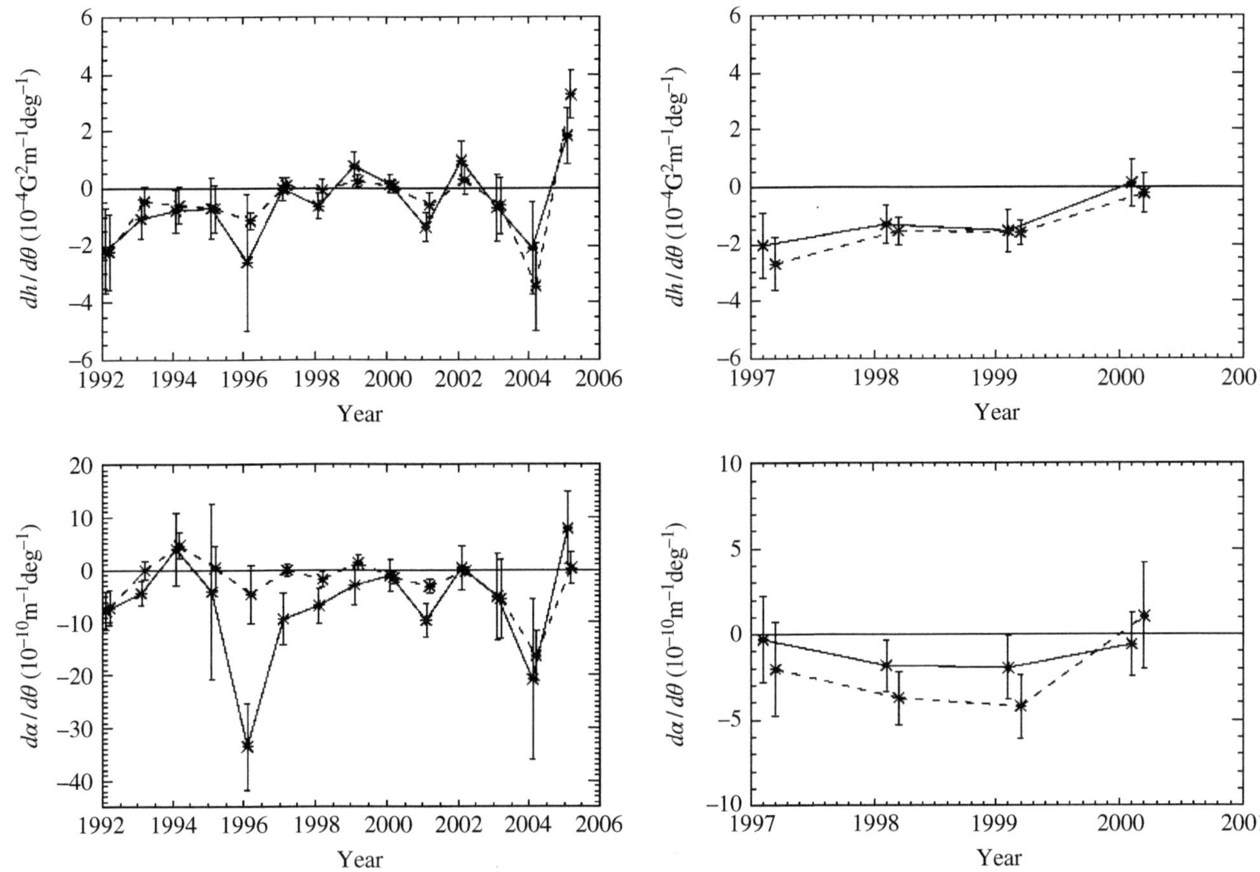

Figure 4.5 Time variation of the mean slope of linear fit calculated from the latitudinal profile of helicity for each year from 1992 to 2005 (left) and 1997 to 2000 (right). The solid (dashed) lines on the left represent Huairou (Mitaka) data, and on the right, Mees (Mitaka) data. The upper panel is $dh_c/d\theta$ and the lower is $d\alpha_{ff}/d\theta$. Error bars represent one sigma uncertainty of the slope of the linear fit. Xu et al., 2007/Elsevier.

variation of the hemispheric helicity rule. Xu et al. (2007) compared a series of vector magnetograms of the same active regions observed by different solar vector magnetographs. They found that a similar tendency of the signs of calculated magnetic helicity from magnetograms was observed at the different observatories in Figure 4.5. However, there were slight differences.

The seemingly contradictory results of the helicity may come primarily from observations of the solar magnetic field and the processing and analysis of the data. Diagnosing solar magnetic fields based on these magnetically sensitive lines is accompanied by approximations of the theoretical analysis, which probably causes the uncertainty in the inversion of spectro-polarimetric observations to obtain magnetic fields. It is important to prevent or avoid the influence of these errors and try to obtain relatively accurate measurements of solar magnetic fields if defects in the different magnetographs can be ignored. There may be additional reasons for the errors observed in vector magnetographs from various observatories (Ai, 1989; Ronan et al., 1992; Wang et al., 1992; Sakurai et al., 1995; Sakurai, 2001; Zhang et al., 2003), and they should be analyzed in detail.

4.3.2. Magnetic Helicity with Solar Cycles

The evolution of helicity with the solar cycle is a notable topic. The mean distribution of the photospheric current helicity density of active regions on the solar surface given by $\overline{h_{cz}} = \overline{(\mathbf{B} \cdot \nabla \times \mathbf{B})_z}$ in equation (4.6) and $\alpha_{av} = \overline{(\nabla \times \mathbf{B})_z / B_z}$ in equation (4.7) in solar cycle 22 was statistically demonstrated by (Bao & Zhang, 1998) and (Zhang & Bao, 1998; Zhang & Bao, 1999). They were inferred from a series of photospheric vector magnetograms recorded at Huairou Solar Observing Station, National Astronomical Observatories of China.

Figure 4.6 shows in a butterfly diagram the distribution of the current helicity of active regions observed at Huairou. Most helicity of active regions follows the hemispheric sign rule, and some super δ active regions are

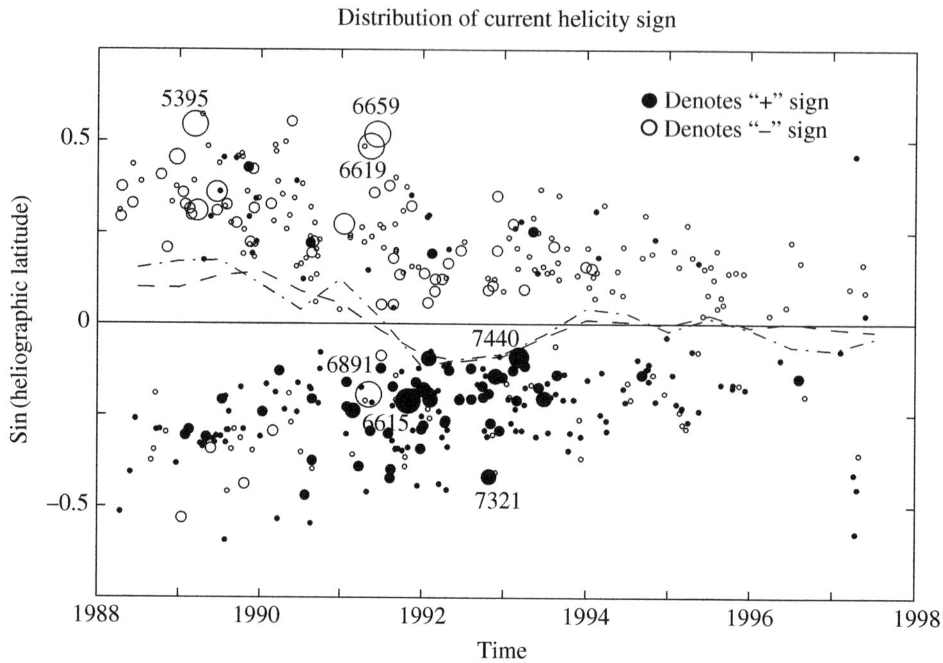

Figure 4.6 Butterfly diagram of the mean current helicity density h_{cz}. The sizes of the circles relate to the grade of the current helicity density: 0, 1, 3, 5, 7 ($10^{-3}G^2m^{-1}$). The white (black) circles indicate a negative (positive) sign for helicity. The dashed-dotted line shows the distributed average value of current helicity, and the dashed line shows the mean value of the imbalance of current helicity after data smoothing. Vertical coordinates represent the sine of latitude, and horizontal coordinates are expressed in years. From Zhang & Bao (1998).

marked in the figure. The high density of current helicity regions (such as NOAA 5395, 6659, and 6619) occurred at high latitudes in the northern hemisphere, while NOAA 7321 and 7440 are in the southern hemisphere. Notably, the active region NOAA 6891 has a reversed sign for helicity due to the hemispheric sign rule.

A subsequent study focused on the distribution of the average helical characteristics of the magnetic field in solar active regions from 1988 to 2005 in the form of butterfly diagrams (latitude–time). The data was observed at Huairou (Zhang et al., 2010b) (Fig. 4.7). It covered most of solar cycles 22 and 23 after removing magneto-optical effects in the measurements of the magnetic field (Su & Zhang, 2004; Gao et al., 2008).

The maximum values of the mean helicity density of active regions at the solar surface tend to occur near the edges of the sunspot butterfly diagram in Fig. 4.7. It is roughly consistent with Fig. 4.6, where some superactive regions are marked. The reversal of the mean current helicity's sign relative to the hemispheric sign rule occurs at the beginning of the butterfly wing in 1997 in Fig. 4.7 (also see Bao et al., 2000). It would be interesting to compare with the traditional idea of the α effect on the solar dynamo theory (such as Parker, 1955).

Kuzanyan et al. (2003), Zhang et al. (2006) identified the longitudinal migration of active regions and the corresponding current helicity with their rotation rates and compared them with the internal differential rotation law in the solar convection zone inferred from helioseismology. They deduced the distribution of the contributed helicity over depth.

Moreover, Zhang (2006) reported that both α factor and current helicity in strong fields have the opposite sign of weak fields. They analyzed large samples of photospheric vector magnetic fields in active regions from 1997 to 2004 observed at Huairou. A similar case with the spectro-polarimeter (SP) onboard the Hinode satellite was demonstrated by Hao & Zhang (2011), who showed that helicity changes the sign from the inner umbra to the outer penumbra. The question is how to express the possible reason and mechanism for opposite signs of the current helicity between the umbra and penumbra in the same sunspots or strong and weak fields in the same active regions.

Following Matthaeus et al. (1982), it is possible to determine the helicity spectrum of magnetic fields in solar active regions from the spectral correlation tensor if we assume that local statistical isotropy exists. The Fourier transform of the two-point correlation tensor, $\langle B_i(\mathbf{x}, t)B_j(\mathbf{x} + \boldsymbol{\xi}, t)\rangle$, can be written as (cf. Moffatt, 1978)

$$\left\langle \hat{B}_i(\mathbf{k}, t)\hat{B}_j^*(\mathbf{k}', t)\right\rangle = \Gamma_{ij}(\mathbf{k}, t)\delta_2(\mathbf{k} - \mathbf{k}'), \qquad (4.8)$$

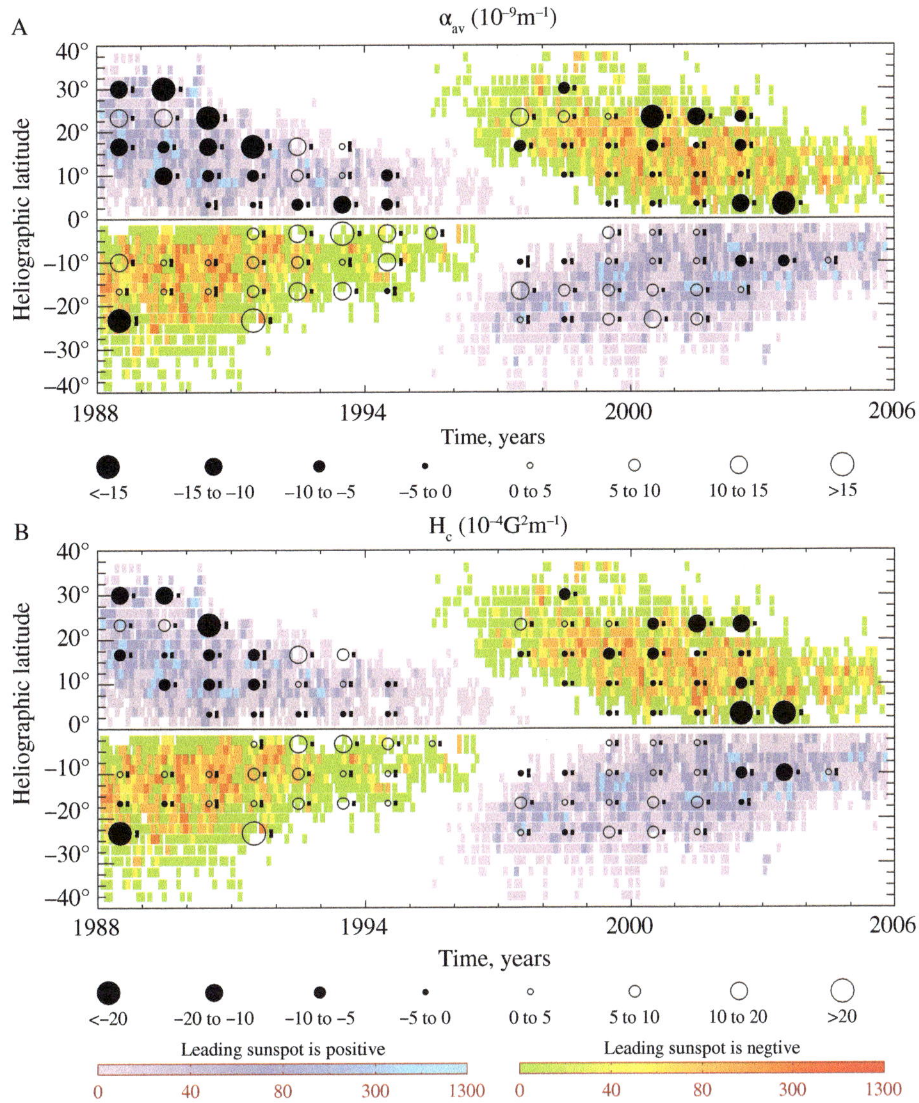

Figure 4.7 (a) Distribution of the averaged twist α_{ff}. (b) Current helicity h_{cz} of solar sunspots in the 22nd and 23rd solar cycles. Superimposed, the underlying grayed butterfly diagram shows how sunspot density varies with latitude over the solar cycle. The open (closed) circles indicate a positive (negative) sign. Zhang et al., 2010b/Oxford University Press.

where $\hat{B}_i(\mathbf{k}, t) = \int B_i(\mathbf{x}, t)\, e^{i\mathbf{k}\cdot\mathbf{x}} d^2x$ is the 2D Fourier transform, subscripts i and j relate to the vector magnetic field components, the asterisk denotes complex conjugation, and ensemble averaging is replaced by averaging over concentric annuli in wavevector space (Zhang, Brandenburg & Sokoloff, 2016). As $\mathbf{k}$ defines the direction in Γ_{ij} and that $k_i\hat{B}_i = 0$, the only possible structure of $\Gamma_{ij}(\mathbf{k}, t)$ takes the form (cf. Moffatt, 1978)

$$\Gamma_{ij}(\mathbf{k}, t) = \frac{2E_M(k, t)}{4\pi k}(\delta_{ij} - \hat{k}_i\hat{k}_j) + \frac{iH_M(k, t)}{4\pi k}\epsilon_{ijk}k_k, \quad (4.9)$$

where $\hat{k}_i = k_i/k$ is the component of the unit vector of $\mathbf{k}$, $k = |\mathbf{k}|$ is the modulus with $k^2 = k_x^2 + k_y^2$, and $E_M(k, t)$ and $H_M(k, t)$ are the magnetic energy and magnetic helicity spectra, respectively.

For example, Fig. 4.8 shows the spectrum of magnetic helicity and energy in an active region; the data was observed by Helioseismic and Magnetic Imager (HMI) onboard the Solar Dynamics Observatory (SDO) satellite. Figure 4.8a shows the sign of the magnetic helicity at different scales after averaging the spectrum over broad, logarithmically spaced wavenumber bins.

Figure 4.9 presents the variation of the slopes of the spectra of magnetic energy and helicities of active regions with solar cycles (Zhang, Brandenburg & Sokoloff, 2016). The magnetic spectra have been taken in the form $k^{-\alpha}$.

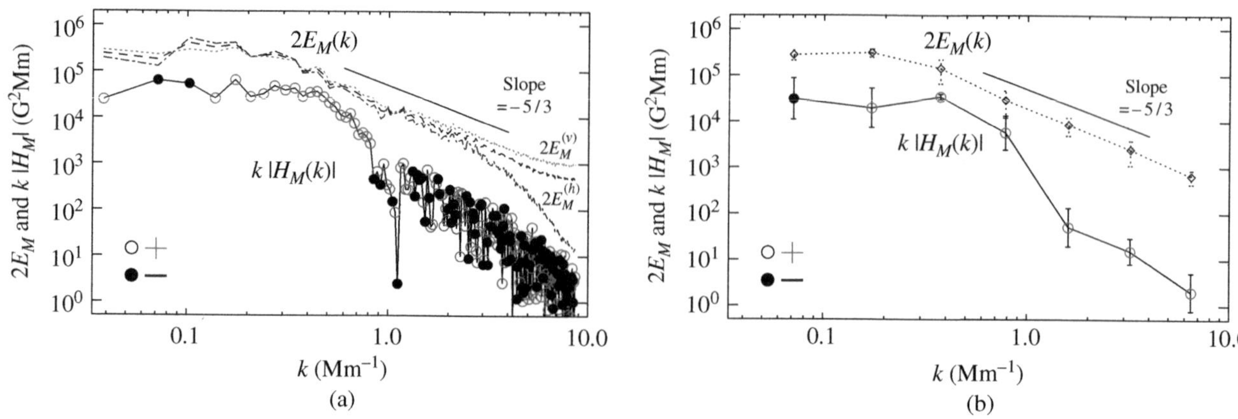

Figure 4.8 (a) $2E_M(k)$ (solid line) and $k|H_M(k)|$ (dotted line) for NOAA 11158 at 23:59:54UT on 13 February 2011. Positive (negative) values of $H_M(k)$ are indicated by open (closed) symbols. $2E_M^{(v)}(k)$ (dotted) and $2E_M^{(h)}(k)$ (dash-dotted) are shown for comparison. (b) Same as (a), but the magnetic helicity is averaged over broad logarithmically spaced wavenumber bins. Zhang, Brandenburg & Sokoloff, 2014/IOP Publishing.

The statistical correlation coefficient between sunspot numbers and slopes α of the magnetic energy is 0.827, and that between sunspot numbers and current helicity is 0.730. The correlation coefficient between sunspot numbers and slopes of current helicity changes to 0.831 as one takes the sunspot numbers to a one-year delay. This is consistent with the observed result in Fig. 4.7, where the maximum value of the mean current helicity of solar sunspots is delayed compared to that of sunspot numbers. Similar evidence is that the complex magnetic configuration of active regions tends to occur in the decaying phase of the solar cycle (Guo et al., 2010).

Another piece of evidence is that one cannot find the significant variation of the mean slopes α of the spectra of magnetic energy and helicities of active regions with the latitudes after a long-term average in Fig. 4.9. This indicates that the mean scale distribution of the magnetic field of solar active regions does not show a significant latitudinal tendency.

A similar study for the active cycle is demonstrated by Gosain & Brandenburg (2019) after computation of the magnetic helicity and energy spectra from magnetic patches on the solar surface using data observed by the Hinode satellite from 2006–2017.

Similar studies of the large-scale current helicity are based on calculating longitudinal full-disk magnetograms obtained by the Michelson Doppler Imager (MDI) as well as the Kitt Peak Vacuum Telescope and the Synoptic Long-term Investigations of the Sun (SOLIS) (Pevtsov & Latushko, 2000; Wang & Zhang, 2010; Pipin & Pevtsov, 2014). The large-scale magnetic fields reflect a clear and consistent current helicity pattern that follows the established hemispheric rule. Pipin et al. (2019) have demonstrated the mean helicity density of the nonaxisymmetric magnetic field of the Sun and separated it into mean large- and small-scale components, which display the hemispheric helicity rule of opposite signs at the beginning of cycle 24.

4.4. MAGNETIC HELICITY INJECTION ON THE SOLAR SURFACE

4.4.1. Magnetic Helicity Injection Inferred from Moving Magnetic Structures

Magnetic helicity in equation (4.1) is not a directly observable quantity in the solar atmosphere. The general definition of magnetic helicity does not satisfy the requirement of gauge invariance, and the concept of relative magnetic helicity has been introduced (Berger & Field, 1984). The injected magnetic helicity from the solar subatmosphere is a quantity that reflects its relative variation. It can be inferred by the motions of foot points of the magnetic fields on the solar surface. The change in magnetic helicity is (Chae, 2001)

$$\frac{dH_m}{dt} = -2\oint_S (\mathbf{A}_p \cdot \mathbf{V}_t) B_n ds + 2\oint_S (\mathbf{A}_p \cdot \mathbf{B}_t) V_n ds, \quad (4.10)$$

where $\mathbf{A}_p$ is the vector potential of the potential field, and subscripts t and n mark the transverse and vertical components of the magnetic field and velocity field, respectively. The first integral on the right-hand side of equation (4.10) is the contribution from twisting or shearing motions of foot points of magnetic fields on the solar surface, and the second integral is from the emergence of magnetic fluxes from the subphotosphere (Antiochos, 1987; Kusano et al., 2002).

According to the analysis of Démoulin & Berger (2003), the injected magnetic helicity can be calculated as (Chae, 2001)

$$\frac{dH_m}{dt} = -2\oint_S (\mathbf{A}_p \cdot \mathbf{U}) B_n ds, \quad (4.11)$$

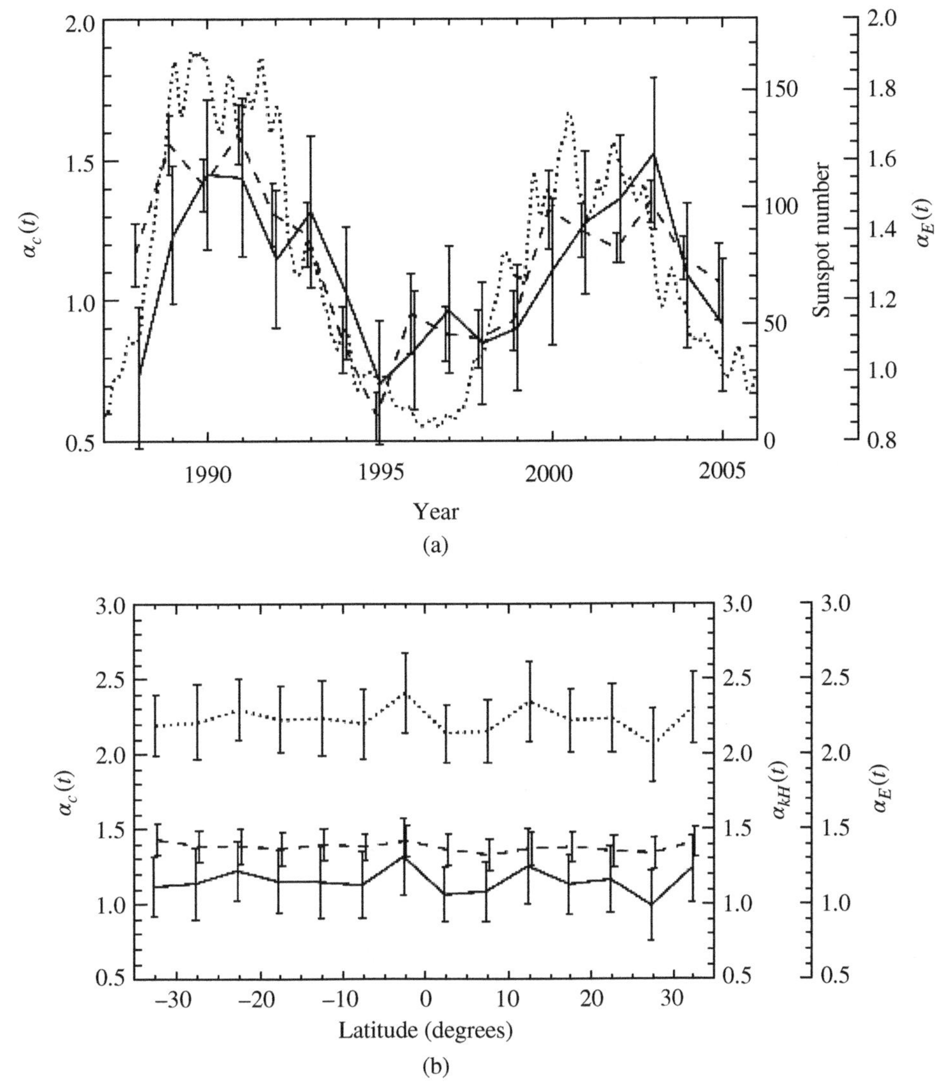

Figure 4.9 (a) Distribution of the average scale exponent α of the current helicity (α_c, solid line) and magnetic energy (α_E, dashed line), with the time inferred by 6629 vector magnetograms of solar active regions from 1988 to 2005. The dotted line is sunspot numbers. (b) Distribution of the average scale exponent α of the magnetic helicity (α_{kH}, dotted line), current helicity (α_c, solid line), and magnetic energy (α_E, dashed line) with the latitude. α_{kH} relates to $k|H_M(k)|$. The error bars are 0.3σ. Zhang, Brandenburg, & Sokoloff, 2016/IOP Publishing.

where

$$\mathbf{U} = \mathbf{V}_t - \frac{V_n}{B_n}\mathbf{B}_t, \tag{4.12}$$

$\mathbf{A}_p$ is the magnetic vector potential of the potential field, and t and n mark the horizontal and vertical components of the magnetic and velocity field, respectively. This implies that one can include the greatest contribution of injective helicity from the horizontal motion of magnetic foot points on the solar surface.

Figure 4.10 shows an example of the injection of magnetic helicity in a newly emerging active region by equation (4.12). Both polarities of the magnetic flux increase very quickly in Fig. 4.10a. Figures 4.10b and 4.10c display the temporal variation of dH_m/dt and accumulated helicity $\Delta H(t)$ separately. The former represents the injective rate of magnetic helicity with the horizontal motion of magnetic fields on the solar surface, and the latter represents the total accumulated helicity.

4.4.2. Magnetic Helicity Injection with Solar Cycles

Berger & Ruzmaikin (2000) evaluated the surface integral using solar magnetogram data and differential

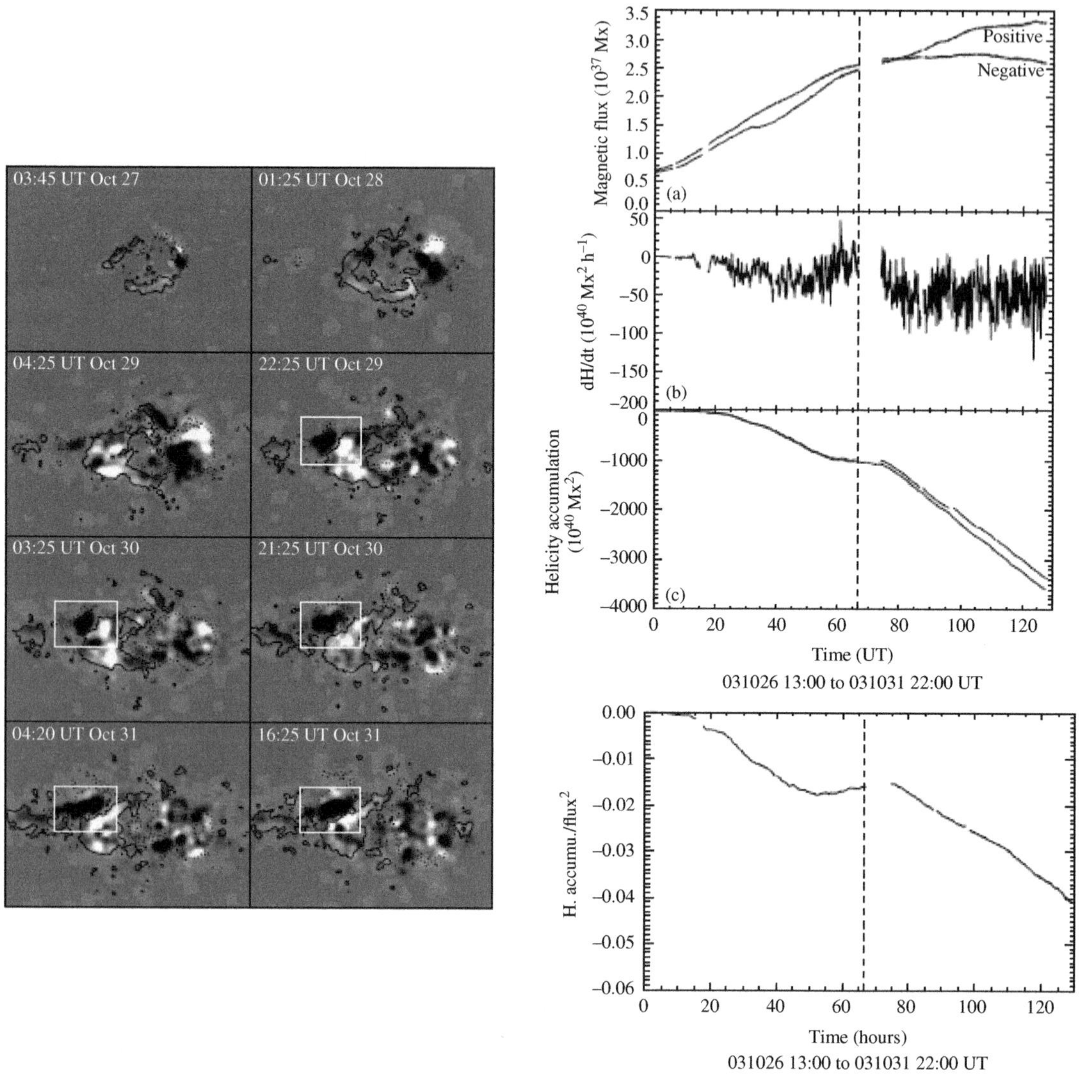

Figure 4.10 Active region NOAA 10488 on 26–31 October 2003. Left: Grayscale maps of $G \equiv -2(\mathbf{u} \cdot \mathbf{A}_p)B_z$ of one-hour averages. White (black)indicate positive (negative) signs. Top-right: (a) Time profile of the longitudinal component of magnetic field flux. (b) Time profile of the injected rate of magnetic helicity by horizontal motion. (c) Accumulated helicity $\Delta H(t)$ calculated from dH/dt. Bottom-right: Ratio of the accumulated coronal helicity to the square of the magnetic flux. Liu & Zhang., 2006/Springer Nature.

rotation curves, and they proposed helicity generation in the solar interior by differential rotation to produce the correct sign compared to observations of coronal structures (Fig. 4.11). They estimated that the net helicity flowing into each hemisphere over this cycle was approximately $4 \times 10^{46}\text{Mx}^2$. Georgoulis et al. (2009) estimated a maximum helicity injection of $6.6 \times 10^{45}\text{Mx}^2$ for solar cycle 23. Hawkes & Berger (2018) used data sets covering 60 years to estimate the magnetic helicity as a predictor of solar cycles, which is an extension following Berger & Ruzmaikin (2000).

According to equation (4.11), the relationship between the mean magnetic helicity density $\overline{h}_m$ and the mean current helicity density $\overline{h}_{cz}$ can be obtained as follows:

$$\overline{h}_m = -\frac{2}{S_c L_c} \int_{T_c} \oint_{\partial S} (\mathbf{U} \cdot \mathbf{A}_p) B_n ds dt \sim k\overline{\mathbf{A} \cdot \mathbf{B}} \sim L_c^2 \overline{h}_{cz}, \tag{4.13}$$

where T_c is the typical relaxation time, S_c and L_c are the typical horizontal and vertical spatial scales of emerging magnetic flux in the solar atmosphere before the transport of helicity into the interplanetary space, and k is a correlation coefficient (about on the order of one).

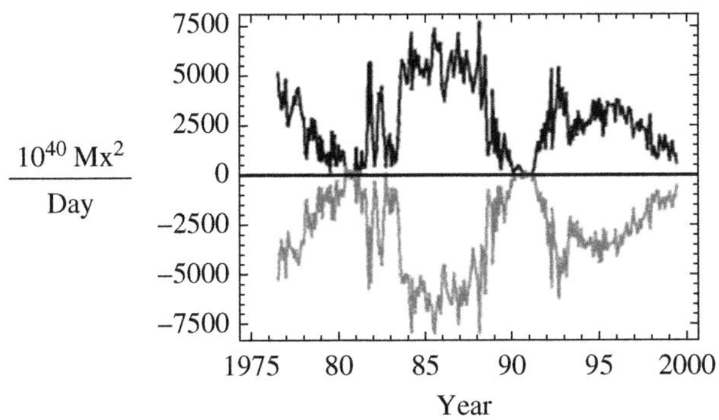

Figure 4.11 Net transfer of helicity into the southern corona and wind $dH_{CS}/dt = \dot{H}(V_S \rightarrow C_S)$ (predominantly positive curve), and into the northern corona and wind $dH_{CN}/dt = \dot{H}(V_N \rightarrow C_N)$ (predominantly negative curve). The units are $10^{40}Mx^2/day$. Berger & Ruzmaikin, 2009/John Wiley & Sons.

In principle, it is difficult to obtain $\overline{h}_m$ directly from observations to compare with the current helicity $\overline{h}_{cz}$ in active regions (in equation (4.6)). The current helicity $\overline{h}_{cz}$ is inferred from the photospheric vector magnetograms shown previously, while equation (4.11) can be used to calculate the injection of magnetic helicity from the subphotosphere. The statistical relationship between magnetic and current helicity with solar cycles is important for understanding the distribution of magnetic chirality in the solar atmosphere and its transformation from the subatmosphere with solar cycles.

To analyze the injection of global magnetic helicity on the solar surface with the solar cycle, MDI full-disk magnetograms were used by Yang et al. (2012) by tracking the MDI synoptic charts and Zhang & Yang (2013) by tracking the full-disk 96-minute magnetograms. The injection of global magnetic helicity on the solar surface can also be calculated or estimated using equations (4.11) and (4.12) from the full-disk magnetograms, as the projective effects of the magnetic field have been neglected nearby the limb of the solar disk. Figure 4.12 shows the mean injection of magnetic helicity flux from 1996 to 2009 in each solar rotation to analyze its long-term evolution on the solar surface. In addition to significantly fluctuating injection of magnetic helicity in the northern and southern hemispheres, we can also find the hemispheric sign rule for the injected magnetic helicity (Zhang & Yang, 2013). It is consistent with the result obtained by photospheric vector magnetograms in the active regions in Fig. 4.7 due to the negligible contribution from the quiet Sun (Welsch & Longcope, 2003). Rough consistency can be estimated using equation (4.13), even though there may be complex internal relationships between different calculated forms of magnetic helicity. We can see from the observations that the total injective helicity in a solar cycle tends to be zero, but the fluctuation is significant.

It is found that the extreme value of negative helicity flux occurred in 1997–2002, after the smoothing of the net magnetic helicity flux in Fig. 4.13. This provides only a rough estimate, in which the extreme value is about $-2 \times 10^{36} Mx^2 s^{-1}$. Figure 4.13 also depicts the injective rate of the total absolute value of magnetic helicity flux

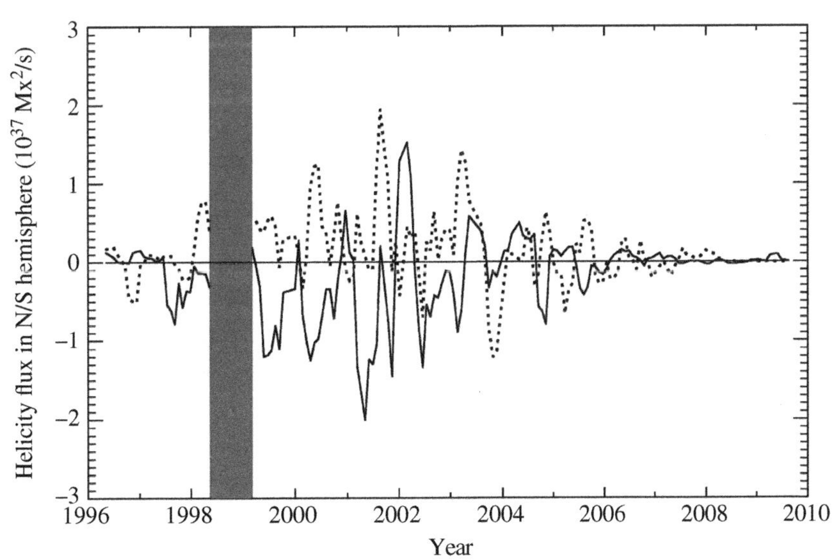

Figure 4.12 Injective magnetic helicity flux from the northern (solid line) and southern (dotted line) hemisphere from 1996 to 2009. The shadow indicates a period without relevant helicity calculations from the magnetograms. Adapted from Zhang & Yang, 2013.

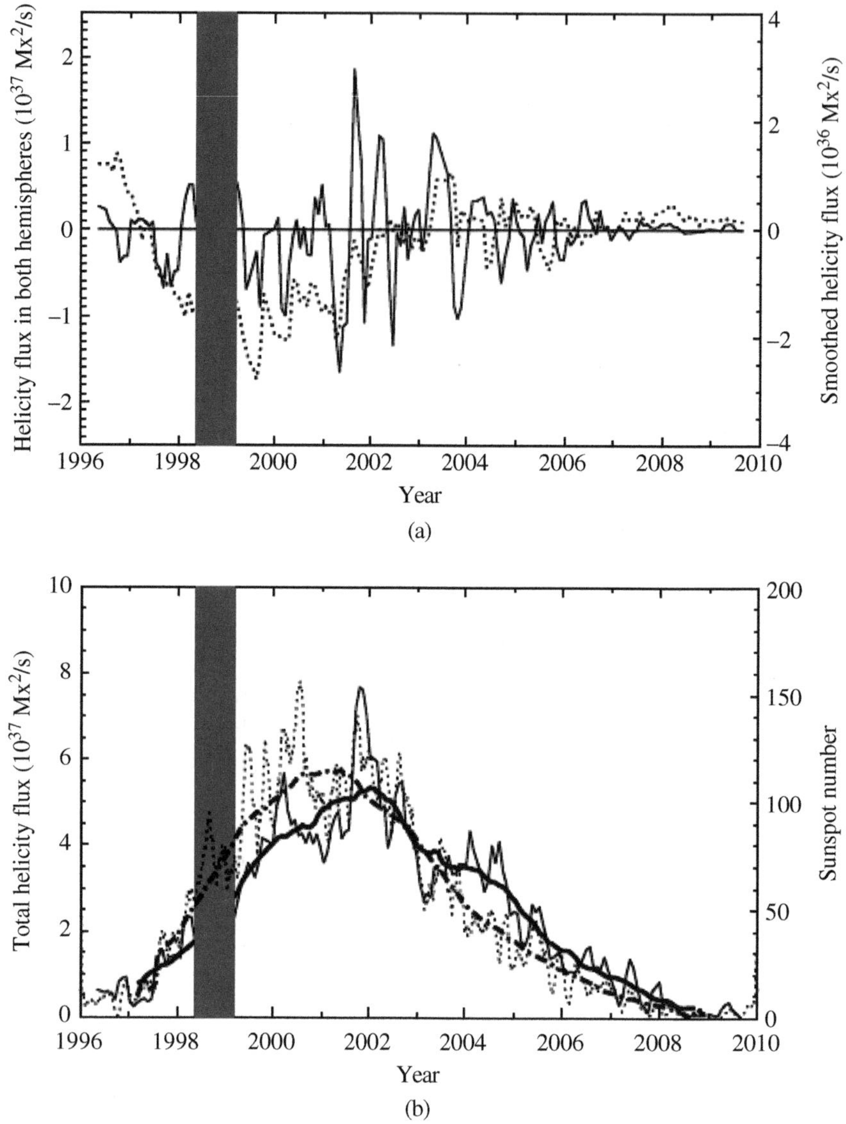

Figure 4.13 (a) Net injective helicity contributed from both hemispheres. The dotted line marks the net injective helicity after smoothing 48 solar rotations. (b) Total magnetic helicity flux (thin solid line) and sunspot numbers (thin dotted line) from 1996 to 2009. The black thick solid (dot-dashed) line shows the total helicity flux (sunspot numbers) after it is smoothed significantly. Adapted from Zhang & Yang, 2013/IOP Publishing.

inferred from the solar surface and the sunspot numbers from 1996 to 2009. The injective rate of the total absolute value of helicity flux is obtained from the absolute values of the positive and negative helicity flux. The mean value of the absolute injective rate of total helicity is $2.40 \times 10^{37} \mathrm{Mx}^2 s^{-1}$ in the calculated solar disk from 1996 to 2009. This provides a basic estimate of the total injective magnetic helicity flux of about $5.0 \times 10^{46} \mathrm{Mx}^2$ in the 23rd solar cycle from both hemispheres; it is of a similar order to that estimated by Berger & Ruzmaikin (2000) and Yang & Zhang (2012). This result may provide a minimum value if the contribution of small-scale helicity flux and the projective effects of the magnetic field on the solar surface are estimated in solar cycle 23.

Also note that the injected magnetic helicity flux tends to be delayed compared to the phase in total sunspot numbers, comparing the smoothed ones in Fig. 4.13. The maximum sunspot numbers occurred in 2001, while the maximum value of current helicity inferred by the vector magnetograms occurred in 2002 after it was smoothed. This is consistent with the maximum butterfly diagram of calculated current helicity of solar sunspot delays compared to sunspot numbers in Fig. 4.7. It also means the sunspot numbers do not entirely reflect the relevant handedness of magnetic fields generated in the solar atmosphere.

4.5. MAGNETIC HELICITY AS AN INDEX IN THE HIGH SOLAR ATMOSPHERE

When we study the upward transport of magnetic helicity from the solar photosphere, the helical characteristics of the coronal magnetic field can give us important information. Under the existing observational conditions, the chirality study of coronal loops is of great significance.

4.5.1. Hemispheric Distribution of Helical Coronal Soft X-Ray Loops

To analyze the magnetic chirality in the solar corona, Fig. 4.14 shows the statistical results of 753 large-scale soft X-ray loops from 1991 to 2001, observed by the Yohkoh satellite. As the unidentified loops are ignored, one can find that 77.3% (81.5%) of the systems in accord with the hemispheric rule are in the northern (southern) hemisphere. This is roughly consistent with the hemispheric sign rule of magnetic helicity in active regions.

Figure 4.14 also depicts the imbalance of chiral soft X-ray loops in both hemispheres. The reverse magnetic helicity occurred with a relatively high tendency in 1991, 1992, and 1995 in the northern hemisphere, while it is insignificant in the southern hemisphere.

Figure 4.15 shows the statistical distribution of helical soft X-ray loops with latitude. The mean latitude of softX-ray loops migrates toward the equator with the solar cycle in the form of a butterfly diagram. It is similar to the distributed form of sunspots. Because there were very few soft X-ray loops in 1991, 1995, and 1996 in our statistics, the deviation from the butterfly diagram in these years can be seen. Most of the large-scale loops follow the helicity hemispheric sign rule, and the statistical distribution of the reverse helical soft X-ray loops relative to the sign rule can also be found in Fig. 4.15.

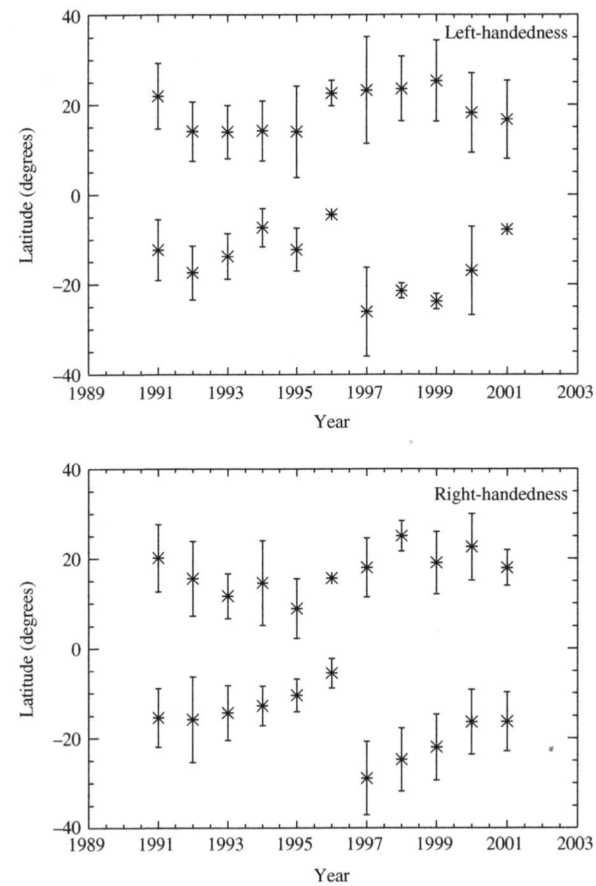

Figure 4.15 The mean latitudinal distribution of soft X-ray loops with left- and right-handedness. σ-error bars are shown with vertical lines. Zhang et al., 2010a/IOP Publishing.

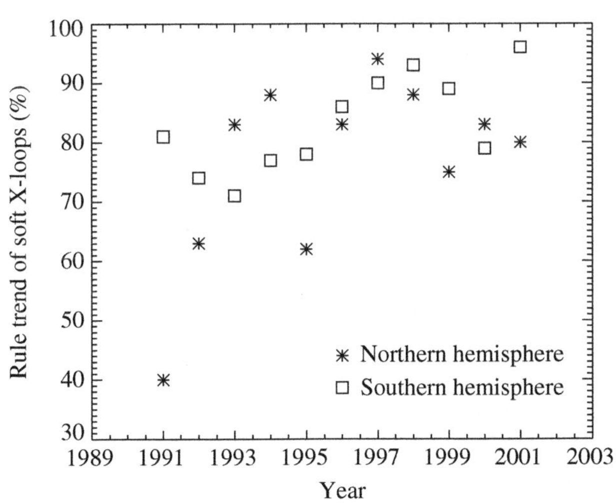

Figure 4.14 The hemispheric handed rule trend of the proportion of soft X-ray loops in the southern and northern hemispheres. Zhang et al., 2010a/IOP Publishing.

Moreover, from the time scale of solar cycles, the transequatorial X-ray loops connect the active solar regions from both hemispheres, and the relevant current helicity of these pair regions has been statistically provided by Chen, Bao & Zhang (2007). Figure 4.16 exhibits the helicity correlation of the 43 pairs of active regions connected by transequatorial loops (Chen, Bao & Zhang, 2007). For the helicity parameter α_{best} ($\sim \overline{(\nabla \times \mathbf{B})_z / B_z}$ in equation (4.7)), 22 pairs (51%) of the active regions show the same helicity patterns, and 21 pairs (49%) of the active regions have the opposite chirality. For the parameter ρ_h ($= \sum h_c(i,j) / \sum |h_c(i,j)|$), 26 pairs (60%) of the active regions have the same sign and 17 pairs (40%)

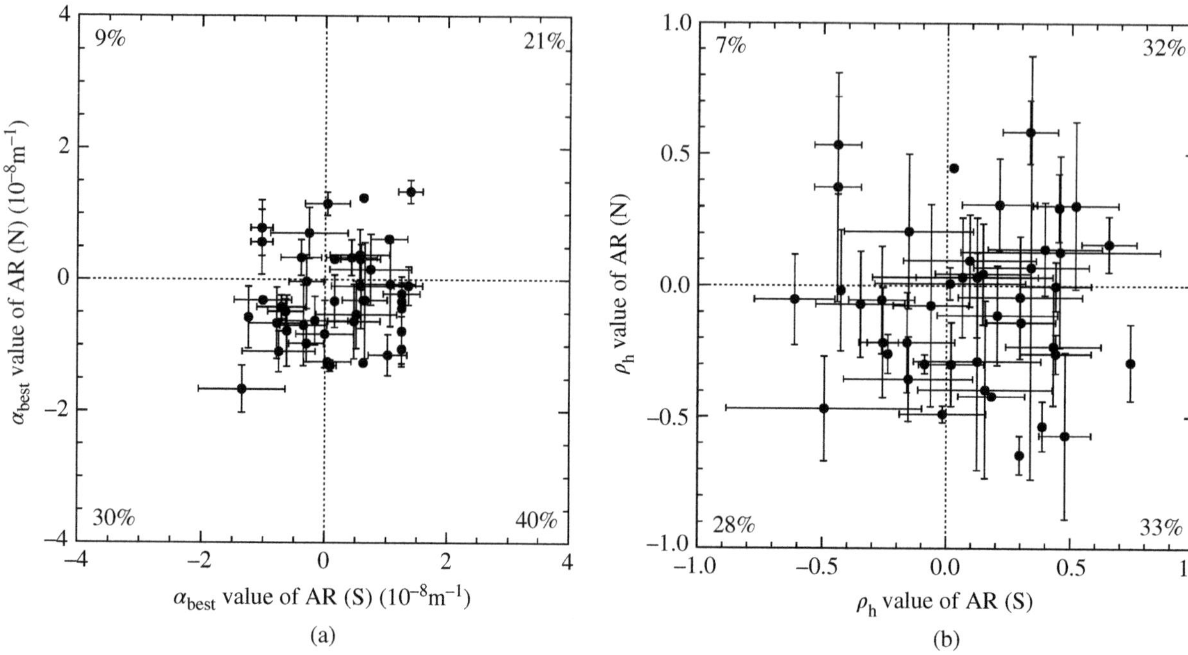

Figure 4.16 (a) Correlation of α_{best} of the active region pairs connected by transequatorial loops. (b) Relationship of ρ_h of the active region pairs connected by transequatorial loops. The horizontal axis presents the helicity values of active regions in the southern hemisphere; the vertical axis shows the values in the northern hemisphere. Error bars (when present) correspond to 1σ of the mean helicity values from multiple magnetograms of the same active region. Points without error bars correspond to active regions re-demonstrated by a single magnetogram. Chen, Bao & Zhang, 2007/Springer Nature.

the opposite sign. The results of both methods indicate that the active region pairs connected by transequatorial loops do not necessarily have the same chirality. This reflects the large-scale poloidal magnetic field in the high solar atmosphere. Whether these loops contain more information about the global transformation of magnetic helicity with solar cycles is worthy of further study.

4.5.2. Helicity with Solar Flare Cycles

The relationship between magnetic helicity and the activity of the flare-coronal mass ejections is a notable topic. It relates to the transformation of magnetic helicity from the solar atmosphere into interplanetary space (cf. Rust, 2002; Yang et al., 2012; Wang & Zhang, 2015; Kim et al., 2017; Park et al., 2021).

For statistical study of the photospheric magnetic nonpotentiality in solar active regions and its relationship with associated flares, Yang et al. (2012) selected 2173 photospheric vector magnetograms from 1106 active regions observed at Huairou from 1988 to 2008, which covers most of the 22nd and 23rd solar cycles. A series of parameters have been calculated, including the mean absolute current helicity density, absolute averaged twist force-free parameter, mean free magnetic energy, and other nonpotential parameters of each vector magnetogram.

Yang et al. (2012) selected as active samples (i.e., flare-productive active regions) those active regions with equivalent flare strength greater than a typical value (such as an M1.0 flare) within the subsequent 24 hours. The rest were quiet samples (i.e., flare-quiet active regions).

In Fig. 4.17, the mean current helicity $\overline{|h_{cz}|}$ of the flare-productive active regions shows statistically higher values than those for flare-quiet active regions in the solar maximum. However, the twist factor $|\alpha_{av}|$ of active regions shows an insignificant difference between quiet and active samples, in comparison with $\overline{|h_{cz}|}$. However, note that the mean helicity density $\overline{|h_{cz}|}$ of the flare-productive active regions shows two delayed peaks relative to the sunspot maximums in 1992 and 2005. A similar case of $|\alpha_{av}|$ can be found in 2005. This is consistent with the results calculated from the magnetic helicity in Figs. 4.9 and 4.13. It most likely indicates that the formation of peaks in flare-productive active regions with strong magnetic helicity tends to be statistically delayed compared to extreme values of sunspot numbers.

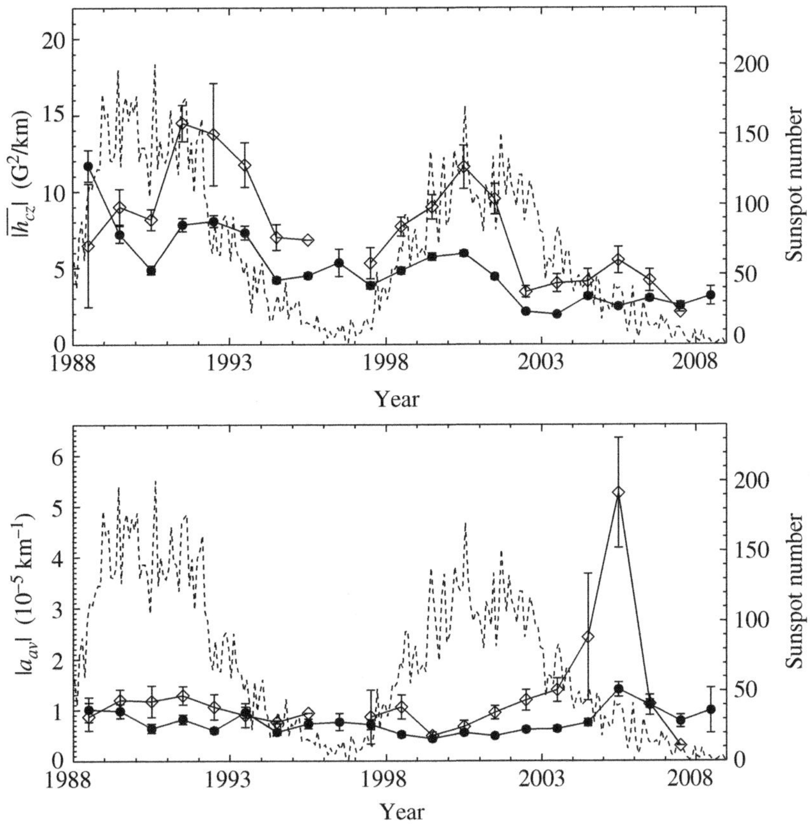

Figure 4.17 Yearly mean values of $\overline{|h_{cz}|}$ and $|\alpha_{av}|$ for samples from active regions from 1988 to 2008. Diamonds represent the yearly mean values of the samples that produced flares with FI $\geq$ 10.0 in the following 24 hours (flare-productive samples). Dots represent the yearly mean values of samples that did not produce flares with FI $\geq$ 10.0 in the following 24 hours (flare-quiet samples). The monthly mean relative sunspot numbers during the same period also overlap (dashed line). Yang et al., 2012/Springer Nature.

4.6. QUESTIONS ABOUT SOLAR DYNAMOS WITH OBSERVATION OF MAGNETIC HELICITY

We note that exploring the variation of magnetic helicity with solar cycles from different perspectives is of great significance for understanding the law of formation of the solar internal magnetic field, which relates to the emergence of twisted magnetic flux in active regions (c.f. Longcope et al., 1998; Fisher et al., 2000; Kleeorin et al., 2020; Kuzanyan et al., 2020).

In comparing observations of magnetic helicity, Choudhuri (2003) argued that an extremely important question is whether, from a solar dynamo perspective, there is a systematic aspect to certain current helicities at certain latitudes and times or whether they are merely statistical fluctuations. This chapter's more extensive data analysis provides a chance to settle this question. On theoretical grounds, they expected that flux tubes might have corresponding current helicity systematically in certain latitudes at certain times. They suggested that flux tubes at the bottom of the convection zone are advected equatorward by the equatorward meridional flow there, whereas the poloidal field at the surface is advected poleward by the poleward meridional flow there. A systematic study of the observed signs of current helicity in different latitudes in different phases of the solar cycle should throw important light on the nature of the solar dynamo. The extended numerical simulation for the solar dynamo with helicity has been demonstrated by Choudhuri et al. (2004).

A similar presentation is that Xu et al. (2007) employed a very simple Parker dynamo model, and direct generalizations of the latitude time distribution of the helicity obey a polarity rule similar to Hale's polarity rule. There is a stable phase shift between the helicity and the toroidal magnetic field in the dynamo model. Their dynamo models are more sophisticated than Parker's simplest model, which could predict phase shifts and butterfly diagrams. It is natural to suppose that a description of the evolution of helicity in terms of a differential equation could considerably modify the phase shift between the helicity and the toroidal magnetic field to compare with observed variations in the magnetic field and helicity with solar cycles (Bao et al., 2000; Zhang et al., 2010b).

We can define the averaged small-scale magnetic helicity as $\overline{\chi} = \overline{\mathbf{a} \cdot \mathbf{b}}$ (**a**, and is the **b** fluctuating part of the magnetic vector potential and magnetic field, respectively. For the isotropic turbulence (Moffatt, 1978), the current helicity is related to magnetic helicity, $h_c = \overline{\mathbf{b} \cdot \nabla \times \mathbf{b}} \sim \overline{\chi}/\ell^2$. The evolution equation for $\overline{\chi}$ can be obtained from the equations that govern the evolution of **a** and **b** (Kleeorin & Rogachevskii, 1999):

$$\frac{\partial \overline{\chi}}{\partial t} = -2\left(\mathcal{E} \cdot \overline{\mathbf{B}}\right) - \frac{\overline{\chi}}{R_m \tau_c} - \nabla \cdot \boldsymbol{F} - \eta \overline{\mathbf{B}} \cdot \overline{\mathbf{J}}, \quad (4.14)$$

where $\mathcal{E} = \alpha_0 \overline{\mathbf{B}} + \left(\overline{\mathbf{V}}^{(p)} \times \overline{\mathbf{B}}\right) - \eta_T \left(\nabla \times \overline{\mathbf{B}}\right)$; $\overline{\mathbf{B}}$, $\overline{\mathbf{J}}$, and $\overline{\mathbf{V}}^{(p)}$ are the mean parts of the magnetic field, current, and velocity fields, respectively; the equipartition value $\overline{B}_{\rm eq} \sim \sqrt{4\pi\overline{\rho}\overline{u^2}}$, R_m is the magnetic Reynolds number; and the helicity flux $\boldsymbol{F} = \overline{\mathbf{a} \times \mathbf{u}} \times \overline{\mathbf{B}} - \overline{\mathbf{a} \times (\mathbf{u} \times \mathbf{b})} - \overline{\mathbf{b}\phi}$, ϕ is an arbitrary scalar function related to a gauge of the vector potential (see Kleeorin et al., 1995, Kleeorin & Rogachevskii, 1999 and references therein). The subsurface kinetic helicity in solar active regions has been inferred from helioseismology (e.g. Gao et al., 2009, 2012; Komm & Gosain, 2019), which tends statistically to have the opposite sign relative to the current helicity (e.g. Gao et al., 2009, 2012) (to the hemispheric current helicity sign rule).

The relatively small time-scale fluctuation or oscillation of mean magnetic helicity in solar cycles has been shown in Figs. 4.7, 4.12, and 4.13. Some of the theoretical explanations for the evolution of magnetic helicity in equation (4.14) in the framework of a mean-field dynamo can be found (Kleeorin et al., 2003; Zhang et al., 2006; Zhang et al., 2012; Yang et al., 2020) in comparing with the observations. These reflect that the generation of the magnetic field is a relatively complex process in the convection zone, and one probably cannot use a simple model to explain the twisting process of the magnetic field inside the Sun. They also provide an important opportunity to study the solar-cycle variation of the magnetic helicity from the perspective of the characteristics of magnetic turbulence based on solar magnetic fields (Hoyng, 1993; Zhang, 2012).

The importance of the magnetic helicity with kinetic helicity and the solar dynamo has been noted (cf. Pouquet et al., 1975; Kleeorin & Ruzmaikin, 1982; Brandenburg & Subramanian, 2005; Rädler & Rheinhardt, 2007). The α effect is produced by the kinetic helicity and current helicity, and it is $\alpha_0 = -\frac{\tau_c}{3}\left(\overline{\mathbf{u} \cdot \nabla \times \mathbf{u}} - \frac{\overline{\mathbf{b} \cdot \nabla \times \mathbf{b}}}{4\pi\overline{\rho}}\right)$. The latter effect is interpreted as resistance of magnetic fields against a twist by helical motions (Vainshtein & Cattaneo, 1992; Rüdiger & Kitchatinov, 1993). This effect has introduced a concept called "catastrophic quenching" of the α effect related to the generated large-scale magnetic field. It was found that

$$\alpha_0\left(\overline{B}\right) = \frac{\alpha_0(0) + \eta R_m \overline{\mathbf{B}} \cdot \overline{\mathbf{J}}}{1 + R_m\left(\overline{B}/\overline{B}_{\rm eq}\right)^2}.$$

In the case of $R_m \gg 1$, the α effect is quickly saturated for a large-scale magnetic field if the strength is much below the equipartition value $\overline{B}_{\rm eq}$ (Ossendrijver et al., 2001).

Furthermore, we have many questions in the solar dynamo mechanism (c.f. Moffatt & Dormy, 2019) from the observational perspective of magnetic helicity, but statistically observed magnetic helicity with the solar cycle provides constraints for this study. A further question is the extent to which we can diagnose the mechanism inside the Sun through the observed transmission and evolution of magnetic helicity in the solar atmosphere.

4.7. DISCUSSION

In this chapter, we have demonstrated some relationships between magnetic helicity and solar cycles. We have introduced magnetic helicity and current helicity and their possible relationship from solar observations. To analyze the relationship between magnetic helicity and solar cycles, we introduced a butterfly diagram of the mean current helicity density inferred from the vector magnetic fields of active regions and the relationship to the sunspot butterfly diagram in Figs. 4.6 and 4.7, and soft X-ray loops in Fig. 4.15. Another notable topic is the relationship between magnetic (current) helicity and solar flare–coronal mass ejections with solar cycles. This was shown in Fig. 4.17. We also demonstrated the injection of magnetic helicity with solar cycles inferred from full-disk magnetograms from space and ground bases, providing an aspect of magnetic helicity that differs from the current helicity calculated by the photospheric vector magnetic field in solar active regions. The message we infer from magnetic helicity with solar cycles is as follows:

1. The twist and helicity butterfly diagrams of active regions are antisymmetric relative to the solar equator. This confirms the hemispheric rule of magnetic helicity of active regions in 11-year cycles (such as Ding et al., 1987; Pevtsov, Canfield, and Metcalf, 1995; Bao & Zhang, 1998; Kleeorin et al., 2003; Zhang & Yang, 2013; Liu et al., 2022). This is revealing regarding the dynamo process of the magnetic field inside the Sun.

2. The study of current helicity on the solar surface in equation (4.6) and injective magnetic helicity through the solar surface in equation (4.11) reflect different aspects of the twisted magnetic field with handedness. Their basic relationship can also be estimated in the simple form of equation (4.13), so one can find a similar tendency in that the maximum magnetic helicity and current helicity are delayed relative to the maximum sunspot number in the solar cycles in Figs. 4.9 and 4.13.

3. Coronal soft X-ray loops reflect the extension of the photospheric magnetic field. It is found that the handedness of most loops tends to follow the hemispheric rule. The interesting question is how the twisted magnetic field extends from the photosphere into the corona within the solar cycles. A similar question relates to the quiescent and active filaments in the solar atmosphere (Martin et al., 1994; Ouyang et al., 2017), how magnetic helicity spreads out from the active region, and whether there are other possible causes.

4. The variation of magnetic helicity with solar-flare activity was reported by Bao et al. (1999). We have demonstrated some results on the statistical correlation between them with solar cycles. However, the complexity between them is not completely clear because more magnetic parameters need to be optimally analyzed for eruptive solar-flare phenomena (Yang et al., 2012). We also note that statistical magnetic (current) helicity related to flares is delayed compared to the sunspot number with solar cycles in Fig. 4.17. Moreover, the relationship between solar flare–coronal mass ejections and magnetic helicity is still a notable question (Georgoulis et al., 2009; Wang & Zhang, 2015; Kim et al., 2017; Park et al., 2021). This involves the basic issues of nonpotential magnetic fields and the relationship with magnetic reconnection, which is beyond the scope of this discussion.

5. Some basic problems related to observing and studying solar magnetic (current) helicity with solar cycles involve measurement and analysis accuracy and completeness of data. It can be found from equation (4.6) that it is impossible to obtain the complete current helicity value in the solar active regions, and it is also difficult to obtain complete data for vector magnetic fields in all the active regions on the solar surface. In calculating magnetic helicity transfer on the solar surface (such as equation (4.11)), there are also problems with the measurement accuracy of the magnetic field and velocity field, and it is difficult to cover the whole surface of the Sun. If we also consider the observation error of the instruments, it is easy to understand that there are many uncertainties in the research and differences in the observations and calculation results of different authors (cf. Ai, 1989; Ronan et al., 1992; Wang et al., 1992; Sakurai et al., 1995; Sakurai, 2001; Zhang et al., 2003; Pevtsov et al., 2005; Xu et al., 2009). Open questions remain.

ACKNOWLEDGMENTS

The authors would like to thank the Huairou Solar Observing Station, the National Astronomical Observatories, the Chinese Academy of Sciences, and the Solar & Heliospheric Observatory (SOHO) for providing observational data. The authors would like especially to thank D. Sokoloff, A. Brandenburg, T. Sakurai, S. Bao, Y. Gao, K. Kuzanyan, and V. Pipin for their kind discussions about the study of magnetic helicity with solar observations. This study is supported by grants from the National Natural Science Foundation (NNSF) of China under project grants 11673033, 11427803, 11427901, 12073040, 12073041, 12003051, and 11973056, and other grants at Huairou Solar Observing Station, National Astronomical Observatories, Chinese Academy of Sciences. S. Y. acknowledges support by grants 11427901, 10921303, 11673033, U1731113, 11611530679, and 11573037 of the National Natural Science Foundation of China and grants XDB09040200, XDA04061002, and XDA15010700 of the Strategic Priority Research Program of the Chinese Academy of Sciences and the Youth Innovation Promotion Association of CAS (2019059).

REFERENCES

Abramenko, V. I. (2005). Relationship between magnetic power spectrum and flare productivity in solar active regions. *Astrophysical Journal*, *629*, 1141–1149.

Abramenko, V. I., Wang, T. J., & Yurchishin, V. B. (1996). Analysis of electric current helicity in active regions on the basis of vector magnetograms. *Solar Physics*, *168*, 75–89.

Ai, G. X. (1989). The solar magnetic telescope at Huairou Solar Observing Station. *Publications of the Yunnan Observatory*, Special Issue, 5.

Arnold, V. I. (1974). The asymptotic Hopf invariant and its applications (in Russian). In *Proceedings of the Summer School in Differential Equations, Erevan, 1974*. Armenian SSR Academy of Sciences.

Antalova, A. (1996). Daily soft X-ray flare index (1969–1972). *Contributions of the Astronomical Observatory Skalnate Pleso*, *26*, 98–120.

Antiochos, S. K. (1987). The topology of force-free magnetic fields and its implications for coronal activity. *Astrophysical Journal*, *312*, 886–894.

Aschwanden, M. J. (2005). *Physics of the solar corona: An introduction with problems and solutions*. Springer.

Bao, S. D., & Zhang, H. Q. (1998). Patterns of current helicity for the twenty-second solar cycle. *Astrophysical Journal*, *496*, L43–L46.

Bao, S. D., Zhang, H. Q., Ai, G. X., & Zhang, M. (1999). A survey of flares and current helicity in active regions. *Astronomy and Astrophysics Supplement Series*, *139*, 311–320.

Bao, S. D., Ai, G. X., & Zhang, H. Q. (2000). The hemispheric sign rule of current helicity during the rising phase of cycle 23. *Journal of Astrophysics and Astronomy*, *21*, 303–306.

Berger, M. A., & Field, G. B. (1984). The topological properties of magnetic helicity. *Journal of Fluid Mechanics*, *147*, 133–148.

Berger, M. A., & Ruzmaikin, A. (2000). Rate of helicity production by solar rotation. *Journal of Geophysical Research*, *105*, 10481–10490.

Bothmer, V., & Schwenn, R. (1998). The structure and origin of magnetic clouds in the solar wind. *Annales Geophysicae*, *16*, 1–24.

Brandenburg, A. (2001). The inverse cascade and nonlinear alpha-effect in simulations of isotropic helical hydromagnetic turbulence. *Astrophysical Journal*, *550*, 824–840.

Brandenburg, A., & Subramanian, K. (2005). Astrophysical magnetic fields and nonlinear dynamo theory. *Solar Physics*, *417*, 1–209.

Chae, J. (2001). Observational determination of the rate of magnetic helicity transport through the solar surface via the horizontal motion of field line footpoints. *Astrophysical Journal*, *560*, L95–L98.

Chen, J., Bao, S., & Zhang, H. (2007). Helicity patterns of the active regions connected by transequatorial loops. *Solar Physics*, *242*, 65–85.

Choudhuri, A. R. (2003). On the connection between mean field dynamo theory and flux tubes. *Solar Physics*, *215*, 31–55.

Choudhuri, A. R., Chatterjee, P., & Nandy, D. (2004). Helicity of solar active regions from a dynamo model. *Astrophysical Journal*, *615*, L57–L60.

Chumak, O. V., Zhang, H. Q., & Guo, J. (2004). Integral properties of the magnetic fields of solar active regions under quiet and flare activity conditions. *Astronomy and Astrophysics Transactions*, *23*, 525–531.

Démoulin, P., & Berger, M. A. (2003). Magnetic energy and helicity fluxes at the photospheric level. *Solar Physics*, *215*, 203–215.

Ding, Y. J., Hong, Q. F., & Wang, H. Z. (1987). A statistical study of the spiral spots on the solar disc. *Solar Physics*, *107*, 221–231.

Fan, Y. (2009). Magnetic fields in the solar convection zone. *Living Reviews in Solar Physics*, *6*, 4.

Feldman, U., Doschek, G. A., & Klimchuk, J. A. (1997). The occurrence rate of soft X-ray flares as a function of solar activity. *Astrophysical Journal*, *474*, 511–517.

Fisher, G. H., Fan, Y., Longcope, D. W., Linton, M. G., & Pevtsov, A. A. (2000). The solar dynamo and emerging flux (invited review). *Solar Physics*, *192*, 119–139.

Gao, Y., Su, J. T., Xu, H., & Zhang, H. (2008). Statistical removal of Faraday rotation in vector magnetograms taken by filter-type magnetographs. *Monthly Notices of the Royal Astronomical Society*, *386*, 1959–1965.

Gao, Y., Zhang, H., & Zhao, J. (2009). Comparison of photospheric current helicity and subsurface kinetic helicity. *Monthly Notices of the Royal Astronomical Society*, *394*, L79–L83.

Gao, Y., Zhao, J., & Zhang, H. (2012). Analysis on correlations between subsurface kinetic helicity and photospheric current helicity in active regions. *Astrophysical Journal*, *761*, L9.

Georgoulis, M. K., Rust, D. M., Pevtsov, A. A., Bernasconi, P. N., & Kuzanyan, K. M. (2009). Solar magnetic helicity injected into the heliosphere: Magnitude, balance, and periodicities over solar cycle 23. *Astrophysical Journal*, *705*, L48–L52.

Gilchrist, S. A., & Wheatland, M. S. (2014). Nonlinear force-free modeling of the corona in spherical coordinates. *Solar Physics*, *289*, 1153–1171.

Gosain, S., & Brandenburg, A. (2019). Spectral magnetic helicity of solar active regions between 2006 and 2017. *Astrophysical Journal*, *882*, 80.

Guo, J., Zhang, H. Q., Chumak, O. V., & Lin, J. B. (2010). The evolution of the quantified magnetic complexity of active regions in the 23rd solar cycle. *Monthly Notices of the Royal Astronomical Society*, *405*, 111–114.

Hagino, M., & Sakurai, T. (2005). Solar-cycle variation of magnetic helicity in active regions. *Publications of the Astronomical Society of Japan*, *57*, 481–485.

Hale, G. E. (1908). On the probable existence of a magnetic field in sun-spots. *Astrophysical Journal*, *28*, 315–343.

Hao, J., & Zhang, M. (2011). Hemispheric helicity trend for solar cycle 24. *Astrophysical Journal*, *733*, L27.

Hawkes, G., & Berger, M. A. (2018). Magnetic helicity as a predictor of the solar cycle. *Solar Physics*, *293*, 109.

Hoyng, P. (1993). Helicity fluctuations in mean field theory: An explanation for the variability of the solar cycle? *Astronomy and Astrophysics*, *272*, 321–339.

Joshi, N. C., Bankoti, N. S., Pande, S., Pande, B., Uddin, W., & Pandey, K. (2010). Statistical analysis of soft X-ray solar flares during solar cycles 21, 22 and 23. *New Astronomy*, *15*, 538–546.

Kim, R. S., Park, S. H., Jang, S., Cho, K. S., & Lee, B. S. (2017). Relation of CME speed and magnetic helicity in CME source regions on the Sun during the early phase of solar cycles 23 and 24. *Solar Physics*, *292*, 66.

Kleeorin, N., & Rädler, K. H. (1980). *Mean-field magnetohydrodynamics and dynamo theory*. Berlin: Akademie-Verlag.

Kleeorin, N. I., & Ruzmaikin, A. A. (1982). Dynamics of the average turbulent helicity in a magnetic field. *Magnetohydrodynamics*, *18*, 116–122.

Kleeorin, N., & Rogachevskii, I. (1999). Magnetic helicity tensor for an anisotropic turbulence. *Physical Review E*, *59*, 6724–6729.

Kleeorin, N., Kuzanyan, K., Moss, D., Rogachevskii, I., Sokoloff, D., & Zhang, H. (2003). Magnetic helicity evolution during the solar activity cycle: Observations and dynamo theory. *Astronomy and Astrophysics*, *409*, 1097–1105.

Kleeorin, N., Safiullin, N., Kuzanyan, K. M., Rogachevskii, I., Tlatov, A., & Porshnev, S. (2020). The mean tilt of sunspot bipolar regions: Theory, simulations and comparison with observations. *Monthly Notices of the Royal Astronomical Society*, *495*, 238–248.

Komm, R., & Gosain, S. (2019). Kinetic helicity and lifetime of activity complexes during solar cycle 24. *Astrophysical Journal*, *887*, 192.

Krause, F., & Rädler, K. H. (1980). *Mean-field magnetohydrodynamics and dynamo theory*. Berlin: Akademie-Verlag.

Kusano, K., Maeshiro, T., Yokoyama, T., & Sakurai, T. (2002). Measurement of magnetic helicity injection and free energy loading into the solar corona. *Astrophysical Journal*, *577*, 501–512.

Kuzanyan, K., Zhang, H., & Bao, S. (2000). Probing signatures of the alpha-effect in the solar convection zone. *Solar Physics*, *191*, 231–246.

Kuzanyan, K. M., Lamburt, V. G., Zhang, H., & Bao, S. (2003). Distribution of helical properties of solar magnetic fields. *Chinese Journal of Astronomy and Astrophysics*, *3*, 257–265.

Kuzanyan, K., Kleeorin, N., Rogachevskii, I., Sokoloff, D., & Zhang, H. (2020). Estimates of current helicity and tilt of solar active regions and Joy's law. *Geomagnetism and Aeronomy*, *60*(8), 1032–1037.

Liu, J. H., & Zhang, H. Q. (2006). The magnetic field, horizontal motion, and helicity in a fast emerging flux region which eventually forms a delta spot. *Solar Physics*, *234*, 21–40.

Liu, J., Liu, Y., Zhang, Y., Huang, J., & Zhang, H. (2022). Helicity observations of active regions during the exchange period of solar cycle 24 and 25. *Monthly Notices of the Royal Astronomical Society*, *509*, 5298–5304.

Longcope, D. W., Fisher, G. H., & Pevtsov, A. A. (1998). Flux-tube twist resulting from helical turbulence: The σ-effect. *Astrophysical Journal*, *507*, 417–432.

Low, B. C. (1996). Solar activity and the corona. *Solar Physics*, *167*, 217–265.

Low, B. C. (2015). Field topologies in ideal and near-ideal magnetohydrodynamics and vortex dynamics. *Science China Physics, Mechanics & Astronomy*, *58*, 5626.

Lynch, B. J., Gruesbeck, J. R., Zurbuchen, T. H., & Antiochos, S. K. (2005). Solar cycle-dependent helicity transport by magnetic clouds. *Journal of Geophysical Research: Space Physics*, *110*, 8107.

Martin, S. F., Bilimoria, R., & Tracadas, P. W. (1994). Magnetic field configurations basic to filament channels and filaments. In R. J Rutten, C. J. Schrijver (Eds), *Solar surface magnetism* (pp. 303).Dordrecht: Springer.

Matthaeus, W. H., Goldstein, M. L., & Smith, C. (1982). Evaluation of magnetic helicity in homogeneous turbulence. *Physical Review Letters*, *48*, 1256.

Moffatt, H. (1969). The degree of knottedness of tangled vortex lines. *Journal of Fluid Mechanics*, *35*, 117–129.

Moffatt, H. K. (1978). *Magnetic field generation in electrically conducting fluids*. Cambridge: Cambridge University Press.

Moffatt, H. K., & Dormy, E. (2019). *Self-exciting fluid dynamos*. Cambridge: Cambridge University Press.

Nindos, A., & Zhang, H. Q. (2002). Photospheric motions and coronal mass ejection productivity. *Astrophysical Journal*, *573*, L133–L136.

Ossendrijver, M., Stix, M., & Brandenburg, A. (2001). Magnetoconvection and dynamo coefficients: dependence of the alpha effect on rotation and magnetic field. *Astronomy and Astrophysics*, *376*, 713–726.

Ouyang, Y., Zhou, Y., Chen, P., & Fang, C. (2017). Chirality and magnetic configurations of solar filaments. *Astrophysical Journal*, *835*, 94.

Park, S. H., Leka, K. D., & Kusano, K. (2021). Magnetic helicity flux across solar active region photospheres. II. Association of hemispheric sign preference with flaring activity during solar cycle 24. *Astrophysical Journal*, *911*, 79.

Parker, E. (1955). Hydromagnetic dynamo models. *Astrophysical Journal*, *122*, 293.

Parker, E. N. (1979). *Cosmical magnetic fields – Their origin and their activity*. Oxford: Oxford University Press.

Pevtsov, A. A., Canfield, R. C., & Metcalf, T. R. (1994). Patterns of helicity in solar active regions. *Astrophysical Journal*, *425*, L117.

Pevtsov, A. A., Canfield, R. C., & Metcalf, T. R. (1995). Latitudinal variation of helicity of photospheric magnetic fields. *Astrophysical Journal*, *440*, L109.

Pevtsov, A. A., & Latushko, S. M. (2000). Current helicity of the large-scale photospheric magnetic field. *Astrophysical Journal*, *528*, 999–1003.

Pevtsov, A. A., Hagyard, M. J., Blehm, Z., Smith, J. E., Canfield, R. C., Sakurai, T., & Hagino, M. (2005). On a cyclic variation of the hemispheric helicity rule. *Highlights of Astronomy*, *13*, 140.

Pevtsov, A. A., & Longcope, D. W. (2007). Helicity as the ultimate test to the surface dynamo problem. In K. Shibata, S. Nagata, & T. Sakurai (Eds.), *New solar physics with Solar-B mission* (pp. 99). San Francisco: Astronomical Society of the Pacific.

Pevtsov, A. A., Canfield, R. C., Sakurai, T., & Hagino, M. (2008). On the solar cycle variation of the hemispheric helicity rule. *Astrophysical Journal*, *667*, 719–722.

Pipin, V. V., & Pevtsov, A. A. (2014). Magnetic helicity of the global field in solar cycles 23 and 24. *Astrophysical Journal*, *789*, 21.

Pipin, V. V., Pevtsov, A. A., Liu, Y., & Kosovichev, A. G. (2019). Evolution of magnetic helicity in solar cycle 24. *Astrophysical Journal*, *877*, L36.

Pouquet, A., Frisch, U., & Léorat, J. (1975). Possibility of an inverse cascade of magnetic helicity in magnetohydrodynamic turbulence. *Journal of Fluid Mechanics*, *68*, 769–778.

Rädler, K. H., & Rheinhardt, M. (2007). Mean-field electrodynamics: Critical analysis of various analytical approaches to the mean electromotive force. *Geophysical & Astrophysical Fluid Dynamics*, *101*, 117–154.

Richardson, R. S. (1941). The nature of solar hydrogen vortices. *Astrophysical Journal*, *93*, 24.

Ronan, R. S., Orrall, F. Q., Mickey, D. L., West, E. A., Hagyard, M. J., & Balasubramaniam, K. S. (1992). A comparison of vector magnetograms from the Marshall Space Flight Center and Mees Solar Observatory. *Solar Physics*, *138*, 49–68.

Rust, D. M. (1994). Spawning and shedding helical magnetic fields in the solar atmosphere. *Geophysical Research Letters*, *21*, 241–244.

Rust, D. M. (2002). Magnetic helicity, coronal mass ejections, and the solar cycle. *ESA Special Publications*, *477*, 39–41).

Rüdiger, G., & Kitchatinov, L. L. A. (1993). Alpha-effect and alpha-quenching. *Astronomy & Astrophysics*, *269*, 581–588.

Sakurai, T., Ichimoto, K., Nishino, Y., Shinoda, K., Noguchi, M., Hiei, E., et al . (1995). Solar flare telescope at Mitaka. *Publications of the Astronomical Society of Japan*, *47*, 81–92.

Sakurai, T. (2001). Stokes profile inversion without resort to specific atmospheric models. In M. Sigwarth (Ed.), *Advanced solar polarimetry – theory, observation, and instrumentation* (pp. 535). San Francisco: Astronomical Society of the Pacific.

Schmieder, B., Archontis, V., & Pariat, E. (2014). Magnetic flux emergence along the solar cycle. *Space Science Reviews*, *186*, 227–250.

Schmieder, B., & Aulanier, G. (2018). Solar active region electric currents before and during eruptive flares. *Geoscience Letters*, *235*, 391–406.

Schwabe, S. H. (1843). Excerpts from solar observations during 1843. *Astronomische Nachrichten*, *20*(495), 234–235.

Seehafer, N. (1990). Electric current helicity in the solar atmosphere. *Solar Physics*, *125*, 219–232.

Seehafer, N. (1994). Alpha effect in the solar atmosphere. *Astronomy & Astrophysics*, *284*, 593.

Sokoloff, D. (2004). The Maunder minimum and the solar dynamo. *Solar Physics*, *224*, 145–152.

Sokoloff, D., Bao, S., Kleeorin, N., Kuzanyan, K., Moss, D., Rogachevskii, I., et al. (2006). The distribution of current helicity at the solar surface at the beginning of the solar cycle. *Astronomische Nachrichten*, *327*(9), 876.

Sokoloff, D., Zhang, H., Kuzanyan, K. M., Obridko, V. N., Tomin, D. N., & Tutubalin, V. N. (2008). Current helicity and twist as two indicators of the mirror asymmetry of solar magnetic fields. *Solar Physics*, *248*, 17–28.

Su, J. T., & Zhang, H. Q. (2004). Diagnosis of Faraday rotation with the video vector magnetograph at Huairou. *Solar Physics*, *222*, 17–33.

Toriumi, S., Takasao, S., Cheung, M. C., Jiang, C., Guo, G., Hayashi, K., & Inoue, S. (2020). Comparative study of data-driven solar coronal field models using a flux emergence simulation as a ground-truth data set. *Astrophysical Journal*, *890*, 103.

Török, T., & Kliem, B. (2005). Confined and ejective eruptions of kink-unstable flux ropes. *Astrophysical Journal*, *630*, L97–L100.

Vainshtein, S., & Cattaneo, F. (1992). Nonlinear restrictions on dynamo action. *Astrophysical Journal*, *393*, 165.

Wang, C., & Zhang, M. (2010). A hemispheric helicity sign rule indicated by large-scale photospheric magnetic fields at three phases of solar cycle 23. *Astrophysical Journal*, *720*, 632–638.

Wang, C., & Zhang, M. (2015). Correlation between CME occurrence rate and current helicity in the global magnetic field of Solar Cycle 23. *Solar Physics*, *290*, 811–818.

Wang, H. M., Varsik, J., Zirin, H., Canfield, R. C., Leka, K. D., & Wang, J. X. (1992). Joint vector magnetograph observations at BBSO, Huairou Station, and Mees Solar Observatory. *Solar Physics*, *142*, 11–20.

Wang, H. N., & Xu, R. L. (Eds.). (2002). *Solar-terrestrial magnetic activity and space environment: proceedings of the COSPAR colloquium on solar-terrestrial magnetic activity and space environment (STMASE), held in the NAOC in Beijing, China, September 10–12, 2001*. Amsterdam: Elsevier Science.

Welsch, B. T., & Longcope, D. W. (2003). Magnetic helicity injection by horizontal flows in the quiet Sun. I. Mutual-helicity flux. *Astrophysical Journal*, *588*, 620.

Woltjer, L. (1958a). A theorem on force-free magnetic fields. *Proceedings of the National Academy of Sciences*, *44*, 489–491.

Woltjer, L. (1958b). On hydromagnetic equilibrium. *Proceedings of the National Academy of Sciences*, *44*, 833–841.

Xu, H. Q., Gao, Y., Zhang, H. Q., Sakurai, T., Pevtsov, A. A., & Sokoloff, D. (2007). Helicity comparison among three magnetographs. *Advances in Space Research*, *39*, 1715–1722.

Xu, H., Gao, Y., Popova, E. P., Nefedov, S. N., Zhang, H., & Sokoloff, D. D. (2009). Magnetic and electric-current helicities in very simple models of the solar dynamo. *Astronomicheskii Zhurnal*, *86*(2), 185–190.

Xu, H., Stepanov, R., Kuzanyan, K., Sokoloff, D., Zhang, H., & Gao, Y. (2015). Current helicity and magnetic field anisotropy in solar active regions. *Monthly Notices of the Royal Astronomical Society*, *454*, 1921–1930.

Yang, S., & Zhang, H. (2012). Large-scale magnetic helicity fluxes estimated from MDI magnetic synoptic charts over the Solar Cycle 23. *Astrophysical Journal*, *758*, 61.

Yang, S., Pipin, V. V., Sokoloff, D. D., Kuzanyan, K. M., & Zhang, H. Q. (2020). The origin and effect of hemispheric helicity imbalance in solar dynamo. *Journal of Plasma Physics*, *86*(3), 775860302.

Yang, X., Zhang, H., Gao, Y., Guo, J., & Lin, G. (2012). A statistical study on photospheric magnetic nonpotentiality of active regions and its relationship with flares during solar cycles 22–23. *Solar Physics*, *280*, 165–181.

Yeates, A. R., Mackay, D. H., & van Ballegooijen, A. A. (2008). Evolution and distribution of current helicity in full-sun simulations. *Astrophysical Journal*, *680*, L165.

Yurchyshyn, V. B., Wang, H., Goode, G., & Deng, Y. (2001). Orientation of the magnetic fields in interplanetary flux ropes and solar filaments. *Astrophysical Journal*, *563*, 381–388.

Zeldovich, Y. B., Ruzmaikin, A. A., & Sokoloff, D. D. (1983). *Magnetic fields in astrophysics*. New York: Gordon and Breach.

Zhang, H. (2012). Reversal magnetic chirality of solar active regions and a possible dynamo model. *Monthly Notices of the Royal Astronomical Society*, *419*, 799–805.

Zhang, H. (2019). From polarimetry to helicity: Studies of solar magnetic fields at the Huairou Solar Observing Station. *Science China-Physics, Mechanics & Astronomy*, *62*(9), 999601.

Zhang, H. (2020). Diagnostic of spectral lines in magnetized solar atmosphere: Formation of the Hβ line in sunspots. *Science China-Physics, Mechanics & Astronomy*, *63*(11), 119611.

Zhang, H., & Bao, S. (1998). Latitudinal distribution of photospheric current helicity and solar activities. *Astrophysical Journal*, *339*, 880–886.

Zhang, H., & Bao, S. (1999). Distribution of photospheric electric current helicity and solar activities. *Astrophysical Journal*, *519*, 876–883.

Zhang, H. Q., Labonte, B., Li, J., & Sakurai, T. (2003). Analysis of vector magnetic fields in solar active regions by Huairou, Mees, and Mitaka vector magnetographs. *Solar Physics*, *213*, 87–102.

Zhang, H., Sokoloff, D., Rogachevskii, I., Moss, D., Lamburt, V., Kuzanyan, K., & Kleeorin, N. (2006). The radial distribution of magnetic helicity in the solar convective zone: Observations and dynamo theory. *Monthly Notices of the Royal Astronomical Society*, *365*, 276–286.

Zhang, H., Yang, S., Gao, Y., Su, J., Sokoloff, D. D., & Kuzanyan, K. (2010a). Large-scale soft X-ray loops and their magnetic chirality in both hemispheres. *Astrophysical Journal*, *719*, 1955–1963.

Zhang, H., Sakurai, T., Pevtsov, A., Gao, Y., Xu, H., Sokoloff, D., & Kuzanyan, K. (2010b). A new dynamo pattern revealed by solar helical magnetic fields. *Monthly Notices of the Royal Astronomical Society*, *402*, L30–L33.

Zhang, H., Moss, D., Kleeorin, N., Kuzanyan, K., Rogachevskii, I., Sokoloff, D., et al. (2012). Current helicity of active regions as a tracer of large-scale solar magnetic helicity. *Astrophysical Journal*, *751*, 47.

Zhang, H., & Yang, S. (2013). Distribution of magnetic helicity flux with solar cycles. *Astrophysical Journal*, *763*, 105.

Zhang, H., Brandenburg, A., & Sokoloff, D. (2014). Magnetic helicity and energy spectra of a solar active region. *Astrophysical Journal*, *784*, L45.

Zhang, H., Brandenburg, A., & Sokoloff, D. (2016). Evolution of magnetic helicity and energy spectra of solar active regions. *Astrophysical Journal*, *819*, 146.

Zhang, M. (2006). Helicity observations of weak and strong fields. *Astrophysical Journal*, *646*, L85–L88.

Zhao, J., Gilchrist, S. A., Aulanier, G., Schmieder, B., Pariat, E., & Li, H. (2016). Hooked flare ribbons and flux-rope-related QSL footprints. *Astrophysical Journal*, *823*, 62.

5

Chirality of Solar Filaments and the Supporting Magnetic Field

Peng-Fei Chen

ABSTRACT

As a perfect tracer of nonpotential solar magnetic field, solar filaments possess both magnetic chirality and structural chirality, which are asymmetric in the two hemispheres of the Sun. It has been believed for decades that the structural chirality of filaments has a one-to-one correspondence to the handedness of the supporting magnetic field; that is, the filaments with left-bearing (right-bearing) barbs have positive (negative) magnetic helicity. Such a well-known notion was recently challenged, and the correspondence is more complicated than previously thought. Instead, we propose that the traditional correspondence is valid only for filaments supported by magnetic flux ropes and is the opposite for those supported by sheared magnetic arcades. In this chapter, we review the progress achieved in the past decade to shed light on the elusive nature of coronal magnetic structures.

5.1. INTRODUCTION

As magnetic fields are generated and amplified in the solar interior by as-yet-unknown dynamo processes, they become unstable and emerge through the solar surface into the solar atmosphere. The effect of the magnetic field in the solar atmosphere is significant: whereas the plasma temperature tends to decrease with height from the solar interior to the photosphere, the plasma temperature increases with height in the chromosphere until the corona. In contrast to the 6,000 K solar surface, the corona reaches 1–2 MK so that the solar corona becomes almost fully ionized. The million-Kelvin corona is maintained by the balance between coronal heating and cooling processes, including radiation, heat conduction, and solar-wind advection. While on large scales, the corona is in dynamic quasi-equilibrium, in some local regions it is not, due to thermal instability (Parker, 1953) or thermal non-equilibrium (Antiochos et al., 2000). This leads to local catastrophic cooling, which results in the formation of condensed cool plasma structures called solar filaments with density ~100 times greater and temperature >100 times lower than the ambient corona. Two factors contribute to the catastrophic cooling: the optically thin radiation of the coronal plasma increases as the temperature decreases; and the optically thin radiation is proportional to the plasma density squared, whereas coronal heating is often proportional to or independent of the plasma density. Hence, a perturbation with a temperature decrease and a density increase would enhance the cooling and condensation.

Such cold, dense plasma absorbs chromospheric lines, so solar filaments appear dark in typical chromospheric filtergrams such as Hα images. When they appear above the solar limb, the ~7,000 K plasma appears bright against the sky even in Hα.

It was revealed that solar filaments are formed above magnetic polarity inversion lines (PILs) between opposite polarities, and the magnetic fields are quasiparallel with the PILs (Martin, 1998; Hanaoka and Sakurai, 2017). Such conditions make solar filaments perfect tracers

Key Lab of Modern Astronomy & Astrophysics, School of Astronomy and Space Science, Nanjing University, Nanjing, China

Helicities in Geophysics, Astrophysics, and Beyond, Geophysical Monograph 283, First Edition.
Edited by Kirill Kuzanyan, Nobumitsu Yokoi, Manolis K. Georgoulis, and Rodion Stepanov.

DOI:10.1002/9781119841715.ch05

of strongly sheared or twisted nonpotential magnetic fields. This is consistent with the facts that more than 70% of coronal mass ejections (CMEs) are associated with filament eruptions, and erupting filaments are the bright core of the resultant CMEs (see Chen, 2011, for a review).

One characteristic of a magnetic field is its chirality: for example, a magnetic field can be left-handed or right-handed. The chirality of the magnetic field can be quantified by either magnetic helicity $H_m = \int \mathbf{A} \cdot \mathbf{B} dV$ or current helicity $H_c = \int \mathbf{B} \cdot \mathbf{J} dV$, where $\mathbf{B}$ is the magnetic field, $\mathbf{A}$ is the magnetic vector potential with $\nabla \times \mathbf{A} = \mathbf{B}$ and $\mathbf{J}$ is the current density (Berger, 1984). For the left-handed magnetic field, the helicity is negative; for the right-handed magnetic field, the helicity is positive. In practice, H_m and H_c have the same sign (van Driel-Gesztelyi et al., 2003). An interesting feature of the solar magnetic field is that the current helicity manifests systematic asymmetry between the northern and southern hemispheres: that is, the helicity is preferentially negative in the northern hemisphere and positive in the southern hemisphere (Seehafer, 1990; Abramenko et al., 1997; Bao and Zhang, 1998; Pevtsov and Balasubramaniam, 2003; Hagino and Sakurai, 2004; Yeates et al., 2007; Lim and Chae, 2009; Wang and Zhang, 2010; Yang and Zhang, 2012). However, because of the 180° ambiguity and uncertainty in measuring the horizontal magnetic components (Xu et al., 2016) and the unavailability of the horizontal component of the current density in the photosphere (except assuming a force-free condition Yang et al., 2014), there is much uncertainty in calculating the helicity. As a result, the strength of such a hemispheric rule is not as great and diverges 60–82% in the existing literature (van Driel-Gesztelyi et al., 2003; Pevtsov et al., 2014; Wang, 2013) (keep in mind that a strength of 50% corresponds to a random distribution without any preference).

The solar atmosphere displays many structural phenomena, such as sunspot whirls, skewed coronal loops or flaring loops, and filaments with barbs (see van Driel-Gesztelyi et al., 2003; Pevtsov et al., 2014, for reviews), which may provide crucial information in deciphering the chirality and magnetic configuration of these structures.

5.2. MAGNETIC CHIRALITY OF SOLAR FILAMENTS

When facing a filament from the positive-polarity side of a filament channel, if the magnetic field is toward the left (right), the magnetic chirality of the filament is called sinistral (dextral), and the magnetic field is right-handed (left-handed), corresponding to positive (negative) helicity, as shown in Table 5.1. Note that sinistral (dextral) is sometimes referred to as a left-handed (right-handed)

Table 5.1 Correspondence between helicity sign and filament chirality.

Filament chirality	Helicity	Handedness
Sinistral	Positive	Right-handed
Dextral	Negative	Left-handed

pattern (Martin et al., 2008); we avoid using these terms because they can cause confusion with left-handed (right-handed) magnetic twist. Rust (1967) used the Hα line to measure 100 quiescent prominences during the declining phase of solar cycle 19 when the north solar pole had negative polarity. It was revealed that the polar crown prominences in the northern hemisphere had their horizontal magnetic field toward the west, implying that the magnetic chirality of these prominences should be dextral (although such a concept did not exist at that time). About one solar cycle later, when the north solar pole had positive polarity, Leroy (1978) measured the horizontal magnetic field of 80 quiescent prominences and found that it was eastward: that is, the horizontal magnetic field was reversed, exactly as anticipated by Rust (1967). Their results indicated that whereas the magnetic polarity orientation reverses in successive solar cycles, the helicity of polar crown prominences does not change sign. To measure the magnetic field of prominences, Rust (1967) used the Zeeman effect, and Leroy (1978) used the Hanle effect. Since such polarimetric measurements were not routinely available, several indirect methods were proposed.

5.2.1. Fibril Orientation

Foukal (1971) found that the Hα fibrils on the opposite sides of a filament channel are antiparallel along the magnetic PIL. Assuming that the Hα fibrils trace the local magnetic field, it was then suggested that we can easily determine the directionality of the axial component of the magnetic field inside the filament channel from the Hα fibril orientation once we know the magnetic polarities of the two sides of the filament channel. Applying this method to a sample of 82 active-region filaments and a sample of 72 filaments across the solar disk, Martin et al. (1994) found that all quiescent filaments in the northern hemisphere are dextral, whereas most of the quiescent filaments in the southern hemisphere are sinistral. However, the active-region filaments do not show noticeable hemispheric preference. It is noted that Foukal's method does not work well for high-altitude filaments with weak magnetic fields or filament channels with intense plages where Hα fibrils are not easy to identify (Martin et al., 1992).

5.2.2. Coronal Cells

Similar to Foukal (1971) using Hα fibrils, Sheeley et al. (2013) proposed that the axial component of the magnetic field of solar filaments can be determined by extreme ultraviolet (EUV) images such as Fe 193 Å or 195 Å. The supergranule-sized coronal cells on the two sides of a filament channel are bent following the local magnetic field, forming antiparallel EUV stalks. Combining the information about magnetic polarities on the two sides of the filament channel, Sheeley et al. (2013) demonstrated that the orientation of the axial magnetic field and hence the chirality of the filament can be determined. Wang (2013) applied this method to the filament channels observed during 1997–2003 and found that over 80% of the filaments obeyed the hemispheric rule: that is, negative/positive helicity in the northern/southern hemisphere.

5.2.3. Filament Drainage Brightening

Once a filament erupts, a pair of conjugate brightenings appears not far from the two endpoints of the filament in the EUV or soft X-ray images. The conjugate brightenings are on the two sides of the filament spine; hence they are either left- or right-skewed with respect to the filament spine. It was proposed that if the conjugate brightenings are left-skewed, the helicity of the filament is negative; if the conjugate brightenings are right-skewed, the helicity of the filament is positive (Wang et al., 2009; Chen et al., 2014).

Regarding the formation of the conjugate brightenings, Wang et al. (2009) proposed that they are due to interchange magnetic reconnection near the foot points of the magnetic field lines supporting the filament. On the other hand, Chen et al. (2014) claimed that the conjugate brightenings are due to filament drainage during the eruption. The two mechanisms may not be exclusive.

Applying the drainage-brightening method to 571 filaments observed by the Atmospheric Imaging Assembly (AIA) telescope onboard the Solar Dynamics Observatory (SDO) satellite from May 2010 to December 2015, Ouyang et al. (2017) found that 91.6% of the filaments followed the hemispheric rule of helicity. In particular, they divided the filaments into quiescent-, intermediate-, and active-region types and found that (1) the strength of the hemispheric preference of the quiescent filaments decreased from ~97% in the ascending phase of solar cycle 24 to ~85% in the declining phase; (2) the strength of the hemispheric preference of the intermediate filaments maintained a high value around $96 \pm 4\%$ during the period; and (3) the strength of the preference of the active-region filaments changed significantly, varying from ~63 to ~95% in the ascending phase of the solar cycle, remaining at $82 \pm 5\%$ during the descending phase. However, during half a year around the solar maximum, the active-region filaments lost hemispheric preference. The helicity signs of the 571 filaments are shown in Fig. 5.1, where the horizontal axis is time and the vertical axis is the latitude of the filaments.

There are other indirect methods to determine the directionality of the axial component of the magnetic field along the filament spine, such as the skew of the coronal arcades or flaring loops (Martin and McAllister, 1996) and the sigmoid shape (Pevtsov and Balasubramaniam, 2003).

5.3. STRUCTURAL CHIRALITY OF SOLAR FILAMENTS

As a three-dimensional structure, a filament displays structural chirality similar to that of typhoons on Earth. Martin et al. (1994) noticed that filaments frequently have barbs, which veer away from the filament spine when viewed from above. The filament barbs are similar to the exits from a highway and hence have two senses of bearing. Highway exits in the UK are left-bearing, and highway exits in China are right-bearing. Martin et al. (1994) found that most filaments in the northern hemisphere have right-bearing barbs, whereas most filaments in the southern hemisphere have left-bearing barbs. More importantly, they claimed that there is a one-to-one correspondence between structural chirality and magnetic chirality: that is, the filaments with right-bearing barbs are dextral, whereas the filaments with left-bearing barbs are sinistral.

Such a relationship made it routinely possible to identify filament helicity by simply checking the Hα images without any extra help from magnetograms. Since it was proposed, many surveys have been taken to investigate the hemispheric preference of the Sun's helicity (Pevtsov and Balasubramaniam, 2003; Yeates et al., 2007; Lim and Chae, 2009), which showed that 76–82% of filaments follow the hemispheric rule. The identification of structural chirality was further realized via an automated detection algorithm by Bernasconi et al. (2005), who found that only 68% of filaments follow the hemispheric rule. According to Hazra et al. (2018), it seems that the current automated detection algorithms are not yet mature. With a larger sample, they found that the visually determined strength of the hemispheric preference is 83%, in contrast to only 58% obtained by the automated detection algorithm.

It should be pointed out that the identification of the structural chirality of solar filaments suffers markedly from projection effects. For example, if a filament is not far from the southeast solar limb, its left-bearing barb on the east side of the filament spine may be identified as a right-bearing barb on the west side of the

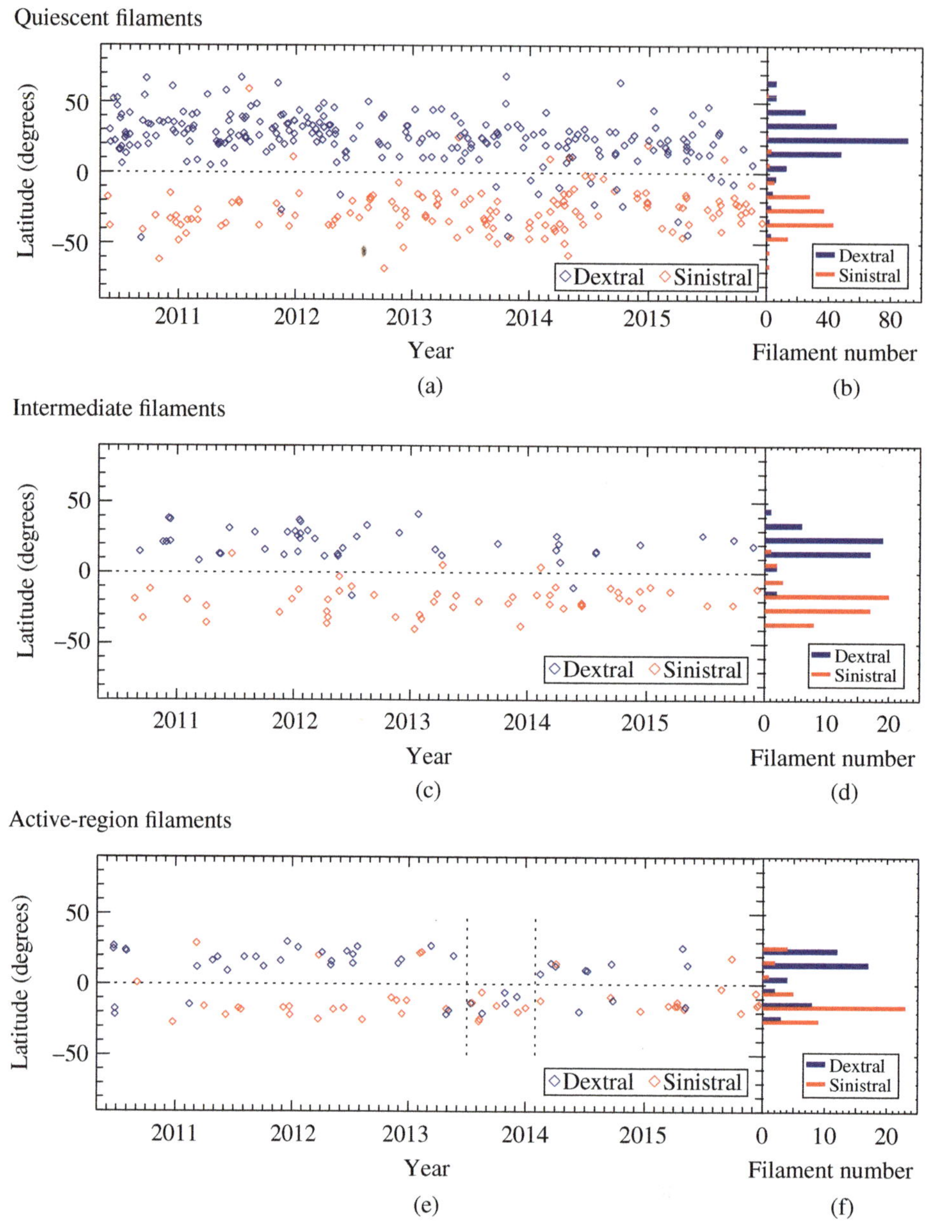

Figure 5.1 Time-longitudinal graphs of the magnetic chirality of three types of solar filaments observed from May 2010 to December 2015. Ouyang et al., 2017 / IOP Publishing.

filament spine. More seriously, Guo et al. (2010) investigated a filament with negative helicity and found that the filament barbs were right-bearing in one segment and left-bearing in the other. With coronal magnetic field extrapolation, they revealed that the segment with right-bearing barbs was supported by a flux rope, whereas the segment with left-bearing barbs was supported by a sheared arcade. Hence, there seems to be no one-to-one correspondence between structural chirality and magnetic chirality as claimed by Martin et al. (1994).

5.4. MAGNETIC CONFIGURATION OF SOLAR FILAMENTS

While magnetic dips are not a necessary condition for filaments (Karpen et al., 2001), most filaments are supported by magnetic dips (Chen et al., 2020). From a theoretical point of view, two types of magnetic configurations have dips: the normal-polarity (Kippenhahn and Schlüter, 1957) and inverse-polarity (Kuperus and Raadu, 1974) configurations, where the poloidal component of the filament's magnetic field is from the positive side to the negative side for the former and from the negative side to the positive side for the latter. Direct measurements of filament magnetic fields confirmed the existence of both types (Leroy et al., 1984; Bommier and Leroy, 1998). Note that because the axial component is dominant in the filament magnetic field, the classical two-dimensional normal-polarity model and inverse-polarity model should be extended to three dimensions, becoming sheared arcades and line-tied flux ropes (van Ballegooijen and Martens, 1989; Aulanier et al., 2006).

Unfortunately, there is a lack of routine observations of the magnetic field for solar filaments and prominences. As a result, the community has been debating for more than two decades whether flux ropes are the progenitors for all CMEs (Cheng et al., 2017; Chen et al., 2020; Patsourakos et al., 2020; Yan et al., 2020; Pan et al., 2021). It would be intriguing if one could determine the magnetic configuration without polarimetric measurements. Chen et al. (2014) pointed out that while the one-to-one correspondence between the structural chirality and magnetic chirality of solar filaments, which we call Martin's rule, is not universal, it can be used to infer the magnetic configuration of a filament without directly measuring the filament's magnetic field. According to Chen et al. (2014), filaments that obey Martin's rule are supported by magnetic flux ropes, and filaments that disobey Martin's rule are supported by sheared arcades. In this updated framework, the relationship between structural and magnetic chirality is not as simple as is sketched in Martin (1998). Instead, Chen et al. (2020) proposed a more complete relationship: as illustrated in Fig. 5.2, if a dextral filament has right-bearing barbs or a sinistral filament has left-bearing barbs, it is supported by a flux rope; if a dextral filament has left-bearing barbs or a sinistral filament has right-bearing barbs, it is supported by a sheared arcade. Notice in Fig. 5.2 that left-skewed coronal loops (or flare loops) and inverse-S-shaped sigmoids (or filament channels) correspond to negative helicity, whereas right-skewed coronal loops (or flare loops) and S-shaped sigmoids (or filament channels) correspond to positive helicity.

As demonstrated in Chen et al. (2014), the procedure to identify the magnetic configuration of a filament consists of two steps:

1. Determine the helicity sign. This can be done by checking the vector magnetograms (Pevtsov et al., 2014),

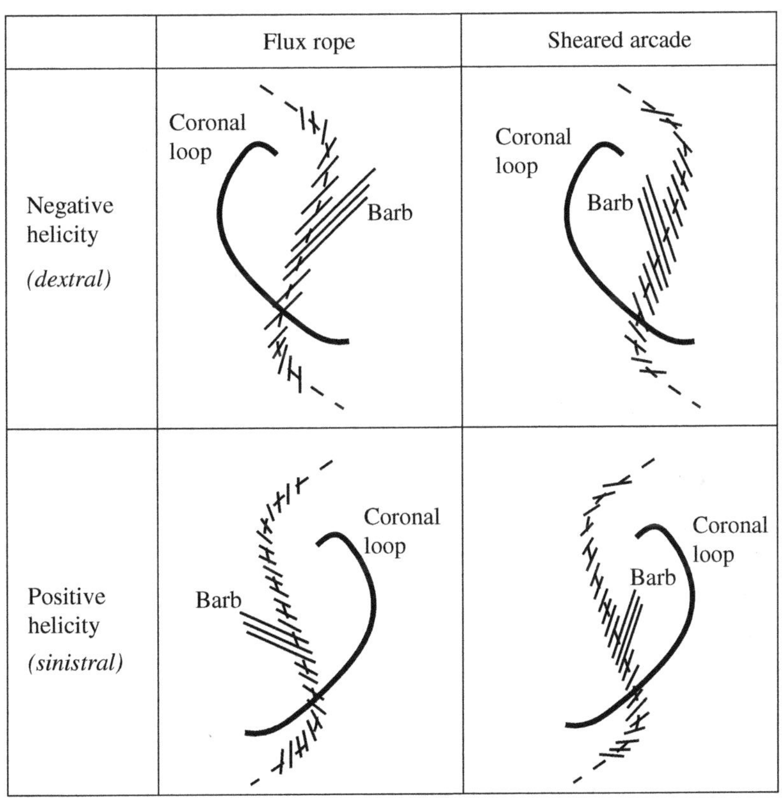

Figure 5.2 Schematic sketch demonstrating the correspondence between the helicity and chirality of filament barbs (longer black lines), threads (shorter black lines) and the overlying coronal loops (thick black lines) in two magnetic configurations: flux ropes and sheared arcades. The dashed lines mark the magnetic neutral lines of filament channels. Chen et al., 2020 / IOP Publishing.

using the orientation of coronal cells plus the longitudinal magnetograms (Sheeley et al., 2013), checking the skew of flaring loops or coronal arcades (Martin and McAllister, 1996), or using the conjugate drainage brightenings (Wang et al., 2009; Chen et al., 2014).

2. Determine the bearing sense of the filament barbs. This can be done by checking the Hα or EUV images. Since barbs may not be easily recognized in some filaments, it was proposed that it is better to check the bearing sense of filament threads (Chen et al., 2014).

Applying this scheme to 564 solar filaments that erupted during 5.5 years from May 2010 to December 2015, we found that ~89% of the filaments were supported by flux ropes and ~11% by sheared arcades (Ouyang et al., 2017). In particular, it was revealed that ~40% of active-region filaments, ~9% of intermediate filaments, and ~4% of quiescent filaments were supported by sheared arcades, which do not obey Martin's rule; hence, using Martin's rule would lead to misidentification of the helicity sign. This may explain the discrepancy between the significant hemispheric preference of active-region magnetograms (Pevtsov et al., 2014) and the absence of hemispheric preference of active-region filaments (Martin et al., 1994). As revealed by Ouyang et al. (2017), with the correct identification of the helicity sign, even active-region filaments have a hemispheric preference strength of 83%, similar to the results obtained via vector magnetograms (Pevtsov et al., 2014). A catalog has been compiled with information about the helicity sign and magnetic configuration for all 564 filaments; it is available at https://github.com/flarechen/Filament-chirality. Only six of the filaments show both left-bearing and right-bearing threads, meaning these filaments are supported partly by sheared arcades and partly by flux ropes.

5.5. FUTURE PROSPECTS

Magnetic fields are the energy source of various solar eruptions, and solar filaments are intimately related to both solar flares and CMEs; hence the magnetic fields of solar filaments are among the most important parameters in solar physics. Sufficient knowledge of the magnetic helicity and configuration before filament eruption may help predict the initiation, propagation characteristics, and geo-effectiveness of the resultant CMEs (Zhang and Low, 2005; Georgoulis et al., 2019; Guo et al., 2019; He et al., 2020; Asvestari et al., 2021).

As seen in this chapter, our understanding of structural and magnetic chirality is constantly updated, and many more questions remain to be solved in the future. These include, for example,

- How are the magnetic and structural chirality of solar filaments formed (Priest et al., 1996; Longcope et al., 1998; DeVore, 2000; Choudhuri, 2003; Mackay, 2015; Knizhnik et al., 2017; Park et al., 2020)?
- How often can a filament have a helicity sign opposite that of the surrounding coronal arcade (Muglach et al., 2009)?
- Ouyang et al. (2017) found that active-region filaments lost their hemispheric preference for half a year during solar maximum in solar cycle 24. Is this a rule or a random phenomenon?
- How might anomalous barbs affect the identification of structural chirality (Filippov, 2017)?
- Is the asymmetry of solar eruptions between the two hemispheres (e.g., Deng et al., 2017) related to the asymmetry of helicity?
- Are erupting filaments with anomalous helicity different in energetics (Zhong et al., 2021)?
- Are the tilt angles significantly different between filaments with normal and anomalous helicity signs (Pevtsov et al., 2014; Hao et al., 2015; Tlatov et al., 2016)?

With the launch of the Chinese Hα Solar Explorer (CHASE, Li et al., 2022), we expect some of these questions to be answered.

ACKNOWLEDGMENTS

The author thanks K. Kuzanyan for the invitation. This research was supported by the National Key R&D Program of China (2020YFC2201200) and the NSFC (12127901, 11961131002, and 11533005).

REFERENCES

Abramenko, V. I., Wang, T., & Yurchishin, V. B. (1997). Electric current helicity in 40 active regions in the maximum of solar cycle 22. *Solar Physics*, *174*, 291–296. https://doi.org/10.1023/A:1004957515498

Antiochos, S. K., MacNeice, P. J., & Spicer, D. S. (2000). The thermal nonequilibrium of prominences. *The Astrophysical Journal*, *536*(1), 494–499. https://doi.org/10.1086/308922

Asvestari, E., Pomoell, J., Kilpua, E., Good, S., Chatzistergos, T., Temmer, M., et al. (2021). Modelling a multi-spacecraft coronal mass ejection encounter with EUHFORIA. *Astronomy and Astrophysics*, *652*, A27. https://doi.org/10.1051/0004-6361/202140315

Aulanier, G., DeVore, C. R., & Antiochos, S. K. (2006). Solar prominence merging. *The Astrophysical Journal*, *646*(2), 1349–1357. https://doi.org/10.1086/505020

Bao, S., & Zhang, H. (1998). Patterns of current helicity for the twenty-second solar cycle. *The Astrophysical Journal Letters*, *496*, L43–L46. https://doi.org/10.1086/311232

Berger, M. A. (1984). Rigorous new limits on magnetic helicity dissipation in the solar corona. *Geophysical and Astrophysical Fluid Dynamics*, *30*(1), 79–104. https://doi.org/10.1080/03091928408210078

Bernasconi, P. N., Rust, D. M., & Hakim, D. (2005). Advanced automated solar filament detection and characterization code: description, performance, and results. *Solar Physics*, *228*, 97–117. https://doi.org/10.1007/s11207-005-2766-y

Bommier, V., & Leroy, J. L. (1998). Global pattern of the magnetic field vectors above neutral lines from 1974 to 1982: pic-du-midi observations of prominences. In D. F. Webb, B. Schmieder, & D. M. Rust (Eds.), *IAU colloq. 167: new perspectives on solar prominences* (p. 434). Astronomical Society of the Pacific.

Chen, P. F. (2011). Coronal mass ejections: models and their observational basis. *Living Reviews in Solar Physics*, *8*. https://doi.org/10.12942/lrsp-2011-1

Chen, P. F., Harra, L. K., & Fang, C. (2014). Imaging and spectroscopic observations of a filament channel and the implications for the nature of counter-streamings. *The Astrophysical Journal*, *784*, 50. https://doi.org/10.1088/0004-637X/784/1/50

Chen, P.-F., Xu, A.-A., & Ding, M.-D. (2020). Some interesting topics provoked by the solar filament research in the past decade. *Research in Astronomy and Astrophysics*, *20*(10), 166. https://doi.org/10.1088/1674-4527/20/10/166

Cheng, X., Guo, Y., & Ding, M. (2017). Origin and structures of solar eruptions I: magnetic flux rope. *Science China Earth Sciences*, *60*, 1383–1407. https://doi.org/10.1007/s11430-017-9074-6

Choudhuri, A. R. (2003). On the connection between mean field dynamo theory and flux tubes. *Solar Physics*, *215*(1), 31–55. https://doi.org/10.1023/A:1024874816178

Deng, L., Zhang, X., An, J., & Cai, Y. (2017). Statistical properties of solar Hα flare activity. *Journal of Space Weather and Space Climate*, *7*, A34. https://doi.org/10.1051/swsc/2017038

DeVore, C. R. (2000). Magnetic helicity generation by solar differential rotation. *The Astrophysical Journal*, *539*(2), 944–953. https://doi.org/10.1086/309274

Filippov, B. P. (2017). Morphological indicators of the chirality of solar filaments. *Astronomy Reports*, *61*(10), 891–900. https://doi.org/10.1134/S1063772917100043

Foukal, P. (1971). Morphological relationships in the chromospheric Hα fine structure. *Solar Physics*, *19*(1), 59–71. https://doi.org/10.1007/BF00148824

Georgoulis, M. K., Nindos, A., & Zhang, H. (2019). The source and engine of coronal mass ejections. *Philosophical Transactions of the Royal Society of London Series A*, *377*(2148), 20180094. https://doi.org/10.1098/rsta.2018.0094

Guo, Y., Schmieder, B., Démoulin, P., Wiegelmann, T., Aulanier, G., Török, T., & Bommier, V. (2010). Coexisting flux rope and dipped arcade sections along one solar filament. *The Astrophysical Journal*, *714*(1), 343–354. https://doi.org/10.1088/0004-637X/714/1/343

Guo, Y., Xia, C., Keppens, R., Ding, M. D., & Chen, P. F. (2019). Solar magnetic flux rope eruption simulated by a data-driven magnetohydrodynamic model. *The Astrophysical Journal Letters*, *870*(2), L21. https://doi.org/10.3847/2041-8213/aafabf

Hagino, M., & Sakurai, T. (2004). Latitude variation of helicity in solar active regions. *Publications of the Astronomical Society of Japan*, *56*, 831–843. https://doi.org/10.1093/pasj/56.5.831

Hanaoka, Y., & Sakurai, T. (2017). Statistical study of the magnetic field orientation in solar filaments. *The Astrophysical Journal*, *851*(2), 130. https://doi.org/10.3847/1538-4357/aa9cf1

Hao, Q., Fang, C., Cao, W., & Chen, P. F. (2015). Statistical analysis of filament features based on the Hα solar images from 1988 to 2013 by computer automated detection method. *The Astrophysical Journal Supplement Series*, *221*(2), 33. https://doi.org/10.1088/0067-0049/221/2/33

Hazra, S., Mahajan, S. S., Douglas, J., Martens, W. K. P. C. H. (2018). Hemispheric preference and cyclic variation of solar filament chirality from 2000 to 2016. *The Astrophysical Journal*, *865*(2), 108. https://doi.org/10.3847/1538-4357/aadb42

He, W., Jiang, C., Zou, P., Duan, A., Feng, X., Zuo, P., & Wang, Y. (2020). Data-driven MHD simulation of the formation and initiation of a large-scale preflare magnetic flux rope in AR 12371. *The Astrophysical Journal*, *892*(1), 9. https://doi.org/10.3847/1538-4357/ab75ab

Karpen, J. T., Antiochos, S. K., Hohensee, M., Klimchuk, J. A., & MacNeice, P. J. (2001). Are magnetic dips necessary for prominence formation? *The Astrophysical Journal Letters*, *553*(1), L85–L88. https://doi.org/10.1086/320497

Kippenhahn, R., & Schlüter, A. (1957). Eine theorie der solaren filamente. Mit 7 textabbildungen. *Zeitschrift für Astrophysik*, *43*, 36.

Knizhnik, K. J., Antiochos, S. K., DeVore, C. R., & Wyper, P. F. (2017). The mechanism for the energy buildup driving solar eruptive events. *The Astrophysical Journal Letters*, *851*(1), L17. https://doi.org/10.3847/2041-8213/aa9e0a

Kuperus, M., & Raadu, M. A. (1974). The support of prominences formed in neutral sheets. *Astronomy and Astrophysics*, *31*, 189.

Leroy, J. L. (1978). On the orientation of magnetic fields in quiescent prominences. *Astronomy and Astrophysics*, *64*(1–2), 247–252.

Leroy, J. L., Bommier, V., & Sahal-Brechot, S. (1984). New data on the magnetic structure of quiescent prominences. *Astronomy and Astrophysics*, *131*, 33–44.

Li, C., Fang, C., & Li, Z., et al. (2022). The Chinese Hα Solar Explorer (CHASE) mission: An overview. *Science China Physics, Mechanics & Astronomy*, *65*(2), 289602. https://doi.org/10.1007/s11433-022-1893-3

Lim, E.-K., & Chae, J. (2009). Chirality of intermediate filaments and magnetic helicity of active regions. *The Astrophysical Journal*, *692*, 104–108. https://doi.org/10.1088/0004-637X/692/1/104

Longcope, D. W., Fisher, G. H., & Pevtsov, A. A. (1998). Flux-tube twist resulting from helical turbulence: The Σ-effect. *The Astrophysical Journal*, *507*(1), 417–432. https://doi.org/10.1086/306312

Mackay, D. H. (2015). Formation and large-scale patterns of filament channels and filaments. In J.-C. Vial & O. Engvold (Eds.), *Solar prominences* (pp. 355). Springer Cham. https://doi.org/10.1007/978-3-319-10416-4_14

Martin, S. F. (1998). Conditions for the formation and maintenance of filaments (invited review). *Solar Physics*, *182*, 107–137. https://doi.org/10.1023/A:1005026814076

Martin, S. F., & McAllister, A. H. (1996). The skew of X-ray coronal loops overlying Hα filaments. In Y. Uchida, T. Kosugi, & H. S. Hudson (Eds.), *Magnetodynamic phenomena in the solar atmosphere – prototypes of stellar magnetic activity* (p. 497). Springer Dordrecht.

Martin, S. F., Marquette, W. H., & Bilimoria, R. (1992). The solar cycle pattern in the direction of the magnetic field along the long axes of polar filaments. In *The solar cycle* (p. 53). Astronomical Society of the Pacific.

Martin, S. F., Bilimoria, R., & Tracadas, P. W. (1994). Magnetic field configurations basic to filament channels and filaments. In R. J. Rutten & C. J. Schrijver (Eds.), *Magnetodynamic phenomena in the solar atmosphere* (p. 303). Springer Dordrecht.

Martin, S. F., Lin, Y., & Engvold, O. (2008). A method of resolving the 180-degree ambiguity by employing the chirality of solar features. *Solar Physics*, *250*, 31–51. https://doi.org/10.1007/s11207-008-9194-8

Muglach, K., Wang, Y. M., & Kliem, B. (2009). Evidence for mixed helicity in erupting filaments. *The Astrophysical Journal*, *703*(1), 976–981. https://doi.org/10.1088/0004-637X/703/1/976

Ouyang, Y., Zhou, Y. H., Chen, P. F., & Fang, C. (2017). Chirality and magnetic configurations of solar filaments. *The Astrophysical Journal*, *835*(1), 94. https://doi.org/10.3847/1538-4357/835/1/94

Pan, H., Liu, R., Gou, T., Kliem, B., Su, Y., Chen, J., & Wang, Y. (2021). Pre-eruption splitting of the double-decker structure in a solar filament. *The Astrophysical Journal*, *909*(1), 32. https://doi.org/10.3847/1538-4357/abda4e

Park, S.-H., Leka, K. D., & Kusano, K. (2020). Magnetic helicity flux across solar active region photospheres. I. Hemispheric sign preference in solar cycle 24. *The Astrophysical Journal*, *904*(1), 6. https://doi.org/10.3847/1538-4357/abbb93

Parker, E. N., (1953). Instability of thermal fields. *The Astrophysical Journal*, *117*, 431, https://doi.org/10.1086/145707

Patsourakos, S., et al. (2020). Decoding the pre-eruptive magnetic field configurations of coronal mass ejections. *Space Science Reviews*, *216*(8), 131. https://doi.org/10.1007/s11214-020-00757-9

Pevtsov, A. A., & Balasubramaniam, K. S. (2003). Helicity patterns on the sun. *Advances in Space Research*, *32*(10), 1867–1874. https://doi.org/10.1016/S0273-1177(03)90620-X

Pevtsov, A. A., Berger, M. A., Nindos, A., Norton, A. A., & van Driel-Gesztelyi, L. (2014). Magnetic helicity, tilt, and twist. *Space Science Reviews*, *186*(1-4), 285–324. https://doi.org/10.1007/s11214-014-0082-2

Priest, E. R., van Ballegooijen, A. A., & Mackay, D. H. (1996). A model for dextral and sinistral prominences. *The Astrophysical Journal*, *460*, 530. https://doi.org/10.1086/176990

Rust, D. M. (1967). Magnetic fields in quiescent solar prominences. I. Observations. *The Astrophysical Journal*, *150*, 313. https://doi.org/10.1086/149333

Seehafer, N. (1990). Electric current helicity in the solar atmosphere. *Solar Physics*, *125*(2), 219–232. https://doi.org/10.1007/BF00158402

Sheeley, N. R. Jr., Martin, S. F., Panasenco, O., & Warren, H. P. (2013). Using coronal cells to infer the magnetic field structure and chirality of filament channels. *The Astrophysical Journal*, *772*, 88. https://doi.org/10.1088/0004-637X/772/2/88

Tlatov, A. G., Kuzanyan, K. M., & Vasil'yeva, V. V. (2016). Tilt angles of solar filaments over the period of 1919 – 2014. *Solar Physics*, *291*(4), 1115–1127. https://doi.org/10.1007/s11207-016-0880-7

van Ballegooijen, A. A., & Martens, P. C. H. (1989). Formation and eruption of solar prominences. *The Astrophysical Journal*, *343*, 971. https://doi.org/10.1086/167766

van Driel-Gesztelyi, L., Démoulin, P., & Mandrini, C. H. (2003). Observations of magnetic helicity. *Advances in Space Research*, *32*(10), 1855–1866. https://doi.org/10.1016/S0273-1177(03)90619-3

Wang, C., & Zhang, M. (2010). A hemispheric helicity sign rule indicated by large-scale photospheric magnetic fields at three phases of solar cycle 23. *The Astrophysical Journal*, *720*(1), 632–638. https://doi.org/10.1088/0004-637X/720/1/632

Wang, Y. M. (2013). On the strength of the hemispheric rule and the origin of active-region helicity. *The Astrophysical Journal Letters*, *775*(2), L46. https://doi.org/10.1088/2041-8205/775/2/L46

Wang, Y. M., Muglach, K., & Kliem, B. (2009). Endpoint brightenings in erupting filaments. *The Astrophysical Journal*, *699*(1), 133–142. https://doi.org/10.1088/0004-637X/699/1/133

Xu, H., Zhang, H., Kuzanyan, K., & Sakurai, T. (2016). On the origin of differences in helicity parameters derived from data of two solar magnetographs. *Solar Physics*, *291*(8), 2253–2267. https://doi.org/10.1007/s11207-016-0975-1

Yan, X., Li, Q., Chen, G., Xue, Z., Feng, L., Wang, J., et al. (2020). Dynamics evolution of a solar active-region filament from a quasi-static state to eruption: Rolling motion, untwisting motion, material transfer, and chirality. *The Astrophysical Journal*, *904*(1), 15. https://doi.org/10.3847/1538-4357/abba81

Yang, S., & Zhang, H. (2012). Large-scale magnetic helicity fluxes estimated from MDI magnetic synoptic charts over the Solar Cycle 23. *The Astrophysical Journal*, *758*(1), 61. https://doi.org/10.1088/0004-637X/758/1/61

Yang, Y.-H., Chen, P. F., Hsieh, M.-S., Wu, S. T., He, H., & Tsai, T.-C. (2014). Characteristics of the photospheric magnetic field associated with solar flare initiation. *The Astrophysical Journal*, *786*(1), 72. https://doi.org/10.1088/0004-637X/786/1/72

Yeates, A. R., Mackay, D. H., & van Ballegooijen, A. A. (2007). Modelling the global solar corona: Filament chirality observations and surface simulations. *Solar Physics*, *245*, 87–107. https://doi.org/10.1007/s11207-007-9013-7

Zhang, M., & Low, B. C. (2005). The hydromagnetic nature of solar coronal mass ejections. *Annual Review of Astronomy and Astrophysics*, *43*(1), 103–137. https://doi.org/10.1146/annurev.astro.43.072103.150602

Zhong, Y., Dai, Y., & Ding, M. D. (2021). Extreme-ultraviolet late phase in homologous solar flares from a complex active region. *The Astrophysical Journal*, *916*(1), 37. https://doi.org/10.3847/1538-4357/ac0430

6

Solar Flares and Magnetic Helicity

Shin Toriumi[1] and Sung-Hong Park[2,3,4]

ABSTRACT

Solar flares and coronal mass ejections are the largest energy-releasing phenomena in the current solar system. They cause drastic enhancements of electromagnetic waves of various wavelengths and sometimes eject coronal material into interplanetary space, disturbing the magnetic surroundings of orbiting planets, including the Earth. It is generally accepted that solar flares are a phenomenon in which magnetic energy stored in the solar atmosphere above an active region is suddenly released through magnetic reconnection. Therefore, to elucidate the nature of solar flares, it is critical to estimate the complexity of the magnetic field and track its evolution. Magnetic helicity, a measure of the twist of coronal magnetic structures, is thus used to quantify and characterize the complexity of flare-productive active regions. This chapter provides an overview of solar flares and discusses how the different concepts of magnetic helicity are used to understand and predict solar flares.

6.1. INTRODUCTION

In astronomy, the term *flare* generally refers to the sudden brightening of electromagnetic waves at various wavelengths. In the Sun, flares (Fig. 6.1) are occasionally observed as brightenings across the spectrum from X-rays to radio waves with time scales of minutes to hours (Fletcher et al., 2011; Benz, 2017). The typical amount of energy released during flares ranges from 10^{28} to 10^{32} erg, making flares the largest energy-releasing phenomena in the current solar system. (The magnitude of solar flares is characterized by the maximum soft X-ray flux of 1-8 Å measured by the Geostationary Operational Environmental Satellite (GOES). The GOES classes are A, B, C, M, and X, corresponding to 10^{-8}, 10^{-7}, 10^{-6}, 10^{-5}, and 10^{-4} W m^{-2} at Earth). Solar flares affect orbiting planets in different ways. For instance, X-ray and ultraviolet emissions enhance the degree of ionization of planetary atmospheres, accelerating the escape of atmospheres from the planets. Coronal mass ejections (CMEs), often accompanying large-scale flares (Chen, 2011), may hit the magnetosphere of the Earth, causing magnetic storms and malfunctions in telecommunication systems. Therefore, in modern civilization with highly advanced information technology, predicting flares and CMEs is an urgent issue (Barnes et al., 2016; Kusano et al., 2021).

Observations have shown that particularly strong flares are typically associated with active regions (ARs), conglomerates of magnetic flux with sunspots, which strongly suggests that flares are a release of the free magnetic energy (ΔE_{mag}) accumulated in the corona during the

[1]*Institute of Space and Astronautical Science, Japan Aerospace Exploration Agency, Sagamihara, Japan*

[2]*Korea Astronomy and Space Science Institute, Daejeon Republic of Korea*

[3]*Institute for Space-Earth Environmental Research, Nagoya University, Nagoya, Japan*

[4]*W. W. Hansen Experimental Physics Laboratory, Stanford University, Stanford, CA, USA*

Helicities in Geophysics, Astrophysics, and Beyond, Geophysical Monograph 283, First Edition.
Edited by Kirill Kuzanyan, Nobumitsu Yokoi, Manolis K. Georgoulis, and Rodion Stepanov.

DOI:10.1002/9781119841715.ch06

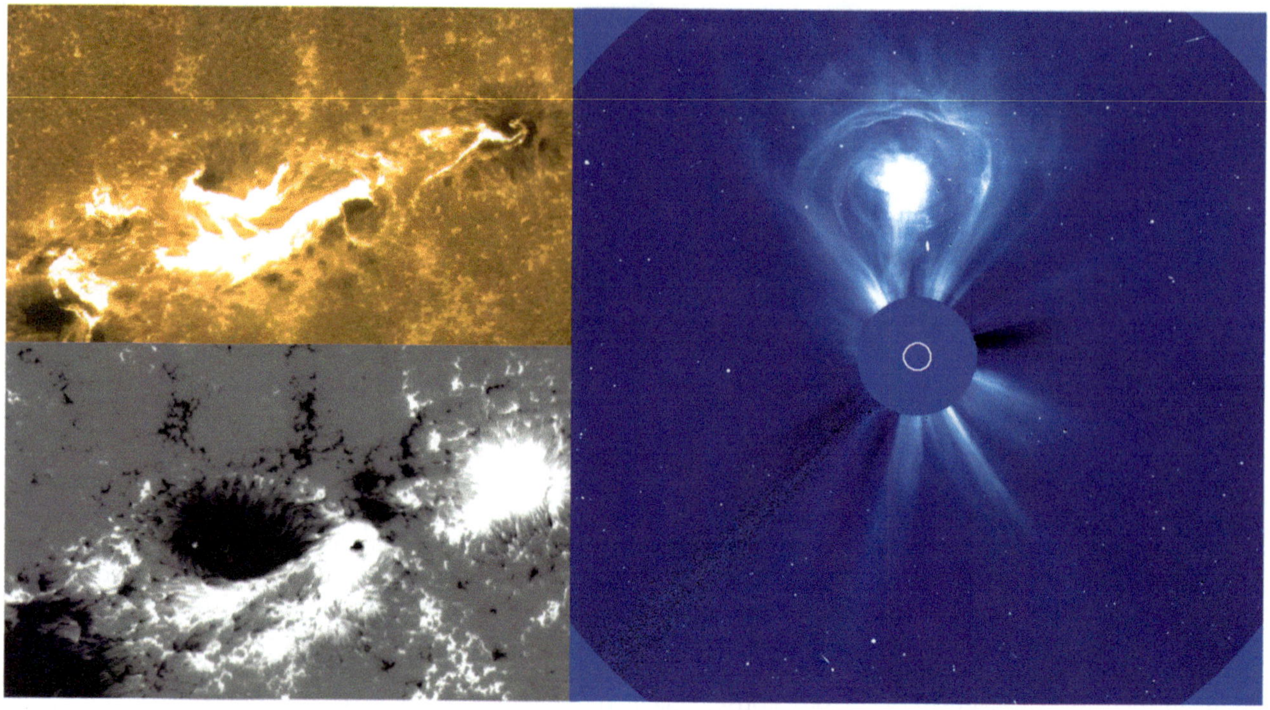

Figure 6.1 Solar flare and CME. (Left) An X2.2-class flare in NOAA AR 11158, as observed by the Hinode satellite. The top panel shows the intensity of the Ca II H line, in which bright flare ribbons expand across the AR, whereas the bottom panel displays the corresponding photospheric magnetogram (white/black corresponds to positive/negative polarity), both detected by Hinode/SOT (Kosugi et al., 2007, Tsuneta et al., 2008). (Right) Lightbulb-shaped CME captured by SOHO (Solar and Heliospheric Observatory)/LASCO (Large Angle and Spectrometric COronagraph experiment) on 27 February 2000. NASA/Public Domain.

evolution of ARs:

$$\Delta E_{\mathrm{mag}} = E_{\mathrm{mag}} - E_{\mathrm{p}} = \int_V \frac{\boldsymbol{B}^2}{8\pi} dV - \int_V \frac{\boldsymbol{B}_{\mathrm{p}}^2}{8\pi} dV, \qquad (6.1)$$

where $\boldsymbol{B}$ is the actual magnetic field and $\boldsymbol{B}_{\mathrm{p}}$ is the potential (current-free) field, $\nabla \times \boldsymbol{B}_{\mathrm{p}} = 0$. Here, the potential field can be proven to be the minimum possible magnetic energy for a given distribution of the normal magnetic field B_{n} on the photospheric surface. (This property is known as Thomson's theorem, which requires the assumption that the magnetic field is solenoidal). In a modern view, flares occur in association with magnetic reconnection, a magnetohydrodynamic (MHD) process that converts (free) magnetic energy into kinetic energy and thermal energy with nonthermal particle acceleration (Priest & Forbes, 2002; Shibata & Magara, 2011).

Statistically, stronger flares occur in sunspot groups not only of larger spot areas or larger amounts of magnetic flux in the photospheric surface (S_{ph}),

$$\Phi = \int_{S_{\mathrm{ph}}} |B_{\mathrm{n}}| \, dS, \qquad (6.2)$$

but also with more complex shapes (Sammis et al., 2000; Toriumi et al., 2017). Most sunspot groups are categorized as β-spots, in which umbrae of the positive and negative polarities have their own penumbrae. However, it is known that the strongest flares tend to occur in sunspot groups called δ-spots, in which the positive and negative polarities are close to each other and surrounded by a single penumbra. In such ARs, magnetic structures often display signs of helical configurations, such as sheared polarity inversion lines (PILs), rotating sunspots, and magnetic flux ropes (Toriumi & Wang, 2019). Therefore, magnetic helicity is widely used as an index to quantitatively measure the degree of twist and complexity of ARs. It is expected that elucidating the relationship between the magnetic helicity supplied to the solar surface and accumulated magnetic energy will lead to an understanding of flares and CMEs (Park et al., 2010b, 2012; Tziotziou et al., 2012; Kim et al., 2017) and give rise to better prediction and forecasting capabilities (Bobra et al., 2014; Bobra & Couvidat, 2015; Bobra & Ilonidis, 2016; Leka et al., 2019).

This chapter discusses the concept of magnetic helicity in the context of AR evolution and its relationship with flare eruptions. For general accounts of flares and CMEs, readers may refer to other reviews by, for example, Priest and Forbes (2002), Fletcher et al. (2011), Shibata and

Magara (2011), and Benz (2017). Regarding the observational characteristics of flare-productive ARs and the attempts to theoretically reproduce them, refer to Toriumi and Wang (2019).

The remainder of this chapter is organized as follows. In section 6.2, we provide an overview of the basic concepts and formulations of magnetic helicity. In section 6.3, we provide methods for measuring the volume helicity and helicity flux. Then, in section 6.4, we review some practical applications of these methods to actual AR data. Section 6.5 describes the numerical reconstruction and modeling of coronal magnetic fields, which are important for accurately measuring the helicity and validating the techniques. Finally, in section 6.6, we summarize this chapter and discuss future perspectives. Although other forms of helicity (e.g., current helicity and kinetic helicity) are also discussed in the context of flare activity in ARs, in this chapter, we keep our primary focus on magnetic helicity.

6.2. BASIC CONCEPTS AND FORMULATIONS

Magnetic helicity is a well-conserved quantity that represents the topology of a magnetic field, such as twists, kinks, and linkages (Elsasser, 1956; Woltjer, 1958; Moffatt, 1969). The helicity H of a magnetic field $\boldsymbol{B}$ contained within a given volume V is defined as

$$H = \int_V \boldsymbol{A} \cdot \boldsymbol{B}\, dV, \tag{6.3}$$

where $\boldsymbol{A}$ denotes the corresponding vector potential of $\boldsymbol{B}$: that is, $\boldsymbol{B} = \nabla \times \boldsymbol{A}$. It is strictly conserved in ideal MHD (Woltjer, 1958). The dissipation is relatively weak even in resistive MHD (Taylor, 1974; Berger & Field, 1984), including typical flares (Berger, 1984), which is one of the reasons magnetic helicity has been used for studying eruptivity.

Equation (6.3) is gauge-invariant only for magnetic fields fully contained in a closed volume. To circumvent this limitation, for open systems such as the solar corona, Berger and Field (1984) and Finn and Antonsen (1985) introduced the *relative helicity*,

$$H_{\mathrm{R}} = \int_V (\boldsymbol{A} + \boldsymbol{A}_0) \cdot (\boldsymbol{B} - \boldsymbol{B}_0)\, dV, \tag{6.4}$$

where $\boldsymbol{B}_0$ is the reference field and $\boldsymbol{A}_0$ is its vector potential, $\boldsymbol{B}_0 = \nabla \times \boldsymbol{A}_0$. For practical purposes, the potential field $\boldsymbol{B}_{\mathrm{p}}$ is often chosen as the reference field,

$$H_{\mathrm{R}} = \int_V (\boldsymbol{A} + \boldsymbol{A}_{\mathrm{p}}) \cdot (\boldsymbol{B} - \boldsymbol{B}_{\mathrm{p}})\, dV. \tag{6.5}$$

The relative helicity (6.5) can be decomposed into two separately gauge-invariant components (Berger, 2003): $H_{\mathrm{R}} = H_{\mathrm{j}} + 2H_{\mathrm{pj}}$ with

$$H_{\mathrm{j}} = \int_V (\boldsymbol{A} - \boldsymbol{A}_{\mathrm{p}}) \cdot (\boldsymbol{B} - \boldsymbol{B}_{\mathrm{p}})\, dV, \tag{6.6}$$

$$H_{\mathrm{pj}} = \int_V \boldsymbol{A}_{\mathrm{p}} \cdot (\boldsymbol{B} - \boldsymbol{B}_{\mathrm{p}})\, dV, \tag{6.7}$$

where H_{j} is the helicity of the current-carrying, or nonpotential, component of the magnetic field, $\boldsymbol{B}_{\mathrm{j}} = \boldsymbol{B} - \boldsymbol{B}_{\mathrm{p}} = \nabla \times (\boldsymbol{A} - \boldsymbol{A}_{\mathrm{p}})$, and H_{pj} is the *mutual helicity* between $\boldsymbol{B}_{\mathrm{j}}$ and $\boldsymbol{B}_{\mathrm{p}}$.

However, care must be taken when considering equation (6.5) because the integrand $(\boldsymbol{A} + \boldsymbol{A}_{\mathrm{p}}) \cdot (\boldsymbol{B} - \boldsymbol{B}_{\mathrm{p}})$ is meaningful only if integrated over the whole domain, and the simple sum of the relative helicity in contiguous subvolumes is not equal to the relative helicity of the entire volume (Berger & Field, 1984; Longcope & Malanushenko, 2008). To overcome this additivity problem, Longcope and Malanushenko (2008) proposed dividing the volume into a set of subvolumes whose boundaries between them are magnetic flux surfaces with no magnetic field penetrating them. The *additive self-helicity* is thus defined as the relative helicity integrated only over the corresponding subvolume with respect to the reference potential field restricted to the subvolume:

$$H_{\mathrm{R}i} = \int_{V_i} (\boldsymbol{A} + \boldsymbol{A}_{\mathrm{p}i}) \cdot (\boldsymbol{B} - \boldsymbol{B}_{\mathrm{p}i})\, dV, \tag{6.8}$$

where V_i is the subvolume bounded by the flux surface, and $\boldsymbol{B}_{\mathrm{p}i}$ and $\boldsymbol{A}_{\mathrm{p}i}$ are its potential field and the vector potential, respectively. The additive self-helicity is integrated to obtain the relative helicity:

$$H_{\mathrm{R}} = \sum_i H_{\mathrm{R}i}, \tag{6.9}$$

where the i indexes denote the subvolumes.

However, in the definition by Longcope and Malanushenko (2008), it is difficult to find a suitable set of subvolumes in practical situations, and the finest possible decomposition may be to take an infinitesimally thin tube around each field line. In this limit, the reference potential field would be the magnetic field itself so that the additive self-helicity would vanish. Yeates and Page (2018) followed a different approach, decomposing the relative helicity into a *relative field-line helicity*, which vanishes along any field line where the original field matches the reference (potential) field. The integral of the relative field-line helicity gives the relative helicity. Although this is also true for the additive self-helicity of equations (6.8) and (6.9), it is only relative to the sum of the local reference fields rather than the global reference field used in the relative helicity.

As described in detail by Pariat et al. (2015), the temporal variation in the relative magnetic helicity H_R in a closed volume V can be expressed as

$$\frac{dH_R}{dt} = 2\int_V \boldsymbol{A}\cdot\frac{\partial \boldsymbol{B}}{\partial t}dV + \int_{\partial V}\left[(\boldsymbol{A}-\boldsymbol{A}_p)\times\frac{\partial(\boldsymbol{A}-\boldsymbol{A}_p)}{\partial t}\right]\cdot d\boldsymbol{S} - 2\int_V \boldsymbol{A}_p\cdot\frac{\partial \boldsymbol{B}_p}{\partial t}dV. \tag{6.10}$$

It should be noted that only the sum of the three integrals on the right-hand side is gauge-invariant, whereas each integral is not. By using Faraday's law and the Gauss divergence theorem and assuming an ideal MHD condition at the boundary of the volume (i.e., $\boldsymbol{E}|_{\partial V} = (-\boldsymbol{v}\times\boldsymbol{B})|_{\partial V}$), dH_R/dt of equation (6.10) can be decomposed into two volume integrals and four flux terms on the surface of V as

$$\frac{dH_R}{dt} = \left.\frac{dH_R}{dt}\right|_{\rm diss} + \left.\frac{dH_R}{dt}\right|_{B_p,\,{\rm var}} + F_{V_n} + F_{B_n} + F_{AA_p} + F_{\Phi}, \tag{6.11}$$

where

$$\left.\frac{dH_R}{dt}\right|_{\rm diss} = -2\int_V \boldsymbol{E}\cdot\boldsymbol{B}dV \tag{6.12}$$

$$\left.\frac{dH_R}{dt}\right|_{B_p,\,{\rm var}} = 2\int_V \frac{\partial\phi}{\partial t}\nabla\cdot\boldsymbol{A}_p dV \tag{6.13}$$

$$F_{V_n} = -2\int_{\partial V}(\boldsymbol{B}\cdot\boldsymbol{A})\boldsymbol{v}\cdot d\boldsymbol{S} \tag{6.14}$$

$$F_{B_n} = 2\int_{\partial V}(\boldsymbol{v}\cdot\boldsymbol{A})\boldsymbol{B}\cdot d\boldsymbol{S} \tag{6.15}$$

$$F_{AA_p} = \int_{\partial V}\left[(\boldsymbol{A}-\boldsymbol{A}_p)\times\frac{\partial(\boldsymbol{A}-\boldsymbol{A}_p)}{\partial t}\right]\cdot d\boldsymbol{S} \tag{6.16}$$

$$F_{\Phi} = -2\int_{\partial V}\frac{\partial\phi}{\partial t}\boldsymbol{A}_p\cdot d\boldsymbol{S} \tag{6.17}$$

and ϕ is the solution of the Laplace equation $\nabla^2\phi = 0$ (i.e., satisfying $\boldsymbol{B}_p = \nabla\phi$). Choosing the Coulomb gauge (CG; i.e., $\nabla\cdot\boldsymbol{A}_p = 0$) for the potential field $\boldsymbol{B}_p$, and having the following two additional constraints for the boundary condition,

$$\boldsymbol{A}|_{\partial V} = \boldsymbol{A}_p|_{\partial V}, \tag{6.18}$$

$$\boldsymbol{A}_p\cdot d\boldsymbol{S}|_{\partial V} = 0, \tag{6.19}$$

we can obtain a simplified expression of equation (6.11):

$$\left.\frac{dH_R}{dt}\right|_{\rm CG} = \left.\frac{dH_R}{dt}\right|_{\rm diss} + F_{V_n} + F_{B_n}. \tag{6.20}$$

In the context of investigating the transport rate of the relative magnetic helicity across photospheric boundary surface of a given AR into the corona, the concept of the *relative magnetic helicity flux* $F_{\rm tot}$ has been used in many previous studies of flaring ARs with photospheric magnetic field observations (e.g., Berger & Field, 1984),

$$F_{\rm tot} = -(F_{V_n} + F_{B_n}) = 2\int_{S_{\rm ph}}[(\boldsymbol{A}_p\cdot\boldsymbol{B}_h)v_n - (\boldsymbol{A}_p\cdot\boldsymbol{v}_h)B_n]\,dS, \tag{6.21}$$

where the subscripts n and h indicate the normal and horizontal components, respectively. In equation (6.21), the F_{V_n} term is associated with the emergence or submergence of a twisted magnetic structure through $S_{\rm ph}$ (often called the *emergence term*). The F_{B_n} term is related to the twisting or untwisting of magnetic field lines by horizontal foot-point motions on $S_{\rm ph}$ (the so-called *shear term*). A negative sign appears in front of the sum of F_{V_n} and F_{B_n} because the surface unit vector $d\boldsymbol{S}$ is chosen outwardly from the volume V in this particular case of helicity injection into the corona. It should be noted, however, that $F_{\rm tot}$ can be attributed to the transport of helicity into either the coronal domain or the subsurface across $S_{\rm ph}$. In Fig. 6.2, the general concept of $F_{\rm tot}$ is

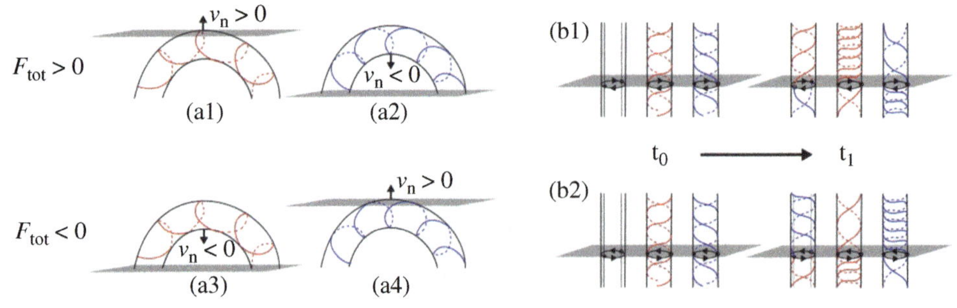

Figure 6.2 Schematic diagram that demonstrates relevant examples of $F_{\rm tot} > 0$ (top) and $F_{\rm tot} < 0$ (bottom) across the solar surface (shown in gray) in the context of the two distinct helicity flux components (a) F_{v_n} and (b) F_{B_n}. The magnetic flux tubes are described by field lines with right-handed (positive; red) and left-handed (negative; blue) helicity. Here, t_0 and t_1 are arbitrarily chosen only to demonstrate the time progress ($t_0 < t_1$).

demonstrated with notable examples of $F_{tot} > 0$ (top) and $F_{tot} < 0$ (bottom) by the two distinct helicity flux terms of F_{V_n} (Fig. 6.2a1–a4) and F_{B_n} (Fig. 6.2b1–b2), respectively.

Once F_{tot} is determined as a function of time, the change in H_R due to helicity transport across S_{ph} over the time interval $\Delta t = t - t_0$ can be obtained:

$$\Delta H_R(t) = \int_{t_0}^{t} F_{tot}\, dt, \tag{6.22}$$

where t_0 is the reference time (often considered as the start time of observations), and t is the time of interest. In the literature, ΔH_R has often been referred to as the *injection* or *accumulation* of helicity into the AR coronal volume through S_{ph}. Nevertheless, strictly speaking, a more appropriate term may be helicity *change* as a result of the helicity flux transport. Again, as described in Fig. 6.2, the transport can be achieved not only by injection into the coronal volume but also by removal from the corona into the subsurface.

6.3. METHODS OF HELICITY ESTIMATION

6.3.1. Volume Helicity

If the coronal magnetic field is obtained as a result of extrapolation based on photospheric magnetic field measurement or numerical simulation (section 6.5), it is possible to directly calculate the helicity in a coronal volume. As mentioned in section 6.2, the relative helicity, $H_R = \int_V (\boldsymbol{A} + \boldsymbol{A}_p) \cdot (\boldsymbol{B} - \boldsymbol{B}_p)\, dV$, has a degree of freedom in the choice of gauge. In many practical applications, the relative helicity is calculated for the magnetic field $\boldsymbol{B} = \nabla \times \boldsymbol{A}$ in a finite volume, often of the 3D computational box with the Cartesian coordinates (x, y, z) above a vector magnetic field on the photospheric surface ($z = 0$). In the past, the CG (i.e., $\nabla \cdot \boldsymbol{A} = 0$) has been used to calculate the volume helicity (e.g., Rudenko and Myshyakov, 2011, Thalmann et al., 2011). However, the CG, when implemented, could be computationally expensive.

To overcome this issue, Valori et al. (2012) adopted a gauge by DeVore (2000), in which one component of the vector potential is 0—for instance, $A_z = 0$—which reduces the calculation cost. For a Cartesian box $V = [x1, x2] \times [y1, y2] \times [z1, z2]$, we are free to select a gauge $\hat{z} \cdot \boldsymbol{A} = 0$ that satisfies

$$\boldsymbol{A} = \boldsymbol{A}_0 - \hat{z} \times \int_{z_1}^{z_2} \boldsymbol{B}(x, y, z')dz', \tag{6.23}$$

where the vector potential $\boldsymbol{A}_0$ is a solution to the z-component of $\boldsymbol{B} = \nabla \times \boldsymbol{A}$ and is described as $\boldsymbol{A}_0 = \boldsymbol{A}(x, y, z_1) = (A_{0x}, A_{0y}, 0)$. We take a simple solution to this:

$$A_{0x} = -\frac{1}{2} \int_{y_1}^{y} B_z(x, y', z = z_1)dy', \tag{6.24}$$

$$A_{0y} = \frac{1}{2} \int_{x_1}^{x} B_z(x', y, z = z_1)dx'. \tag{6.25}$$

Similarly, the vector potential for the reference potential field, $\boldsymbol{A}_p$, is calculated using the same $\boldsymbol{A}_0$ as

$$\boldsymbol{A}_p = \boldsymbol{A}_0 - \hat{z} \times \int_{z_1}^{z_2} \boldsymbol{B}_p(x, y, z')dz'. \tag{6.26}$$

The relative helicity for the finite volume, H_R, can be computed using the set $\boldsymbol{B}$, $\boldsymbol{A}$, $\boldsymbol{B}_p$, and $\boldsymbol{A}_p$.

It should be noted here that the calculation results of the relative helicity may differ depending not only on the choice of gauge (CG or DeVore gauge) but also on the implementation method. Valori et al. (2016) systematically compared the relative helicity with multiple calculation methodologies and investigated the difference in calculation results and the dependence on spatial resolution. As a result, it was found, for instance, that the accuracy in computing H_R is found to vary with different methods, especially the CG methods, and that the spread in H_R estimates between different methods converge in general at higher spatial resolutions.

6.3.2. Magnetic Helicity Flux

To determine the relative magnetic helicity in a coronal volume, it is essential to have a 3D coronal magnetic field, which requires complicated modeling efforts based on either the photospheric magnetic field observation or the polarization of coronal lines. On the other hand, it is straightforward to determine the time series of the magnetic helicity flux F_{tot} and examine the temporal variation and total amount of helicity transported across the AR photosphere.

One of the first practical methods for estimating F_{tot} in ARs was developed by Chae (2001, hereafter F_{tot}^{JC}), using line-of-sight magnetograms obtained by the Michelson Doppler Imager (MDI; Scherrer et al., 1995) onboard the Solar and Heliospheric Observatory (SOHO):

$$F_{tot}^{JC} = -2 \int_{S_{ph}} (\boldsymbol{A}_p \cdot \boldsymbol{v}_{LCT}) B_n dS, \tag{6.27}$$

where $\boldsymbol{v}_{LCT}$ is the horizontal velocity of the *apparent* motions of magnetic elements in the photosphere, which is determined using the technique of local correlation tracking (LCT; November & Simon, 1988). $\boldsymbol{A}_p$ is

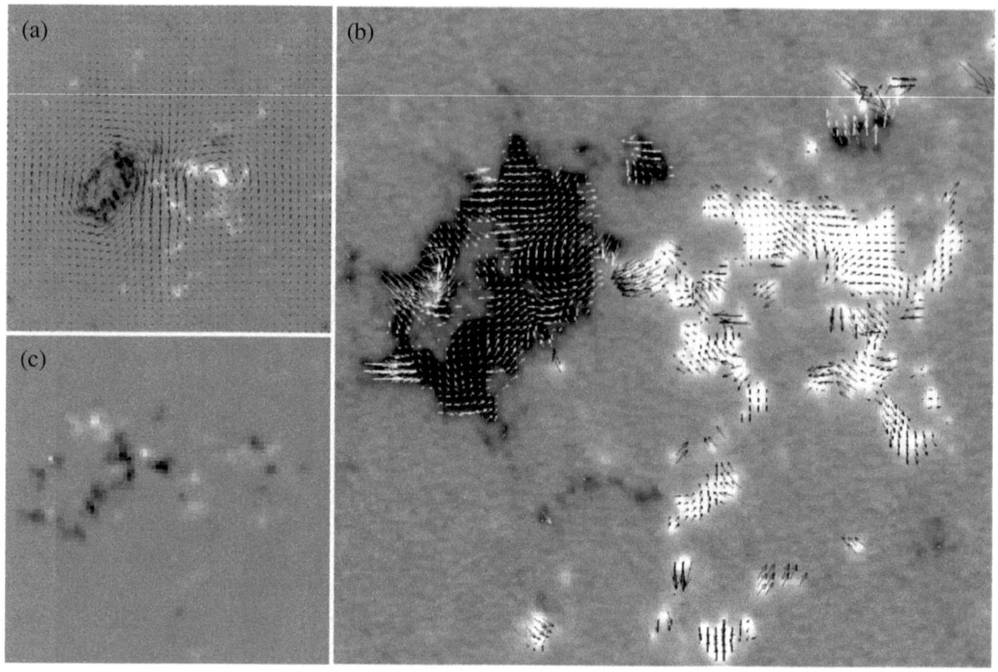

Figure 6.3 An example of estimating F_{tot} for NOAA AR 8011 with the F_{tot}^{JC} method developed by Chae (2001). The line-of-sight magnetic field B_{LOS} was obtained by SOHO/MDI when the AR was located near the center of the solar disk at 17:00 UT on 17 January 1997. Therefore, it can be well approximated to the normal component B_n. The grayscale image of B_{LOS} is overplotted with (a) $\boldsymbol{A}_p$ (arrows) and (b) $\boldsymbol{v}_{LCT}$ (arrows). (c) F_{tot} is estimated by integrating proxy values: that is, $-2(\boldsymbol{A}_p \cdot \boldsymbol{v}_{LCT})B_n$ (grayscale image), for the relative magnetic helicity flux density over the entire AR photospheric surface under consideration. Reproduced with permission from Chae., 2001/IOP Publishing.

calculated from B_n using the fast Fourier transform (FFT) as

$$A_{p,x} = \mathrm{FFT}^{-1}\left(\frac{jk_y}{k_x^2+k_y^2}\mathrm{FFT}(B_n)\right),$$
$$A_{p,y} = \mathrm{FFT}^{-1}\left(-\frac{jk_x}{k_x^2+k_y^2}\mathrm{FFT}(B_n)\right). \quad (6.28)$$

F_{tot}^{JC} is a good proxy of F_{tot} under the condition that $\boldsymbol{v}_{LCT}$ reasonably represents the magnetic field-line foot-point velocity (sometimes called the flux transport velocity) $\boldsymbol{u} \equiv \boldsymbol{v}_t - (v_n/B_n)\boldsymbol{B}_t$. However, we caution here that $\boldsymbol{v}_{LCT}$ is not valid in places where newly emerging or completely submerging flux patches appear. Figure 6.3 shows an example of the implementation of the F_{tot}^{JC} method for NOAA AR 8011 observed by SOHO/MDI on 17 January 1997.

Another method of estimating F_{tot} was developed by Kusano et al. (2002, hereafter F_{tot}^{KK}), applying the LCT technique and an inversion from the normal component of the magnetic induction equation, respectively, to infer $\boldsymbol{v}_h$ and v_n, and using the helicity flux formula of equation (6.21). Several follow-up studies were carried out using the same F_{tot}^{KK} method, in particular, considering the temporal variations of F_{V_n} and F_{B_n} separately (e.g., Kusano et al., 2003, Yokoyama et al., 2003). Since then, F_{tot} has been estimated using several different methods to infer the plasma flow velocity from a sequence of (vector) magnetic field data (e.g., Welsch et al., 2004; Longcope, 2004; Schuck, 2006, 2008; Chae & Sakurai, 2008).

In addition, there was an approach by Pariat et al. (2005, hereafter F_{tot}^{EP}) to determine F_{tot} explicitly using a specific form of $\boldsymbol{A}_p$ that satisfies the CG (i.e., $\nabla \cdot \boldsymbol{A}_p = 0$),

$$\boldsymbol{A}_p = \frac{1}{2\pi}\hat{\boldsymbol{n}} \times \int_{S'_{ph}} B_n(x')\frac{\boldsymbol{r}}{r^2}dS'. \quad (6.29)$$

By inserting $\boldsymbol{A}_p$ of equation (6.29) into equation (6.21), F_{tot}^{EP} can be expressed as

$$F_{tot}^{EP} = -\frac{1}{2\pi}\int_{S_{ph}}\int_{S'_{ph}} \frac{\boldsymbol{r} \times [\boldsymbol{u} - \boldsymbol{u}']}{r^2}\bigg|_n B_n B'_n \, dS \, dS'. \quad (6.30)$$

As discussed in Pariat et al. (2005), the F_{tot}^{EP} method was shown to be more efficient in reducing spurious signals in the map of the relative magnetic helicity flux density (i.e., the integrand in equation (6.30)) compared to that from the F_{tot}^{JC} method (i.e., the integrand in equation (6.27)).

6.4. PRACTICAL APPLICATIONS

6.4.1. Temporal Variation

Since several practical methods were developed in the 2000s to estimate F_{tot} across the AR photosphere from solar magnetic field observations (refer to section 6.3.2), there have been many studies to find any temporal variations in the magnetic helicity over various timescales in relation to flaring activity of the source ARs, particularly for comparing differences of flaring versus nonflaring ARs. In the context of the timescales under consideration, time series analyses of F_{tot} can be generally classified into two categories: (1) short-term (tens of minutes to a few hours) and (2) long-term (a few hours to several days). In the case of short-term variation, for example, there were reports of a large, impulsive variation in the time series of F_{tot} over the course of the flare (e.g., Moon et al., 2002a,b, Hartkorn and Wang, 2004, Romano et al., 2005, Bi et al., 2018). The impulsive variation in F_{tot} typically occurs during the rise phase of the flare X-ray flux, and the magnitude of the impulsive helicity variation (i.e., $\Delta H_{\mathrm{R}} = \int_{t_0}^{t_1} F_{\mathrm{tot}}\, dt$, with t_0 and t_1 as the flare start and end times) correlates with the flare X-ray peak flux. Although this impulsive variation in F_{tot} was found for multiple flare events, we should be cautious because photospheric magnetograms derived from spectropolarimetric observations (in particular, pixels where white-light flare kernels are located) can be significantly affected by flare emission, as mentioned by Hartkorn and Wang (2004). (An example of such disturbed polarimetric signals can be found in the magnetogram in Fig. 6.1, in which the core of the positive (negative) polarity sunspot shows a negative (positive) signal). We note that a detailed investigation of short-term variations in F_{tot} may help us understand the flare-trigger process (e.g., tether-cutting magnetic reconnection, rapidly growing MHD instabilities) and the relevant flare-associated evolution of the photospheric magnetic field.

With respect to flare-associated variations in F_{tot} on a long-term scale, one of the most remarkable observational findings is that flaring ARs show a large, continuous increase in ΔH_{R} (i.e., the time integral of F_{tot}) over an interval of several hours to a few days before the flare onset. Figure 6.4 presents an example of this long-term increasing trend of ΔH_{R} (cross-solid line in Fig. 6.4c) in NOAA AR 9672, which produced an X1.3 flare (dotted line in Fig. 6.4c) on 25 October 2001. Interestingly, ΔH_{R} increased monotonically during a preflare phase of ~24 hours (marked by the dark gray bar and labeled I) and then remained almost constant for the post-flare phase (labeled II). On the other hand, the AR's total unsigned magnetic flux (diamond-solid line in Fig. 6.4c) changed little over the entire investigation period, making it difficult to find any flare-associated temporal variations. This trend of increase in ΔH_{R} was found in many other large flaring ARs (e.g., Park et al., 2008, 2010a,b, 2012, 2013; Kusano et al., 2002; Magara & Tsuneta, 2008; Vemareddy et al., 2012; Romano et al., 2014; Vemareddy,

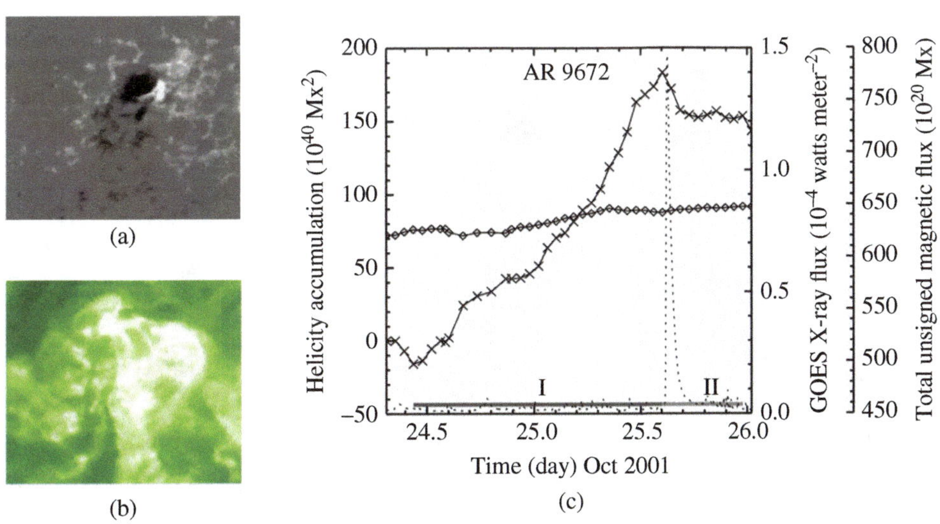

Figure 6.4 Long-term trend of increase in ΔH_{R} across the photospheric surface of NOAA AR 9672 before the onset of an X1.3 flare on 25 October 2001. (a, b) MDI line-of-sight magnetic field and extreme ultraviolet imaging telescope (EIT) 195 Å intensity images, respectively of the AR taken at ~16:00 UT. (c) Temporal variations in ΔH_{R} (cross-solid line), total unsigned flux (diamond-solid line), and GOES soft X-ray flux (dotted line), together with the two distinct (i.e., increase and steady) phases in the observed profile of ΔH_{R}. Park et al. 2008/IOP Publishing.

2015; Romano et al., 2015), whereas no characteristic evolution of ΔH_R was observed in flare-quiet ARs. Some statistical studies of this increase in helicity in relation to the flaring activity are discussed in section 6.4.2.

In the same context, as shown in Fig. 6.5, a long-term increase in helicity was also found in the time profile of the relative magnetic helicity H_R in the coronal volume of ARs that produced large flares (e.g., Park et al., 2010a; Jing et al., 2012, 2015; Thalmann et al., 2019). The observed increasing patterns of both ΔH_R and H_R prior to flares support the idea that H_R in the AR corona is mainly supplied by F_{tot} through the bottom surface (i.e., the AR photosphere). It is also thought that H_R is well preserved over timescales of a few hours to several days: for example, as shown by Park et al. (2010a) based on the comparison of approximately five-day profiles between H_R and ΔH_R for NOAA AR 10930. Moreover, over the course of AR evolution, the preflare increasing trend of the helicity may reflect a complicated process of building up the free magnetic energy in the corona by the emergence of twisted magnetic flux tubes and/or foot-point shear motions of the magnetic field line.

One of the most important aspects of studying the magnetic helicity is that it measures the handedness of twisted magnetic field lines: that is, *positive* helicity in the case of right-handed twist and *negative* helicity for the left-handed case. As shown in Fig. 6.6, Kusano et al. (2002) examined the spatial distribution and evolution of the relative magnetic helicity flux density across the entire photospheric surface of NOAA AR 8100 during a five-day interval from 1 to 5 November 1997. They found that negative helicity transport was predominantly caused by F_{V_n} over the first 2.5 days of the investigated interval, while helicities with positive and negative signs were transported by F_{V_n} and F_{B_n}, respectively, for the rest of the interval. In the latter phase, multiple flares, consisting of three M-class flares and one X2.1 flare, occurred in the region (labeled G in Fig. 6.6b), where negative helicity was consistently transported primarily owing to the shear motion along the main PIL. Such a characteristic follow-up transport of helicity via a localized area (typically around PILs) with a sign opposite that of the preloaded helicity in the AR corona has been reported in many other observations of flaring ARs (e.g., Yokoyama et al., 2003; Wang et al., 2004; Park et al., 2010a; Vemareddy et al., 2012; Jing et al., 2015). When the subsequent opposite-sign helicity flux in the localized areas of the AR becomes larger than the helicity flux in the rest of the areas, the sign of F_{tot} is reversed over the course of its measurement, which is called *helicity sign reversal*.

Park et al. (2012) performed time series analysis of F_{tot} for a set of different AR samples, each of which produced eruptive flares and accompanying CMEs. They found that all the investigated eruptive flares occurred following a remarkable increase in ΔH_R first and the subsequent reversal of the helicity sign (see Fig. 6.7). Here the helicity sign reversal is shown as the slope of the time profile of ΔH_R changes its sign. In contrast, examining temporal variations of F_{tot} for NOAA AR 12257, Vemareddy (2021) reported that there were only small C-class flares without CMEs even when the helicity sign reversal proceeded. However, it should be noted that in the case of NOAA AR 12257, there was no significant transport

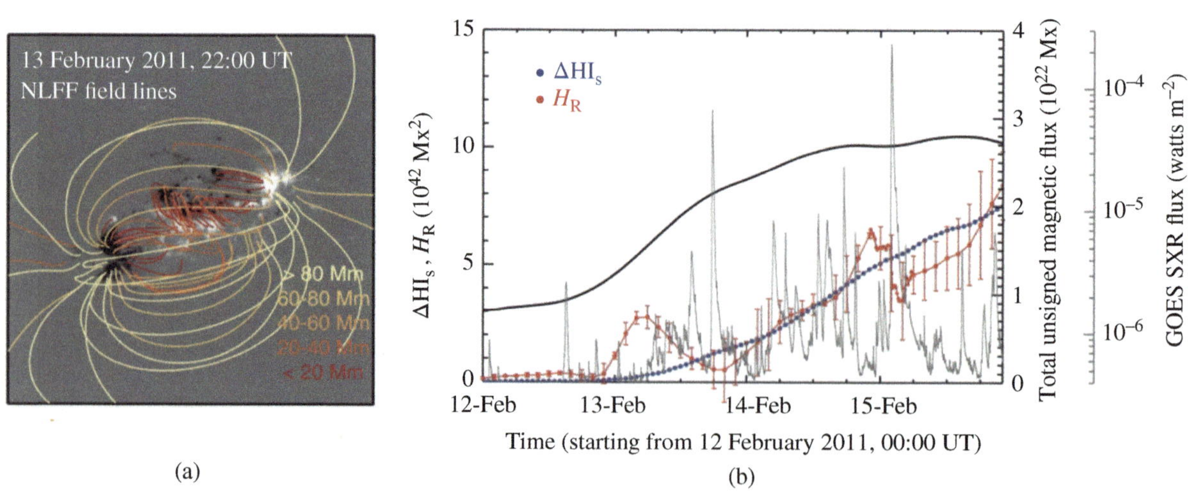

Figure 6.5 Long-term increasing trend of H_R in the coronal volume of NOAA AR 11158 before an X2.2 flare on 15 February 2011. (a) Selected set of coronal magnetic field lines extrapolated from the photospheric vertical magnetic field (grayscale image) with different colors according to their maximum altitude. (b) Temporal variations of H_R (red line), ΔH_R (blue line), the total unsigned magnetic flux (black line), and the GOES soft X-ray flux (gray line). Jing et al., 2012/IOP Publishing.

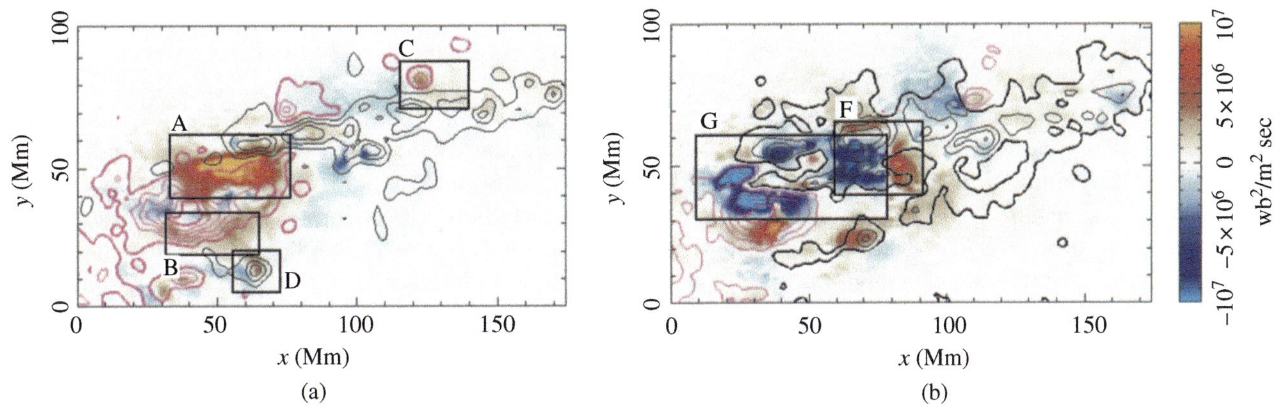

Figure 6.6 Maps of the relative magnetic helicity flux density for NOAA AR 8100, observed at (a) 00:49 UT and (b) 23:11 UT on 3 November 1997 in the blue and red color scale. The AR produced an X2.1 flare at 05:52 UT on 4 November. The positive and negative vertical magnetic fields are indicated by pink and black contours, respectively. Some regions of interest are marked by rectangular boxes with the following labels. A: the main PIL; B: a southern part of the main positive polarity sunspot; C and D: satellite patches in which the magnetic polarity is opposite to that of the nearby main spot; F: a region with strong upflows on the photospheric surface; G: a region in which the helicity flux density is predominantly negative in sign. Kusano et al. 2002/IOP Publishing.

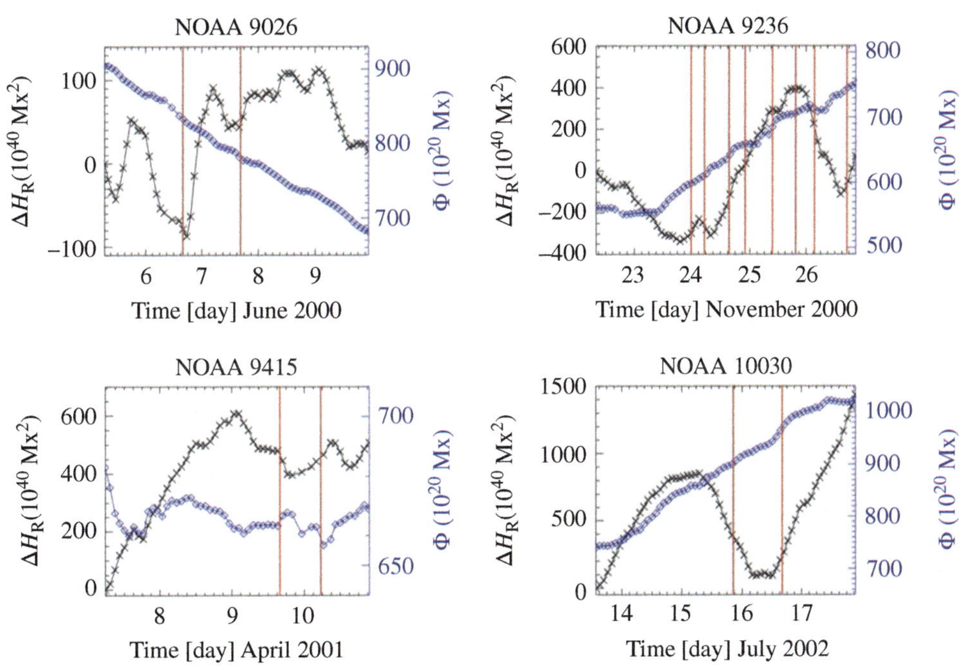

Figure 6.7 Time profiles of ΔH_R (black line) and Φ (blue line) for four ARs that produced multiple eruptive flares with CMEs. The red vertical lines indicate when the CMEs first appeared in the LASCO/C2 field of view. Park et al. 2012/IOP Publishing.

of helicity before the sign reversal, while a large amount of helicity transport was made for the ARs under study in Park et al. (2012). A careful investigation is needed when interpreting the helicity flux density map; for example, artificial signals may be generated due to various limitations in observational data and practical methods, as indicated by Pariat et al. (2005, 2006). Meanwhile, Jing et al. (2015) conducted an interesting comparison study of eruptive versus confined flares by analyzing the evolution of H_R in the AR corona. They found that H_R decreased remarkably ~4 hours prior to the eruptive X2.2 flare on 15 February 2011, whereas no such decrease in

H_R was found in the case of the confined X3.1 flare on 24 October 2014.

It is noteworthy that the observed signatures of (1) the helicity sign reversal and (2) the sudden decrease in H_R prior to eruptive flares can be explained by the flare-trigger MHD simulation model of Kusano et al. (2012), in which, as a flare trigger, a small-scale bipolar magnetic structure supplies helicity with the opposite sign into the preexisting sheared arcade system.

There have been longstanding discussions about the role of magnetic helicity as a sufficient indicator of solar eruptions. Based on practical methods of estimating the magnetic helicity (refer to section 6.3), many trials have been conducted to determine the *magic* parameter that can provide a threshold above which flares and CMEs occur. In this respect, Pariat et al. (2017) proposed a new parameter for the solar eruptivity criterion: the ratio of the magnetic helicity of the current-carrying magnetic field to the total relative helicity, $|H_j|/|H_R|$. For NOAA AR 12673, which produced the largest X-class flare of solar cycle 24, Moraitis et al. (2019) found that the helicity ratio increased and then relaxed to lower values before and after, respectively, each of the two major X2.2 and X9.3 flares on 6 September 2017. Through analysis of the observations and coronal field extrapolations of two flaring ARs with eruptive versus confined flares, as shown in Fig. 6.8, Thalmann et al. (2019) found a similar increasing/decreasing trend of $|H_j|/|H_R|$ before and after the major eruptive flares in NOAA AR 11158. However, this was not observed in NOAA AR 12192, which produced only confined flares without CMEs. Another follow-up study by Gupta et al. (2021), with a set of 10

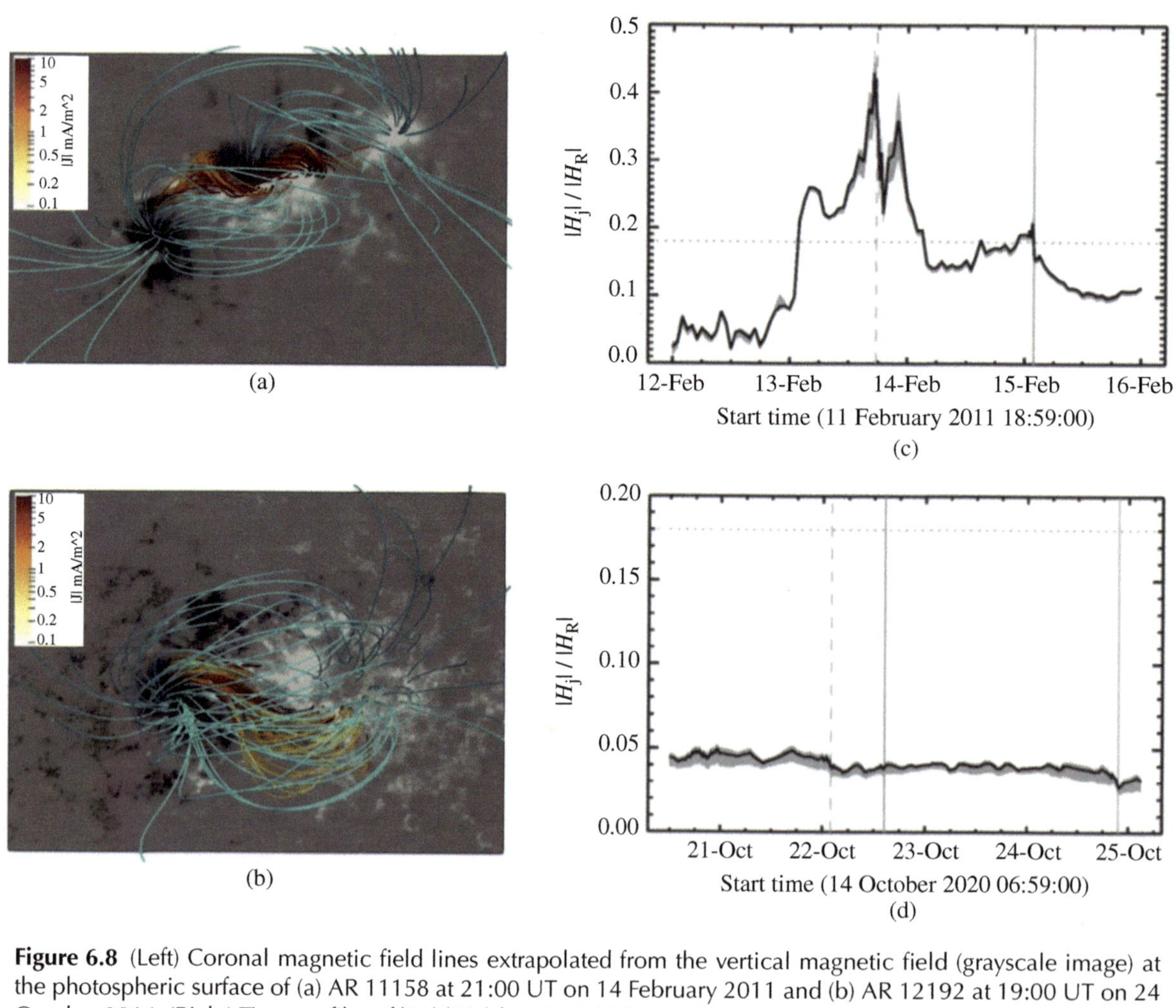

Figure 6.8 (Left) Coronal magnetic field lines extrapolated from the vertical magnetic field (grayscale image) at the photospheric surface of (a) AR 11158 at 21:00 UT on 14 February 2011 and (b) AR 12192 at 19:00 UT on 24 October 2014. (Right) Time profiles of $|H_j|/|H_R|$ for (c) AR 11158 and (d) AR 12192. In (c) and (d), the shaded areas represent the spread of the values of $|H_j|/|H_R|$ from the three different methods used to estimate the volume helicity. The vertical dashed and solid lines indicate the GOES peak times of the M- and X-class flares, respectively. The horizontal dashed line marks the value of $|H_j|/|H_R| = 0.17$ as a reference. Thalmann et al., 2019/IOP Publishing.

different NOAA-numbered AR samples, shows results supporting that the helicity ratio has a strong ability to indicate the eruptive potential of an AR, albeit with a few exceptions. A further statistical study with more samples is needed.

Then why does the helicity ratio, $|H_j|/|H_R|$, seem to predict flare eruptions well? There have been intensive studies on critical conditions for a current-carrying structure (e.g., twisted flux ropes) to successfully erupt against its surrounding confinement field, as represented by the torus instability (Kliem & Török, 2006; Démoulin & Aulanier, 2010) and the double-arc instability (Ishiguro & Kusano, 2017). In fact, various "relative" quantities have been proposed in the framework of instabilities for a current system in the corona and tested with observations of the eruptive and confined flares (Sun et al., 2015, Toriumi et al., 2017, Lin et al., 2020, Li et al., 2020, Kusano et al., 2020, Kazachenko et al., 2022). In this respect, $|H_j|/|H_R|$ may reflect the magnetic relationship between the current-carrying structure (the flux rope) and the entire system (the whole AR), which is one of the key factors that determine these instabilities.

Recently, using wavelet analysis, attempts have been made to determine any characteristic periodicities in the time series of F_{tot} for flaring ARs. For example, Korsós et al. (2020) and Soós et al. (2021) reported that there are some differences in the identified periodicities between flaring ARs with large M- and X-class flares and ARs with smaller B- and C-class flares. In this context, more detailed studies in the future may help us better understand whether the evolution of the magnetic field on the photosphere is inherently different between large flaring ARs and the others, as well as predict the likelihood of flare occurrence for a target AR.

6.4.2. Statistical Trends

Examination of the magnetic helicity on many sample ARs has revealed the statistical tendency that flare-rich ARs harbor a larger amount of magnetic helicity or helicity flux. For instance, LaBonte et al. (2007) surveyed the helicity injections for 48 X-flaring ARs and 345 reference regions without X-flares. They derived an empirical threshold for the occurrence of an X-class flare that the peak helicity flux, $\max|\langle F_{tot}\rangle|$, given by equation (6.27), should exceed a magnitude of 6×10^{36} $Mx^2\ s^{-1}$.

Park et al. (2010b) expanded the analysis to 378 ARs and investigated the correspondence between the 24-hour-averaged helicity flux, $|\langle F_{tot}\rangle|$, based on equation (6.27), and the next-24-hours flare index (i.e., the sum of GOES soft X-ray peak intensities of flares that occurred in the 24 hours after the helicity measurement). They demonstrated that for 91 subsamples with an unsigned magnetic flux of $(3-5)\times10^{22}$ Mx, the helicity

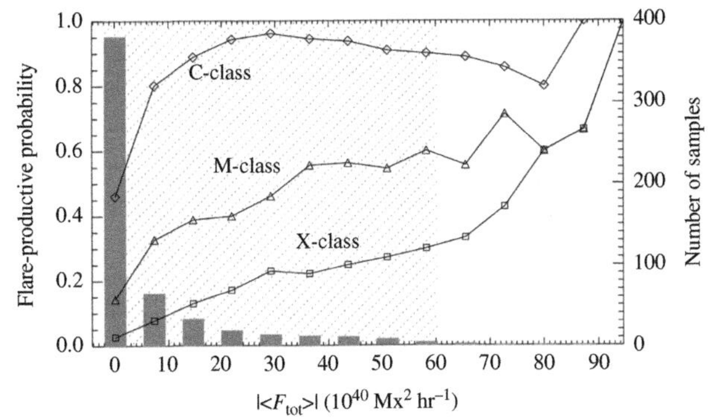

Figure 6.9 Flare-productive probability as a function of $|\langle F_{tot}\rangle|$, respectively, for C-class (diamonds), M-class (triangles), and X-class (squares). The gray bars indicate the total number of AR samples in the defined bins. Park et al. 2010b/IOP Publishing.

flux for the flaring ARs was approximately twice that of the quiescent ARs. On the other hand, 118 ARs with large helicity rates did not show a significant difference in magnetic flux between the flaring and quiescent groups. As shown in Fig. 6.9, these authors also found that the larger the value of $|\langle F_{tot}\rangle|$ an AR has, the greater the flaring activity of the AR in all cases of C-, M-, and X-class flares.

Regarding the differences between CME-eruptive and non-eruptive flares, Nindos and Andrews (2004) examined the coronal magnetic helicity, derived under the assumption that the coronal field of each AR has a constant force-free α (section 6.5.1), for 133 M-class events from 78 ARs. They revealed the statistical tendency that the preflare α values and coronal helicity of the non-eruptive flares are smaller than those of the eruptive flares. Park et al. (2012) examined 47 CMEs in 28 ARs and showed that there was also good correlation between $|\langle F_{tot}\rangle|$ and the CME speed. Interestingly, by tracing the helicity evolution, the CMEs analyzed were divided into two groups: one in which helicity increased monotonically with one sign of helicity, and another in which the sign of helicity reverses after a significant helicity injection (reversed shear; Kusano et al., 2004, 2012).

Tziotziou et al. (2012) investigated the relationship between the free magnetic energy (ΔE_{mag}) and relative helicity (H_R) on 42 ARs, finding that the flaring ARs possess higher free energy and relative helicity. As demonstrated in Fig. 6.10, monotonic scaling exists between the two variables, with a threshold for M-class events of 4×10^{31} erg and 2×10^{42} Mx^2 for the energy and helicity, respectively. This energy-helicity diagram was further examined by Tziotziou et al. (2014). By using 3D numerical simulations of eruptive and non-eruptive cases and two observed eruptive and non-eruptive ARs, they

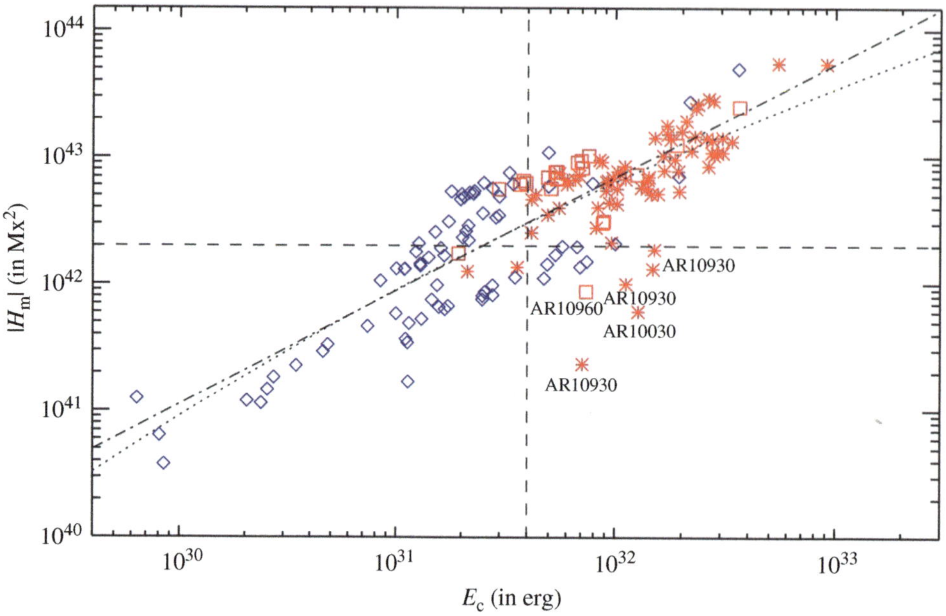

Figure 6.10 Scatter plots of ARs as functions of free magnetic energy (denoted as E_c in this diagram) and absolute relative magnetic helicity ($|H_m|$ here). The blue diamonds, red squares, and red asterisks denote the nonflaring, M-class flaring, and X-class flaring ARs, respectively. The dashed horizontal and vertical lines indicate the thresholds for M-class events (2×10^{42} Mx^2 and 4×10^{31} erg). The dotted and dash-dotted lines are the least-square best fit and least-square best logarithmic fit for the data points, respectively. Tziotziou et al. 2012/IOP Publishing.

reconstructed the energy-helicity diagram and found a consistent monotonic scaling law.

However, care must be taken when considering such statistical results because magnetic helicity is homogeneous to the square of the magnetic flux. As shown in section 6.1, ARs with larger areas or magnetic fluxes are more prone to flares, which may be included in the statistical results using extensive parameters (those scaling with AR size) including magnetic helicity (Welsch et al., 2009; Bobra & Couvidat, 2015; Sun et al., 2015; Toriumi & Takasao, 2017). Therefore, to assess the eruptivity of ARs purely owing to structural complexity and not to the AR area, it is worth considering the helicity normalized by the flux squared, H_R/Φ^2 (Démoulin & Pariat, 2009). On the other hand, the results of statistical studies clearly demonstrate quantitative differences between flare-productive and quiescent ARs. This suggests that helicity-related variables may be useful for predicting and forecasting imminent flare occurrences.

6.4.3. Magnetic Tongues: A Proxy for the Helicity Sign

It has been suggested that *magnetic tongues* in ARs, the elongated yin-yang-shaped patches of positive and negative magnetic polarities on the solar surface, can be used as a proxy for the magnetic helicity (López Fuentes et al., 2000). It is thought that the positive and negative polarities extended on both sides of the PIL in longitudinal (or line-of-sight) magnetograms reflect the vertical projection of the poloidal (azimuthal) component of the emerging magnetic flux tube. When an arched flux tube rises and penetrates the photospheric surface, the horizontal cross section of the tube displays a pair of opposite-polarity patches (magnetic tongues) with the PIL in between. If the flux tube is twisted or writhed – that is, if the emerging flux has a nonzero magnetic helicity (the situation in Fig. 6.2a1 or a4) – the tongue structure has an axisymmetric pattern, and the PIL deviates from an orthogonal angle to the main bipolar axis. Therefore, the layout of the tongues and the direction of the PIL were used to determine the sign of the magnetic helicity of the ARs. Flux-emergence simulations that assume twisted flux tubes often display magnetic tongues in the photosphere (e.g., Archontis and Hood, 2010, Toriumi et al., 2011).

Chandra et al. (2009) analyzed the evolution of the magnetic field in NOAA AR 10365 in May 2003 and found that it had a positive magnetic helicity. This AR produced an M1.6-class flare leaving a pair of J-shaped flare ribbons, another indication of twisting in ARs.

Mandrini et al. (2014) analyzed the force-free parameter α (section 6.5.1) for NOAA ARs 11121 and 11123 and confirmed that the positive value for α agrees with the observed magnetic tongue.

Statistical analysis of magnetic tongues by Luoni et al. (2011) on 40 ARs showed that the helicity signs determined from the tongues were consistent with those derived from the photospheric helicity fluxes. Poisson et al. (2015a,b) systematically surveyed 41 bipolar ARs and demonstrated that the twist estimated from the tongues in general matched the force-free α derived from the coronal field extrapolations. By analyzing 187 ARs, Poisson et al. (2016) further investigated the dependence of tongues on the activity cycle. They found that the angle of the PIL relative to the bipolar axis (AR tilt) has only a weak sign dominance in each solar hemisphere ($\geq$50%). In addition, the characteristics of the tongues – not only the PIL angle but also the amount of magnetic flux, polarity size, latitude, emergence rate, etc. – are not dependent on the periods of the cycle.

6.5. NUMERICAL MODELS

6.5.1. Coronal Field Reconstructions

Owing to the limitations of magnetic field measurements in the solar corona, we rely on coronal field models to estimate the magnetic field energy and helicity. The simplest approach is to extrapolate the coronal field based on the observed photospheric magnetic field under the assumption that nonmagnetic forces (e.g., pressure gradient and gravity) are negligible and the Lorentz force vanishes: that is,

$$\boldsymbol{j} \times \boldsymbol{B} = 0, \tag{6.31}$$

where $\boldsymbol{j} = c/(4\pi)(\nabla \times \boldsymbol{B})$ is the electric current. This force-free condition can also be written as

$$\nabla \times \boldsymbol{B} = \alpha \boldsymbol{B}, \tag{6.32}$$

where α is the force-free parameter. If $\alpha = 0$, or equivalently $\boldsymbol{j} = 0$ (current-free), the coronal field is the potential

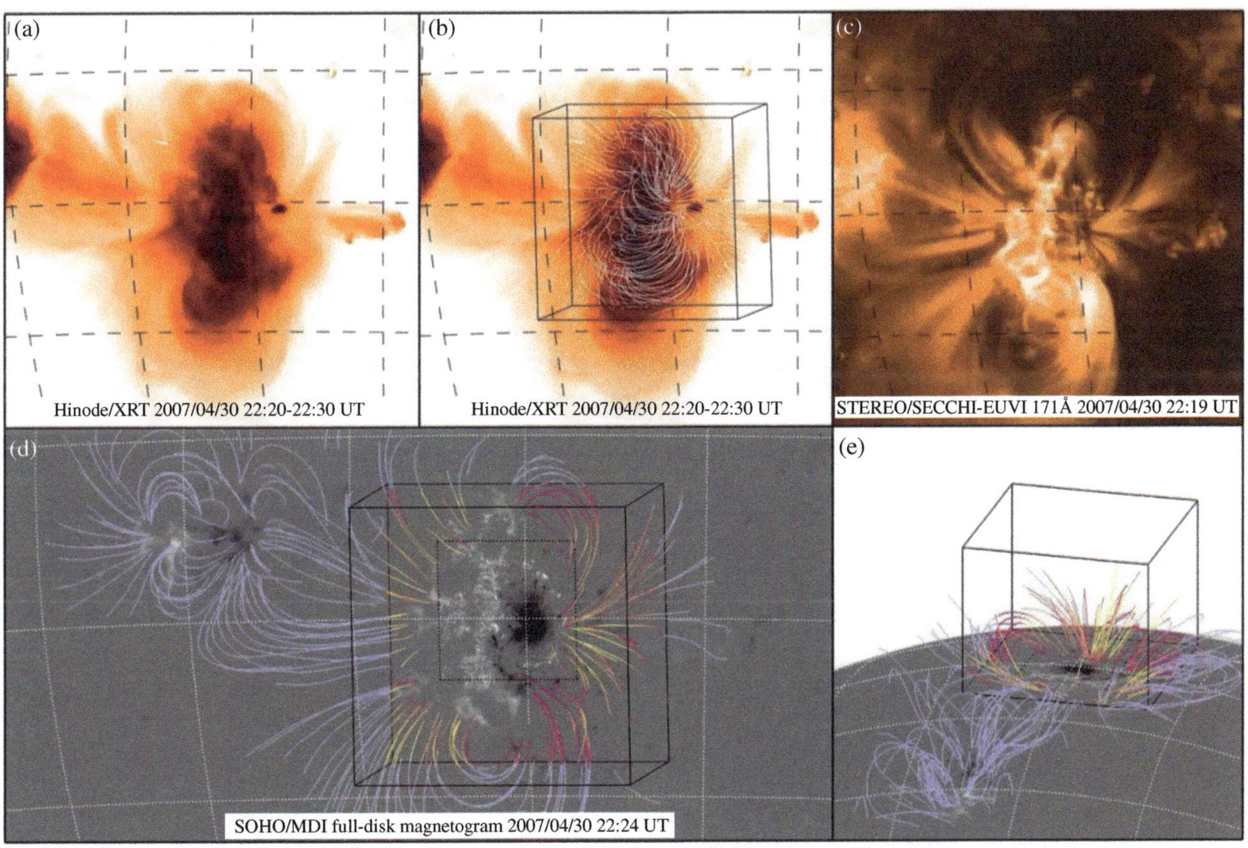

Figure 6.11 NLFFF extrapolation on NOAA AR 10953. (a, c) Coronal observations in X-ray and extreme ultraviolet. (b) Best-fit NLFFF result, shown as white lines. (d, e) Stereoscopically reconstructed field lines; solid black cubes outline the NLFFF computational domain. The loops inside the box are colored according to their misalignment angle from the NLFFF solution (yellow to orange corresponds to 5° to 45°). De Rosa et al., 2009/IOP Publishing.

field ($\boldsymbol{B}_\text{p}$), which is often used as the reference field when estimating the relative magnetic helicity (as in equation (6.5)). If α is nonzero and constant everywhere in the coronal volume under consideration, the magnetic field is called a linear force-free field (LFFF). If α is non-uniform, the field is a nonlinear force-free field (NLFFF; Figs. 6.5, 6.8, and 6.11).

Here, only the vertical component of the photospheric magnetic field (B_n) is required as the bottom boundary condition to obtain the potential field, whereas all components of the vector magnetogram ($\boldsymbol{B}$) are used for the LFFF and NLFFF. Extrapolation models, particularly NLFFF methods, have been used extensively for the direct determination of volume helicity in the AR corona (see the discussions in the previous sections). Readers are referred to, for example, Inoue (2016) and Wiegelmann and Sakurai (2021) for detailed accounts of extrapolation methods.

These reconstruction models, however, only provide steady-state coronal fields and may not be applicable to flare-productive ARs, which are highly dynamic in nature. To overcome this issue, temporally evolving models have been used. One approach of such coronal field reconstructions is data-constrained models, where the initial

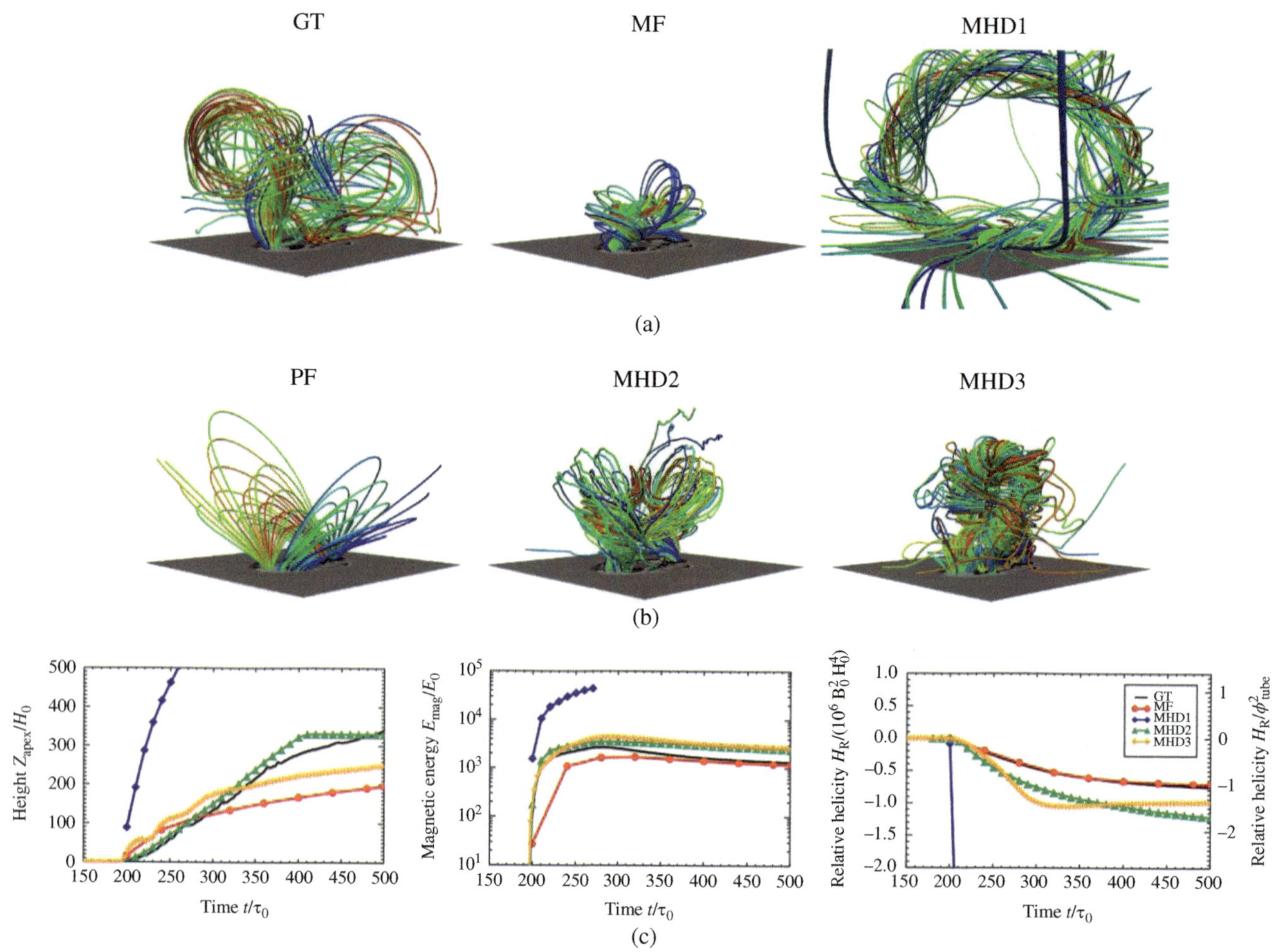

Figure 6.12 (a, b) Magnetic fields for ground-truth flux-emergence simulation (GT), potential field extrapolation (PF), and four data-driven coronal field reconstructions (MF, MHD1, MHD2, and MHD3). The bottom boundary shows the vertical GT magnetic field (i.e., the photospheric magnetogram). The colored tubes indicate the field lines, where the tubes with reddish (bluish) colors are integrated from the seeds placed in the positive (negative) polarity. The seeds were identical in all six cases. The time for MHD1 is 1000 s after the photospheric appearance of the emerging flux, while all other times are 4000 s. (c) Temporal evolution of the apex height z_apex, magnetic energy E_mag, and relative magnetic helicity H_R for GT and the four coronal field models MF, MHD1, MHD2, and MHD3. The normalization factor for time τ_0 corresponds to 25 s. Toriumi et al., 2020/IOP Publishing.

coronal field is prepared by extrapolations based on a snapshot magnetogram, and the subsequent dynamical evolution is obtained by solving the MHD equations.

Data-driven models, in which the coronal field evolves in response to sequentially updated bottom boundary magnetograms, may provide even more realistic coronal reconstructions. To investigate how accurately different data-driven models can reproduce a coronal field and quantities such as the magnetic helicity, Toriumi et al. (2020) used a flux-emergence simulation as the ground truth (GT) data set (Toriumi and Takasao, 2017). In this GT simulation, a twisted flux tube initially placed in the convection zone rises into the atmosphere and eventually builds up coronal loops. These authors compared the GT coronal field with those reconstructed using multiple data-driven models based on sequential GT magnetograms. As illustrated in Fig. 6.12, the helical flux rope structure was reproduced in all coronal field models. On the other hand, model-dependent results were obtained for quantitative comparison, with the magnetic energies (E_{mag} and ΔE_{mag}) and relative magnetic helicity (H_R) varying from cases almost comparable to GT to those differing by orders of magnitude. The magnetic helicity was derived by the method in DeVore (2000), following Valori et al. (2012). The observed model discrepancies were attributed to a highly non-force-free input bottom boundary (GT magnetograms) and to modeling treatments of the background atmosphere, bottom boundary, and spatial resolution.

6.5.2. Idealized Simulations

Numerical models of flux rope eruptions (i.e., CMEs) have also been employed to test the concept of magnetic helicity. For instance, Malanushenko et al. (2009) used eruption models to show that the additive self-helicity (H_{Ri}, equation (6.8)) is useful for determining the threshold beyond which a flux rope becomes unstable. The additive self-helicity divided by the square of the magnetic flux through a cross section of the tube, H_{Ri}/Φ^2, can be considered a generalization of the twist number (Berger and Field, 1984), which has been used to discuss the threshold of kink instability. The authors investigated the evolution of H_{Ri}/Φ^2 in the simulations, in which a twisted flux rope kinematically inserted into the computational domain interacted with a preexisting coronal arcade and deformed into a writhed configuration owing to the kink instability (Fan & Gibson, 2003). Over time, the (absolute) magnitude of H_{Ri}/Φ^2 increased until it reached a threshold value of ~1.5 for the instability to occur.

Another simulation model employed to validate the usage of helicity was given by Leake et al. (2013, 2014), where a magnetic flux emerges into a magnetized corona, which may or may not erupt depending on the direction of the coronal arcade (Fig. 6.13). Pariat et al. (2017) investigated the temporal evolution of several global scalar quantities to determine whether they could discriminate between eruptive and non-eruptive cases. It was found that the helicity ratio $|H_j|/|H_R|$ can determine the

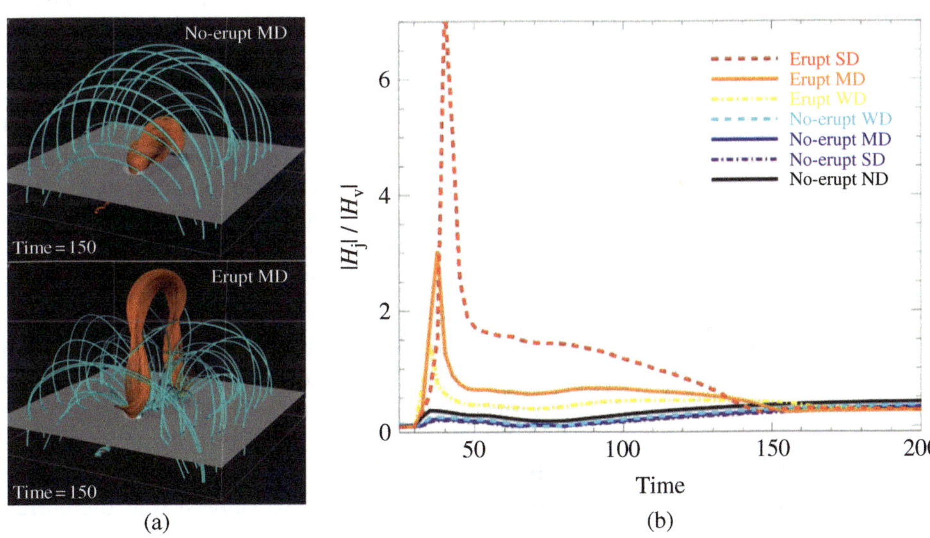

Figure 6.13 Examination of helicity ratio ($|H_j|/|H_R|$) by Pariat et al. (2017). (a) Snapshots of two simulation cases (non-eruptive and eruptive) with medium arcade strength. (b) Time evolution of the helicity ratio for the seven parameter cases. The eruptive cases show higher values above 0.45. Note that the relative helicity H_R is denoted in the figure as H_V. Corresponding to the magnetic field strengths of the coronal arcades, the simulation cases are referred to as strong arcade (SD), medium arcade (MD), weak arcade (WD), and no arcade (ND). E. Pariat, et al., 2017/ European Southern Observatory.

eruptivity, whereas other quantities such as the magnetic flux (Φ), magnetic energies (ΔE_{mag}, $\Delta E_{mag}/E_{mag}$, etc.), and total relative helicity ($|H_R|$) cannot (see section 6.4.1 for the details).

6.6. SUMMARY AND DISCUSSION

In this chapter, we have reviewed the following: (1) solar flares and CMEs tend to occur in ARs with complex structures; (2) therefore, various concepts of magnetic helicity, which is the physical quantity that measures the complexity of magnetic fields, help to understand the occurrence of flares and CMEs; and (3) based on the statistical findings, these concepts may be useful for predicting and forecasting such eruptive events. Magnetic helicity is not limited to mere theoretical research, but its practicality has been proven by a variety of studies that exploit actual observational data. However, there still remain important topics that cannot be addressed in detail in this chapter.

For instance, there is an issue that the absolute value of magnetic helicity may vary depending on which gauge is employed for both A and A_p, as discussed in Pariat et al. (2015) and Schuck and Antiochos (2019). This gauge issue can be problematic, particularly when comparing results obtained from different methods of estimating helicity and investigating a critical value of helicity for flare- or CME-trigger instabilities. In addition, as in equation (6.11), the definition of the temporal variation of the relative magnetic helicity dH_R/dt and its flux components on the boundary surface may not be fully consistent between previous studies with somewhat different gauge choices. We also note that each flux component may not be gauge-invariant; rather, only their sum is. On the other hand, different ways of implementing helicity formulas and preprocessing observational data can result in inconsistent results due to numerical errors and artificial noise. Future community-wide efforts, such as Helicity Thinkshops and Helicity 2020, may be needed to discuss and achieve a general consensus on the issues addressed here.

It is important to note that we still need to resolve various sources of uncertainty in determining the evolution of magnetic helicity for ARs. First, as a problem on the observational side, there are inaccuracies in photospheric magnetograms. For instance, if there are magnetic field structures with sizes smaller than the spatial resolution, magnetic helicity may be underestimated. Other causes of inaccuracies include the 180° ambiguity of the transverse field, general difficulties in calibration, improperly assumed atmospheres in the Stokes inversions, corrugation of line formation heights, and inadequate sampling of spectral profiles (see, for example, del Toro Iniesta 2003; some of these issues will be discussed later in this section). Second, there is a difficulty of using the photospheric data in estimating the relative helicity and/or helicity flux for the corona. The dynamic chromosphere lying between the photosphere and the corona may have an effect on coronal energetics that cannot be perceived by photosphere observation alone, and this may also be one cause of misassessing the magnetic helicity of the corona. Third, as mentioned in the previous paragraph, because the relative helicity is not invariant to gauge selection, it is critically important to confirm that the applied gauges are the same when comparing different relative helicity estimates. Otherwise, the comparison will lead to a wrong conclusion about the AR evolution and its relationship with eruptivity (see sections 2 and 3.1 for details).

To understand the flaring activity of ARs, most magnetic helicity studies thus far have focused on the temporal evolution of magnetic helicity with a set of *individual* ARs over a span (typically, a few days to two weeks) of their (partial) passages on the solar disk. On the other hand, very recently, a novel approach was designed by Park et al. (2021) to examine whether there are any flare-associated differences in heliographic regions on a larger scale with respect to the degree of complying with the hemispheric sign preference (HSP) of magnetic helicity. The HSP of magnetic helicity has a unique cycle-independent tendency, in which various features in the northern hemisphere of the Sun tend to exhibit negative or left-handed helicity, while they are positive in the southern hemisphere. In the study of Park et al. (2021), heliographic regions were defined in the Carrington longitude-latitude plane with a grid spacing of 45° in longitude and 15° in latitude. The authors examined the relative magnetic helicity flux F_{tot} across the photospheric surface for 4,802 samples of 1,105 unique ARs observed during solar cycle 24. They found that heliographic regions with lower degrees of HSP compliance are likely to show higher levels of flaring activity and larger values of average magnetic flux. What does this observational result tell us? Based on simulations of the rise of magnetic flux tubes deep in the convection zone by Manek and Brummell (2021), we can deduce that if the helicity sign of a rising flux tube is against the HSP, the tube's field strength is relatively larger for its successful rise compared to a tube with a helicity sign following the HSP. This speculation assumes that the background poloidal field strength is quasi-isotropic in the tachocline. The association of the lower HSP with greater flaring activity may also suggest that there are localized regions in the convection zone where large-scale downflows and strong turbulence are present. Rising flux tubes therein may have a sign opposite that of magnetic helicity as compared to the sign expected from the HSP, may possess greater magnetic complexity (such as δ-spots), and may produce large flares. Such HSP studies may offer an indirect understanding of the elusive magnetic nature of the solar interior as well as the interaction between flux tubes and convection flows, which may be the origin of bearing flare-productive ARs (Toriumi & Wang, 2019). Moreover, realistic dynamo simulations and helioseismology studies will help us better understand the formation and evolution of flaring ARs before their emergence on the solar surface.

In recent decades, especially since sequential (vector) magnetograms became available from space, we have deepened our understanding of the usefulness of magnetic helicity. While difficulties remain in accurately estimating magnetic helicity, such as those caused by the ambiguity mentioned earlier in deriving the transverse magnetic field in the photosphere, we expect that such difficulties can be overcome or at least mitigated in the near future. For example, methods that employ stereoscopic disambiguation may be helpful (e.g., with the Solar Orbiter; Rouillard et al., 2020; Valori et al., 2022). The photospheric magnetic field, from which the coronal field is reconstructed, deviates significantly from the force-free state. However, it may be possible to improve the coronal-field reconstruction by incorporating the chromospheric field, which yields better force-freeness (e.g., with the Daniel K. Inouye Solar Telescope and Sunrise III; Wiegelmann et al., 2008, Fleishman et al., 2019). Numerical simulations have been used to validate the helicity estimation methods, but by utilizing more realistic flux-emergence simulations that have appeared in recent years, it will be possible to examine these methods with even greater accuracy (Cheung et al., 2019; Toriumi & Hotta; 2019; Hotta & Iijima, 2020). These advanced studies are expected to lead to further understanding and better prediction of flares and CMEs in the near future.

ACKNOWLEDGMENTS

S.-H. Park was supported by basic research funding from the Korea Astronomy and Space Science Institute (KASI2023185007).

REFERENCES

Archontis, V., & Hood, A. W. (2010). Flux emergence and coronal eruption. *Astronomy & Astrophysics*, *514*, A56. https://doi.org/10.1051/0004-6361/200913502

Barnes, G., Leka, K. D., Schrijver, C. J., Colak, T., Qahwaji, R., Ashamari, O. W., et al. (2016). A comparison of flare forecasting methods. I. Results from the "All-Clear" Workshop. *The Astrophysical Journal*, *829*(2), 89. https://doi.org/10.3847/0004-637X/829/2/89

Benz, A. O. (2017). Flare observations. *Living Reviews in Solar Physics*, *14*(1), 2. https://doi.org/10.1007/s41116-016-0004-3

Berger, M. A. (1984). Rigorous new limits on magnetic helicity dissipation in the solar corona. *Geophysical and Astrophysical Fluid Dynamics*, *30*(1), 79–104. https://doi.org/10.1080/03091928408210078

Berger, M. A., & Field, G. B. (1984). The topological properties of magnetic helicity. *Journal of Fluid Mechanics*, *147*, 133–148. https://doi.org/10.1017/S0022112084002019

Berger, M. A. (2003). Topological quantities in magnetohydrodynamics. In A. Ferriz-Mas, M. Nunez (Eds.), *Advances in nonlinear dynamos* (pp. 345–374). London: CRC Press. https://doi.org/10.1201/9780203493137.ch10

Bi, Y., Liu, Y. D., Liu, Y., Yang, J., Xu, Z., & Ji, K. (2018). A survey of changes in magnetic helicity flux on the photosphere during relatively low-class flares. *The Astrophysical Journal*, *865*(2), 139. https://doi.org/10.3847/1538-4357/aada7e

Bobra, M. G., & Couvidat, S. (2015). Solar flare prediction using SDO/HMI vector magnetic field data with a machine-learning algorithm. *The Astrophysical Journal*, *798*(2), 135. https://doi.org/10.1088/0004-637X/798/2/135

Bobra, M. G., & Ilonidis, S. (2016). Predicting coronal mass ejections using machine learning methods. *The Astrophysical Journal*, *821*(2), 127. https://doi.org/10.3847/0004-637X/821/2/127

Bobra, M. G., Sun, X., Hoeksema, J. T., Turmon, M., Liu, Y., Hayashi, K., et al. (2014). The Helioseismic and Magnetic Imager (HMI) Vector Magnetic Field Pipeline: SHARPs - Space-Weather HMI Active Region Patches. *Solar Physics*, *289*(9), 3549–3578. https://doi.org/10.1007/s11207-014-0529-3

Chae, J. (2001). Observational determination of the rate of magnetic helicity transport through the solar surface via the horizontal motion of field line footpoints. *The Astrophysical Journal Letters*, *560*(1), L95–L98. https://doi.org/10.1086/324173

Chae, J., & Sakurai, T. (2008). A test of three optical flow techniques – LCT, DAVE, and NAVE. *The Astrophysical Journal*, *689*(1), 593–612. https://doi.org/10.1086/592761

Chandra, R., Schmieder, B., Aulanier, G., & Malherbe, J. M. (2009). Evidence of magnetic helicity in emerging flux and associated flare. *Solar Physics*, *258*(1), 53–67. https://doi.org/10.1007/s11207-009-9392-z

Chen, P. F. (2011). Coronal mass ejections: Models and their observational basis. *Living Reviews in Solar Physics*, *8*(1), 1. https://doi.org/10.12942/lrsp-2011-1

Cheung, M. C. M., Rempel, M., Chintzoglou, G., Chen, F., Testa, P., Martínez-Sykora, J., et al. (2019). A comprehensive three-dimensional radiative magnetohydrodynamic simulation of a solar flare. *Nature Astronomy*, *3*, 160–166. https://doi.org/10.1038/s41550-018-0629-3

De Rosa, M. L., Schrijver, C. J., Barnes, G., Leka, K. D., Lites, B. W., Aschwanden, M. J., et al. (2009). A critical assessment of nonlinear force-free field modeling of the solar corona for active region 10953. *The Astrophysical Journal*, *696*(2), 1780–1791. https://doi.org/10.1088/0004-637X/696/2/1780

del Toro Iniesta, J. C. (2003). *Introduction to spectropolarimetry*. Cambridge University Press.

Démoulin, P., & Aulanier, G. (2010). Criteria for flux rope eruption: Non-equilibrium versus torus instability. *The Astrophysical Journal*, *718*(2), 1388–1399. https://doi.org/10.1088/0004-637X/718/2/1388

Démoulin, P., & Pariat, E. (2009). Modelling and observations of photospheric magnetic helicity. *Advances in Space Research, 43*(7), 1013–1031. https://doi.org/10.1016/j.asr.2008.12.004

DeVore, C. R. (2000). Magnetic helicity generation by solar differential rotation. *The Astrophysical Journal, 539*(2), 944–953. https://doi.org/10.1086/309274

Elsasser, W. M. (1956). Hydromagnetic dynamo theory. *Reviews of Modern Physics, 28*(2), 135–163. https://doi.org/10.1103/RevModPhys.28.135

Fan, Y., & Gibson, S. E. (2003). The emergence of a twisted magnetic flux tube into a preexisting coronal arcade. *The Astrophysical Journal Letters, 589*(2), L105-L108. https://doi.org/10.1086/375834

Finn, J. M., & Antonsen Jr., T. M. (1985). Magnetic helicity: What is it and what is it good for? *Comments on Plasma Physics and Controlled Fusion, 9*, 111–126. https://doi.org/10.1088/0004-637X/696/2/1780

Fleishman, G., Mysh'yakov, I., Stupishin, A., Loukitcheva, M., & Anfinogentov, S. (2019). Force-free field reconstructions enhanced by chromospheric magnetic field data. *The Astrophysical Journal, 870*(2), 101. https://doi.org/10.3847/1538-4357/aaf384

Fletcher, L., Dennis, B. R., Hudson, H. S., Krucker, S., Phillips, K., Veronig, A., et al. (2011). An observational overview of solar flares. *Space Science Reviews, 159*(1–4), 19–106. https://doi.org/10.1007/s11214-010-9701-8

Gupta, M., Thalmann, J. K., & Veronig, A. M. (2021). Magnetic helicity and energy budget around large confined and eruptive solar flares. *Astronomy & Astrophysics, 653*, A69. https://doi.org/10.1051/0004-6361/202140591

Hartkorn, K., & Wang, H. (2004). Magnetic helicity change rate associated with X-class and M-class flares. *Solar Physics, 225*(2), 311–324. https://doi.org/10.1007/s11207-004-5699-y

Hotta, H., & Iijima, H. (2020). On rising magnetic flux tube and formation of sunspots in a deep domain. *Monthly Notices of the Royal Astronomical Society, 494*(2), 2523–2537. https://doi.org/10.1093/mnras/staa844

Inoue, S. (2016). Magnetohydrodynamics modeling of coronal magnetic field and solar eruptions based on the photospheric magnetic field. *Progress in Earth and Planetary Science, 3*(1), 19. https://doi.org/10.1186/s40645-016-0084-7

Ishiguro, N., & Kusano, K. (2017). Double arc instability in the solar corona. *The Astrophysical Journal, 843*(2), 101. https://doi.org/10.3847/1538-4357/aa799b

Jing, J., Park, S.-H., Liu, C., Lee, J., Wiegelmann, T., Xu, Y., et al. (2012). Evolution of relative magnetic helicity and current helicity in NOAA active region 11158. *The Astrophysical Journal Letters, 752*(1), L9. https://doi.org/10.1088/2041-8205/752/1/L9

Jing, J., Xu, Y., Lee, J., Nitta, N. V., Liu, C., Park, S.-H., et al. (2015). Comparison between the eruptive X2.2 flare on 2011 February 15 and confined X3.1 flare on 2014 October 24. *Research in Astronomy and Astrophysics, 15*(9), 1537. https://doi.org/10.1088/1674-4527/15/9/010

Kazachenko, M. D., Lynch, B. J., Savcheva, A., Sun, X., & Welsch, B. T. (2022). Toward improved understanding of magnetic fields participating in solar flares: Statistical analysis of magnetic fields within flare ribbons. *The Astrophysical Journal, 926*(1), 56. https://doi.org/10.3847/1538-4357/ac3af3

Kim, R. S., Park, S. H., Jang, S., Cho, K. S., & Lee, B. S. (2017). Relation of CME speed and magnetic helicity in CME source regions on the Sun during the early phase of solar cycles 23 and 24. *Solar Physics, 292*(4), 66. https://doi.org/10.1007/s11207-017-1079-2

Kliem, B., & Török, T. (2006). Torus instability. *Physical Review Letters, 96*(25), 255002. https://doi.org/10.1103/PhysRevLett.96.255002

Korsós, M. B., Romano, P., Morgan, H., Ye, Y., Erdélyi, R., & Zuccarello, F. (2020). Differences in periodic magnetic helicity injection behavior between flaring and non-flaring active regions: Case study. *The Astrophysical Journal Letters, 897*(2), L23. https://doi.org/10.3847/2041-8213/ab9d7a

Kosugi, T., Matsuzaki, K., Sakao, T., Shimizu, T., Sone, Y., Tachikawa, S., et al. (2007). The Hinode (Solar-B) mission: An overview. *Solar Physics, 243*(1), 3–17. https://doi.org/10.1007/s11207-007-9014-6

Kusano, K., Maeshiro, T., Yokoyama, T., & Sakurai, T. (2002). Measurement of magnetic helicity injection and free energy loading into the solar corona. *The Astrophysical Journal, 577*(1), 501–512. https://doi.org/10.1086/342171

Kusano, K., Maeshiro, T., Yokoyama, T., & Sakurai, T. (2003). Measurement of magnetic helicity flux into the solar corona. *Advances in Space Research, 32*(10), 1917–1922. https://doi.org/10.1016/S0273-1177(03)90626-0

Kusano, K., Maeshiro, T., Yokoyama, T., & Sakurai, T. (2004). The trigger mechanism of solar flares in a coronal arcade with reversed magnetic shear. *The Astrophysical Journal, 610*(1), 537–549. https://doi.org/10.1086/421547

Kusano, K., Bamba, Y., Yamamoto, T. T., Iida, Y., Toriumi, S., & Asai, A. (2012). Magnetic field structures triggering solar flares and coronal mass ejections. *The Astrophysical Journal, 760*(1), 31. https://doi.org/10.1088/0004-637X/760/1/31

Kusano, K., Iju, T., Bamba, Y., & Inoue, S. (2020). A physics-based method that can predict imminent large solar flares. *Science, 369*(6503), 587–591. https://doi.org/10.1126/science.aaz2511

Kusano, K., Ichimoto, K., Ishii, M., Miyoshi, Y., Yoden, S., Akiyoshi, H., et al. (2021). PSTEP: Project for solar-terrestrial environment prediction. *Earth, Planets and Space, 73*(1), 159. https://doi.org/10.1186/s40623-021-01486-1

LaBonte, B. J., Georgoulis, M. K., & Rust, D. M. (2007). Survey of magnetic helicity injection in regions producing X-class flares. *The Astrophysical Journal, 671*(1), 955–963. https://doi.org/10.1086/522682

Leake, J. E., Linton, M. G., & Török, T. (2013). Simulations of emerging magnetic flux. I. The formation of stable coronal flux ropes. *The Astrophysical Journal, 778*(2), 99. https://doi.org/10.1088/0004-637X/778/2/99

Leake, J. E., Linton, M. G., & Antiochos, S. K. (2014). Simulations of emerging magnetic flux. II. The formation of unstable coronal flux ropes and the initiation of coronal mass ejections. *The Astrophysical Journal, 787*(1), 46. https://doi.org/10.1088/0004-637X/787/1/46

Leka, K. D., Park, S.-H., Kusano, K., Andries, J., Barnes, G., Bingham, S., et al. (2019). A comparison of flare forecasting methods. II. Benchmarks, metrics, and performance results for operational solar flare forecasting systems. *The*

Astrophysical Journal Supplement Series, *243*(2), 36. https://doi.org/10.3847/1538-4365/ab2e12
Li, T., Hou, Y., Yang, S., Zhang, J., Liu, L., & Veronig, A. M. (2020). Magnetic flux of active regions determining the eruptive character of large solar flares. *The Astrophysical Journal*, *900*(2), 128. https://doi.org/10.3847/1538-4357/aba6ef
Lin, P. H., Kusano, K., Shiota, D., Inoue, S., Leka, K. D., & Mizuno, Y. (2020). A new parameter of the photospheric magnetic field to distinguish eruptive-flare producing solar active regions. *The Astrophysical Journal*, *894*(1), 20. https://doi.org/10.3847/1538-4357/ab822c
Longcope, D. W. (2004). Inferring a photospheric velocity field from a sequence of vector magnetograms: The minimum energy fit. *The Astrophysical Journal*, *612*(2), 1181–1192. https://doi.org/10.1086/422579
Longcope, D. W., & Malanushenko, A. (2008). Defining and calculating self-helicity in coronal magnetic fields. *The Astrophysical Journal*, *674*(2), 1130–1143. https://doi.org/10.1086/524011
López Fuentes, M. C., Demoulin, P., Mandrini, C. H., & van Driel-Gesztelyi, L. (2000). The counterkink rotation of a non-Hale active region. *The Astrophysical Journal*, *544*(1), 540–549. https://doi.org/10.1086/317180
Luoni, M. L., Démoulin, P., Mandrini, C. H., & van Driel-Gesztelyi, L. (2011). Twisted flux tube emergence evidenced in longitudinal magnetograms: Magnetic tongues. *Solar Physics*, *270*(1), 45–74. https://doi.org/10.1007/s11207-011-9731-8
Magara, T., & Tsuneta, S. (2008). Hinode's observational result on the saturation of magnetic helicity injected into the solar atmosphere and its relation to the occurrence of a solar flare. *Publications of the Astronomical Society of Japan*, *60*, 1181. https://doi.org/10.1093/pasj/60.5.1181
Malanushenko, A., Longcope, D. W., Fan, Y., & Gibson, S. E. (2009). Additive self-helicity as a kink mode threshold. *The Astrophysical Journal*, *702*(1), 580–592. https://doi.org/10.1088/0004-637X/702/1/580
Mandrini, C. H., Schmieder, B., Demoulin, P., Guo, Y., & Cristiani, G. D. (2014). Topological analysis of emerging bipole clusters producing violent solar events. *Solar Physics*, *289*(6), 2041–2071. https://doi.org/10.1007/s11207-013-0458-6
Manek, B., & Brummell, N. (2021). On the origin of solar hemispherical helicity rules: Simulations of the rise of magnetic flux concentrations in a background field. *The Astrophysical Journal*, *909*(1), 72. https://doi.org/10.3847/1538-4357/abd859
Moffatt, H. K. (1969). The degree of knottedness of tangled vortex lines. *Journal of Fluid Mechanics*, *35*, 117–129. https://doi.org/10.1017/S0022112069000991
Moon, Y. J., Chae, J., Choe, G. S., Wang, H., Park, Y. D., Yun, H. S., et al. (2002a). Flare activity and magnetic helicity injection by photospheric horizontal motions. *The Astrophysical Journal*, *574*(2), 1066–1073. https://doi.org/10.1086/340975
Moon, Y. J., Chae, J., Wang, H., Choe, G. S., & Park, Y. D. (2002b). Impulsive variations of the magnetic helicity change rate associated with eruptive flares. *The Astrophysical Journal*, *580*(1), 528–537. https://doi.org/10.1086/343130
Moraitis, K., Sun, X., Pariat, E., & Linan, L. (2019). Magnetic helicity and eruptivity in active region 12673. *Astronomy & Astrophysics*, *628*, A50. https://doi.org/10.1051/0004-6361/201935870
Nindos, A., & Andrews, M. D. (2004). The association of big flares and coronal mass ejections: What is the role of magnetic helicity? *The Astrophysical Journal*, *616*(2), L175-L178. https://doi.org/10.1086/426861
November, L. J., & Simon, G. W. (1988). Precise proper-motion measurement of solar granulation. *Astrophysical Journal*, *333*, 427. https://doi.org/10.1086/166758
Pariat, E., Démoulin, P., & Berger, M. A. (2005). Photospheric flux density of magnetic helicity. *Astronomy & Astrophysics*, *439*(3), 1191–1203. https://doi.org/10.1051/0004-6361:20052663
Pariat, E., Valori, G., Démoulin, P., & Dalmasse, K. (2015). Testing magnetic helicity conservation in a solar-like active event. *Astronomy & Astrophysics*, *580*, A128. https://doi.org/10.1051/0004-6361/201525811
Pariat, E., Leake, J. E., Valori, G., Linton, M. G., Zuccarello, F. P., & Dalmasse, K. (2017). Relative magnetic helicity as a diagnostic of solar eruptivity. *Astronomy & Astrophysics*, *601*, A125. https://doi.org/10.1051/0004-6361/201630043
Park, S.-H., Lee, J., Choe, G. S., Chae, J., Jeong, H., Yang, G., et al. (2008). The variation of relative magnetic helicity around major flares. *The Astrophysical Journal*, *686*(2), 1397–1403. https://doi.org/10.1086/591117
Park, S.-H., Chae, J., Jing, J., Tan, C., & Wang, H. (2010a). Time evolution of coronal magnetic helicity in the flaring active region NOAA 10930. *The Astrophysical Journal*, *720*(2), 1102–1107. https://doi.org/10.1088/0004-637X/720/2/1102
Park, S.-H., Chae, J., & Wang, H. (2010b). Productivity of solar flares and magnetic helicity injection in active regions. *The Astrophysical Journal*, *718*(1), 43–51. https://doi.org/10.1088/0004-637X/718/1/43
Park, S.-H., Cho, K.-S., Bong, S.-C., Kumar, P., Chae, J., Liu, R., & Wang, H. (2012). The occurrence and speed of CMEs related to two characteristic evolution patterns of helicity injection in their solar source regions. *The Astrophysical Journal*, *750*(1), 48. https://doi.org/10.1088/0004-637X/750/1/48
Park, S.-H., Kusano, K., Cho, K.-S., Chae, J., Bong, S.-C., Kumar, P., et al. (2013). Study of magnetic helicity injection in the active region NOAA 9236 producing multiple flare-associated coronal mass ejection events. *The Astrophysical Journal*, *778*(1), 13. https://doi.org/10.1088/0004-637X/778/1/13
Park, S.-H., Leka, K. D., & Kusano, K. (2021). Magnetic helicity flux across solar active region photospheres. II. Association of hemispheric sign preference with flaring activity during Solar Cycle 24. *The Astrophysical Journal*, *911*(2), 79. https://doi.org/10.3847/1538-4357/abea13
Poisson, M., Mandrini, C. H., Démoulin, P., & López Fuentes, M. (2015a). Evidence of twisted flux-tube emergence in active regions. *Solar Physics*, *290*(3), 727–751. https://doi.org/10.1007/s11207-014-0633-4
Poisson, M., López Fuentes, M., Mandrini, C. H., & Démoulin, P. (2015b). Active-region twist derived from magnetic tongues and linear force-free extrapolations. *Solar Physics*, *290*(11), 3279–3294. https://doi.org/10.1007/s11207-015-0804-y
Poisson, M., Démoulin, P., López Fuentes, M., & Mandrini, C. H. (2016). Properties of magnetic tongues over a solar

cycle. *Solar Physics*, *291*(6), 1625–1646. https://doi.org/10.1007/s11207-016-0926-x

Priest, E. R., & Forbes, T. G. (2002). The magnetic nature of solar flares. *The Astronomy and Astrophysics Review*, *10*(4), 313–377. https://doi.org/10.1007/s001590100013

Romano, P., Contarino, L., & Zuccarello, F. (2005). Observational evidence of the primary role played by photospheric motions in magnetic helicity transport before a filament eruption. *Astronomy & Astrophysics*, *433*(2), 683–690. https://doi.org/10.1051/0004-6361:20041807

Romano, P., Zuccarello, F. P., Guglielmino, S. L., & Zuccarello, F. (2014). Evolution of the magnetic helicity flux during the formation and eruption of flux ropes. *The Astrophysical Journal*, *794*(2), 118. https://doi.org/10.1088/0004-637X/794/2/118

Romano, P., Zuccarello, F., Guglielmino, S. L., Berrilli, F., Bruno, R., Carbone, V., et al. (2015). Recurrent flares in active region NOAA 11283. *Astronomy & Astrophysics*, *582*, A55. https://doi.org/10.1051/0004-6361/201525887

Rouillard, A. P., Pinto, R. F., Vourlidas, A., De Groof, A., Thompson, W. T., Bemporad, A., et al. (2020). Models and data analysis tools for the Solar Orbiter mission. *Astronomy & Astrophysics*, *642*, A2. https://doi.org/10.1051/0004-6361/201935305

Rudenko, G. V., & Myshyakov, I. I. (2011). Gauge-invariant helicity for force-free magnetic fields in a rectangular box. *Solar Physics*, *270*(1), 165–173. https://doi.org/10.1007/s11207-011-9743-4

Sammis, I., Tang, F., & Zirin, H. (2000). The dependence of large flare occurrence on the magnetic structure of sunspots. *The Astrophysical Journal*, *540*(1), 583–587. https://doi.org/10.1086/309303

Scherrer, P. H., Bogart, R. S., Bush, R. I., Hoeksema, J. T., Kosovichev, A. G., Schou, J., et al. (1995). The Solar Oscillations Investigation - Michelson Doppler Imager. *Solar Physics*, *162*(1–2), 129–188. https://doi.org/10.1007/BF00733429

Schuck, P. W. (2006). Tracking magnetic footpoints with the magnetic induction equation. *The Astrophysical Journal*, *646*(2), 1358–1391. https://doi.org/10.1086/505015

Schuck, P. W. (2008). Tracking vector magnetograms with the magnetic induction equation. *The Astrophysical Journal*, *683*(2), 1134–1152. https://doi.org/10.1086/589434

Schuck, P. W., & Antiochos, S. K. (2019). Determining the transport of magnetic helicity and free energy in the Sun's atmosphere. *The Astrophysical Journal*, *882*(2), 151. https://doi.org/10.3847/1538-4357/ab298a

Shibata, K., & Magara, T. (2011). Solar flares: Magnetohydrodynamic processes. *Living Reviews in Solar Physics*, *8*(1), 6. https://doi.org/10.12942/lrsp-2011-6

Soós, S., Korsós, M. B., Morgan, H., & Erdélyi, R. (2021). On the differences in the periodic behaviour of magnetic helicity flux in flaring active regions with and without X-class events. *arXiv e-prints*, art. arXiv:2112.05933

Sun, X., Bobra, M. G., Hoeksema, J. T., Liu, Y., Li, Y., Shen, C., et al. (2015). Why is the Great Solar Active Region 12192 Flare-rich but CME-poor? *The Astrophysical Journal Letters*, *804*(2), L28. https://doi.org/10.1088/2041-8205/804/2/L28

Taylor, J. B. (1974). Relaxation of toroidal plasma and generation of reverse magnetic fields. *Physical Review Letters*, *33*(19), 1139–1141. https://doi.org/10.1103/PhysRevLett.33.1139

Thalmann, J. K., Inhester, B., & Wiegelmann, T. (2011). Estimating the relative helicity of coronal magnetic fields. *Solar Physics*, *272*(2), 243–255. https://doi.org/10.1007/s11207-011-9826-2

Thalmann, J. K., Moraitis, K., Linan, L., Pariat, E., Valori, G., & Dalmasse, K. (2019). Magnetic helicity budget of solar active regions prolific of eruptive and confined flares. *The Astrophysical Journal*, *887*(1), 64. https://doi.org/10.3847/1538-4357/ab4e15

Toriumi, S., & Hotta, H. (2019). Spontaneous generation of δ-sunspots in convective magnetohydrodynamic simulation of magnetic flux emergence. *The Astrophysical Journal Letters*, *886*(1), L21. https://doi.org/10.3847/2041-8213/ab55e7

Toriumi, S., & Takasao, S. (2017). Numerical simulations of flare-productive active regions: δ-sunspots, sheared polarity inversion lines, energy storage, and predictions. *The Astrophysical Journal*, *850*(1), 39. https://doi.org/10.3847/1538-4357/aa95c2

Toriumi, S., & Wang, H. (2019). Flare-productive active regions. *Living Reviews in Solar Physics*, *16*(1), 3. https://doi.org/10.1007/s41116-019-0019-7

Toriumi, S., Miyagoshi, T., Yokoyama, T., Isobe, H., & Shibata, K. (2011). Dependence of the magnetic energy of solar active regions on the twist intensity of the initial flux tubes. *Publications of the Astronomical Society of Japan*, *63*, 407–415. https://doi.org/10.1093/pasj/63.2.407

Toriumi, S., Schrijver, C. J., Harra, L. K., Hudson, H., & Nagashima, K. (2017). Magnetic properties of solar active regions that govern large solar flares and eruptions. *The Astrophysical Journal*, *834*(1), 56. https://doi.org/10.3847/1538-4357/834/1/56

Toriumi, S., Takasao, S., Cheung, M. C. M., Jiang, C., Guo, Y., Hayashi, K., et al. (2020). Comparative study of data-driven solar coronal field models using a flux emergence simulation as a ground-truth data set. *The Astrophysical Journal*, *890*(2), 103. https://doi.org/10.3847/1538-4357/ab6b1f

Tsuneta, S., Ichimoto, K., Katsukawa, Y., Nagata, S., Otsubo, M., Shimizu, T., et al. (2008). The Solar Optical Telescope for the Hinode Mission: An Overview. *Solar Physics*, *249*(2), 167–196. https://doi.org/10.1007/s11207-008-9174-z

Tziotziou, K., Moraitis, K., Georgoulis, M. K., & Archontis, V. (2014). Validation of the magnetic energy vs. helicity scaling in solar magnetic structures. *Astronomy & Astrophysics*, *570*, L1. https://doi.org/10.1051/0004-6361/201424864

Tziotziou, K., Georgoulis, M. K., & Raouafi, N.-E. (2012). The magnetic energy-helicity diagram of solar active regions. *The Astrophysical Journal Letters*, *759*(1), L4. https://doi.org/10.1088/2041-8205/759/1/L4

Valori, G., Démoulin, P., & Pariat, E. (2012). Comparing values of the relative magnetic helicity in finite volumes. *Solar Physics*, *278*(2), 347–366. https://doi.org/10.1007/s11207-012-9951-6

Valori, G., Pariat, E., Anfinogentov, S., Chen, F., Georgoulis, M. K., Guo, Y., et al. (2016). Magnetic helicity estimations in models and observations of the solar magnetic field. Part I: Finite volume methods. *Space Science Reviews*, *201*(1–4), 147–200. https://doi.org/10.1007/s11214-016-0299-3

Valori, G., Löschl, P., Stansby, D., Pariat, E., Hirzberger, J., & Chen, F. (2022). Disambiguation of vector magnetograms by stereoscopic observations from the Solar Orbiter (SO)/Polarimetric and Helioseismic Imager (PHI) and the Solar Dynamic Observatory (SDO)/Helioseismic and Magnetic Imager (HMI). *Solar Physics*, *297*(1), 12. https://doi.org/10.1007/s11207-021-01942-x

Vemareddy, P. (2015). Investigation of helicity and energy flux transport in three emerging solar active regions. *The Astrophysical Journal*, *806*(2), 245. https://doi.org/10.1088/0004-637X/806/2/245

Vemareddy, P. (2021). Successive injection of opposite magnetic helicity: evidence for active regions without coronal mass ejections. *Monthly Notices of the Royal Astronomical Society*, *507*(4), 6037–6044. https://doi.org/10.1093/mnras/stab2401

Vemareddy, P., Ambastha, A., Maurya, R. A., & Chae, J. (2012). On the injection of helicity by the shearing motion of fluxes in relation to flares and coronal mass ejections. *The Astrophysical Journal*, *761*(2), 86. https://doi.org/10.1088/0004-637X/761/2/86

Wang, J., Zhou, G., & Zhang, J. (2004). Helicity patterns of coronal mass ejection-associated active regions. *The Astrophysical Journal*, *615*(2), 1021–1028. https://doi.org/10.1086/424584

Welsch, B. T., Fisher, G. H., Abbett, W. P., & Regnier, S. (2004). ILCT: Recovering photospheric velocities from magnetograms by combining the induction equation with local correlation tracking. *The Astrophysical Journal*, *610*(2), 1148–1156. https://doi.org/10.1086/421767

Welsch, B. T., Li, Y., Schuck, P. W., & Fisher, G. H. (2009). What is the relationship between photospheric flow fields and solar flares? *The Astrophysical Journal*, *705*(1), 821–843. https://doi.org/10.1088/0004-637X/705/1/821

Wiegelmann, T., Thalmann, J. K., Schrijver, C. J., De Rosa, M. L., & Metcalf, T. R. (2008). Can we improve the preprocessing of photospheric vector magnetograms by the inclusion of chromospheric observations? *Solar Physics*, *247*(2), 249–267. https://doi.org/10.1007/s11207-008-9130-y

Wiegelmann, T., & Sakurai, T. (2021). Solar force-free magnetic fields. *Living Reviews in Solar Physics*, *18*(1), 1. https://doi.org/10.1007/s41116-020-00027-4

Woltjer, L. (1958). A theorem on force-free magnetic fields. *Proceedings of the National Academy of Science*, *44*(6), 489–491. https://doi.org/10.1073/pnas.44.6.489

Yeates, A. R., & Page, M. H. (2018). Relative field-line helicity in bounded domains. *Journal of Plasma Physics*, *84*(6), 775840602. https://doi.org/10.1017/S0022377818001204

Yokoyama, T., Kusano, K., Maeshiro, T., & Sakurai, T. (2003). Relation between magnetic helicity injection and flare activities in active region NOAA 8100. *Advances in Space Research*, *32*(10), 1949–1952. https://doi.org/10.1016/S0273-1177(03)90631-4

7

Magnetic Helicity Measurements in the Solar Wind

Yasuhito Narita

ABSTRACT

The concept of magnetic helicity plays an important role in studies of solar wind turbulence. Various methods have been developed to evaluate magnetic helicity, some for multipoint measurements and others for single-point measurements. Observational studies show increasing evidence that the helical sense of the magnetic field changes the sign nearly randomly around zero helicity in the spectral domain on spatial scales where the picture of magnetohydrodynamic turbulence is considered valid (typically at 1,000 km or more). On short-wavelength, ion-kinetic scales where length is on the order of the ion gyroradius or inertial length (100–1,000 km or less), the helical sense of the magnetic field tends to shift to nonzero values. Nonzero helicity in the ion-kinetic domain is the signature of linear-mode wave excitation, such as kinetic Alfvén waves and whistler waves. There is also a more systematic trend in which the helicity changes sign between the northern and southern hemispheres, and a weak trend in which helicity evolves in the same fashion as the solar cycle. Extensive studies of magnetic helicity in the solar wind are expected to lead to a breakthrough in solving transport problems related to the Sun, such as the amount of helicity, magnetic flux, and energy.

7.1. INTRODUCTION

The concept of magnetic helicity has drawn increasing attention from researchers in the field of solar wind turbulence over the last four decades. In ideal magnetohydrodynamics (MHD), total energy and magnetic and cross helicities are known as rugged invariants (Biskamp, 2003). As such, they determine the final state of nondissipative dynamics toward which the system evolves. The final state determines how the energy cascade of each quantity can predict things such as the power spectrum in dissipative MHD. Like fluctuation energies (magnetic energy, kinetic energy), fluctuation helicities (magnetic helicity, kinetic helicity, and cross helicity) and electromotive force are second-order statistical quantities in MHD and are observationally derived from covariance matrices (in the spatial or temporal domain) or spectral matrices (in the spectral domain). While information about magnetic energy is contained in the diagonal elements in the covariance matrix, information about magnetic helicity is contained in the off-diagonal elements of the matrix. Hence, naively speaking, the MHD turbulence system can be determined when both the energies and helicities are set. Another quantity closely related to magnetic helicity is current helicity, which plays an important role in dynamo mechanisms and solar magnetism studies (Yokoi, 2006, 2013).

One formulation of magnetic helicity is that it is a measure of the asymmetry of the energy between a right-hand spatially rotating field and a left-hand field. The concept of helicity is a spatial definition. See, for example, Stix (1992) and Gary (1993) about the difference between spatial rotation (measured by helicity) and temporal rotation (measured by polarization). In contrast to magnetic helicity, cross-helicity can be

Institute for Theoretical Physics, Technical University of Braunschweig, Braunschweig, Germany

Helicities in Geophysics, Astrophysics, and Beyond, Geophysical Monograph 283, First Edition.
Edited by Kirill Kuzanyan, Nobumitsu Yokoi, Manolis K. Georgoulis, and Rodion Stepanov.

DOI:10.1002/9781119841715.ch07

formulated to measure the energy asymmetry between parallel-propagating components (Alfvén waves, for example) and antiparallel-propagating components.

The solar wind is a natural laboratory for studying the detailed evolution of a collisionless plasma into turbulence by exciting various kinds of fluctuations such as linear-mode waves, nonlinear waves, pumped waves, eddies, and coherent structures. Observational access to magnetic helicity is often limited, mostly due to the nature of the single-point measurements in space lacking in finding the spatial gradients. Helicity can nevertheless be determined from observations: for example, by incorporating Taylor's frozen-in-flow hypothesis (Taylor, 1938) and an assumption of homogeneity and integrating the 3D spectrum over the two directions perpendicular to the flow velocity (Matthaeus and Smith, 1981; Matthaeus et al., 1982; Matthaeus and Goldstein, 1982). Evaluating helicity then involves converting the measured temporal field rotation around the flow direction into the spatial rotation advected by the flow. When the mean magnetic field is aligned with the flow direction, the evaluated helicity is equivalent to the polarization (the temporal field rotation sense about the mean magnetic field).

This chapter reviews the observational studies of magnetic helicity in the turbulent solar wind. While the interplanetary magnetic field inherits the helical structures of the solar magnetic field to some extent, turbulence excited in the solar wind plays an important role in transporting and distributing helicity over different scales—for example, by turbulent spectral transport and the inverse cascade. After reviewing the large-scale helicity distribution for the Parker solar wind model (section 7.2), the methods of evaluating magnetic helicity are presented (section 7.3). The helicity profile obtained from the solar wind studies is reviewed in section 7.4. The chapter concludes with a summary and an outlook for future helicity study directions (section 7.5). The emphasis of the chapter is on turbulent solar wind. Of course, large-scale magnetic helicity is nonzero in solar transient events such as magnetic clouds, flux ropes, flux tubes, coronal mass ejections, and solar flares. Magnetic helicity is a useful quantity or tool for analyzing the magnetic field structure of those transient events (concisely reviewed in section 7.5).

7.2. THE PARKER MODEL AS A GUIDE

It is worth noting that the Parker spiral, the large-scale Archimedean winding magnetic field pattern in the solar wind, has nonzero values of magnetic helicity density, and the sign of the helicity depends on the hemisphere. The analytic form of the large-scale magnetic helicity for the Parker spiral field was obtained by Bieber et al. (1987a). The Parker field model is expressed as

$$B_r = B_{\odot r}\frac{R_\odot^2}{r^2}\,\mathrm{sgn}(\cos\theta) \tag{7.1}$$

$$B_\theta = 0 \tag{7.2}$$

$$B_\phi = -B_{\odot\phi}\frac{R_\odot}{r}\,\sin\theta\,\mathrm{sgn}(\cos\theta) \tag{7.3}$$

where r denotes the radial distance from the Sun, θ the heliographic latitude, ϕ the azimuthal (longitudinal) angle around the solar rotation axis, $R_\odot$ the solar radius, and $B_{\odot r}$ and $B_{\odot\phi}$ the radial and azimuthal components of the nominal solar surface magnetic field, respectively. The magnetic helicity density (hereafter, simply helicity) corresponding to equations (7.1)–(7.3) is

$$h_\mathrm{m}^\mathrm{P} = \boldsymbol{A}\cdot\boldsymbol{B} \tag{7.4}$$

$$= -\frac{B_{\odot r}B_{\odot\phi}R_\odot^3}{3r^2}\left[1+2|\cos\theta|\,\ln(1+|\cos\theta|)\right] \times \mathrm{sgn}(\cos\theta). \tag{7.5}$$

Equation (7.5) indicates that helicity decreases by r^2 at larger distances from the Sun. The profile of the Parker helicity density (equation (7.5)) can be obtained by normalizing into $\tilde{h}_\mathrm{m}^\mathrm{P}$ (or eliminating the surface field components ($B_{\odot r}$ and $B_{\odot\phi}$) and solar radial dependence involving $R_\odot$ and r) in the following fashion,

$$\tilde{h}_\mathrm{m}^\mathrm{P} = -\frac{1}{3}\left[1+2|\cos\theta|\,\ln(1+|\cos\theta|)\right]\,\mathrm{sgn}(\cos\theta). \tag{7.6}$$

which is graphically plotted in Fig. 7.1 as a function of heliographic latitude. The Parker field lines are confined in the plane perpendicular to the solar rotation axis. The helicity of the Parker spiral has a total value of zero when integrated over the whole solid angles. Magnetic helicity peaks at the solar poles and reverses

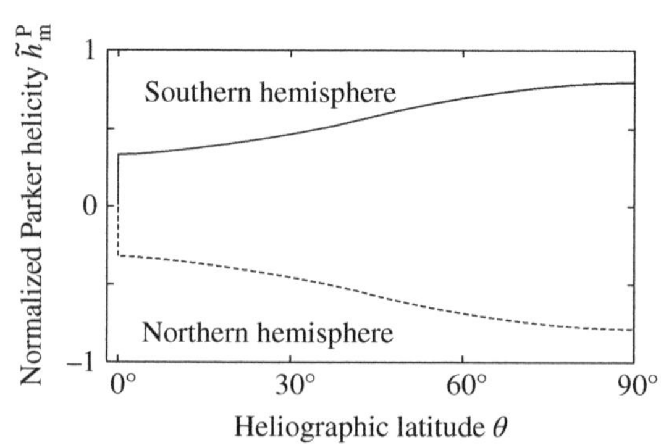

Figure 7.1 Normalized magnetic helicity density $\tilde{h}_\mathrm{m}^\mathrm{P}$ as a function of the heliographic latitude θ for the Parker solar wind model.

sign at the equator. The helicity sign reversal between the hemispheres was observed using Ulysses spacecraft data (Brandenburg et al., 2011). Podesta & Gary (2011) found the change in helicity at small scales associated with dissipation; the change is caused by reversal of the local mean magnetic field between the two hemispheres.

Parker spiral helicity is subject to distortion by large-scale processes such as stream interaction. Naively speaking, one may expect that large-scale helicity in interplanetary space inherits the helicity at the solar surface or coronal region. On intermediate to smaller scales (compared to the size of the Parker spiral), nonlinear turbulent processes affect helicity, such as the helicity in the energy-containing range (at spacecraft-frame frequencies below 10^{-4} Hz.

7.3. EVALUATING MAGNETIC HELICITY

7.3.1. Uncurling the Vector Potential

It is straightforward to evaluate magnetic helicity in the Fourier domain, making extensive use of spectral analysis founded by Matthaeus & Smith (1981). By uncurling the vector potential $\boldsymbol{a}$ (in the Fourier domain) under the Coulomb gauge $\nabla \cdot \boldsymbol{A} = 0$ or, equivalently, $\mathrm{i}\boldsymbol{k} \cdot \boldsymbol{a} = 0$ when assuming a set of plane waves in a homogeneous background such as $\boldsymbol{A} = \boldsymbol{A}_0 + \boldsymbol{a}\exp(\mathrm{i}\boldsymbol{k} \cdot \boldsymbol{r})$, the fluctuating vector potential $\boldsymbol{a}$ can then be explicitly expressed as a combination of the wave vector $\boldsymbol{k}$ and the associated magnetic field amplitude $\boldsymbol{b}(\boldsymbol{k})$:

$$\boldsymbol{a} = \frac{\mathrm{i}}{k^2}\boldsymbol{k} \times \boldsymbol{b}(\boldsymbol{k}). \tag{7.7}$$

We now compute the scalar product with the magnetic field (with the averaging denoted by $\langle \cdots \rangle$) and obtain the magnetic helicity:

$$\langle \boldsymbol{a}^\dagger \cdot \boldsymbol{b} \rangle = -\frac{\mathrm{i}}{k^2} \sum_i^{x,y,z} \sum_j^{x,y,z} \sum_k^{x,y,z} \epsilon_{ijk} k_i \langle b_j^* b_k \rangle \tag{7.8}$$

$$= -\frac{\mathrm{i}}{k^2} \left[k_x \left(\langle b_y^* b_z \rangle - \langle b_z^* b_y \rangle \right) + k_y \left(\langle b_z^* b_x \rangle - \langle b_x^* b_z \rangle \right) + k_z \left(\langle b_x^* b_y \rangle - \langle b_y^* b_x \rangle \right) \right], \tag{7.9}$$

where k is the magnitude of the wave vector, $k = |\boldsymbol{k}|$, and the epsilon symbol ϵ_{ijk} denotes the Levi-Civita tensor with the properties $\epsilon_{xyz} = \epsilon_{yzx} = \epsilon_{zxy} = 1$, $\epsilon_{zyx} = \epsilon_{xzy} = \epsilon_{yxz} = -1$, and $\epsilon_{ijk} = 0$ otherwise. The minus sign before the factor i/k^2 in equations (7.8) and (7.9) comes from the Hermitian conjugation of the vector potential. Equations (7.8) and (7.9) are gauge independent. Adding a gauge term $\nabla\Phi$ (with a scalar function Φ) in the vector potential ends up with a term $i\tilde{\Phi} \sum_i^{x,u,z} k_i B_i$ in the spectral domain (where $\tilde{\Phi}$ is the Fourier transform of Φ), which vanishes because of the divergence-free nature of the magnetic field, $\sum_i^{x,y,z} k_i B_i = 0$. Expressions in equations (7.8) and (7.9) can be used in multipoint observations where fluctuating magnetic fields can be evaluated directly in the wave vector domain. Equation (7.9) indicates that magnetic is a measure of asymmetry in the off-diagonal elements of the covariant or spectral matrix, in contrast to energy, which represents the diagonal elements of the spectral matrix. In the case of three-dimensional isotropic helical turbulence, the spectral matrix for the magnetic field can be explicitly decomposed into the energy term and the helicity term (Brandenburg et al., 2011).

The helicity evaluation (equation (7.9)) can readily be applied to observational data when the three-dimensional wave vector $\boldsymbol{k}$ and associated magnetic field amplitude (as a function of the wave vectors) are available. Such a three-dimensional helicity evaluation is possible when using multispacecraft data. It was shown, for example, that data from only four spacecraft in a tetrahedral array configuration is sufficient to evaluate magnetic helicity as a function of the frequencies and wave vectors (Narita et al., 2009). The wave vector was treated as a free parameter in the multispacecraft data analysis (Glassmeier et al., 2001), In many cases of in situ measurements in the solar wind, however, observational of magnetic helicity is limited to time-domain or frequency-domain analysis because of the lack of a spatial gradient in single-point measurements. The fluctuation amplitude (and phase) can only be determined as a function of the frequencies in the spacecraft frame, not as a function of the wave vectors. Still, observational access to magnetic helicity is possible in the one-dimensional and one-component sense when Taylor's frozen-in hypothesis (Taylor, 1938) is used and the helicity study is limited to the helical sense around the flow direction. Taylor's hypothesis relabels or maps the time stamps in the data into length scales along the flow, which is radial and outward from the Sun. In the spectral domain, the frequencies in the spacecraft frame are transformed into stream-wise wave numbers (i.e., wave numbers along the flow). The stream-wise wave numbers are linearly proportional to the frequencies (in the spacecraft frame) under Taylor's hypothesis, and the mapping into the stream-wise wave numbers k_R reads

$$k_\mathrm{R} \simeq \frac{\omega_\mathrm{sc}}{U_\mathrm{R}}. \tag{7.10}$$

where ω_sc is the spacecraft-frame (angular) frequency and U_R the flow speed in the stream direction. Here we use the notation for the RTN coordinate system spanning the radial direction from the Sun (R), the tangential direction (T), and the northward direction (N).

7.3.2. Reduced Magnetic Helicity

The one-dimensional magnetic helicity is as a function of the stream-wise wave numbers using Taylor's hypothesis. In equations (7.8) and (7.9), we highlight the helical field around the radial direction from the Sun and obtain the helicity as (Matthaeus and Smith, 1981; Matthaeus et al., 1982)

$$h_{\mathrm{m}} = -\mathrm{i}\frac{k_{\mathrm{R}}}{k^2}\left(\langle b_{\mathrm{T}}^* b_{\mathrm{N}}\rangle - \langle b_{\mathrm{N}}^* b_{\mathrm{T}}\rangle\right), \tag{7.11}$$

where b_{T} and b_{N} are the T and N components of the magnetic field, respectively. The wave numbers are assumed to be radial such that $k = k_{\mathrm{R}}$ holds. The difference in the brackets in equation (7.11), $\langle b_{\mathrm{T}}^* b_{\mathrm{N}}\rangle - \langle b_{\mathrm{N}}^* b_{\mathrm{T}}\rangle$, is purely imaginary if the two fluctuating components form a around the reference axis (around the direction of the R component). In the simplest case, one may set a sense in the T–N plane (right-hand spatially rotating for increasing values of r) with the unit amplitude, such as $b_{\mathrm{T}} = \exp\left[\mathrm{i}k_{\mathrm{R}}r\right]$ and $b_{\mathrm{N}} = \exp\left[\mathrm{i}\left(k_{\mathrm{R}}r - \pi/2\right)\right]$. The helicity component around the radial direction is then evaluated using equation (7.11): $h_{\mathrm{m}} = -\mathrm{i}\left(-\mathrm{i} - \mathrm{i}\right)/k_{\mathrm{R}} = -2/k_{\mathrm{R}}$.

A right-hand spatially rotating field around the R axis is characterized by a negative value of h_{m} when $k_{\mathrm{R}} > 0$ in equation (7.11).

Another useful representation of the helicity component around the k_i axis, $-\mathrm{i}k_i(\langle b_j^* b_k\rangle - \langle b_k^* b_j\rangle)/k^2$, is obtained when introducing two complex-number (or phase-extended) variables b_+ and b_- defined as

$$b_+ = \frac{1}{\sqrt{2}}\left(b_j + \mathrm{i}b_k\right) \tag{7.12}$$

$$b_- = \frac{1}{\sqrt{2}}\left(b_j - \mathrm{i}b_k\right). \tag{7.13}$$

Here, b_+ represents the in the b_j–b_k plane. The right-hand spatially rotating field, for example, is expressed by $b_j = \exp(\mathrm{i}k_i r_i)$ and $b_k = \exp(\mathrm{i}(k_i r_i - \pi/2))$, yielding $b_+ = \sqrt{2}\exp(\mathrm{i}k_i r_i)$ and $b_- = 0$. Likewise, b_- represents the left-hand spatially rotating field in the b_j–b_k plane. A prefactor of $1/\sqrt{2}$ comes from energy conservation such that $|b_+|^2 + |b_-|^2 = |b_j|^2 + |b_k|^2$ holds. Equation (7.11) is then expressed (when interpreting the j component as the T component and the k component as the N component) as the energy difference between the right-hand spatially rotating field $|b_+|^2$ and the left-hand spatially rotating field $|b_-|^2$:

$$h_{\mathrm{m}} = -\frac{k_{\mathrm{R}}}{k^2}\left(|b_+|^2 - |b_-|^2\right). \tag{7.14}$$

See also the Stokes V parameter method adapted from the research field of optics and spectroscopy (Comişel et al., 2016). Again, the dominance of the right-hand spatially rotating field results in a negative value of h_{m} due to the minus sign in equation (7.14).

In the observational field, since helicity extracts the antisymmetric part of the off-diagonal elements (e.g., $\langle b_{\mathrm{T}}^* b_{\mathrm{N}}\rangle$), one may refer directly to the imaginary part of the off-diagonal elements and multiply by a factor of 2 instead of computing the antisymmetric difference as in equation (7.11). The following formula is often used in the literature (Matthaeus et al., 1982; Podesta & Gary, 2011):

$$h_{\mathrm{m}} = \frac{2}{k_{\mathrm{R}}}\,\mathrm{Im}\left[\left\langle b_{\mathrm{T}}^* b_{\mathrm{N}}\right\rangle\right] \tag{7.15}$$

The helicity spectrum in the form of equation (7.15) has been widely applied to various single-point magnetic field data in the solar wind (Matthaeus et al., 1982; Smith, 2003; Podesta and Gary, 2011; Iovieno et al., 2016) such that the off-diagonal element $\left\langle b_{\mathrm{T}}^* b_{\mathrm{N}}\right\rangle$ is evaluated in the spacecraft-frame frequency domain and k_{R} is from the frequencies using Taylor's hypothesis.

Observational studies show that helicity values can be both positive and negative (around the flow direction) and oscillate between the two signs in a wider range of frequencies (see section 7.4 for details). The helicity magnitude falls off by nearly a power law as a function of the frequencies or stream-wise wave numbers. Naively speaking, when one applies the inertial-range energy spectrum for isotropic, homogeneous turbulence to the off-diagonal element of the spectral matrix—that is, $|\left\langle b_{\mathrm{T}}^* b_{\mathrm{N}}\right\rangle| \propto k_{\mathrm{R}}^{-5/3}$—the helicity magnitude falls off as $|h_{\mathrm{m}}| \propto k^{-8/3}$, which is steeper than the inertial-range energy spectrum, as confirmed observationally by Matthaeus et al. (1982).

It is also useful to normalize the helicity (equation (7.11) or (7.15)) as σ_{m} in the spacecraft data analysis by referring to the magneticnetic energy (diagonal elements of the spectral matrix) at the same frequency or stream-wise wave number. Note that the units are different for helicity and energy. By compensating (i.e., multiplying) the helicity by the wave number, the normalized (or reduced) magnetic helicity is constructed as

$$\sigma_{\mathrm{m}} = \frac{k_{\mathrm{R}} h_{\mathrm{M}}}{\mathcal{E}_{\mathrm{m}}} \tag{7.16}$$

$$-1 \leq \sigma_{\mathrm{m}} \leq 1, \tag{7.17}$$

where $\mathcal{E}_{\mathrm{m}}$ denotes the trace of the

$$\mathcal{E}_{\mathrm{m}} = \langle b_{\mathrm{R}}^2\rangle + \langle b_{\mathrm{T}}^2\rangle + \langle b_{\mathrm{N}}^2\rangle. \tag{7.18}$$

In the practical case, normalized helicity can be evaluated first in the frequency domain as

$$\sigma_{\mathrm{m}} = \frac{2\mathrm{Im}\left[\left\langle b_{\mathrm{T}}^* b_{\mathrm{N}}\right\rangle\right]}{\langle b_{\mathrm{R}}^2\rangle + \langle b_{\mathrm{T}}^2\rangle + \langle b_{\mathrm{N}}^2\rangle}, \tag{7.19}$$

or by limiting the analysis to the two field components (the T and N components) as

$$\sigma_{\mathrm{m}} = \frac{2\mathrm{Im}\left[\langle b_{\mathrm{T}}^{*} b_{\mathrm{N}}\rangle\right]}{\langle b_{\mathrm{T}}^{2}\rangle + \langle b_{\mathrm{N}}^{2}\rangle}. \tag{7.20}$$

Then helicity is mapped onto the stream-wise wave numbers using the flow speed.

7.3.3. Caveats

The measured polarization depends on the reference frame, but helicity does not. Polarization is a temporal measurement, whereas helicity is a spatial measurement. When properly measured, helicity is Galilean invariant to the reference frame. However, the direction of the wave vector needs to be known or assumed when evaluating helicity in the spectral domain. For example, the assumption of radially outward wave vectors from the Sun must be evaluated by other means.

The applicability or validity of Taylor's hypothesis has been studied extensively by various means, including deviation from the frozen-in-flow condition due to non-negligible intrinsic frequencies and propagation speeds (Howes et al., 2014), inaccuracy due to random sweeping by large-scale flow variation and counter-propagating Alfvén waves (Narita, 2017; Perez, 2021), inaccuracy through accumulating errors caused by fluctuating flow velocity (Treumann et al., 2019), and decorrelation effects (Bourouaine and Perez, 2018).

Turbulence studies based on single-spacecraft and multispacecraft data analyses shows that magnetic field fluctuations are associated with perpendicular wave vectors to the mean or large-scale magnetic field in a wider range of frequencies, from magnetohydrodynamic scales (1,000 km and greater) down to ion-kinetic scales (100–1,000 km and less) (Bieber et al., 1996; Leamon et al., 1998b; Hamilton et al., 2008; Narita, 2011; Pine et al., 2020b). Care must be exercised with the assumption of radially pointing wave vectors when using the single-point method; one-dimensional helicity spectra (along the flow) obtained from the solar wind intervals or samples with larger Parker spiral angles are more representative of the full three-dimensional spectrum.

From the point of view of using Taylor's hypothesis, when combined with the observational conclusion that wave vectors are nearly perpendicular to the large-scale magnetic field in the solar wind, the reduced helicity is more representative of the 3D helicity at larger distances from the Sun (where the large-scale magnetic field makes a larger angle to the flow direction according to the Parker spiral structure). The reduced spectrum is an integration across the two directions perpendicular to the flow. While the integration does move the signal to higher wave numbers, the reduced helicity is more representative of the 3D helicity because the radial component of the wave vector becomes more representative of the 3D wave vector perpendicular to the large-scale field.

On the other hand, magnetic field fluctuations in the solar wind are incompressible, so the fluctuation appears mostly perpendicular to the large-scale field. From the point of view of studying the helical sense of the magnetic field, the reduced helicity analysis is representative of the three-dimensional helicity at smaller distances from the Sun (where the large-scale magnetic field is more radial) because the off-diagonal element $\langle b_{\mathrm{T}} * b_{\mathrm{N}}\rangle$ and its Hermitian conjugate are the dominating elements in the spectral matrix.

7.4. HELICITY PROFILE IN THE SOLAR WIND

7.4.1. Reduced Helicity Spectra

The solar wind is considered a nearly ideal laboratory for studying the role of magnetic helicity in collisionless plasma, particularly plasma evolution into a turbulent state. The solar wind also exhibits highly helical structures associated with flux ropes and coronal mass ejections. Although not as extensively studied as magnetic energy spectra, reduced magnetic helicity has been observed in various solar-wind regions. Helicity measurements in the ecliptic plane were performed using data from Voyager (Matthaeus et al., 1982; Iovieno et al., 2016), STEREO (Podesta and Gary, 2011; Podesta et al., 2013; He et al., 2011), the Advanced Composition Explorer (ACE) (Smith, 2003; Hamilton et al., 2008), Wind (Leamon et al., 1998b; Dasso et al., 2003, 2006; Markovskii et al., 2016; Vasquez et al., 2018; Zhao et al., 2021), Cluster, Helios (Carbone and Bruno, 1997; Carbone et al., 2010), and the Parker Solar Probe (Pecora et al., 2021). Measurements out of the ecliptic plane (high-latitude regions) were performed using Ulysses data (Brandenburg et al., 2011; Klein et al., 2014). Magnetic helicity is also studied using numerical simulations (Stribling et al., 1995; Vasquez et al., 2022).

Observational studies of magnetic helicity in the solar wind began with analyzing Voyager data at a distance of about 2.8 AU (astronomical units) from the Sun (Matthaeus et al., 1982). The magnetic helicity was in the spacecraft-frame frequency ranges from about 10^{-5} Hz to about 10^{-2} Hz. It was found that magnetic helicity (around the radial direction from the Sun) randomly changes sign around the zero mean and that the helicity magnitude falls off like a power-law spectrum with a spectral index of about $-8/3$, in correspondence with the inertial-range energy spectrum with an index of $-5/3$.

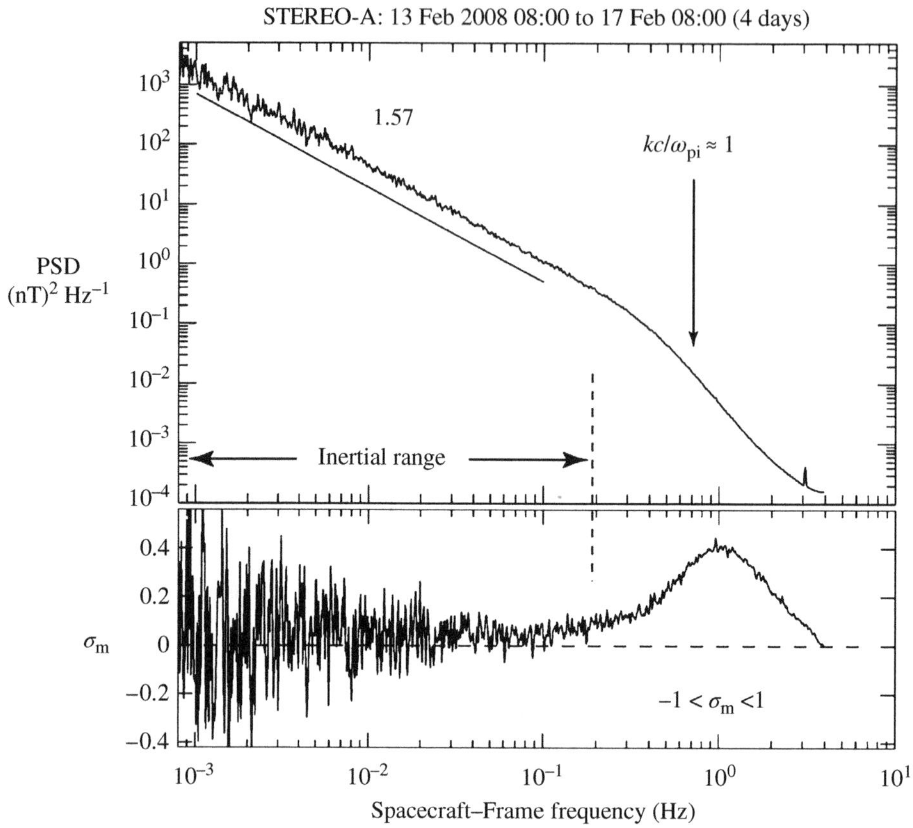

Figure 7.2 Power spectrum (top panel) and reduced magnetic helicity spectrum (bottom panel) in the frequency domain. Magnetic field data from STEREO-A spacecraft. Podesta et al., 2013/Springer Nature.

Reduced helicity analysis has been applied extensively to turbulent fluctuations in the solar wind (Matthaeus et al., 1982; Matthaeus and Goldstein, 1982; Gary, 1986; Leamon et al., 1998a; Goldstein et al., 1994; Hamilton et al., 2008; Howes and Quataert, 2010). Figure 7.2 displays the magnetic energy spectrum (top panel) and the reduced helicity spectrum (bottom panel) in the spacecraft-frame frequency domain, covering both the intermediate-wavelength fluid scale (below 10^{-1} Hz) and the short-wavelength ion-kinetic scale (above 10^{-1} Hz). The spectra are derived from STEREO-A magnetic field data in the solar wind at a distance of about 1 AU from the Sun (Podesta et al., 2013). Pine et al. (2020a) extended the analysis of helicity associated with the inertial and dissipation ranges in the Voyager data to a greater heliocentric distance and found that reduced helicity increases (systematically deviates from zero) as the dissipation range spectral index steepens.

In three dimensions, magnetic helicity is a signed quantity, so the scaling behavior needs to be treated adequately. Magnetic helicity fluctuates quasirandomly around zero in a turbulent environment. However, some studies have reported that the scaling of cancellation between the positive and negative parts follows a well-defined scaling law. The scaling of cancellations is characterized by the cancellation exponent (Carbone and Bruno, 1997; Carbone et al., 2010).

The frequencies are proportional to the stream-wise wave numbers when applying Taylor's hypothesis, $\omega_{sc} \simeq \boldsymbol{k} \cdot \boldsymbol{U}_0$, where ω_{sc} is the spacecraft-frame frequency, $\boldsymbol{k}$ the wave vector, and $\boldsymbol{U}_0$ the mean flow velocity vector. The features of the energy spectrum and reduced helicity are as follows:

- The fluid-scale energy spectrum is well-fitted by a power law with an index close to that of the inertial-range spectrum $-5/3$. The fluid-scale helicity changes sign around the zero mean, $\langle \sigma_m \rangle = 0$, and the helicity oscillation amplitude in the spectral domain becomes smaller at higher frequencies.
- The ion-kinetic energy spectrum exhibits a steeper spectrum than that of the fluid scale. The transition into the ion-kinetic range occurs at the scale of ion inertial length (the wave number of the ion inertial length is denoted by $kc/\omega_{pi} \approx 1$ in the figure at a frequency of about 7×10^{-1} Hz). The reduced helicity has

a systematic trend to the positive-value side of helicity, which peaks at $\sigma_m \sim 0.4$ at a frequency of 1 Hz.

Extended helicity studies by Podesta & Gary (2011), He et al. (2011), and Klein et al. (2014) using Ulysses data show the existence of two distinct populations of reduced helicity around the ion inertial length in the two-dimensional domain spanning the wave period (using the wavelet analysis) and the angle of the solar wind direction to the large-scale magnetic field. One is a population of positive-value helicities with larger angles (nearly 90° to the large-scale field), and the other are negative-value helicities with smaller angles (nearly 0° to the mean field).

7.4.2. Wave Picture of Magnetic Helicity

Various waves, structures, and fluctuation types may exist or coexist in turbulence. Interpretation of the linear-mode waves is presented as a "wave picture."

Five linear modes are relevant to ion-kinetic range turbulence when interpreting turbulent fields representing a set of linear-mode waves in plasma (the linear Vlasov model):

- The Alfvén wave group contains ion-cyclotron waves and kinetic Alfvén waves. Ion-cyclotron waves are left-hand polarized and damp at the ion-cyclotron frequency when propagating parallel to the mean magnetic field. As the propagation direction becomes increasingly oblique to the mean field (say, at angles of 70° to 80°, a transition occurs such that the ion-cyclotron waves flip the spatial field rotation and are less damped to the ion cyclotron resonance. The reversal of spatial field rotation is caused by a change in the electric current (or electric field) contributing to wave motion from Hall current (yielding left-hand polarization) to diamagnetic current (yielding right-hand polarization) (Narita et al., 2020).
- The fast-mode wave group contains whistler and ion-Bernstein waves. In contrast to ion-cyclotron waves, whistler waves are right-hand polarized (when propagating parallel to the mean magnetic field) and can thus resonate with electrons. A transition occurs at larger propagation angles, and whistler waves become ion-Bernstein waves that resonate at the ion-cyclotron frequency and its harmonics. Ion-Bernstein waves are nearly linearly polarized yet can exhibit a moderate left-hand polarization at the wave vectors around the transition from whistler waves.
- Kinetic slow waves are short-wavelength extensions of slow magnetosonic mode (or long-wavelength extensions of ion-acoustic mode) (Zhao et al., 2014; Narita and Marsch, 2015). A kinetic slow wave is more electrostatic and nearly linearly polarized.

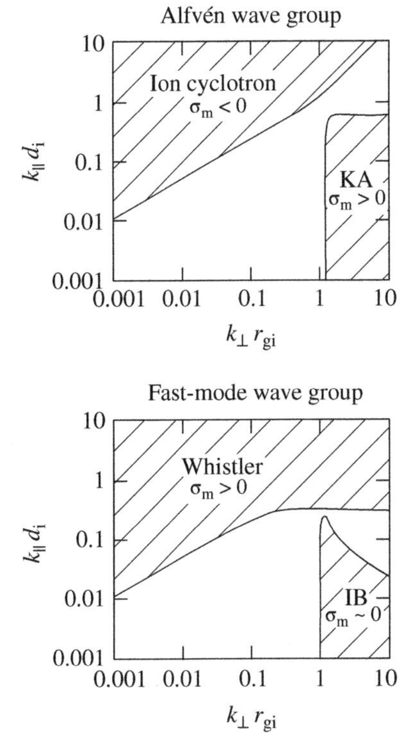

Figure 7.3 Schematic diagram of reduced magnetic helicity with clear signatures for the linear-mode waves in the Vlasov plasma model based on the calculation by Klein et al. (2014): ion-cyclotron waves, kinetic Alfvén waves (KA), whistler waves, and ion-Bernstein waves (IB). The perpendicular components of the wave vectors are normalized to the ion-gyroradius r_{gi}. The parallel components of the wave vectors are normalized to the ion inertial length d_i.

Figure 7.3 shows a schematic diagram of reduced helicity for the Alfvén wave group and the fast-mode wave group.

The signature of small-angle negative helicities is statistically weak or infrequent such that the averaging procedure over the angles masks that negative-helicity population, and only large-angle positive helicities are emphasized.

7.4.3. Scale and Hemispheric Dependencies

Large-scale magnetic helicity exhibits a more systematic trend. Brandenburg et al. (2011) analyzed Ulysses magnetic field data in high-latitude regions of the heliosphere from 1993 to 1996 and found sign reversal of reduced helicity occurring at a stream-wise wave number of about 2 AU^{-1} (about 10^{-8} km^{-1}) at distances less than 2.8 AU from the Sun and at a wave number of 30 AU^{-1} (about 10^{-7} km^{-1}) at larger distances. The transition is such that magnetic helicity turns positive on smaller scales (i.e., higher wave numbers) in the northern heliosphere and negative in the southern heliosphere. The sign reversal

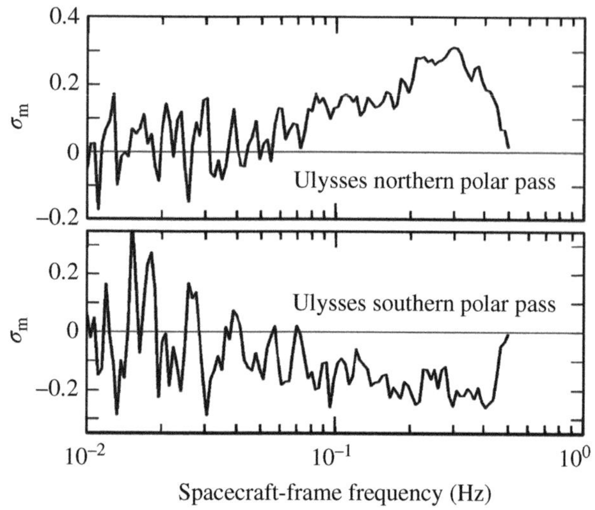

Figure 7.4 Reduced magnetic helicity in the solar polar regions evaluated from the Ulysses polar pass data. Data from Podesta and Gary, 2011.

is interpreted as the result of the reversal of the mean magnetic field between the two hemispheres. The transition or sign reversal of magnetic helicity occurring at a streamwise-wave number of about 10^{-8} km^{-1} is also found by analyzing Voyager-2 data (Iovieno et al., 2016).

Energy dissipation by ions is polarization dependent. Polarization in the plasma frame translates into helicity according to the mean-field direction (used to evaluate polarization as temporal rotation) and propagation direction (used to evaluate helicity as spatial rotation). The hemispheric dependence of the small-scale magnetic helicity sign is also found using Ulysses magnetic field data from 1993 to 1995 (Podesta and Gary, 2011). Figure 7.4 displays the reduced helicity in the spacecraft-frame frequency domain up to about 0.5 Hz for the northern polar pass (top panel) and the southern polar pass (bottom panel). The helicity sign is positive at frequencies above 0.1 Hz in the northern hemisphere and negative in the southern hemisphere. The radial component of the large-scale magnetic field in the solar wind was outward in the northern hemisphere (i.e., a positive sense of polarity) during the analyzed time interval. Again, the change of helicity sign is interpreted as the result of a change of mean magnetic field between the two hemispheres. The sign change in reduced helicity spectra associated with magnetic sectors in the ecliptic plane has also been reported by Goldstein et al. (1994), Leamon et al. (1998b), and He et al. (2011).

7.4.4. Solar Cycle Dependence

There are indications that magnetic helicity observed in the solar wind evolves with the solar cycle such that one may expect an oscillatory variation as demonstrated for net magnetic flux and helicity using solar magnetogram data by Berger and Ruzmaikin (2000).

The solar cycle dependence of helicity may be associated with how the turbulence level in the solar wind varies varies with the solar cycle, thereby changing the rate of dissipation and the strength of the dissipation signal. A caveat: the steepening of magnetic energy spectra may represent not only dissipation (energy conversion from magnetic field fluctuations into thermal energy of plasmas) but also dispersion (energy conversion from magnetic field fluctuations into electric field energy due to excitation of dispersive waves like whistler, kinetic Alfvén, or ion-Bernstein wave when limiting our interest to the linear-mode waves).

The top panel of Fig. 7.5 shows the variation of sunspot numbers from 1960 (the end of cycle 19) to 2000 (the beginning of cycle 23). Bieber et al. (1987b) presented the long-term evolution of magnetic helicity using hourly averaged magnetic field data at 1 AU for the interval from 1965 to 1984 (OMNItape data). Helicity is determined separately for the radially inward sector (toward-sector) and the radially outward sector (away-sector). The yearly averaged helicity is displayed in the middle panel of Fig. 7.5. The difference in helicity between the two sectors, defined as $\Delta H_{\mathrm{m}} = H_{\mathrm{t}}^{r-} - H_{\mathrm{m}}^{r+}$ using the helicity in the inward sector H_{m}^{r-} and that in the outward sector H_{m}^{r+}, is found to be time-dependent in

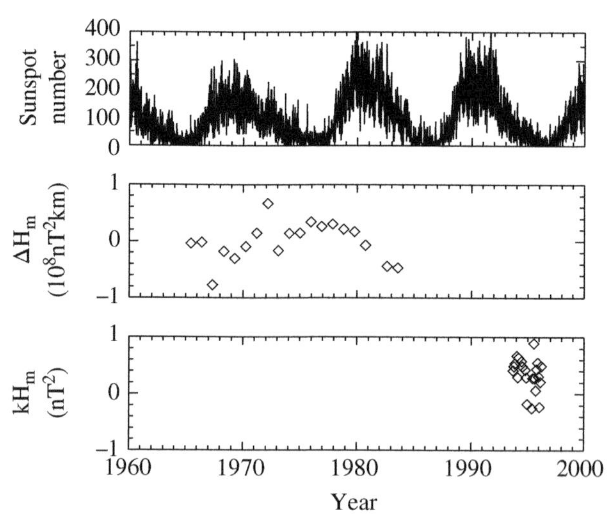

Figure 7.5 (top) Sunspot number variation. Data courtesy of the World Data Center SILSO, Royal Observatory of Belgium, Brussels. (middle) Difference in magnetic helicity between the two sectors (toward-sector and away-sector with respect to the Sun). Data from OMNItape from 1965 to 1984 (Bieber et al., 1987b). (bottom) Scale-compensated magnetic helicity at a length scale of 10^6 km. Data from Ulysses from 1993 to 1996 (Brandenburg et al., 2011).

the sense that the helicity in the inward sector dominates ($\Delta H_m > 0$) during the solar minimum and in the growing phase of the solar cycle (first half of the cycle), outward sector dominates ($\Delta H_m < 0$) in the decaying phase (second half) of the solar cycle. The asymmetric distribution of magnetic helicity between the two sectors (or the two hemispheres) may be related to cosmic ray modulation in the heliosphere. Brandenburg et al. (2011) determined the compensated magnetic helicity kH_m (compensated for the length scale by multiplying by the wave number k) for a shorter interval from 1993 to 1996 using Ulysses data in high-latitude regions (bottom panel of Fig. 7.5). There is a weak tendency for magnetic helicity at a scale of 10^6 km to decrease or even change sign toward the solar minimum. The trend is unclear at larger spatial scales (about 10^8 km).

7.5. SUMMARY AND OUTLOOK

Magnetic helicity and reduced helicity can be determined using spacecraft data and serve as useful tools to advance our understanding of large-scale solar wind structures and turbulent fluctuations and dissipation mechanisms therein, in association with the Sun. Helicity can be evaluated by integrating over the two perpendicular directions around the flow. Field rotation can be studied not only in terms of polarization (temporal rotation around the mean magnetic field) but also in terms of helicity (spatial rotation around the wave vector). The two concepts are related as shown by Smith et al. (1984) and Goldstein et al. (1984):

$$(\boldsymbol{B}_0)_z \sigma_m(k_z) \frac{\omega}{k_z} > 0 \qquad (7.21)$$

for right-hand polarization, and

$$(\boldsymbol{B}_0)_z \sigma_m(k_z) \frac{\omega}{k_z} < 0 \qquad (7.22)$$

for left-hand polarization. Here, ω/k_z is the component of phase velocity in the z direction. Of course, the assumption of Taylor's hypothesis—the assumption of homogeneous turbulence or isotropic turbulence— must also be justified or validated by other means. The reduced helicity method was applied to magnetic field data in the ecliptic plane and high-latitude regions.

Overall, reduced helicity in the solar wind has the following behaviors: (1) nonzero helicity on larger scales (but smaller than the scale of the Parker field), which presumably inherits the helicity profile of the solar surface (the Parker field helicity results from the solar dipolar field and the solar wind expansion); (2) zero-mean randomly oscillating helicity on the intermediate fluid scale (where turbulence is considered to be magnetohydrodynamic); and (3) deviation from zero-mean helicity in the short-wavelength, ion-kinetic range, with the possibility of linear-mode wave excitation.

Linear-mode waves may serve as sources and sinks of the helicity budget. The existence of left-hand polarized, ion-cyclotron waves has been indicated by the pitch-angle scattering of protons in Helios data (Marsch, 2018). Polarization reversal can occur within the Alfvén and fast-mode wave groups. Wave-wave couplings generate not only linear-mode waves but also sideband waves and nonlinear or pumped waves, and the polarization changes when the sign of the frequency or wave number changes. Perhaps the Alfvén and fast-mode wave groups develop independently into different cascades such that whistler waves survive on electron scales (about 10 km in interplanetary space).

Helical field structures are also observed in transient events such as flux ropes (Dasso et al., 2003), twisted flux tubes (Zaqarashvili et al., 2014), and solar flares (Gupta et al., 2021). Dasso et al. (2003) proposed that magnetic helicity (per unit length) is a useful quantity to reconstruct twisted magnetic fields observed by the Wind spacecraft using various theoretical models (the force-free field, non-force-free field, and uniformly twisted field models). A possible link of magnetic helicity with magnetic reconnection between the flux tube field and the solar wind magnetic field was also proposed by Dasso et al. (2006). The solar corona region also exhibits magnetic helicity sign reversal (Bourdin et al., 2018), which may spatially extend or convect to interplanetary space. Tthree spacecraft continue to measure the magnetic field in the inner heliosphere (Parker Solar Probe, Solar Orbiter, and BepiColombo's cruise), so it may be possible to map magnetic helicity from the solar wind down onto the solar corona or the solar surface.

Several ideas are proposed to improve the accuracy of evaluating magnetic helicity. First, it is helpful to decompose the turbulent field into the plus-helical component b_+ and the minus-helical component b_- in the helicity analysis. The representation of the helicity spectra is no longer oscillatory or discrete in the spectral domain, but the helicity spectra can be derived more continuously such that the spectral slope, spectral break, and cross-talk between the two spectra become more visible. Second, it is helpful to incorporate the results from multispacecraft data analyses into helicity studies. In particular, the direction of wave vectors is nearly perpendicular in a wide range of scales (both magnetohydrodynamic turbulence scales and ion-kinetic turbulence scales) (Narita, 2011). The perpendicular wave number can then be from single-point measurements as

$$k_\perp = \frac{\omega_{sc}}{U_R \sin\theta_{BU}}. \qquad (7.23)$$

Here, the stream-wise wave number (or Taylor-mapped wave number) $k_T = \omega_{sc}/U_R$ is inversely projected onto the perpendicular axis to the local mean magnetic field (in the plane spanning the mean field and flow direction). The projection of the perpendicular wave number $k_\perp$ onto the Taylor-mapped wave number k_T is expressed by $k_\mathrm{T} = k_\perp \cos\theta_{kU} = k_\perp \sin\theta_{BU}$ using the relation $\theta_{kU} = \frac{\pi}{2} - \theta_{BU}$, where θ_{kU} is the angle between the wave vector and the flow direction, and θ_{BU} is the angle between the large-scale magnetic field and the flow direction.

ACKNOWLEDGEMENTS

Discussions with U. Motschmann, G. Kleindienst, K.-H. Glassmeier, E. Marsch, Ph.-A. Bourdin, S. P. Gary, and N. Yokoi were greatly appreciated.

REFERENCES

Berger, M. A., & Ruzmaikin, A. (2000). Rate of helicity production by solar rotation. *Journal of Geophysical Research: Space Physics*, *105*, 10,481–10,490. https://doi.org/10.1029/1999JA900392

Bieber, J. W., Evenson, P., & Matthaeus, W. H. (1987a). Magnetic helicity of the Parker spiral. *The Astrophysical Journal*, *315*, 700–705. https://doi.org/10.1086/165171

Bieber, J. W., Evenson, P., & Matthaeus, W. H. (1987b). Magnetic helicity of the IMF and the solar modulation of cosmic rays. *Geophysical Research Letters*, *14*, 864–867.

Bieber, J. W., Wanner, W., & Matthaeus, W. H. (1996). Dominant two-dimensional solar wind turbulence with implications for cosmic ray transport. *Journal of Geophysical Research: Space Physics*, *101*, 2511–2522. https://doi.org/10.1029/95JA02588

Biskamp, D. (2003). *Magnetohydrodynamic turbulence*. Cambridge: Cambridge University Press.

Bourdin, P., Singh, N. K., & Brandenburg, A. (2018). Magnetic helicity reversal in the corona at small plasma beta. *The Astrophysical Journal*, *869*, 2. https://doi.org/10.3847/1538-4357/aae97a

Bourouaine, S., & Perez, J. C. (2018). On the limitations of Taylor's hypothesis in Parker Solar Probe's measurements near the Alfvén critical point. *The Astrophysical Journal*, *858*, L20. https://doi.org/10.3847/2041-8213/aabccf

Brandenburg, A., Subramanian, K., Balogh, A., & Goldstein, M. L. (2011). Scale dependence of magnetic helicity in the solar wind. *The Astrophysical Journal*, *734*, 9. https://doi.org/10.1088/0004-637X/734/1/9

Carbone, V., & Bruno, R. (1997). Sign singularity of the magnetic helicity from in situ solar wind observations. *The Astrophysical Journal*, *488*, 482–487. https://doi.org/10.1086/304670

Carbone, V., Perri, S., Yordanova, E., Veltri, P., Bruno, R., Khotyaintsev, Y., & André, M. (2010). Sign-singularity of the reduced magnetic helicity in the solar wind plasma. *Physical Review Letters*, *104*, 181101. https://doi.org/10.1103/PhysRevLett.104.181101

Comişel, H., Nariyuki, Y., Narita, Y., & Motschmann, U. (2016). On the role of ion-scale whistler waves in space and astrophysical plasma turbulence. *Annales Geophysicae*, *34*, 975–984. https://doi.org/10.5194/angeo-34-975-2016

Dasso, S., Mandrini, C. H., Démoulin, P., & Farrugia, C. J. (2003). Magnetic helicity analysis of an interplanetary twisted flux tube. *Journal of Geophysical Research*, *108*, 1362. https://doi.org/10.1029/2003JA009942

Dasso, S., Mandrini, C. H., Démoulin, P., & Luoni, M. L. (2006). A new model-independent method to compute magnetic helicity in magnetic clouds. *Astronomy & Astrophysics*, *455*, 349–359. https://doi.org/10.1051/0004-6361:20064806

Gary, S. P. (1986). Low-frequency waves in a high-beta collisionless plasma: polarization, compressibility and helicity. *Journal of Plasma Physics*, *35*, 431–447. https://doi.org/10.1017/S0022377800011442

Gary, S. P. (1993). *Theory of space plasma microinstabilities*. Cambridge: Cambridge University Press. https://doi.org/10.1017/CBO9780511551512

Glassmeier, K.-H., Motschmann, U., Dunlop, M., Balogh, A., Acu na, M. H., Carr, C., et al. (2001). Cluster as a wave telescope – first results from the fluxgate magnetometer. *Annales Geophysicae*, *19*, 1439–1447. https://doi.org/10.5194/angeo-19-1439-2001(correctionin21,1071,2003)

Goldstein, M. L., Smith, C. W., & Matthaeus, W. H. (1984). Correction to "Large amplitude MHD waves upstream of the Jovian bow shock," *Journal of Geophysical Research: Space Physics*, *89*, 9161. https://doi.org/10.1029/JA089iA10p09161

Goldstein, M. L., Roberts, D. A., & Fitch, C. A. (1994). Properties of the fluctuating magnetic helicity in the inertial and dissipation ranges of solar wind turbulence. *Journal of Geophysical Research: Space Physics*, *99*, 11,519–11,538. https://doi.org/10.1029/94JA00789

Goldstein, M. L., Wicks, R. T., Perri, S., & Sahraoui, F. (2015). Kinetic scale turbulence and dissipation in the solar wind: key observational results and future outlook. *Philosophical Transactions of the Royal Society A*, *373*, 20140147. http://dx.doi.org/10.1098/rsta.2014.0147

Guo, Z. B., Diamond, P. H., & Wang, X. G. (2012). Magnetic reconnection, helicity dynamics, and hyper-diffusion. *Astrophysical Journal*, *757*, 173. http://doi.org/10.1088/0004-637X/757/2/173

Guo, Y., Pariat, E., Valori, G., Anfinogentov, S., Chen, F., Georgoulis, M. K., et al. (2017). Magnetic helicity estimations in models and observations of the solar magnetic field. III. Twist number method. *Astrophysical Journal*, *840*, 40. https://doi.org/10.3847/1538-4357/aa6aa8

Gupta, M., Thalmann, J. K., & Veronig, A. M. (2021). Magnetic helicity and energy budget around large confined and eruptive solar flares. *Astronomy & Astrophysics*, *653*, A69. https://doi.org/10.1051/0004-6361/202140591

Hamilton, K., Smith, C. W., Vasquez, B. J., & Leamon, R. J. (2008). Anisotropies and helicities in the solar wind inertial and dissipation ranges at 1 AU. *Journal of Geophysical Research: Space Physics*, *113*, A01106. https://doi.org/10.1029/2007JA012559

He, J., Marsch, E., Tu, C., Yao, S., & Tian, H. (2011). Possible evidence of Alfvén-cyclotron waves in the angle distribution of magnetic helicity of solar wind turbulence. *Astrophysical Journal*, *731*, 85. https://doi.org/10.1088/0004-637X/731/2/85

Hoshino, M., & Goldstein, M. L. (1989). Time evolution from linear to nonlinear stages in magnetohydrodynamic parametric instabilities. *Physics of Fluids B*, *1*, 1405–1415. https://doi.org/10.1063/1.858971

Howes, G. G., & Quataert, E. (2010). On the interpretation of magnetic helicity signatures in the dissipation range of solar wind turbulence. *Astrophysical Journal Letters*, *709*, L49. https://doi.org/10.1088/2041-8205/709/1/L49

Howes, G. G., Klein, K. G., & TenBarge, J. M. (2014). Validity of the Taylor hypothesis for linear kinetic waves in the weakly collisional solar wind. *Astrophysical Journal*, *789*, 106. https://doi.org/10.1088/0004-637X/789/2/106

Iovieno, M., Gallana, L., Fraternale, F., Richardson, J. D., Opher, M., & Tordella, D. (2016). Cross and magnetic helicity in the outer heliosphere from Voyager 2 observations. *European Journal of Mechanics - B/Fluids*, *55*, 394–401. https://doi.org/10.1016/j.euromechflu.2015.08.009

Klein, K. G., Howes, G. G., TenBarge, J. M., & Podesta, J. J. (2014). Physical interpretation of the angle-dependent magnetic helicity spectrum in the solar wind: The nature of turbulent fluctuations near the proton gyroradius scale. *Astrophysical Journal*, *785*, 138. https://doi.org/10.1088/0004-637X/785/2/138

Leamon, R. J., Matthaeus, W. H., Smith, C. W., & Wong, H. K. (1998a). Contribution of cyclotron-resonant damping to kinetic dissipation of interplanetary turbulence. *Astrophysical Journal*, *507*, L181–L184. https://doi.org/10.1086/311698

Leamon, R. J., Smith, C. W., Ness, N. F., Matthaeus, W. H., & Wong, H. K. (1998b). Observational constraints on the dynamics of the interplanetary magnetic field dissipation range. *Journal of Geophysical Research: Space Physics*, *103*, 4775–4787. https://doi.org/10.1029/97JA03394

Markovskii, S. A., Vasquez, B. J., & Smith, C. W. (2016). The effect of electron thermal pressure on the observed magnetic helicity in the solar wind. *Astrophysical Journal*, *833*, 212. https://doi.org/10.3847/1538-4357/833/2/212

Marsch, E. (2018). Solar wind and kinetic heliophysics. *Annals of Geophysics*, *36*, 1607–1630. https://doi.org/10.5194/angeo-36-1607-2018

Matthaeus, W. H. and Smith, C. W. (1981). Structure of correlation tensors in homogeneous anisotropic turbulence. *Physical Review A*, *24*, 2135. https://doi.org/10.1103/PhysRevA.24.2135

Matthaeus, W. H., Goldstein, M. L., & Smith, C. (1982). Evaluation of magnetic helicity in homogeneous turbulence. *Physical Review Letters*, *48*, 1256–1259. https://doi.org/10.1103/PhysRevLett.48.1256

Matthaeus, W. H., & Goldstein, M. L. (1982). Measurement of the rugged invariants of magnetohydrodynamic turbulence in the solar wind. *Journal of Geophysical Research*, *87*, 6011–6028. https://doi.org/10.1029/JA087iA08p06011

Narita, Y., Kleindienst, G., & Glassmeier, K.-H. (2009). Evaluation of magnetic helicity density in the wave number domain using multi-point measurements in space. *Annals of Geophysics*, *27*, 3967–2976. https://doi.org/10.5194/angeo-27-3967-2009

Narita, Y., Gary, S. P., Saito, S., Glassmeier, K.-H., & Motschmann, U. (2011). Dispersion relation analysis of solar wind turbulence. *Geophysical Research Letters*, *38*, L05101. https://doi.org/10.1029/2010GL046588

Narita, Y., & Marsch, E. (2015). Kinetic slow mode in the solar wind and its possible role in turbulence dissipation and ion heating. *Astrophysical Journal*, *805*, 24. https://doi.org/10.1088/0004-637X/805/1/24

Narita, Y. (2017). Error estimate of Taylor's frozen-in flow hypothesis in the spectral domain. *Annals of Geophysics*, *35*, 325–331. https://doi.org/10.5194/angeo-35-325-2017

Narita, Y., Roberts, O. W., Vörös, Z., & Hoshino, M. (2020). Transport ratios of the kinetic Alfvén mode in space plasmas. *Frontiers in Physics*, *8*, 166. https://doi.org/10.3389/fphy.2020.00166

Pecora, F., Servidio, S., Greco, A., & Matthaeus, W. H. (2021). Identification of coherent structures in space plasmas: the magnetic helicity–PVI method. *Astronomy and Astrophysics*, *650*, A20. https://doi.org/10.1051/0004-6361/202039639

Perez, J. C., Bourouaine, S., Chen, C. H. K., & Raouafi, E. (2021). Applicability of Taylor's hypothesis during Parker Solar Probe perihelia. *Astronomy and Astrophysics*, *650*, A22. https://doi.org/10.1051/0004-6361/202039879

Pine, Z. B., Smith, C. W., Hollick, S., Argall, M. R., Vasquez, B. J., Isenberg, P. A., et al. (2020). Solar wind turbulence from 1 to 45 au. I. Evidence for dissipation of magnetic fluctuations using Voyager and ACE observations. *Astrophysical Journal*, *900*, 91. https://doi.org/10.3847/1538-4357/abab10

Pine, Z. B., Smith, C. W., Hollick, S., Argall, M. R., Vasquez, B. J., Isenberg, P. A., et al. (2020). Solar wind turbulence from 1 to 45 au. III. Anisotropy of magnetic fluctuations in the inertial range using Voyager and ACE observations. *Astrophysical Journal*, *900*, 93. https://doi.org/10.3847/1538-4357/abab11

Podesta, J. J., & Gary, S. P. (2011). Magnetic helicity spectrum of solar wind fluctuations as a function of the angle with respect to the local mean magnetic field. *Astrophysical Journal*, *734*, 15. https://doi.org/10.1088/0004-637X/734/1/15

Podesta, J. J. (2013). Evidence of kinetic Alfvén waves in the solar wind at 1 AU. *Solar Physics*, *286*, 529. https://doi.org/10.1007/s11207-013-0258-z

Smith, C. W., Goldstein, M. L., Matthaeus, W. H., & Vi nas, A. F. (1984). Correction to "Turbulence analysis of the Jovian upstream 'wave' phenomenon," *Journal of Geophysical Research*, *89*, 9159–9160. https://doi.org/10.1029/JA089iA10p09159

Smith, C. W. (2003). Magnetic helicity in the solar wind. *Advances in Space Research*, *32*, 1971–1980. https://doi.org/10.1016/S0273-1177(03)00770-1

Stix, T. H. (1992). *Waves in plasmas*. Melville, NY: American Institute of Physics.

Stribling, T., Matthaeus, W. H., & Oughton, S. (1995). Magnetic helicity in magnetohydrodynamic turbulence with a mean magnetic field. *Physics of Plasmas*, *2*, 1437–1452. https://doi.org/10.1063/1.87135

Taylor, G. I. (1938). The spectrum of turbulence. *Proceedings of the Royal Society of London A*, *164*, 476–490. https://doi.org/10.1098/rspa.1938.0032

Treumann, R. A., Baumjohann, W., & Narita, Y. (2019). On the applicability of Taylor's hypothesis in streaming magnetohydrodynamic turbulence. *Earth, Planets and Space*, *71*, 41. https://doi.org/10.1186/s40623-019-1021-y

Vasquez, B. J., Markovskii, S. A., & Smith, C. W. (2018). The turbulence magnetic helicity signature in the interplanetary medium: A Blackman-Tukey and Morlet wavelet analysis. *Astrophysical Journal*, *855*, 121. https://doi.org/10.3847/1538-4357/aaad0d

Vasquez, B. J., Markovskii, S. A., & Smith, C. W. (2022). Three-dimensional hybrid simulation results of a variable magnetic helicity signature at proton kinetic scales. *Astrophysical Journal*, *924*, 41. https://doi.org/10.3847/1538-4357/ac3bbc

Wiegelmann, T., & Büchner, J. (2001). Evolution of magnetic helicity in the course of kinetic magnetic reconnection. *Nonlinear Processes in Geophysics*, *8*, 127–140. https://doi.org/10.5194/npg-8-127-2001

Yeates, A. R., & Hornig, G. (2016). The global distribution of magnetic helicity in the solar corona. *Astronomy and Astrophysics*, *594*, A98. https://doi.org/10.1051/0004-6361/201629122

Yeates, A. R., & Hornig, G. (2017). The global distribution of magnetic helicity in the solar corona (Corrigendum). *Astronomy and Astrophysics*, *603*, C2. https://doi.org/10.1051/0004-6361/201629122e

Yokoi, N. (2006). Modeling of the turbulent magnetohydrodynamic residual-energy equation using a statistical theory. *Physics of Plasmas*, *13*, 062306. https://doi.org/10.1063/1.2209232

Yokoi, N. (2013). Cross helicity and related dynamo. *Geophysical & Astrophysical Fluid Dynamics*, *107*, 114–184. https://doi.org/10.1080/03091929.2012.754022

Zaqarashvili, T. V., Vörös, Z., Narita, Y., & Bruno, R. (2014). Twisted magnetic flux tubes in the solar wind. *Astrophysical Journal Letters*, *783*, L19. https://doi.org/10.1088/2041-8205/783/1/L19

Zhao, J. S., Voitenko, Y., Yu, M. Y., Lu, J. Y., & Wu, D. J. (2014). Properties of short-wavelength oblique Alfvén and slow waves. *Astrophysical Journal*, *793*, 107. https://doi.org/10.1088/0004-637X/793/2/107

Zhao, G. Q., Lin, Y., Wang, X. Y., Feng, H. Q., Wu, D. J., Li, H. B., et al. (2021). Magnetic helicity signature and its role in regulating magnetic energy spectra and proton temperatures in the solar wind. *Astrophysical Journal*, *906*, 123. https://doi.org/10.3847/1538-4357/abca3b

8

Magnetic Helicity in Rotating Neutron Stars

Maxim Dvornikov

ABSTRACT

We study the evolution of magnetic helicity in neutron stars (NSs). First, we analyze how the surface terms affect the conservation law for the sum of the chiral imbalance of charged particle densities and the density of magnetic helicity. Our results are applied to a system with a finite volume, which can be a magnetized NS. We show that the contribution of these surface terms can lead to the reconnection of magnetic field lines followed by X-ray or gamma bursts in magnetars. Comparing the new quantum surface term with the classical contribution known in standard magnetic hydrodynamics, we find that it dominates the classical term only for NSs with rigid rotation. Second, we study the dynamics of chiral electrons, which interact electroweakly with background fermions having a velocity, arbitrarily depending on spatial coordinates and nonuniform density. We derive the kinetic equations and effective action for right and left particles. We obtain the correction to the Adler anomaly and anomalous electric current for rotating matter. Then we apply the results to magnetic helicity flow inside a magnetized rotating NS. We compute the characteristic time of the helicity change and demonstrate that it coincides with the magnetic cycle period of certain pulsars.

8.1. INTRODUCTION

Studies of various phenomena taking place in systems of chiral fermions, apart from a theoretical interest, have applications in astroparticle physics, cosmology, and solid state and accelerator physics. By chiral media, we mean situations when the temperature or chemical potential is much greater than the particle mass.

The chiral magnetic effect (CME) is the generation of anomalous electric current along an external magnetic field. Such a current can exist only in matter with a nonzero chiral imbalance: that is, the chemical potentials of left and right fermions should be different. We also recall that an ultrarelativistic fermion has a certain projection of its spin $\mathbf{s}$ on the particle momentum $\mathbf{p}$. If $\mathbf{s} \uparrow\uparrow \mathbf{p}$, we use the term *right particles*. The opposite situation of $\mathbf{s} \uparrow\downarrow \mathbf{p}$ corresponds to *left fermions*. An anomalous current $\mathbf{j} \parallel \mathbf{B}$ in a vacuum is forbidden by parity conservation in electrodynamics. The CME was first predicted by Vilenkin (1980a) and then reconsidered by Fukushima et al. (2008).

The next important chiral phenomenon is the chiral vortical effect (CVE): the excitation of an electric current along the vorticity of matter in a rotating system of ultrarelativistic fermions. The CVE is also forbidden in a vacuum. The existence of the CVE was also predicted theoretically by Vilenkin (1980b). We also recall the chiral separation effect (CSE), which produces an axial current along an external magnetic field. Other interesting chiral phenomena are summarized in the review by Kharzeev et al. (2016).

Kharzeev et al. (2016); Zhao and Wang (2019) suggested that one can explain the asymmetry in collisions

Pushkov Institute of Terrestrial Magnetism, Ionosphere and Radiowave Propagation (IZMIRAN), Russian Academy of Sciences, Moscow, Russia

Helicities in Geophysics, Astrophysics, and Beyond, Geophysical Monograph 283, First Edition.
Edited by Kirill Kuzanyan, Nobumitsu Yokoi, Manolis K. Georgoulis, and Rodion Stepanov.

DOI:10.1002/9781119841715.ch08

of heavy ions based on the CME and CVE. The CME and CSE were reported to be observed in magnetized atomic gases (Huang, 2016) and in Weyl and Dirac semimetals (Armitage et al., 2018).

There are numerous applications of chiral phenomena in cosmology. For instance, Boyarsky et al. (2012) showed that the evolution of magnetic fields in a primordial plasma is strongly affected by the quantum chiral anomaly at temperatures $T > 10\,\mathrm{MeV}$. Boyarsky et al. (2012) also found that the magnetic helicity density in plasma of the early universe becomes long scale after the electroweak phase transition. This process leads to the production of lepton asymmetry in primordial plasma cooling to $T \sim 10\,\mathrm{MeV}$. Giovannini and Shaposhnikov (1998) assumed that a plasma before the electroweak phase transition ($T > 100\,\mathrm{GeV}$) can be permeated by a primordial hypermagnetic field. This field can be transformed to the fermionic number densities due to the anomalous coupling. The influence of primordial magnetic fields on the electroweak phase transition was analyzed. The production of baryon asymmetry in some specific configurations of the magnetic field was studied.

There are applications of the CME and CVE in astrophysics. They are related to the possible explanation for strong magnetic fields in some compact stars. Kaspi and Beloborodov (2017) showed that magnetic fields with strength $B > 10^{15}\,\mathrm{G}$ are observed in magnetars. Yamamoto (2016) accounted for the chirality of neutrinos in conventional hydrodynamics and kinetic theory. It was suggested that a dense gas of chiral neutrinos could be important in generating a strong helical magnetic field in a core-collapsing supernova. Yamamoto (2016) provided a mechanism for transforming the gravitational energy of a collapsing star to electromagnetic energy and explained strong magnetic fields in magnetars. Masada et al. (2018) used chiral magnetohydrodynamics to describe the macroscopic evolution of relativistic charged matter with a chiral imbalance. Real-time evolution of magnetic fields and the properties of chiral MHD turbulence were studied numerically. The inverse cascade of magnetic energy was observed. The application of the results for generating strong magnetic fields in magnetars was discussed by Masada et al. (2018). Kaminski et al. (2016) studied neutrino emission along the vorticity or magnetic field in dense matter of a neutron star (NS), taking into account the effects of the anomalous hydrodynamics. The results were applied to account for the great linear velocities $\sim 10^3\,\mathrm{km \cdot s^{-1}}$ observed in some pulsars. Sigl and Leite (2016) examined the effect of the chiral instability of the spectra of magnetic energy density and helicity density during the early cooling phase of a hot protoneutron star. The generation of a magnetic field with strength close to that observed in magnetars was predicted. The density fluctuations and turbulence spectrum of matter were found to affect the generation of the magnetic field. The scale of the magnetic field was obtained from the submillimeter to centimeter range.

Dvornikov and Semikoz (2015a,b,c) suggested that the instability of a magnetic field caused by electroweak correction to the CME leads to the generation of strong fields observed in magnetars. The analogous mechanism was used by Dvornikov (2016a, 2018) in quark stars. Dvornikov (2020) accounted for the electroweak interaction between chiral electrons and background matter in a rotating NS. Dvornikov and Semikoz (2018, 2019) studied the evolution of magnetic helicity and magnetic fields in NSs with inhomogeneous matter using anomalous magnetohydrodynamics (AMHD) and applied the results to various problems of magnetars. Dvornikov et al. (2020) applied the CME and CSE in the nonuniformly heated medium of a protoneutron star to explain the strong magnetic fields of magnetars.

In this chapter, we continue our previous research on the evolution of magnetic fields in compact stars. We tackle two main problems. First, we examine how AMHD is modified when we account for a nonzero electron mass in a magnetized plasma. Second, we study the corrections to the Adler-Bell-Jackiw anomaly due to the electroweak interaction of ultrarelativistic electrons with rotating background matter. Both problems are connected to the evolution of magnetic helicity in magnetized compact stars.

One assumes zero electron mass, $m = 0$, to support the CME based on the anomalous current (Fukushima et al., 2008) (see equations (8.7) and (8.49), as well as works by Brandenburg et al. (2017), Boyarsky et al. (2012)) in the AMHD. Note that the particles are massive, $m_e \neq 0$, in realistic cosmological and astrophysical media, such as in the ultrarelativistic degenerate electron gas in a supernova, where $p_{\mathrm{F}_e} \gg m_e$, or in hot primordial plasma of the early universe with $T \gg m_e$. Dvornikov (2016b) demonstrated that this results in very rapid washing out of a chiral imbalance, $\mu_5 \to 0$, because of the spin-flip in particle collisions.

If we consider the nonzero mass $m_e \neq 0$ for left and right electrons or positrons, the CME is no longer valid. Nevertheless, Ioffe (2006) showed that the Adler-Bell-Jackiw anomaly still takes place since it is also valid for massive particles. Thus, the standard MHD and AMHD are compatible. The compatibility of MHD and AMHD takes place under the following additional condition: a change in the magnetic helicity within a domain volume is given by the new quantum effect consisting of the nonzero flux of the averaged spin through the surface of that domain. The evolution of magnetic helicity accounts for this new term, which can be comparable to known helicity losses at the surface in classical MHD (Priest, 2016). We recall that the evolution of the magnetic helicity at a stellar surface is important for reconfiguring magnetic field lines there. It can lead to efficient conversion of magnetic energy into kinetic and thermal energies of plasma, causing the strong electromagnetic emission of magnetars (Kaspi and Beloborodov, 2017).

As mentioned earlier, in this chapter we also study the behavior of ultrarelativistic electrons in plasma, accounting for their electroweak interaction with nonrelativistic background matter composed of protons and neutrons. The electroweak interaction is treated within the Fermi approximation. We assume that the characteristics of this matter, such as the macroscopic velocity and number density, can depend on spatial coordinates. We take that the electron gas is degenerate: that is, the Fermi momentum of electrons $p_{\mathrm{F}e}$ is much greater than the mass of an electron m_e. The existence of such a system is possible in NSs. One can apply the chiral phenomena to such fermions since these particles are ultrarelativistic: $p_{\mathrm{F}e} \gg m_e$. The mixing of left and right fermions mentioned previously depends on the spin-flip rate in particle collisions. Anomalous electric currents of protons are negligible since protons are nonrelativistic in NSs. Hydrodynamic currents of neutrinos, which are ultrarelativistic particles, can receive an anomalous contribution. However, the fluxes of neutrinos are small for an old NS studied here. Thus, we neglect the neutrino contribution.

This chapter is organized as follows. First, in section 8.2.1, following Dvornikov and Semikoz (2018), we recall some known results for spin magnetization in plasma and formulas describing the electron density in the relativistic degenerate electron gas of a magnetized NS. Then, in section 8.3.1, we derive from the Adler-Bell-Jackiw anomaly the surface terms for the modified CME accounting for all spin terms, including that originated by the pseudoscalar ($\sim \bar{\psi}\gamma_5\psi$). Here, $\gamma_5 = \mathrm{i}\gamma^0\gamma^1\gamma^2\gamma^3$ and $\gamma^\mu = (\gamma^0, \boldsymbol{\gamma})$ are the Dirac matrices. In section 8.3.2, we derive the new magnetic helicity evolution equation, where, neglecting chiral anomaly $n_{\mathrm{R}} - n_{\mathrm{L}} = 0$ due to the spin flip, we find the contribution of the mean spin flux through the boundary of the volume to the dynamics of the magnetic helicity known in classical MHD. In section 8.3.3, we compare estimates of magnetic helicity losses given by classical MHD (Priest, 2016) and our new (quantum) contribution given by that mean spin flux.

Then, in section 8.2.2, based on the results of Dvornikov (2020), we formulate the dynamics of chiral fermions in dense inhomogeneous matter, accounting for the electroweak interaction. The evolution of chiral fermions was described approximately based on the Berry phase evolution. We find the effective actions and kinetic equations for chiral fermions in section 8.2.2. Then, in section 8.2.3, we derive the corrections to the anomalous electric current and the Adler anomaly from the electroweak interaction with inhomogeneous matter. Finally, we apply our results in section 8.3.4 to describe the evolution of magnetic helicity in a rotating NS and compare our prediction for the period of the magnetic cycle with astronomical observations. We conclude in section 8.4.

8.2. THEORETICAL BACKGROUND

In this section, we summarize the theoretical background necessary for applying chiral phenomena to the evolution of magnetic fields in NSs. First, in section 8.2.1, we calculate the mean spin of the relativistic electron gas in an external magnetic field. Using the Berry phase approximation in section 8.2.2, we derive the kinetic equation for electrons in rotating matter, accounting for the electroweak interaction between fermions. Based on the results of section 8.2.2, we derive the correction to the electric currents and the Alder anomaly in section 8.2.3.

8.2.1. Mean Spin of the Electron Gas in an External Magnetic Field

Here, we recall the basic properties of plasma in an external magnetic field. The motion of a relativistic charged particle in a magnetic field obeys the Dirac equation. The energy levels of a 1/2-spin fermion with electric charge q in an external magnetic field $\mathbf{H} = (0, 0, H)$ have the form (Berestetskiĭ et al., 1998, pp. 121–122)

$$\mathcal{E}(p_z, n, \lambda) = \sqrt{m_e^2 + p_z^2 + |q|H(2n+1) - qH\lambda}, \quad (8.1)$$

where p_z is the conserved projection of the fermion momentum along the magnetic field, $n = 0, 1, \ldots$ is the discrete main quantum number, and $\lambda = \pm 1$ is the eigenvalue of the matrix $\Sigma_z = \gamma^5\gamma^0\gamma^3$, which appears in the squared Dirac equation. (Recall that the commutator of the matrix Σ_z with the Hamiltonian is nonzero for a moving Dirac particle (Berestetskiĭ et al., 1998, p. 86). However, we use the quantity λ in the energy levels in equation (8.1)). Negatively charged particles (electrons) should have $q = -e$ and positively charged ones (positrons) $q = +e$. Here, $e > 0$ is the absolute value of the elementary charge.

Consider $e^\pm$ plasma in the external magnetic field. Semikoz and Valle (1997) found that the main Landau level with $n = 0$ contributes the mean spin of electrons and positrons, which has the form

$$\mathcal{S} = \langle \psi_e^\dagger \boldsymbol{\Sigma} \psi_e \rangle_0 = -\frac{2e\mathbf{H}}{(2\pi)^2}$$

$$\times \int_0^\infty \mathrm{d}p_z \left\{ \frac{1}{\exp\left[(\sqrt{p_z^2 + m_e^2} - \mu_e)/T\right] + 1} \right.$$

$$-\frac{1}{\exp\left[(\sqrt{p_z^2+m_e^2}+\mu_e)/T\right]+1}\Bigg\}, \tag{8.2}$$

where ψ_e is the bispinor exactly accounting for the contribution of the external magnetic field, $\mathbf{\Sigma} = \gamma^5\gamma^0\boldsymbol{\gamma}$, μ_e is the chemical potential of relativistic electrons, and T is the temperature of plasma.

We neglect the electron mass in a chiral plasma, $m_e \to 0$. Thus, the integral in the last line is proportional to the chemical potential μ_e. This result does not depend on the temperature. Therefore, the mean spin in a relativistic electron-positron plasma takes the form

$$S = -\frac{e\mu_e\mathbf{H}}{2\pi^2} = -\left(\frac{\alpha_{em}}{\pi}\right)\left(\frac{2\mu_e}{e}\right)\mathbf{H}, \tag{8.3}$$

where $\alpha_{em} = e^2/4\pi \approx 7.3\times10^{-3}$ is the fine structure constant.

We can also consider a nonrelativistic degenerate electron gas, such as forming plasma in a metal. In this situation, using the chemical potential $\mu_e = m_e + p_{F_e}^2/2m_e$ in equation (8.2), one obtains the paramagnetic magnetization term $\mathbf{M} = -\mu_B S$

$$\frac{\mathbf{M}}{\mu_B} = \frac{ep_{F_e}}{2\pi^2}\mathbf{H}, \tag{8.4}$$

where $\mu_B = e/2m_e > 0$ is the Bohr magneton. (Here, we account for the negative magnetic moment of an electron: $-\mu_B$). This result means the time-independent paramagnetic susceptibility is equal to the known value $\chi = \alpha_{em}v_{F_e}/\pi \ll 1$, where $v_{F_e} = p_{F_e}/m_e$ is the Fermi velocity. This leads to the standard definitions $\mathbf{M} = \chi\mathbf{H}$ and $\mathbf{B} = \mu\mathbf{H} = (1+\chi)\mathbf{H}$ (Landau and Lifshitz, 1980, see equation (59.5) on p. 173). Here, μ is the magnetic permeability of the electron gas. The approximation $\mathbf{B} \approx \mathbf{H}$ remains valid with a good accuracy since $\chi \ll 1$.

Only electrons, which are on the main Landau level $n = 0$, contribute to the mean spin:

$$S = -n_0\hat{\mathbf{n}}_B, \quad \hat{\mathbf{n}}_B = \frac{\mathbf{B}}{B}, \quad n_0 = \frac{eBp_{F_e}}{2\pi^2}. \tag{8.5}$$

This fact is valid for a degenerate electron gas, both ultrarelativistic, as in equation (8.3), and nonrelativistic, as in equation (8.4). Nunokawa et al. (1997) found that n_0 is part of the total electron density in the degenerate electron gas in equation (8.5),

$$n_e = \frac{eBp_{F_e}}{2\pi^2} + \sum_{n=1}^{n_{max}}\frac{2eB\sqrt{p_{F_e}^2 - 2eBn}}{2\pi^2}, \tag{8.6}$$

where the sum is up to a maximal value $n_{max} = [p_{F_e}^2/(2eB)]$ and $[\dots]$ remains for the integer part of $p_{F_e}^2/(2eB)$.

The sum in equation (8.6) is vanishing for strong magnetic fields, $2eB > p_{F_e}^2$. Thus all electrons populate the main Landau level, $n_e = n_0$. Then, if we consider ultrarelativistic or massless electrons, the total electric current becomes anomalous. It is the manifestation of the CME (Fukushima et al., 2008; Kharzeev et al., 2016),

$$\mathbf{j} = \alpha\mathbf{B} = \frac{\alpha_{em}\mu_5}{\pi}\mathbf{B}, \tag{8.7}$$

where $\mu_5 = (\mu_R - \mu_L)/2$ is the chiral imbalance.

The generation of the mean spin along the external field in equation (8.5) is identical to the CSE (Kharzeev et al., 2016), which consists of the excitation of the axial current along an external magnetic field, $\langle\mathbf{j}_5\rangle = \overline{\psi}\boldsymbol{\gamma}\gamma_5\psi \propto \mathbf{B}$. Using the algebra of the Dirac matrices, one gets that $\langle\mathbf{j}_5\rangle = \psi^\dagger\mathbf{\Sigma}\psi = S$.

Only electrons on the main Landau level contribute to the CME's vector current and the CSE's axial current. However, Vilenkin (1980a) found that for the CME, these electrons should be ultrarelativistic. Vilenkin (1980a) also demonstrated that the susceptibility of the CME is proportional to the chiral imbalance, as represented in equation (8.7). Thus, the anomalous vector current $\mathbf{j}$ in equation (8.7) due to the CME vanishes for massive electrons in the equilibrium when $\mu_R = \mu_L$.

This current in equation (8.7) is known to produce a forceless magnetic field, $\mathbf{j}\times\mathbf{B} = 0$. The explicit form of such a field in three dimensions, which is the solution of the equation $\nabla\times\mathbf{B} = \alpha\mathbf{B}$, was found by Chandrasekhar and Kendall (1957).

If we deal with a magnetic field obeying $eB > p_{F_e}^2/2$, a realistic assumption $p_{F_e} \sim 100\,\text{MeV}$ is hard to implement in the core of a NS. This is because the magnetic field, necessary for the condition $n_e = n_0$, is strong: $B > 10^4 B_{cr}/2 \sim 2.2\times10^{17}\,\text{G}$. Here, $B_{cr} = m_e^2/e = 4.4\times10^{13}\,\text{G}$ is the critical magnetic field.

We can take a moderately strong magnetic field, $m_e^2 \ll 2eB \ll p_{F_e}^2$. This situation can be implemented in magnetars, where $B \sim 10^{15}\,\text{G}$. Both electrons and protons in a NS have $p_{F_e} = p_{F_p} \sim 100\,\text{MeV}$ in a electroneutral plasma, $n_e = n_p$. In this case, the sum for Landau levels $1 \le n \le n_{max}$ in equation (8.6) makes a greater contribution to the total current than to the anomalous current. Therefore, the usual transverse components $\mathbf{B} \perp \mathbf{j}$ become greater, leading to a Lorentz force. This means the CME is negligible since the population of the main Landau level is small compared to the total (Nunokawa et al., 1997):

$$n_e \approx \frac{p_{F_e}^3}{3\pi^2}\left[1 + \frac{3eB}{2p_{F_e}^2}\right]. \tag{8.8}$$

The correction to the electron number density from the magnetic field in equation (8.8) results in the dependence of the chemical potential on the magnetic field; also see equation (8.2). (We demonstrate shortly that such a small correction leads to the very important term for magnetic helicity dissipation in the rigid rotation of a magnetar).

8.2.2. Approximate Description of Chiral Plasma Based on Berry Phase Evolution

In this section, we study the motion of a chiral fermion in background matter, accounting for the electroweak interaction. We will add the contribution of the electromagnetic field later. One has to consider the Dirac equation for an electron in the form

$$i\dot{\psi} = \left[(\boldsymbol{\alpha}\cdot\hat{\mathbf{p}}) + V_R^\mu \gamma^0\gamma_\mu P_R + V_L^\mu \gamma^0\gamma_\mu P_L\right]\psi, \quad (8.9)$$

where $\hat{\mathbf{p}} = -i\nabla$ is the operator of the electron momentum, ψ is the wave function of an electron, $\boldsymbol{\alpha} = \gamma^0\boldsymbol{\gamma}$ are the Dirac matrices, $P_{L,R} = (1 \pm \gamma^5)/2$ are the projectors to the chiral states, and $V^\mu_{L,R} = V^\mu_{L,R}(\mathbf{x}) = (V^0_{L,R}, \mathbf{V}_{L,R})$ are the potentials of the interaction of the left and right chiral states with the inhomogeneous matter.

Dvornikov and Semikoz (2015a) found the explicit expressions for $V^0_{L,R} \sim G_F n_f$ for electrons. Here, n_f is the density of background fermions and $G_F = 1.17 \times 10^{-5}\,\text{GeV}^{-2}$ is the Fermi constant. Dvornikov and Studenikin (2002) derived that $\mathbf{V}_{L,R} = V^0_{L,R}\mathbf{v}$ for arbitrarily moving unpolarized background matter. Here, $\mathbf{v}(\mathbf{x})$ is the velocity of plasma, in which all background fermions are taken to move as a whole. We can consider the situation when the density of background matter is also a function of coordinates $n_f = n_f(\mathbf{x})$. Since electroweak forces violate parity, then $V^\mu_R \neq V^\mu_L$.

We write ψ in the form of two chiral projections $\psi^T = (\psi_R, \psi_L)$ if we adopt the chiral representation for the Dirac matrices. Using equation (8.9), one obtains the wave equations for $\psi_{L,R}$,

$$i\dot{\psi}_{L,R} = \left[\mp(\boldsymbol{\sigma}\cdot\hat{\mathbf{p}}) + V^0_{L,R} \pm (\boldsymbol{\sigma}\cdot\mathbf{V}_{L,R})\right]\psi_{L,R}, \quad (8.10)$$

where $\boldsymbol{\sigma}$ are the Pauli matrices.

One cannot find the solution of equation (8.10) if the effective potentials arbitrarily depend on coordinates, $V^\mu_{L,R}(\mathbf{x})$. However, we can apply the quasiclassical method, based on the concept of the Berry phase (Berry, 1984), to analyze the evolution of $\psi_{L,R}$. For this purpose, we replace the field theory approach with classical mechanics. Namely, we take that particles move along certain trajectories. Thus we should choose canonical coordinates, $\{\mathbf{x}(t), \mathbf{p}(t)\}$. These coordinates can be taken in numerous ways. Therefore, we replace the wave functions $\psi_{L,R}(\mathbf{x}, t)$, depending on time t and coordinates $\mathbf{x}$, with effective functions depending only on t, $\psi_{L,R}(t)$: $\psi_{L,R}(\mathbf{x}, t) \to \exp(i\Phi)\psi_{L,R}(t)$. The dependence on the canonical coordinates is in the phase $\Phi = \Phi(\mathbf{x}, \mathbf{p})$, which is not yet fixed.

Based on equation (8.10), we find that the new spinors $\psi_{L,R} = \psi_{L,R}(t)$ obey the following equation:

$$i\frac{d}{dt}\psi_{L,R} = H_{L,R}\psi_{L,R}, \quad H_{L,R} = V^0_{L,R} \mp (\boldsymbol{\sigma}\boldsymbol{\Pi}), \quad (8.11)$$

where $\boldsymbol{\Pi} = \nabla\Phi - \mathbf{V}_{L,R}$. The Hamiltonians $H_{L,R}$ in equation (8.11) depend only on time since we assume that $\mathbf{x} = \mathbf{x}(t)$.

Now we take that $H_{L,R}$ change adiabatically. In this situation, Vinitskiĭ et al. (1990) found that

$$\psi_{L,R}(t) = \exp(-i\Theta_{L,R})u_{L,R}, \quad (8.12)$$

where $\Theta_{L,R} = \Theta_{L,R}(t)$ is the phase introduced by Berry (1984). In our case, the Berry phase differs for left and right electrons. In equation (8.12), $u_{L,R}$ are the constant spinors. Following Vinitskiĭ et al. (1990), we impose constraints on $\psi_{L,R}$,

$$\psi^\dagger_{L,R}\dot{\psi}_{L,R} = 0, \quad \psi^\dagger_{L,R}\psi_{L,R} = 1, \quad (8.13)$$

which fix their amplitudes and phases.

Vinitskiĭ et al. (1990) suggested solving the Schrödinger equation

$$H_{L,R}u_{L,R}(\xi) = E_{L,R}u_{L,R}(\xi) \quad (8.14)$$

to find $\Theta_{L,R}$. In equation (8.14), the canonical variable $\xi_a = \{\mathbf{x}, \mathbf{p}\}$, $a = 1, \ldots, 6$, is taken to be constant: $\xi_a = \text{const}$. The solution of equation (8.14) has the form

$$E_{L,R} = |\boldsymbol{\Pi}| + V^0_{L,R}, \quad u_L = \begin{pmatrix} -e^{-i\varphi/2}\sin\frac{\theta}{2} \\ e^{i\varphi/2}\cos\frac{\theta}{2} \end{pmatrix},$$

$$u_R = \begin{pmatrix} e^{-i\varphi/2}\cos\frac{\theta}{2} \\ e^{i\varphi/2}\sin\frac{\theta}{2} \end{pmatrix}, \quad (8.15)$$

where the angles θ and φ determine the direction of the vector $\boldsymbol{\Pi}$.

Using the results of Vinitskiĭ et al. (1990) and equations (8.11)–(8.14), we obtain that the Berry phases $\Theta_{L,R}$ satisfy the equation

$$\dot{\Theta}_{L,R} = -iu^\dagger_{L,R}\frac{\partial u_{L,R}}{\partial\xi_a}\dot{\xi}_a. \quad (8.16)$$

Vinitskiĭ et al. (1990) found that the solution of equation (8.16) has the form

$$\Theta_{L,R}(t) = \int_C A_a^{(L,R)}(\xi)d\xi_a, \quad (8.17)$$

where C is the trajectory of an electron in the phase space and

$$\mathrm{i}A_a^{(\mathrm{L,R})} = u_{\mathrm{L,R}}^\dagger \frac{\partial u_{\mathrm{L,R}}}{\partial \xi_a} \tag{8.18}$$

is the Berry connection.

Using two three-dimensional vectors, we rewrite the Berry connection in the form $A_a^{(\mathrm{L,R})} = (\mathcal{B}_{\mathrm{L,R}}, \mathcal{A}_{\mathrm{L,R}})$. Based on equation (8.18), we get that

$$\mathcal{B}_i^{(\mathrm{L,R})} = -\mathrm{i}u_{\mathrm{L,R}}^\dagger \frac{\partial u_{\mathrm{L,R}}}{\partial x_i} = -\mathrm{i}u_{\mathrm{L,R}}^\dagger \frac{\partial u_{\mathrm{L,R}}}{\partial \Pi_j}\frac{\partial \Pi_j}{\partial x_i},$$
$$\mathcal{A}_i^{(\mathrm{L,R})} = -\mathrm{i}u_{\mathrm{L,R}}^\dagger \frac{\partial u_{\mathrm{L,R}}}{\partial p_i} = -\mathrm{i}u_{\mathrm{L,R}}^\dagger \frac{\partial u_{\mathrm{L,R}}}{\partial \Pi_j}\frac{\partial \Pi_j}{\partial p_i}. \tag{8.19}$$

Thus, we replace the integration along the particle trajectory with the time integration. Finally, we rewrite equation (8.17) as

$$\Theta_{\mathrm{L,R}}(t) = \int_{t_0}^{t} \left(\mathcal{B}_{\mathrm{L,R}}\dot{\mathbf{x}} + \mathcal{A}_{\mathrm{L,R}}\dot{\mathbf{p}}\right) \mathrm{d}t$$
$$= \int_{t_0}^{t} \mathcal{A}_{\Pi j}^{(\mathrm{L,R})} \left[\frac{\partial \Pi_j}{\partial p_i}\dot{p}_i + \frac{\partial \Pi_j}{\partial x_i}\dot{x}_i\right] \mathrm{d}t, \tag{8.20}$$

where

$$\mathrm{i}\mathcal{A}_{\Pi j}^{(\mathrm{L,R})} = u_{\mathrm{L,R}}^\dagger \frac{\partial u_{\mathrm{L,R}}}{\partial \Pi_j}, \tag{8.21}$$

which stems from equation (8.19).

According to equation (8.12), the wave function evolves as $\psi_{\mathrm{L,R}}(t) \sim \exp(-\mathrm{i}\Theta_{\mathrm{L,R}})$. If we recall that the spinors $u_{\mathrm{L,R}}$ correspond to the energies $E_{\mathrm{L,R}}$ in equation (8.15), as well as accounting for equation (8.20), we can introduce the effective energies

$$E_{\mathrm{eff}} = \mathcal{A}_{\Pi j}^{(\mathrm{L,R})} \left[\frac{\partial \Pi_j}{\partial p_i}\dot{p}_i + \frac{\partial \Pi_j}{\partial x_i}\dot{x}_i\right] + E_{\mathrm{L,R}}. \tag{8.22}$$

Now we can consider the electromagnetic interaction of chiral electrons. We assume that their electric charge is e, and one has the external electromagnetic field with the vector potential $A^\mu = (A_0, \mathbf{A})$. Based on equation (8.22), we get the effective actions for right and left electrons:

$$S_{\mathrm{eff}}^{(\mathrm{L,R})}[\mathbf{x},\mathbf{p}] = \int_{t_0}^{t} \mathrm{d}t \left\{\dot{\mathbf{x}}\left[\mathbf{p} + e\mathbf{A}(\mathbf{x})\right]\right.$$
$$- \mathcal{A}_{\Pi j}^{(\mathrm{L,R})} \left[\frac{\partial \Pi_j}{\partial p_i}\dot{p}_i + \frac{\partial \Pi_j}{\partial x_i}\dot{x}_i\right]$$
$$\left. - \epsilon_{\mathrm{L,R}}(\mathbf{x}) - eA_0(\mathbf{x})\right\}, \tag{8.23}$$

where, following Son and Yamamoto (2013), we account for the dependence of the electron energy on coordinates $\epsilon_{\mathrm{L,R}}(\mathbf{x})$.

We have not yet specified the canonical coordinates and $\mathbf{\Pi}$. We take $\{\mathbf{x},\mathbf{p}\}$ to have $\mathbf{\Phi} = (\mathbf{px})$. In this situation,

$$\mathbf{\Pi} = \mathbf{p} - \mathbf{V}_{\mathrm{L,R}}, \quad \frac{\partial \Pi_k}{\partial p_i} = \delta_{ik}, \quad \frac{\partial \Pi_k}{\partial x_i} = -\frac{\partial V_k^{(\mathrm{L,R})}}{\partial x_i}, \tag{8.24}$$

and $\epsilon_{\mathbf{P}}^{(\mathrm{L,R})}(\mathbf{x}) = V_{\mathrm{L,R}}^0 + |\mathbf{P}| + \epsilon_0$. Here, $\epsilon_0 = \epsilon_0(\mathbf{x})$ due to the coordinate dependence of the electromagnetic field.

The effective action in equation (8.23) has the form

$$S_{\mathrm{eff}}^{(\mathrm{L,R})} = \int_{t_0}^{t} \mathrm{d}t \left\{\dot{\mathbf{x}}\left[\mathbf{p} + e\mathbf{A}(\mathbf{x})\right] - \mathcal{A}_{\mathrm{L,R}}\left[\dot{\mathbf{p}} - (\dot{\mathbf{x}}\nabla)\mathbf{V}_{\mathrm{L,R}}\right]\right.$$
$$\left. - \epsilon_{\mathbf{P}}^{(\mathrm{L,R})}(\mathbf{x}) - eA_0(\mathbf{x})\right\}, \tag{8.25}$$

where the index $\mathbf{P}$ in the Berry connection is omitted: $\mathcal{A}_{\mathbf{P}}^{(\mathrm{L,R})} \to \mathcal{A}_{\mathrm{L,R}}$. Because $\dot{\mathbf{P}} = \dot{\mathbf{p}} - (\dot{\mathbf{x}}\nabla)\mathbf{V}_{\mathrm{L,R}}$, we can recast equation (8.25) as

$$S_{\mathrm{eff}}^{(\mathrm{L,R})}[\mathbf{x},\mathbf{P}] = \int_{t_0}^{t} \mathrm{d}t \left\{\dot{\mathbf{x}}\left[\mathbf{P} + e\mathbf{A}_{\mathrm{eff}}(\mathbf{x})\right] - \mathcal{A}_{\mathrm{L,R}}\dot{\mathbf{P}}\right.$$
$$\left. - \epsilon_{\mathbf{P}}^{(\mathrm{L,R})}(\mathbf{x}) - eA_0(\mathbf{x})\right\}, \tag{8.26}$$

where $\mathbf{A}_{\mathrm{eff}} = \mathbf{A} + \mathbf{V}_{\mathrm{L,R}}/e$. Now we can utilize the new set of canonical variables $\Xi_a = \{\mathbf{x},\mathbf{P}\}$. These canonical variables were used by Silva et al. (1999) to construct kinetic equations of fermions electroweakly interacting with inhomogeneous background matter.

The Hamilton equations, which are based on the effective actions in equation (8.26), correspond to noncommutative dynamics. This means the Poisson brackets of coordinates of a test fermion are nonzero, $\{x_i, x_j\} \neq 0$. These Poisson brackets are proportional to the Berry curvature. This fact was noticed by Son and Yamamoto (2013), who considered an analogous system without the contribution of electroweak interaction with background matter.

Son and Yamamoto (2013) generalized the Liouville theorem and obtained the kinetic equations for the distribution functions of left and right chiral fermions. The form of the actions in equation (8.26) formally coincides with that derived by Son and Yamamoto (2013). Thus, we can apply their approach to get the kinetic equation for the distribution functions of left and right fermions, $f_{\mathrm{L,R}}(\mathbf{x},\mathbf{P},t)$, in the form

$$\frac{\partial f_{\mathrm{L,R}}}{\partial t} + \frac{1}{1 + e\mathbf{B}_{\mathrm{eff}}\mathbf{\Omega}} \left\{\left[\mathbf{v} + e(\tilde{\mathbf{E}} \times \mathbf{\Omega}_{\mathbf{P}}) + e(\mathbf{v}\cdot\mathbf{\Omega}_{\mathbf{P}})\mathbf{B}_{\mathrm{eff}}\right]\frac{\partial f_{\mathrm{L,R}}}{\partial \mathbf{x}}\right.$$
$$\left. + \left[e\tilde{\mathbf{E}} + e(\mathbf{v}\times\mathbf{B}_{\mathrm{eff}}) + e^2(\tilde{\mathbf{E}}\cdot\mathbf{B}_{\mathrm{eff}})\mathbf{\Omega}_{\mathbf{P}}\right]\frac{\partial f_{\mathrm{L,R}}}{\partial \mathbf{P}}\right\} = 0, \tag{8.27}$$

where $\tilde{\mathbf{E}} = \mathbf{E}_{\mathrm{eff}} - e^{-1}\partial\epsilon_{\mathbf{P}}/\partial\mathbf{x}$, $\mathbf{v} = \partial\epsilon_{\mathbf{P}}/\partial\mathbf{P}$, $\mathbf{E}_{\mathrm{eff}} = -\nabla A_0 - \nabla V_{\mathrm{L,R}}^0/e$, $\mathbf{B}_{\mathrm{eff}} = (\nabla \times \mathbf{A}_{\mathrm{eff}})$, and $\mathbf{\Omega}_{\mathbf{P}} = (\nabla_{\mathbf{P}} \times \mathcal{A}_{\mathrm{L,R}})$ is the Berry curvature. To derive equation (8.27), we take $\epsilon_{\mathbf{P}}^{(\mathrm{L,R})} = V_{\mathrm{L,R}}^0 + \epsilon_{\mathbf{P}}$, where $\epsilon_{\mathbf{P}}(\mathbf{x}) = |\mathbf{P}| + \epsilon_0(\mathbf{x})$.

Note that if $\epsilon_0(\mathbf{x}) = 0$, equation (8.27) coincides with that derived by Dvornikov and Semikoz (2017), Semikoz and Dvornikov (2018) for a neutrino electroweakly interacting with electron-positron plasma since $\tilde{\mathbf{E}} = \mathbf{E}_{\text{eff}} = \mathbf{E} + \mathbf{E}_{\text{EW}}$ and $\mathbf{B}_{\text{eff}} = \mathbf{B} + \mathbf{B}_{\text{EW}}$, where $\mathbf{E}_{\text{EW}} = -e^{-1}\nabla V^0_{\text{L,R}}$ and $\mathbf{B}_{\text{EW}} = e^{-1}(\nabla \times \mathbf{V}_{\text{L,R}})$ are the effective corrections induced by the electroweak interaction to the conventional electromagnetic field $F_{\mu\nu} = (\mathbf{E}, \mathbf{B})$.

8.2.3. Electric Current and the Adler Anomaly in Inhomogeneous Matter Accounting for Electroweak Interaction

In this section, we derive the corrections to the anomalous current and the Adler anomaly arising from the electroweak interaction of chiral fermions with inhomogeneous background matter. We consider the particular case of rotating matter.

Son and Yamamoto (2013) defined the number densities

$$n_{\text{L,R}} = \int \frac{d^3P}{(2\pi)^3} \left[1 + e\left(\mathbf{B}_{\text{eff}} \cdot \mathbf{\Omega}_{\mathbf{P}}\right)\right] f_{\text{L,R}} \tag{8.28}$$

and the currents

$$\mathbf{j}_{\text{L,R}} = -\int \frac{d^3P}{(2\pi)^3} \left[\epsilon_{\mathbf{P}} \frac{\partial f_{\text{L,R}}}{\partial \mathbf{P}} + e\left(\mathbf{\Omega}_{\mathbf{P}} \cdot \frac{\partial f_{\text{L,R}}}{\partial \mathbf{P}}\right) \epsilon_{\mathbf{P}} \mathbf{B}_{\text{eff}} + \epsilon_{\mathbf{P}} \left(\mathbf{\Omega}_{\mathbf{P}} \times \frac{\partial f_{\text{L,R}}}{\partial \mathbf{x}}\right) + e\left(\mathbf{E}_{\text{eff}} \times \mathbf{\Omega}_{\mathbf{P}}\right) f_{\text{L,R}}\right]. \tag{8.29}$$

Dvornikov and Semikoz (2017), Semikoz and Dvornikov (2018) found that the quantities in equations (8.28) and (8.29) satisfy the equation

$$\partial_t n_{\text{L,R}} + (\nabla \cdot \mathbf{j}_{\text{L,R}}) = -e^2 \int \frac{d^3P}{(2\pi)^3} \left(\mathbf{\Omega}_{\mathbf{P}} \cdot \frac{\partial f_{\text{L,R}}}{\partial \mathbf{P}}\right) \times \left(\mathbf{E}_{\text{eff}} \cdot \mathbf{B}_{\text{eff}}\right). \tag{8.30}$$

Dvornikov and Semikoz (2017), Semikoz and Dvornikov (2018) implied the validity of the Maxwell equations for the effective electromagnetic fields $(\nabla \cdot \mathbf{B}_{\text{eff}}) = 0$ and $\partial_t \mathbf{B}_{\text{eff}} + (\nabla \times \mathbf{E}_{\text{eff}}) = 0$ to derive equation (8.30).

We will consider the homogeneous distribution functions: $\partial f_{\text{L,R}}/\partial \mathbf{x} = 0$. In this situation, $\epsilon_{\mathbf{P}} = \mu_{\text{L,R}} + \cdots$, where $\mu_{\text{L,R}}$ are the chemical potentials of right and left electrons. Moreover, the zero-order terms in the Fermi constant G_{F} are kept. One gets that the anomalous part of the current $\mathbf{J}_{\text{L,R}} = e\mathbf{j}_{\text{L,R}}$ is expressed as

$$\begin{aligned}
\mathbf{J}_{\text{L,R}} &= -\mu_{\text{L,R}} e^2 \mathbf{B}_{\text{eff}} \int \frac{d^3P}{(2\pi)^3} \left(\mathbf{\Omega}_{\mathbf{P}} \cdot \frac{\partial f_{\text{L,R}}}{\partial \mathbf{P}}\right) \\
&= \mp \frac{e^2}{4\pi^2} \mu_{\text{L,R}} \mathbf{B}_{\text{eff}} \\
&= \mp \frac{e^2}{4\pi^2} \mu_{\text{L,R}} \left[\mathbf{B} + \frac{1}{e}(\nabla \times \mathbf{V}_{\text{L,R}})\right].
\end{aligned} \tag{8.31}$$

The first term in equation (8.31), namely $\mathbf{J}_{\text{L,R}} \propto \mu_{\text{L,R}}\mathbf{B}$, reproduces the CME (Fukushima et al., 2008). We take the following distribution function

$$\begin{aligned}
f_{\text{L,R}}(\mathbf{P}) &= \frac{1}{\exp[\beta(|\mathbf{P}| - \mu_{\text{L,R}})] + 1}, \\
\frac{\partial f_{\text{L,R}}}{\partial \mathbf{P}} &= -\frac{\mathbf{P}}{|\mathbf{P}|} \frac{\beta \exp[\beta(|\mathbf{P}| - \mu_{\text{L,R}})]}{\left\{\exp[\beta(|\mathbf{P}| - \mu_{\text{L,R}})] + 1\right\}^2}
\end{aligned} \tag{8.32}$$

to derive equation (8.31). Here, $\beta = 1/T$ is the inverted temperature. We consider the degenerate plasma with $\beta \to \infty$ and $\nabla_{\mathbf{P}} f_{\text{L,R}} \to \delta(|\mathbf{P}| - \mu_{\text{L,R}})$ in equations (8.31) and (8.32). Moreover, we consider the result of Dvornikov and Semikoz (2017) that $\mathbf{\Omega}_{\mathbf{P}} = \mp\mathbf{P}/2|\mathbf{P}|^3$.

The correction to the Adler anomaly can be calculated analogously to equation (8.31) using equation (8.30). First, one assumes that the density of background fermions is homogeneous: that is, $\nabla V^0_{\text{L,R}} = 0$. In this situation, $\mathbf{E}_{\text{eff}} = \mathbf{E}$, and

$$\begin{aligned}
\partial_t n_{\text{L,R}} + (\nabla \cdot \mathbf{j}_{\text{L,R}}) &= \mp \frac{e^2}{4\pi^2} \left(\mathbf{E} \cdot \mathbf{B}_{\text{eff}}\right) \\
&= \mp \frac{e^2}{4\pi^2} \left[(\mathbf{E} \cdot \mathbf{B}) + \frac{1}{e}\left(\mathbf{E} \cdot \nabla \times \mathbf{V}_{\text{L,R}}\right)\right].
\end{aligned} \tag{8.33}$$

Integrating this expression and neglecting the boundary effects, we get that

$$\frac{dn_5}{dt} = \frac{e^2}{2\pi^2 V} \int \left[(\mathbf{E} \cdot \mathbf{B}) - \frac{1}{e}\left(\mathbf{E} \cdot \nabla \times \mathbf{V}_5\right)\right] d^3x - \Gamma n_5, \tag{8.34}$$

where $n_5 = n_{\text{R}} - n_{\text{L}}$ is the imbalance of chiral electrons and $\mathbf{V}_5 = (\mathbf{V}_{\text{L}} - \mathbf{V}_{\text{R}})/2$. Moreover, one has that $(\nabla \times \mathbf{V}_5) = V_5 \boldsymbol{\omega}$, where $V_5 = (V^0_{\text{L}} - V^0_{\text{R}})/2$. Here, $\boldsymbol{\omega}$ is the angular velocity.

We take into account the fact that electrons are ultrarelativistic but not massless particles in equation (8.34). This means their helicity is changed in collisions with other fermions. Thus we introduce the term $-\Gamma n_5$ in the right-hand side (r.h.s.) of equation (8.34). Dvornikov (2016b) computed the spin-flip rate $\Gamma \sim 10^{11}\,\text{s}^{-1}$ in degenerate matter of old NSs.

The correction to the anomalous current due to the electroweak interaction has the form, $\delta\mathbf{J}_{\text{anom}} = (\mathbf{J}_{\text{L}} + \mathbf{J}_{\text{R}}) \propto \nabla \times (\mathbf{V}_{\text{R}} - \mathbf{V}_{\text{L}}) \sim V_5\boldsymbol{\omega}$. This expression in equation (8.31) reproduces the result of Dvornikov (2015) except the factor $1/2\pi$ missed there. Dvornikov (2015) named the generation of such anomalous current the galvano-rotational effect.

8.3. ASTROPHYSICAL APPLICATIONS

In this section, we apply the results of section 8.2 to study the evolution of magnetic fields in compact stars. In sections 8.3.1–8.3.3, we analyze the magnetic helicity evolution in a NS, considering the polarization of relativistic electrons. In section 8.3.4, we study the flux of magnetic helicity through the equator of a rotating NS and propose a possible explanation for the observed magnetic cycles of some pulsars.

8.3.1. Quantum Surface Correction to the Chiral Anomaly in a Finite Volume of a Neutron Star

In this section, we analyze the system of AMHD equations for chiral fermions. Then we study the modification of these equations when one accounts for a particle mass. In particular, we consider the contribution of the nonzero fermion mass to the Adler anomaly.

Pavlović et al. (2017) formulated the full set of AMHD equations:

$$\rho\left[\frac{\partial \mathbf{v}}{\partial t}+(\mathbf{v}\nabla)\mathbf{v}-\nu\nabla^2\mathbf{v}\right]=-\nabla p+\sigma_{\text{cond}}\times[(\mathbf{E}\times\mathbf{B})+(\mathbf{v}\times\mathbf{B})\times\mathbf{B}],$$

$$\frac{\partial\rho}{\partial t}+\nabla(\rho\mathbf{v})=0, \tag{8.35}$$

$$\frac{\partial \mathbf{B}}{\partial t}=-(\nabla\times\mathbf{E}), \tag{8.36}$$

$$(\nabla\times\mathbf{B})=\sigma_{\text{cond}}\left[\mathbf{E}-\frac{e^2}{2\pi^2\sigma_{\text{cond}}}\mu_5\mathbf{B}+(\mathbf{v}\times\mathbf{B})\right], \tag{8.37}$$

$$\frac{\mathrm{d}\mu_5}{\mathrm{d}t}=-\frac{e^2}{4\mu_e^2}\frac{\mathrm{d}h}{\mathrm{d}t}, \tag{8.38}$$

where ρ is the density of matter, $\mathbf{v}$ is the velocity of plasma, p is the pressure, σ_{cond} is the electric conductivity, $\mu_5=(\mu_{eR}-\mu_{eL})/2$ is the imbalance between chiral right and left electrons, h is the density of the magnetic helicity, μ_e is the mean chemical potential of the electron gas, and ν is the coefficient of the viscosity.

Continuity and Navier-Stokes equations are given in equation (8.35). In equation (8.37), the anomalous current due to the CME is taken into account in the Maxwell equation. The displacement current is omitted there. The induction equation from equations (8.36) and (8.37) is changed due to the CME:

$$\frac{\partial \mathbf{B}}{\partial t}=\nabla\times(\mathbf{v}\times\mathbf{B})+\frac{1}{\sigma_{\text{cond}}}\nabla^2\mathbf{B}-\frac{e^2}{2\pi^2\sigma_{\text{cond}}}\nabla\times(\mu_5\mathbf{B}). \tag{8.39}$$

Equation (8.38) results from the Adler-Bell-Jackiw anomaly, discussed shortly, for massless particles. Note that we assume that the electron gas is degenerate in equation (8.38). We will change equation (8.38) to take into account the nonzero mass of electrons.

We will examine which terms are important for the Adler anomaly in a finite volume: for instance, a NS. In particular, we compute the correction due to the averaged mass term $2im_e\langle\overline{\psi}\gamma_5\psi\rangle$ to the Adler-Bell-Jackiw anomaly in quantum electrodynamics (QED) known as nonconservation of the axial current:

$$\frac{\partial}{\partial x^\mu}\overline{\psi}\gamma^\mu\gamma^5\psi=\frac{\partial j^\mu_{\text{R}}}{\partial x^\mu}-\frac{\partial j^\mu_{\text{L}}}{\partial x^\mu}=2im_e\overline{\psi}\gamma_5\psi+\frac{2e^2}{16\pi^2}F_{\mu\nu}\tilde{F}^{\mu\nu}, \tag{8.40}$$

where left and right currents are not conserved separately, even for massless fermions, due to the triangle anomaly in the presence of an electromagnetic field,

$$\frac{\partial j^\mu_{\text{L,R}}}{\partial x^\mu}=\mp im_e\overline{\psi}\gamma_5\psi\mp\frac{e^2}{16\pi^2}F_{\mu\nu}\tilde{F}^{\mu\nu}. \tag{8.41}$$

Using the fact that $\gamma^0\boldsymbol{\gamma}\gamma^5=\boldsymbol{\Sigma}$ and averaging the left-hand side (l.h.s.) in equation (8.40), one obtains

$$\int\frac{\mathrm{d}^3x}{V}\frac{\partial}{\partial x^\mu}\langle\overline{\psi}\gamma^\mu\gamma^5\psi\rangle=\frac{\mathrm{d}}{\mathrm{d}t}(n_{\text{R}}-n_{\text{L}})+\frac{1}{V}\oint\mathrm{d}^2S(\mathcal{S}\cdot\mathbf{n}), \tag{8.42}$$

where $\mathcal{S}$ is the averaged spin given in equation (8.3) for massless fermions. Averaging the r.h.s. of equation (8.40), we get

$$\cdots=-\frac{1}{V}\oint_S(\mathcal{S}_5\cdot\mathbf{n})\mathrm{d}^2S+\frac{2\alpha_{\text{em}}}{\pi}\int\frac{\mathrm{d}^3x}{V}(\mathbf{E}\cdot\mathbf{B}), \tag{8.43}$$

where the first term $\sim\mathcal{S}_5$ is determined by the averaged pseudoscalar

$$2im_e\int\frac{\mathrm{d}^3x}{V}\langle\overline{\psi}\gamma_5\psi\rangle=-\int\frac{\mathrm{d}^3x}{V}(\nabla\cdot\mathcal{S}_5(\mathbf{x},t)). \tag{8.44}$$

The second term in equation (8.43) gives the dissipation of the magnetic helicity according to standard MHD (Priest, 2016):

$$2\int\frac{\mathrm{d}^3x}{V}(\mathbf{E}\cdot\mathbf{B})=-\frac{\mathrm{d}h}{\mathrm{d}t}-\oint\left(\mathbf{n}\cdot\left[\mathbf{B}A_0+\mathbf{E}\times\mathbf{A}\right]\right)\frac{\mathrm{d}^2S}{V}. \tag{8.45}$$

Equation (8.44) results from the spin distribution functions in a weakly inhomogeneous electron-positron plasma:

$$\mathcal{S}_5(\mathbf{x},t)=-\int\frac{\mathrm{d}^3p}{\gamma(2\pi)^3}\left(\mathbf{S}^{(e)}(\mathbf{p},\mathbf{x},t)-\mathbf{S}^{(\bar{e})}(\mathbf{p},\mathbf{x},t)\right)\times\left(\frac{2}{3}+\frac{1}{3\gamma}\right). \tag{8.46}$$

Dvornikov and Semikoz (2018) provided the detailed derivation of equation (8.46).

Silin (1968) derived the equilibrium part of the total Wigner spin distribution function of electrons resulting from the paramagnetic contribution in an inhomogeneous magnetic field:

$$\mathbf{S}^{(e)}_{\text{eq}}(\varepsilon_p,\mathbf{x},t)=\frac{\mu_{\text{B}}\mathbf{H}(\mathbf{x},t)}{\gamma}\frac{\mathrm{d}f^{(e)}_{\text{eq}}(\varepsilon_p)}{\mathrm{d}\varepsilon_p},$$

$$\varepsilon_p = \sqrt{p^2 + m_e^2} = \gamma m_e, \tag{8.47}$$

which are used in equation (8.46): $\mathbf{S}^{(e)}(\mathbf{p}, \mathbf{x}, t) = \mathbf{S}^{(e)}_{\text{eq}}(\varepsilon_p, \mathbf{x}, t) + \delta\mathbf{S}^{(e)}(\mathbf{p}, \mathbf{x}, t)$.

Based on equations (8.42) and (8.43), and taking into account equation (8.45) for the magnetic helicity density $h(t) = \int \frac{\mathrm{d}^3x}{V}(\mathbf{A}\cdot\mathbf{B})$, we obtain the master equation

$$\frac{\mathrm{d}}{\mathrm{d}t}\left(n_{\mathrm{R}} - n_{\mathrm{L}} + \frac{\alpha_{\mathrm{em}}}{\pi}h\right) = -\frac{\alpha_{\mathrm{em}}}{\pi V}\oint_S([\mathbf{E}\times\mathbf{A} + A_0\mathbf{B}]\cdot\mathbf{n})\mathrm{d}^2S - \oint_S([\mathcal{S} + \mathcal{S}_5]\cdot\mathbf{n})\frac{\mathrm{d}^2S}{V}. \tag{8.48}$$

If one neglects the surface terms in a plasma of chiral massless particles, we obtain the conservation law for the sum of the chiral imbalance, $n_{\mathrm{R}} - n_{\mathrm{L}}$, and the density of the magnetic helicity h, accounting for the factor α_{em}/π:

$$\frac{\mathrm{d}}{\mathrm{d}t}\left(n_{\mathrm{R}} - n_{\mathrm{L}} + \frac{\alpha_{\mathrm{em}}}{\pi}h\right) = 0. \tag{8.49}$$

Equation (8.49) is crucial for AMHD since it describes the evolution of the chiral imbalance, $\sim \mathrm{d}\mu_5/\mathrm{d}t$, in a magnetized plasma (Boyarsky et al., 2012; Fukushima et al., 2008): a change in the imbalance results in the production of magnetic helicity and vice versa.

Equation (8.48) includes the surface term known in classical MHD. It contains electromagnetic fields only. The new quantum correction results from the sum of the spin terms, $\mathcal{S}_{\text{eff}} = \mathcal{S} + \mathcal{S}_5$. Here, we will examine the latter term.

It is worth mentioning that the equilibrium distribution function of spin for positron gas has the same positive sign as that for electrons in equation (8.47):

$$\mathbf{S}^{(\bar{e})}_{\text{eq}}(\varepsilon_p, \mathbf{x}, t) = \frac{\mu_{\mathrm{B}}\mathbf{H}(\mathbf{x}, t)}{\gamma}\frac{\mathrm{d}f^{(\bar{e})}_{\text{eq}}(\varepsilon_p)}{\mathrm{d}\varepsilon_p}. \tag{8.50}$$

Thus the pseudovector in equation (8.46) is given by the difference of the particle and antiparticle contributions due to the permutation of the operators $\hat{d}\hat{d}^+ \to -\hat{d}^+\hat{d}$ while deriving equation (8.46):

$$\mathcal{S}_5(\mathbf{x}, t) = -\frac{\mu_{\mathrm{B}}m_e^2\mathbf{H}(\mathbf{x}, t)}{2\pi^2}\int\frac{p^2\mathrm{d}p}{\varepsilon_p^2}\left(\frac{\mathrm{d}f^{(e)}_{\text{eq}}}{\mathrm{d}\varepsilon_p} - \frac{\mathrm{d}f^{(\bar{e})}_{\text{eq}}}{\mathrm{d}\varepsilon_p}\right) \times\left(\frac{2}{3} + \frac{1}{3\gamma}\right). \tag{8.51}$$

If we integrate by parts in equation (8.51) and separate the Lande factor $g_s = 2$ from the Fermi distribution functions $f^{(e,\bar{e})}(\varepsilon_p) = g_s[\exp(\varepsilon_p \mp \mu_e)/T + 1]^{-1}$, we can rewrite the effective averaged spin $\mathcal{S}_{\text{eff}} = \mathcal{S} + \mathcal{S}_5$ contributing the surface term in equation (8.48):

$$\mathcal{S}_{\text{eff}} = -\frac{e\mathbf{H}}{2\pi^2}\int_0^\infty \mathrm{d}p\left[1 - \frac{1}{3\gamma^2}\left(\frac{2}{\gamma} + \frac{2}{\gamma^2} - 1\right)\right] \times\left(\frac{1}{\exp[(\varepsilon_p - \mu_e)/T] + 1} - \frac{1}{\exp[(\varepsilon_p + \mu_e)/T] + 1}\right). \tag{8.52}$$

One has that $\varepsilon_p = \gamma m_e \approx m_e$ in a nonrelativistic plasma. Thus, the total spin effect is vanishing, $\mathcal{S}_{\text{eff}} = \mathcal{S} + \mathcal{S}_5 \to 0$, since the spin terms cancel each other. For instance, one gets that $\mathcal{S} = -\mathcal{S}_5 = -2em_e v_{\mathrm{F}_e}\mathbf{B}/4\pi^2$, where $v_{\mathrm{F}_e} = p_{\mathrm{F}_e}/m_e \ll 1$ in a degenerate electron gas, where the positron contribution is absent. The first correction to the last equality appears due to the decomposition $\gamma^{-1} \approx 1 - v^2/2$. It is quite small: $\mathcal{S}_{\text{eff}} = em_e v^3_{\mathrm{F}_e}\mathbf{B}/3\pi^2$. This correction is valid for weak magnetic fields, $m_e^2 \gg p^2_{\mathrm{F}_e} \gg eB$, where the Wentzel-Kramers-Brillouin (WKB) approximation with large Landau numbers, $n \gg 1$, used to calculate $\mathcal{S}_5$, remains correct.

In the case of ultrarelativistic chiral plasma with $\gamma \gg 1$, one gets from equation (8.52) that $\mathcal{S}_{\text{eff}} \approx \mathcal{S} = -e\mu_e(r, \theta)\mathbf{B}/2\pi^2$; see also equation (8.3). Hence the effect of magnetization becomes great under the condition $\mu_e(r, \theta) \gg m_e$, which can be implemented, for instance, in the NS core. Note that here we use an inhomogeneous chemical potential $\mu_e(r, \theta)$ that corresponds to a realistic profile of the spherically symmetric electron density in a NS (Lattimer and Prakash, 2001), $n_e(r) = n_{\text{core}}Y_e[1 - r^2/R^2_{\mathrm{NS}}]$, where $Y_e = 0.04$ is the electron abundance and $n_{\text{core}} \simeq 10^{38}\,\mathrm{cm}^{-3}$ is the density in the center of a star.

For example, the anisotropy for the chemical potential results from the magnetic field correction proportional to the total magnetic field entering Landau levels $B(r, \theta) = \sqrt{B_p^2(r, \theta) + B_t^2(r, \theta)}$, in rotating a star, which is axially symmetric. Here, $B_{t,p}$ are the toroidal and poloidal components of the magnetic field (see section 8.3.3). (We take a slightly nonuniform magnetic field: that is, it can be uniform at microscopic scales that are smaller than the mean distance between particles in matter, $L < (n_e)^{-1/3}$. This is required to receive the energy levels in equation (8.1). Nevertheless, $\mathbf{B}$ is considered nonuniform at macroscopic length scales similar to R_{NS}). This anisotropy is derived from inverting equation (8.8):

$$\mu_e(r, \theta) = [3\pi^2 n_e(r)]^{1/3}\left[1 - \frac{eB(r, \theta)}{2(3\pi^2 n_e(r))^{2/3}}\right]. \tag{8.53}$$

(Meaning the cubic equation (8.8) written as $\mu_e^3 + 3eB\mu_e/2 - 3\pi^2 n_e = 0$, where $\mu_e \equiv p_{\mathrm{F}_e}$. The parameter $D = q^2 + p^3 = (eB)^3/8 + 9\pi^4 n_e^2/4 > 0$ is positive for the Cartan solution. Thus, we have one real root of this equation given in equation (8.53), as well as two complex self-adjoint roots). Otherwise, the surface integral

$$\oint_S(\mathcal{S}\cdot\mathbf{n})\frac{\mathrm{d}^2S}{V} = -\frac{e}{2\pi^2 V}\oint_S\mu_e(r, \theta)(\mathbf{B}\cdot\mathbf{n})\mathrm{d}^2S \tag{8.54}$$

remaining in equation (8.48) for a NS is vanishing for the uniform spherically symmetric $\mu_e(r)_{r=R_s} = \text{const}$ because of Gauss's law, $\int d^3x(\nabla \cdot \mathbf{B}) = \oint (\mathbf{B} \cdot \mathbf{n}) d^2S = 0$.

8.3.2. Magnetic Helicity Evolution as the Averaged Spin Flux through the Boundary of a Domain

We consider a realistic situation when the chiral imbalance is vanishing, $n_R - n_L \to 0$, in relativistic plasmas, taking into account the nonzero electron mass, $m_e \neq 0$, because of the spin flip. It is necessary to reconcile the evolution of the magnetic helicity density in equation (8.45) and the statistically averaged Adler anomaly in chiral medium given in equation (8.48). Dvornikov (2016b), Dvornikov and Semikoz (2015a) showed that this situation happens very soon in the core of a young NS, during $\sim 10^{-12}$ s. It takes place even when an initial difference is positive $n_R(t_0) - n_L(t_0) > 0$, which can arise in a protoneutron star due to the direct Urca process, $p + e^-_L \to n + \nu_{eL}$. Boyarsky et al. (2012) found similar decay of the chiral imbalance $2\mu_5 = \mu_R - \mu_L \to 0$ due to the spin flip in the cooling universe with temperatures below $T = 10\,\text{MeV}$.

Then, at time $t \gg t_0$, we modify the standard MHD equation (8.45) due to equation (8.48), accounting for the magnetic flux through the volume surface weighted by the nonuniform chemical potential $\mu_e(r, \theta)$ in equation (8.53) that enters the quantum (magnetization) term in equation (8.54):

$$\oint d^2S\,(S_{\text{eff}} \cdot \mathbf{n}) \approx -\frac{e}{2\pi^2}\oint_S \mu_e(r,\theta)(\mathbf{B}\cdot\mathbf{n})d^2S = \frac{\alpha_{\text{em}}}{\pi[3\pi^2 n_e(R)]^{1/3}}\oint B(R,\theta)(\mathbf{B}\cdot\mathbf{n})d^2S. \tag{8.55}$$

Here, the nonuniform electron density $n_e(R) = n_{\text{core}} Y_e(1 - R^2/R^2_{\text{NS}})$ should be large enough to obey the inequality $2eB(R,\theta) \ll [3\pi^2 n_e(R)]^{2/3}$ at the surface with radius $R < R_{\text{NS}}$ since the decomposition is made over the small parameter $2eB/\mu_e^2 \ll 1$ in equation (8.8). Then we can generalize standard equation (8.45) due to the Adler-Bell-Jackiw anomaly, accounting for the additional quantum correction in equation (8.55):

$$\frac{dH}{dt} = -2\int_V d^3x(\mathbf{E}\cdot\mathbf{B}) - \oint_S (\mathbf{n}\cdot[A_0\mathbf{B} + \mathbf{E}\times\mathbf{A}])d^2S + \frac{1}{\mu_e(R)}\oint_S B(R,\theta)B_r(R,\theta)d^2S, \tag{8.56}$$

where $H = Vh$ is the total helicity of a star and $B(R,\theta) = \sqrt{B_r^2 + B_\theta^2 + B_\varphi^2}$ is the total magnetic field strength entering Landau levels. We also use the short notation $\mu_e(R) = [3\pi^2 n_e(R)]^{1/3}$.

Priest (2016) claims that the volume term on the l.h.s. in equation (8.45) can be neglected at evolution times smaller than the diffusion one: $t \ll \tau_D \sim L^2\sigma_{\text{cond}}$. Thus the behavior of the magnetic helicity is driven only by the second term in equation (8.56) or the last term in equation (8.45). Schmitt and Shternin (2018) found that the diffusion time is great, $\tau_D \simeq 30\,\text{yr}\,(L/\text{cm})^2$, due to the high electric conductivity in a NS, $\sigma_{\text{cond}} \sim 10^9\,\text{MeV}$, for $T = 10^8\,\text{K}$. For example, the diffusion time is longer than the age of the universe, $\tau_D \approx 3\times 10^{13}\,\text{yr} \gg t_{\text{Univ}} = 1.4\times 10^{10}\,\text{yr}$, for the maximal length scale $L = R_{\text{NS}} = 10^6\,\text{cm}$. Hence, for such large scales, there is no reason to take into account both the magnetic helicity diffusion and the quantum term in equation (8.55), since both could be essential only at times $t > \tau_D$. However, at small scales $L \ll R_{\text{NS}}$, the magnetic helicity diffusion time is less than the age of young magnetars $\sim 10^3\,\text{yr}$: for example, for $L = 1\,\text{cm}$, $\tau_D \sim 30\,\text{yr} \ll t \sim 10^3\,\text{yr}$. Therefore, the quantum contribution to the evolution equation (8.56), missing in the classical approach (Priest, 2016), can be essential for small-scale magnetic fields in a NS at times $t > \tau_D$.

8.3.3. Evolution of Magnetic Helicity in NSs

The study of the evolution of magnetic helicity is important for a possible reconnection of magnetic field lines near the NS surface, mostly outside the crust in the NS magnetosphere. This process, in turn, could explain the gamma or X-ray flares that Kaspi and Beloborodov (2017) reported take place in magnetars. However, it is interesting to study how magnetic helicity dissipation proceeds inside a NS, below the crust.

Using the gauge $A_0 = 0$ and $(\nabla \cdot \mathbf{A}) = 0$ in equation (8.45), valid in classical MHD, and substituting Ohm's law $\mathbf{E} = -\mathbf{v}\times\mathbf{B} + \mathbf{j}/\sigma_{\text{cond}}$, one can rederive the result of Priest (2016)

$$\frac{dH}{dt} = -2\sigma^{-1}_{\text{cond}}\int \mathbf{j}\cdot\mathbf{B}d^3x - \oint_S [(\mathbf{B}\cdot\mathbf{A})(\mathbf{v}\cdot\mathbf{n}) - (\mathbf{v}\cdot\mathbf{A})(\mathbf{B}\cdot\mathbf{n})]d^2S, \tag{8.57}$$

which is generalized in equation (8.56) in this gauge:

$$\frac{dH}{dt} = -2\sigma^{-1}_{\text{cond}}\int \mathbf{j}\cdot\mathbf{B}d^3x - \oint_S [(\mathbf{B}\cdot\mathbf{A})(\mathbf{v}\cdot\mathbf{n}) - (\mathbf{v}\cdot\mathbf{A})(\mathbf{B}\cdot\mathbf{n})]d^2S + \frac{1}{\mu_e(R)}\oint_S B(R,\theta)(\mathbf{B}\cdot\mathbf{n})d^2S. \tag{8.58}$$

The last term on the r.h.s. of equation (8.58) results from the quantum contribution of the averaged spin flux through the surface $R_{\text{NS}} > R$. We will show that it is essential at small scales. Notice that the magnetic helicity in equation (8.58) is conserved, $\mathrm{d}H/\mathrm{d}t = 0$, in a closed volume, where $(\mathbf{v}\cdot\mathbf{n}) = 0$ and $(\mathbf{B}\cdot\mathbf{n}) = 0$. This happens when one neglects volume losses for an ideal plasma, $\sigma_{\text{cond}} \to \infty$. Both quantum and classical surface terms are different from zero if a magnetic field penetrates the domain volume: that is, when $(\mathbf{B}\cdot\mathbf{n}) \neq 0$.

Let us estimate the size of the new quantum contribution: that is, the last term in equation (8.58). This term does not explicitly depend on the NS rotation. We can take the axially symmetric magnetic field consisting of the toroidal and quadrupole poloidal fields,

$$\mathbf{B}(r,\theta) = B_p(r)[\cos 2\theta\mathbf{e}_r + \sin 2\theta\mathbf{e}_\theta] + B_\varphi(r)\cos\theta\mathbf{e}_\varphi, \tag{8.59}$$

proposed by Moss et al. (2008). Here, the angle θ is measured from the NS equator, corresponding to $\theta = 0$. Notice that the magnetic field components are nonzero at the NS equator. The structure of this magnetic field is schematically shown in Fig. 8.1.

The factors $(\mathbf{B}\cdot\mathbf{n}) = B_r = B_p\cos 2\theta$ at the spherical surface, where $\mathbf{n} = \mathbf{e}_r$, and $B(R,\theta) = \sqrt{B_p^2 + B_\varphi^2\cos^2\theta}$, are used in the last term in equation (8.58). Then one obtains such a surface term in the form

$$\begin{aligned}\left(\frac{\mathrm{d}H}{\mathrm{d}t}\right)_{\text{quant}} &= \frac{1}{\mu_e(R)}\oint_S B(R,\theta)(\mathbf{B}\cdot\mathbf{n})\mathrm{d}^2S \\ &= -\left(\frac{2\pi R^2 B_p}{\mu_e}\right)\int_{-1}^{1}(2x^2-1)\sqrt{B_p^2 + B_\varphi^2 x^2}\mathrm{d}x \\ &= \left(\frac{2\pi R^2 B_p B_\varphi}{\mu_e}\right)\times A \sim \frac{B^2R^2}{\mu_e},\end{aligned} \tag{8.60}$$

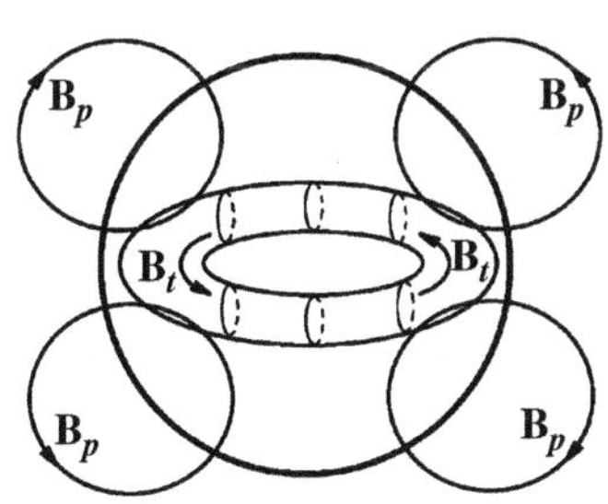

Figure 8.1 The structure of the magnetic field presented in equation (8.59). The poloidal component of the magnetic field reads $\mathbf{B}_p = B_p(r)[\cos 2\theta\mathbf{e}_r + \sin 2\theta\mathbf{e}_\theta]$. The toroidal component is $\mathbf{B}_t = B_\varphi(r)\cos\theta\mathbf{e}_\varphi$. Dvornikov and Semikoz, 2018/IOP Publishing.

where $\mathrm{d}^2S = 2\pi R^2\sin\theta\mathrm{d}\theta$. The factor A

$$\begin{aligned}A(B_p/B_t) = &-\frac{1}{2}\left(\frac{B_p}{B_\varphi}\right)^2\sqrt{1+\left(\frac{B_p}{B_\varphi}\right)^2} + \left(\frac{B_p}{B_\varphi}\right)^2 \\ &\times\left[1+\frac{1}{2}\left(\frac{B_p}{B_\varphi}\right)^2\right] \\ &\times\ln\left(\frac{B_\varphi}{B_p} + \frac{B_\varphi}{B_p}\sqrt{1+\left(\frac{B_p}{B_\varphi}\right)^2}\right)\end{aligned} \tag{8.61}$$

depends on the ratio B_p/B_t. For example, for $B_p \sim B_\varphi$, one gets $A \approx -0.05$. Thus, the last estimate in equation (8.60) is valid: $|(\mathrm{d}H/\mathrm{d}t)_{\text{quant}}| \sim B^2R^2/\mu_e$.

The factor A can be rewritten in the case when $B_\varphi \gg B_p$:

$$A = A(B_p/B_t) = \left(\frac{B_p}{B_\varphi}\right)^2\left[\ln\left(\frac{2B_\varphi}{B_p}\right) - \frac{1}{2}\right]. \tag{8.62}$$

In the opposite case, when $B_p \gg B_\varphi$, one gets

$$\left(\frac{\mathrm{d}H}{\mathrm{d}t}\right)_{\text{quant}} = \frac{4\pi R^2 B_p^2}{3\mu_e}, \tag{8.63}$$

which can be obtained from equation (8.60).

We compare the result in equation (8.60) with the classical surface term:

$$\begin{aligned}\left(\frac{\mathrm{d}H}{\mathrm{d}t}\right)_{\text{class}} &= -\oint_S[(\mathbf{B}\cdot\mathbf{A})(\mathbf{v}\cdot\mathbf{n}) - (\mathbf{v}\cdot\mathbf{A})(\mathbf{B}\cdot\mathbf{n})]\mathrm{d}^2S \\ &= \oint(vA_\varphi)(\mathbf{B}\cdot\mathbf{n})\mathrm{d}^2S \sim -B_p^2R^3\langle v\rangle,\end{aligned} \tag{8.64}$$

where we take the azimuthal rotation velocity $\mathbf{v} = v\mathbf{e}_\varphi$: that is, the first term in equation (8.64) is negligible, $(\mathbf{v}\cdot\mathbf{n}) = 0$. The rotation velocity $v = \Omega R\cos\theta$ and $\langle v\rangle \approx \Omega R$ corresponds to the mean velocity after the integration over the polar angle in equation (8.64). Notice that one can estimate the azimuthal potential $A_\varphi \sim -B_pR\sin 2\theta$ at the sphere $R < R_{\text{NS}}$ for the poloidal component in equation (8.59) $B_\theta = -r^{-1}\partial_r(rA_\varphi) = B_p\sin 2\theta$.

If we compare the classical and quantum surface terms in the total sum,

$$\left(\frac{\mathrm{d}H}{\mathrm{d}t}\right) = \left(\frac{\mathrm{d}H}{\mathrm{d}t}\right)_{\text{class}} + \left(\frac{\mathrm{d}H}{\mathrm{d}t}\right)_{\text{quant}}, \tag{8.65}$$

we find that the quantum term is dominant only in the corotational reference frame, $\langle v\rangle - \Omega R = \delta v \ll 1$, for $R < \mu_e^{-1}/\delta v$. Using $R = 10^5\,\text{cm}$ and $\mu_e^{-1} = 2\times 10^{-13}\,\text{cm}$ for $\mu_e = 100\,\text{MeV}$, one obtains $\delta v < 10^{-18}$. This result correspond to the rigid rotation, $\partial_\theta\Omega = \partial_r\Omega = 0$: that is, $\Omega = \text{const}$. In this situation, $\langle v\rangle - \Omega R \to 0$. Yakovlev and Pethick (2008) found that the neutron component in a NS, which can be superfluid, has some deviations from rigid rotation, $\delta v_n \neq 0$, unlike the proton component, $\delta v_p = 0$. Thus, our assumption of the absence of any differential rotation is invalid in such NSs.

The result in equation (8.60) can be interpreted as an intertwining of two thin tubes with magnetic fields having small base areas $S_p = \pi R_p^2$ and $S_t = \pi R_t^2$. Such tubes are situated at the sphere with radius $R \gg R_{p,t}$ (Priest, 2016):

$$\left(\frac{\mathrm{d}H}{\mathrm{d}t}\right)_{\text{quant}} = \dot{\theta}_{\text{pt}} F_p F_t, \tag{8.66}$$

where $F_t = B_t S_t$ and $F_p = B_p S_p$ are the fluxes of the magnetic field that tear off the two different toroids resulting from the toroidal and quadrupole poloidal components in equation (8.59). They penetrate the spherical surface $R < R_{\text{NS}}$, floating up. The parameter

$$\dot{\theta}_{\text{pt}} = \frac{A}{\pi^2} \times \left(\frac{R}{R_p}\right)^2 \left(\frac{1}{R_t^2 \mu_e}\right) \tag{8.67}$$

provides the angular velocity of the twisting of the magnetic loop bases around each other, causing the interlacing of the flux tubes. For instance, one can evaluate the parameter in equation (8.66) as $\dot{\theta}_{\text{pt}} \sim 10^7 \times (6A/\pi^2)\,\text{s}^{-1}$ for $R_p = R_t = 1\,\text{cm}$, $R = 10^5\,\text{cm}$. Thus, within the time $t \sim 3 \times 10^{-6}\,\text{s} \gg 10^{-12}\,\text{s}$, the magnetic helicity evolution in equation (8.67) leads to a flux tangling with the linkage (topology) number $L_{\text{pt}} = \int \dot{\theta}_{\text{pt}} dt = L_{12} \sim 1$ in the famous Gauss formula $H = 2L_{12}F_1F_2$, where $L_{12} = L_{\text{pt}} = 1$ is conserved afterward (Priest, 2016). (We substitute estimates $A = -0.05$ for $B_p = B_t$ in equation (8.61) when $n_{\text{R}} = n_{\text{L}}$. This means the CME is irrelevant in a NS, $\mu_5 = 0$).

8.3.4. Magnetic Helicity Flow through the Equator of a Rotating NS

Now we turn to the problem of the evolution of magnetic helicity in a rotating NS, accounting for the electroweak interaction between fermions. For this purpose, we use equation (8.34). We mentioned in section 8.2.3 that the spin-flip rate is huge in a NS. Hence, $\dot{n}_5 \to 0$ in equation (8.34). Using equation (8.34), we find that the chiral imbalance reaches the saturated value

$$n_5 \to n_5^{(\text{sat})} = \frac{e^2}{2\pi^2 VT} \int \left[(\mathbf{E}\cdot\mathbf{B}) - \frac{1}{e}\left(\mathbf{E}\cdot\nabla\times\mathbf{V}_5\right)\right] \mathrm{d}^3x. \tag{8.68}$$

We will demonstrate that we can neglect $n_5^{(\text{sat})}$ in an old star. Thus, the saturated densities of left and right fermions are equal.

If both n_5 and $\dot{n}_5$ vanish in equation (8.34), then, taking into account the fact that

$$\frac{\mathrm{d}H}{\mathrm{d}t} = -2\int (\mathbf{E}\cdot\mathbf{B})\,\mathrm{d}^3x, \tag{8.69}$$

where

$$H = \int (\mathbf{A}\cdot\mathbf{B})\,\mathrm{d}^3x \tag{8.70}$$

is the magnetic helicity, we obtain the contribution of the electroweak interaction to the evolution of the magnetic helicity in the form

$$\begin{aligned}\left(\frac{\mathrm{d}H}{\mathrm{d}t}\right)_{\text{EW}} &= -2\frac{V_5}{e}\int (\mathbf{E}\cdot\boldsymbol{\omega})\,\mathrm{d}^3x \\ &= -2\frac{V_5}{e\sigma}\int (\nabla\times\mathbf{B}\cdot\boldsymbol{\omega})\,\mathrm{d}^3x.\end{aligned} \tag{8.71}$$

Here, we account for the Maxwell equations and the fact that $\mathbf{E} = \mathbf{J}/\sigma = (\nabla\times\mathbf{B})/\sigma$, where $\mathbf{J}$ is the ohmic electric current and σ is the conductivity. Note that the analogous consideration was made in section 8.3.3.

Equation (8.71) should be included in the contribution of classical MHD for the behavior or magnetic helicity (see equation (8.72)):

$$\begin{aligned}\left(\frac{\mathrm{d}H}{\mathrm{d}t}\right)_{\text{class}} &= -\frac{2}{\sigma}\int (\mathbf{J}\cdot\mathbf{B})\,\mathrm{d}^3x \\ &\quad + 2\oint \left[(\mathbf{B}\cdot\mathbf{A})(\mathbf{v}\cdot\mathbf{n}) - (\mathbf{v}\cdot\mathbf{A})(\mathbf{B}\cdot\mathbf{n})\right]\mathrm{d}^2S,\end{aligned} \tag{8.72}$$

where $\mathbf{n}$ is the external normal to the surface of a star. The total rate of the helicity variation is $\dot{H} = (\dot{H})_{\text{class}} + (\dot{H})_{\text{EW}}$, where the electroweak and classical MHD contributions are present in equations (8.71) and (8.72). We omit the subscript EW since we are primarily interested in the $(\dot{H})_{\text{EW}}$ term.

Equation (8.71) represents the new mechanism for helicity evolution due to the electroweak interaction of chiral electrons with inhomogeneous background matter. The inhomogeneity of the interaction of chiral fermions is present in the dependence of $\mathbf{V}_{\text{L,R}} = V^0_{\text{L,R}}\mathbf{v}$ on spatial coordinates. It can be implemented if we consider rotating matter with velocity $\mathbf{v}(\mathbf{x}) = \boldsymbol{\omega}\times\mathbf{x}$. The anomalous contribution to the helicity change in equation (8.71) takes place if $(\nabla\times\mathbf{B}) \parallel \boldsymbol{\omega}$. This can be the case in a rotating NS with a toroidal magnetic field component.

We assume that a compact star has a dipole configuration for the magnetic field: the two tori of the toroidal magnetic field $\mathbf{B}_t$ and poloidal field component $\mathbf{B}_p$. The directions of the field in the tori are different in the opposite hemispheres of a star, as schematically depicted in Fig. 8.2. Braithwaite and Nordlund (2006) showed that a stellar magnetic field, which has only a toroidal or poloidal component, is unstable; see also Fig. 8.1.

Let us assume that the toroidal field in one hemisphere of the star produces magnetic field rings or vortices, which disattach from the corresponding torus. Andersson et al. (2007) found that certain turbulence processes, which occur inside some compact rotating stars, can form such

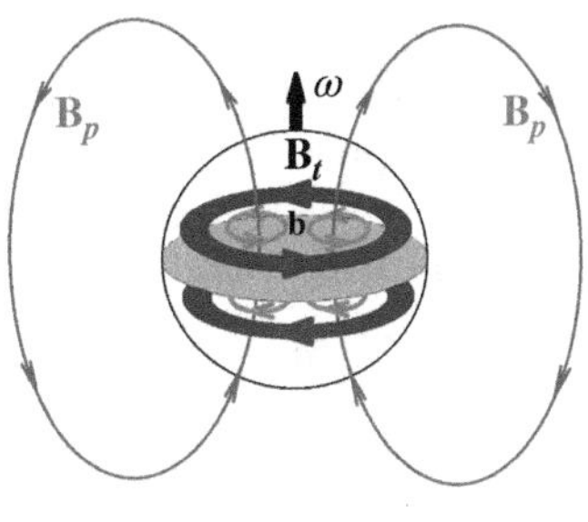

Figure 8.2 The schematic structure of the stellar magnetic field under consideration. The toroidal field $\mathbf{B}_t$ is placed within two blue tori with opposite directions above and below the equator of a star. The poloidal field $\mathbf{B}_p$, which can be nonzero outside the star, is depicted by red lines. The fluctuations of the magnetic field are shown with small green rings. They can exist in both hemispheres of a star. The direction of the magnetic field **b** in such rings coincides with that for the toroidal component. Dvornikov, 2020/Elsevier/Licensed under CC BY 4.0.

rings. These vortices are shown in green in Fig. 8.2. Notice that the direction of the magnetic field in a ring coincides with that in a torus, in which the toroidal component is concentrated. The electroweak mechanism, given in equation (8.71), acts on these vortices.

This process modifies the star's magnetic field: magnetic helicity begins to flow through the stellar equator. Indeed, $\dot{H}$ has a different sign in opposite hemispheres since $\mathbf{B}_t$, and rings have opposite directions in the southern and northern hemispheres. Thus total helicity is conserved: $\dot{H}_{\text{tot}} = \dot{H}_{\text{S}} + \dot{H}_{\text{N}} = 0$. Here, $H_{\text{S,N}}$ are the helicities of the southern and northern hemispheres.

We take that the mean strength of the magnetic field in a ring is **b**, and the mean width of a flux tube of a vortex is r. Assuming that the entire magnetic field in the northern hemisphere is modified into rings, and taking into account the conservation of the magnetic flux, we obtain that $br^2N = B_tR_t^2$, where N is the total number of rings and R_t is the torus radius. Thus we can evaluate $(\nabla \times \mathbf{B})$ in equation (8.71):

$$(\nabla \times \mathbf{B}) \sim N\frac{b}{r} \sim \frac{B_tR_t^2}{r^3}. \tag{8.73}$$

Let us consider, for instance, the northern hemisphere. Using equation (8.71), we have the rate of the helicity change in the form

$$\dot{H}_{\text{N}} \sim -\frac{4}{3}\pi B_t R_t^2 \omega \frac{V_5}{e\sigma}\frac{R^3}{r^3}, \tag{8.74}$$

where $R \sim 10\,\text{km}$ is the radius of the NS. The typical time of the helicity change in the hemisphere is

$$\tau \sim \frac{H_{\text{N}}}{|\dot{H}_{\text{N}}|} = \frac{3}{4\pi}\frac{eB_p\sigma r^3}{\omega V_5 R}. \tag{8.75}$$

Berger (1999) proposed an alternative definition for helicity than that in equation (8.70): $H = 2L\Phi_1\Phi_2$, where $\Phi_{1,2}$ are the linked magnetic fluxes and $L = 0, \pm 1, \ldots$ is the linkage number (see also section 8.3.3). This definition is used in equation (8.75). We take that $H_{\text{N}} \sim B_tR_t^2B_pR^2$ is the total helicity in the northern hemisphere in equation (8.75).

We apply the mechanism for the helicity change, proposed earlier, inside a NS. Dvornikov and Semikoz (2015a) found that $V_5 = G_{\text{F}}n_n/2\sqrt{2} = 6\,\text{eV}$ in this case. Here, $n_n = 1.8 \times 10^{38}\,\text{cm}^{-3}$ is the density of neutrons in a NS. Using chiral phenomena for relativistic electrons in dense matter of a NS is justified because the Fermi momentum of such particles $p_{\text{F}e} = (3\pi^2n_e)^{1/3} \sim 10^2\,\text{MeV}$ is much greater than their masses. Here, we take that the density of electrons n_e is about 5% of n_n inside a NS.

We assume that $B_p = 10^{12}\,\text{G}$ is the typical magnetic field of a pulsar and $\omega = 10^3\,\text{s}^{-1}$ is the angular velocity of a millisecond pulsar. Schmitt and Shternin (2018) found that the electric conductivity of NS matter having temperature $T \sim 10^7\,\text{K}$ is $\sigma = 10^6\,\text{GeV}$. Gusakov and Dommes (2016) predicted the existence of magnetic flux tubes with $r \sim 6 \times 10^{-5}\,\text{cm}$ in the background matter of certain NSs. We obtain that $\tau = 1.3 \times 10^{11}\,\text{s} = 4 \times 10^3\,\text{yr}$ if we use the previous values in equation (8.75).

We have described a new possibility for magnetic helicity to flow through the equator of a NS driven by the electroweak interaction of chiral fermions with rotating matter. Del Sordo et al. (2013) found that helicity flux through the equator is related to the periodic variation of the stellar magnetic field. For instance, it can be the known solar cycle with a period of 22 yr. Therefore, we can conclude that the obtained $\tau \sim 10^3\,\text{yr}$ is related to a periodic variation of the magnetic field in pulsars. Contopoulos (2007) observed analogous cycles in certain NSs with $10^2\,\text{yr} < \tau < 10^4\,\text{yr}$ by studying pulsars spin-down. The mechanism predicted in this chapter is a possible explanation for the observations of Contopoulos (2007).

Now we will show that we can omit $n_5^{(\text{sat})}$ in equation (8.68). It was mentioned earlier that the total magnetic helicity of a NS is conserved. The second term in equation (8.68), $\sim (\mathbf{E} \cdot \nabla \times \mathbf{V}_5)$, is important while turbulent vortices of the magnetic field are produced. Thus we can compute the contribution to $n_5^{(\text{sat})}$ at the beginning of the helicity change from the first term in the integrand, $\sim (\mathbf{E} \cdot \mathbf{B})$. Given that $\mathbf{E} = (\nabla \times \mathbf{B})/\sigma$, we get that $n_5^{(\text{sat})} \sim e^2B^2/2\pi^2R\sigma\Gamma$. If one takes the typical magnetic field of a pulsar $B = 10^{12}\,\text{G}$, the radius of a NS $R = 10\,\text{km}$, the rate of the spin flip $\Gamma = 10^{11}\,\text{s}^{-1}$, and the electric conductivity $\sigma = 10^6\,\text{GeV}$, we obtain that $n_5^{(\text{sat})} \sim 10^{12}\,\text{cm}^{-3}$.

We can compare the value of $n_5^{(\text{sat})}$ estimated with the initial chiral imbalance $n_5(0)$, which is produced at the formation of the NS. Dvornikov and Semikoz (2015a) proposed that a seed chiral imbalance can be created in parity-violating direct Urca processes, washing left electrons out of the NS. The energy scale of Urca processes is $\mu_5(0) = m_n - m_p \sim 1\,\text{MeV}$, where $\mu_5(0) = (\mu_R - \mu_L)/2 = \pi^2 n_5(0)/p_{Fe}^2$ and $m_{p,n}$ are the masses of a proton and a neutron. Dvornikov (2016a), Dvornikov and Semikoz (2015a), Dvornikov et al. (2020) used this value of $\mu_5(0)$ to study the problem of magnetars with the help of chiral phenomena. If $p_{Fe} \sim 10^2\,\text{MeV}$, one obtains that $n_5(0) \sim 10^{33}\,\text{cm}^{-3}$. Therefore, one can see that $n_5^{(\text{sat})} \ll n_5(0)$. Hence, we can neglect both n_5 and $\dot{n}_5$ in equation (8.34) in an old NS.

Nevertheless, one can have a backreaction to the chiral imbalance from the magnetic field. Dvornikov et al. (2020) studied this process recently and found a spike in the dependence of n_5 on time associated with the change of the polarity of a young NS by considering the full set of AMHD equations. Concerning the problem discussed in this chapter, n_5 can grow sharply within the time τ evaluated in equation (8.75). However, one needs a separate study to clarify this issue.

8.4. CONCLUSION

In this chapter, we have studied two possibilities for the evolution of magnetic helicity in a compact star. First, we derived the new equation (8.48) accounting for the new quantum surface term in equation (8.52) with spin effects in plasma by averaging the Adler anomaly in equation (8.40). This new term is important for the evolution of magnetic helicity at the spherical surface around a finite volume of a dense NS.

The evolution of magnetic helicity could cause magnetic field lines to reconnect at the surface and result in flares from a star's outer boundary in the stellar magnetosphere. We did not solve this problem in this chapter. We are only trying to find how strong the new magnetization effect is deep within the NS core where our approximations are valid. For this purpose, we considered small base areas $L^2 = (R\Delta\theta)^2 \ll R^2$ at the surface with radius $R < R_{NS}$ for the corresponding thin magnetic tubes intersecting such a surface for both poloidal and toroidal components in equation (8.59). Because both the magnetic helicity diffusion $-2\int_V d^3x(\mathbf{E}\cdot\mathbf{B})$ and the new quantum term in equation (8.55) are important only at small scales $L \ll R$, when the evolution time is greater than a long diffusion time for the high conductivity of matter in a NS, $t > \tau_D = L^2\sigma_{\text{cond}}$. Notice that the contribution of the new term is not compared with the diffusion losses since we considered an ideal plasma in the limit $\sigma_{\text{cond}} \to \infty$.

We find that the inequalities $m_e^2 \ll eB \ll p_{F_e}^2$ are fulfilled in magnetars with strong magnetic fields $B \sim 10^{15}\,\text{G}$. Moreover, $p_{F_e} = 100\,\text{MeV}$ within a core of these NSs with ultrarelativistic electrons. Thus, the CME contribution is small due to a small population of electrons at the main Landau level $n = 0$; see also equation (8.8). Nevertheless, these electrons provide magnetic helicity diffusion through the new quantum contribution in the evolution equation (8.56).

Note that the WKB approximation with $n \gg 1$ in equation (8.8), when the paramagnetic (spin) contribution is a small correction in the Landau spectrum equation (8.1), simplifies the derivation of the pseudoscalar term $2im_e\langle\bar{\psi}\gamma_5\psi\rangle = \mu_B^{-1}(\nabla\cdot S_5)$ made by Dvornikov and Semikoz (2018).

The WKB approximation is invalid in the NS crust, where degenerate electrons become nonrelativistic, $eB \gg m_e^2 \gg p_{F_e}^2$. This means they populate only the main Landau level $n = 0$. In this situation, one can calculate the pseudovector S_5 in such a medium. It is expected to be comparable with the standard magnetization in equation (8.4). This case is of special interest since it corresponds to the outer NS surface where the magnetic helicity evolution can lead to gamma or X-ray bursts in magnetars.

The second task, considered in this chapter, is related to chiral phenomena in spatially inhomogeneous matter accounting for the electroweak interaction between chiral electrons and background fermions. In section 8.2.2, we derived the wave equation for chiral electrons electroweakly interacting with background matter having a nonuniform density and an arbitrary velocity. We analyzed this equation based on the concept of the Berry phase. In section 8.2.2, we derived the kinetic equations for left and right particles and the effective actions that generalize the results of Dvornikov and Semikoz (2017), Semikoz and Dvornikov (2018).

Then, in section 8.2.3, we derived the corrections to the Adler-Bell-Jackiw anomaly and the anomalous current from the electroweak interaction of chiral electrons with inhomogeneous matter. In this situation, we considered a particular example of rotating matter. The contribution to the anomalous current that we obtained is consistent with the result of Dvornikov (2015).

In section 8.3.4, we applied our results to describe the evolution of magnetic fields in the dense matter of a compact star. We obtained the contribution to the evolution of the magnetic helicity of a rotating NS due to the electroweak interaction of chiral fermions with background neutrons by supposing that the chiral imbalance vanishes in collisions between particles. This effect generates the magnetic helicity flux through the equator of a compact star. We assumed that magnetic vortices in the form of rings are formed because of magnetic turbulence in the NS. Then we computed the typical time of the change in magnetic helicity in one hemisphere of a compact star. We demonstrated that the total magnetic helicity of a NS is constant.

Del Sordo et al. (2013) showed that the flux of the magnetic helicity through the stellar equator is related to the periodic cycle of the magnetic activity of a star, analogous to the known solar cycle. The typical time, obtained in section 8.3.4, was compared to the period of the cyclic electromagnetic activity of certain pulsars observed by Contopoulos (2007). Therefore, the mechanism developed in this chapter is a feasible explanation for the results of Contopoulos (2007).

REFERENCES

Andersson, N., Sidery, T., & Comer, G. L. (2007). Superfluid neutron star turbulence. *Monthly Notices of the Royal Astronomical Society*, *381*, 747–756. astro-ph/0703257

Armitage, N. P., Mele, E. J., & Vishwanath, A. (2018). Weyl and Dirac semimetals in three-dimensional solids. *Reviews of Modern Physics*, *90*, 15001. arXiv:1705.01111

Berestetskiĭ, V. B., Lifshitz, E. M., & Pitaevskiĭ, L. P. (1998). *Quantum electrodynamics* (2nd ed.).Oxford: Pergamon Press.

Berger, M. A. (1999). Introduction to magnetic helicity. *Plasma Physics and Controlled Fusion*, *41*, B167.

Berry, M. V. (1984). Quantal phase factors accompanying adiabatic changes. *Proceedings of the Royal Society of London. Series A, mathematical and physical sciences*, *392*(1802), 45–57.

Braithwaite, J., & Nordlund, A. (2006). Stable magnetic fields in stellar interiors. *Astronomy and Astrophysics*, *450*, 1077–1095. astro-ph/0510316.

Brandenburg, A., Schober, J., Rogachevskii, I., Kahniashvili, T., Boyarsky, A., Fröhlich, J., et al. (2017). Stable magnetic fields in stellar interiors. *The Astrophysical Journal Letters*, *845*, L21. arXiv:1707.03385.

Boyarsky, A., Fröhlich, J., & Ruchayskiy, O. (2012). Self-consistent evolution of magnetic fields and chiral asymmetry in the early Universe. *Physical Review Letters*, *108*, 031301. arXiv:1109.3350.

Chandrasekhar, S., & Kendall, P. C. (1957). On force-free magnetic fields. *The Astrophysical Journal*, *126*, 457–460.

Contopoulos, I. (2007). A note on the cyclic evolution of the pulsar magnetosphere. *Astronomy and Astrophysics*, *475*, 639–642. arXiv:0709.3957.

DelSordo, F., Guerrero, G., & Brandenburg, A. (2013). Turbulent dynamos with advective magnetic helicity flux. *Monthly Notices of the Royal Astronomical Society*, *429*, 1686–1694. arXiv:1205.3502.

Dvornikov, M. (2015). Galvano-rotational effect induced by electroweak interactions in pulsars. *Journal of Cosmology and Astroparticle Physics*, *05*, 037. arXiv:1503.00608.

Dvornikov, M. (2016a). Generation of strong magnetic fields in dense quark matter driven by the electroweak interaction of quarks. *Nuclear Physics B*, *913*, 79–92. arXiv:1608.04946.

Dvornikov, M. (2016b). Relaxation of the chiral imbalance and the generation of magnetic fields in magnetars. *Journal of Experimental and Theoretical Physics*, *123*, 967–978. arXiv:1510.06228.

Dvornikov, M. (2018). Magnetic fields in turbulent quark matter and magnetar bursts. *International Journal of Modern Physics D*, *27*, 1750184. arXiv:1612.06540.

Dvornikov, M. (2020). Magnetic helicity in plasma of chiral fermions electroweakly interacting with inhomogeneous matter. *Nuclear Physics B*, *955*, 115049. arXiv:2005.09358.

Dvornikov, M., & Semikoz, V. B. (2015a). Magnetic field instability in a neutron star driven by the electroweak electron-nucleon interaction versus the chiral magnetic effect. *Physical Review D*, *91*, 061301. arXiv:1410.6676.

Dvornikov, M., & Semikoz, V. B. (2015b). Generation of the magnetic helicity in a neutron star driven by the electroweak electron-nucleon interaction. *Journal of Cosmology and Astroparticle Physics*, *05*, 032. arXiv:1503.04162.

Dvornikov, M., & Semikoz, V. B. (2015c). Energy source for the magnetic field growth in magnetars driven by the electron-nucleon interaction. *Physical Review D*, *92*, 083007. arXiv:1507.03948.

Dvornikov, M. S., & Semikoz, V. B. (2017). Nonconservation of lepton current and asymmetry of relic neutrinos. *Journal of Experimental and Theoretical Physics*, *124*, 731–739. arXiv:1603.07946.

Dvornikov, M., & Semikoz, V. B. (2018). Magnetic helicity evolution in a neutron star accounting for the Adler-Bell-Jackiw anomaly. *Journal of Cosmology and Astroparticle Physics*, *08*, 021. arXiv:1805.04910.

Dvornikov, M., & Semikoz, V. B. (2019). Permanent mean spin source of the chiral magnetic effect in neutron stars. *Journal of Cosmology and Astroparticle Physics*, *06*, 053. arXiv:1904.05768.

Dvornikov, M., & Studenikin, A. (2002). Neutrino spin evolution in presence of general external fields. *Journal of High Energy Physics*, *09*, 016. hep-ph/0202113.

Dvornikov, M., Semikoz, V. B., & Sokoloff, D. D. (2020). Generation of strong magnetic fields in a nascent neutron star accounting for the chiral magnetic effect. *Physical Review D*, *101*, 083009. arXiv:2001.08139.

Fukushima, K., Kharzeev, D. E., & Warringa, H. J. (2008). The chiral magnetic effect. *Physical Review D*, *78*, 074033. arXiv:0808.3382.

Giovannini, M., & Shaposhnikov, M. E. (1998). Primordial hypermagnetic fields and triangle anomaly. *Physical Review D*, *57*, 2186–2206. hep-ph/9710234.

Gusakov, M. E., & Dommes, V. A. (2016). Relativistic dynamics of superfluid-superconducting mixtures in the presence of topological defects and an electromagnetic field with application to neutron stars. *Physical Review D*, *94*, 083006. arXiv:1607.01629.

Huang, X.-G. (2016). Simulating chiral magnetic and separation effects with spin-orbit coupled atomic gases. *Scientific Reports*, *6*, 20601. arXiv:1506.03590.

Ioffe, B. L. (2006). Axial anomaly: The modern status. *International Journal of Modern Physics A*, *21*, 6249–6266. hep-ph/0611026.

Kaminski, M., Uhlemann, C. F., Bleicher, M., & Schaner-Bielich, J. (2016). Anomalous hydrodynamics kicks neutron stars. *Physics Letters B*, *760*, 170–174. arXiv:1410.3833.

Kaspi, V. M., & Beloborodov, A. M. (2017). Magnetars. *Annual Review of Astronomy and Astrophysics*, *55*, 261. arXiv:1703.00068.

Kharzeev, D. E., Liao, J., Voloshin, S. A., & Wang, G. (2016). Chiral magnetic and vortical effects in high-energy nuclear collisions – A status report. *Progress in Particle and Nuclear Physics*, *88*, 1–28. arXiv:1511.04050.

Landau, L. D., & Lifshitz, E. M. (1980). *Statistical physics* (3rd ed.) Oxford: Pergamon Press.

Lattimer, J. M., & Prakash, M. (2001). Neutron star structure and the equation of state. *The Astrophysical Journal*, *550*, 426. astro-ph/0002232.

Masada, Y., Kotake, K., Takiwaki, T., & Yamamoto, N. (2018). Chiral magnetohydrodynamic turbulence in core-collapse supernovae. *Physical Review D*, *98*, 083018. arXiv:1805.10419.

Moss, D., Saar, S. H., & Sokoloff, D. (2008). What can we hope to know about the symmetry of stellar magnetic fields? *Monthly Notices of the Royal Astronomical Society*, *388*, 416–420.

Nunokawa, H., Semikoz, V. B., Smirnov, A. Yu., & Valle, J. W. F. (1997). Neutrino conversions in a polarized medium. *Nuclear Physics B*, *501*, 17–40. hep-ph/9701420.

Pavlović, P., Leite, N., & Sigl, G. (2017). Chiral magnetohydrodynamic turbulence. *Physical Review D*, *96*, 023504. arXiv:1612.07382.

Priest, E. (2016). MHD structures in three-dimensional reconnection. In W. Gonzales, E. Parker (Eds.), *Magnetic reconnection: Concepts and applications, astrophysics and space science library* (Vol. 427, pp. 101–142). Cham: Springer.

Schmitt, A., & Shternin, P. (2018). Reaction rates and transport in neutron stars. In L. Rezzolla, P. Pizzochero, D. Jones, N. Rea, I. Vidaña (Eds.), *The physics and astrophysics of neutron stars, astrophysics and space science library* (Vol. 457, pp. 455–574). Cham: Springer. arXiv:1711.06520.

Semikoz, V. B., & Dvornikov, M. (2018). Generation of the relic neutrino asymmetry in a hot plasma of the early universe. *International Journal of Modern Physics D*, *27*, 1841008. arXiv:1712.06565.

Semikoz, V. B., & Valle, J. W. F. (1997). Erratum: Nucleosynthesis constraints on active sterile neutrino conversions in the early universe with random magnetic field. *Nuclear Physics B*, *485*, 545–547. hep-ph/9607208.

Sigl, G., & Leite, N. (2016). Chiral magnetic effect in protoneutron stars and magnetic field spectral evolution. *Journal of Cosmology and Astroparticle Physics*, *01*, 025. arXiv:1507.04983.

Silin, V. P. (1968). Spin waves in non-ferromagnetic metals. In A. I. Akhiezer, V. G. Bary'achtar, S. V. Peletminsky (Eds.), *Spin waves*. Amsterdam: North-Holland Publishers Co.

Silva, L. O., Bingham, R., Dawson, J. M., Mendonça, J. T., & Shukla, P. K. (1999). Neutrino driven streaming instabilities in a dense plasma. *Physical Review Letters*, *83*, 2703–2706.

Son, D. T., & Yamamoto, N. (2013). Kinetic theory with Berry curvature from quantum field theories. *Physical Review D*, *87*, 085016. arXiv:1210.8158.

Vilenkin, A. (1980a). Equilibrium parity violating current in a magnetic field. *Physical Review D*, *22*, 3080–3084.

Vilenkin, A. (1980b). Quantum field theory at finite temperature in a rotating system. *Physical Review D*, *21*, 2260–2269.

Vinitskiĭ, S. I., Derbov, V. L., Dubovik, V. M., Markovski, B. L., & Stepanovskiĭ, Yu. P. (1990). Topological phases in quantum mechanics and polarization optics. *Physics–Uspekhi*, *33*, 403–428.

Yakovlev, D. G., & Pethick, C. J. (2008). Neutron star cooling. *Annual Review of Astronomy and Astrophysics*, *42*, 169–210. astro-ph/0402143.

Yamamoto, N. (2016). Chiral transport of neutrinos in supernovae: Neutrino-induced fluid helicity and helical plasma instability. *Physical Review D*, *93*, 065017. arXiv:1511.00933.

Zhao, J., & Wang, F. (2019). Experimental searches for the chiral magnetic effect in heavy-ion collisions. *Progress in Particle and Nuclear Physics*, *107*, 200–236. arXiv:1906.11413.

9

Writhing From Biophysics to Solar Physics and Back

Christopher Prior and Arron N. Bale

ABSTRACT

We introduce a brief chronological history of the development and interplay between the fields of solar physics and biophysics of the concept of writhing. Writhing quantifies a structure's global self-entanglement (knotting) and plays a fundamental role in DNA compactification (supercoiling). We discuss how the measure arose initially in a biophysical context and how mathematical developments in this field were used in solar physics to advance the study of magnetic helicity. We then highlight how subsequent mathematical advances in a solar physics context have recently provided solutions to long-standing mathematical issues with the quantity's use in biophysical applications. Finally, we provide preliminary evidence that bounds derived for magnetic helicity can be used to provide insight into the understanding of protein structures through their writhing measures.

9.1. INTRODUCTION

Writhing is a measure of the global entanglement of a three-dimensional curve, first introduced by Georges Călugăreanu [1959, 1961] (see Moffatt and Ricca (1995) for a historical context). It rose to prominence as an important geometrical quantity in DNA models, forming part of an invariant sum with the helical coiling of the molecule's dipeptide backbone. A famous pair of articles by Fuller (1971, 1978) popularized the theorem and give it its modern notation:

$$Lk = Wr + Tw. \tag{9.1}$$

The meaning of the theorem is best introduced visually, as shown in Fig. 9.1. First, in panel (a), we see a *ribbon structure*, the mathematical object to which Călugăreanu's theorem applies. An *axis curve* $\mathbf{x}(s)$ (parameterized by its arclength) is surrounded by a unit vector field $\mathbf{V}(s)$, centered on its axis, and normal to the unit tangent curve of the axis: $\mathbf{T}_{\mathbf{x}} = \mathrm{d}\mathbf{x}/\mathrm{d}s$. This vector field rotates, drawing out a second curve $\mathbf{y}$ winding around $\mathbf{x}$. Formally we have

$$\mathbf{y}(s) = \mathbf{x}(s) + \epsilon \mathbf{V}(s), \tag{9.2}$$

with ϵ the ribbon's width. The quantity Lk represents the integer number of times the two curves $\mathbf{x}$ and $\mathbf{y}$ interlink. This quantity is a topological invariant; that is to say, it remains unchanged if the structure distorts without crossing itself. Formally it is defined by the Gauss linking integral:

$$Lk = \frac{1}{4\pi}\oint_x\oint_y \mathbf{T}_{\mathbf{x}}(s)\times\mathbf{T}_{\mathbf{y}}(t)\cdot\frac{\mathbf{x}(s)-\mathbf{y}(t)}{\|\mathbf{x}(s)-\mathbf{y}(t)\|^3}\,\mathrm{d}s\,\mathrm{d}t. \tag{9.3}$$

A geometrical/topological interpretation of this double integral is illustrated in Fig. 9.2. In short, we can calculate the linking of the two curves by projecting them onto a plane and counting the apparent signed crossings (it is orientation dependent) (Fig. 9.2a). This sign is related to the sign of the vector $\mathbf{T}_{\mathbf{x}}\times\mathbf{T}_{\mathbf{y}}\cdot(\mathbf{x}-\mathbf{y})$. Of course, one can do this for any choice of projection; two other examples are shown in Fig. 9.2b,c. Interestingly, despite the number of crossings changing between the projections indicated in Fig. 9.2b,c, the signed sum does not. In fact, this projection invariance is a result of the degree theorem

Department of Mathematical Sciences, Durham University, Durham, UK

Helicities in Geophysics, Astrophysics, and Beyond, Geophysical Monograph 283, First Edition.
Edited by Kirill Kuzanyan, Nobumitsu Yokoi, Manolis K. Georgoulis, and Rodion Stepanov.

DOI:10.1002/9781119841715.ch09

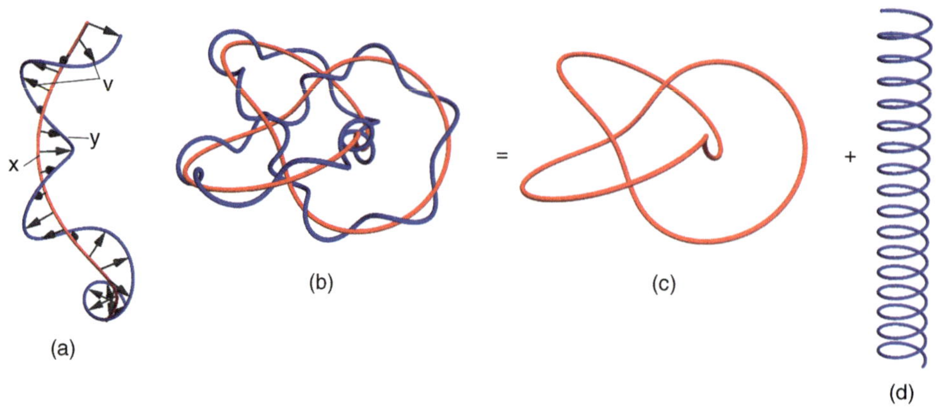

Figure 9.1 A visualization of the Călugăreanu theorem. (a) A ribbon structure, the mathematical object to which the theorem applies. (b) A closed ribbon (whose axis is a trefoil knot). The two curves are interlinked, as measured by the linking quantity *Lk*. (c) The axis curve alone. Its self-entanglement, writhe *Wr*, accounts for the global entanglement of the structure in (b) and changes when the axis (the overall shape) deforms. (d) The twisting *Tw* of the blue curve in (b) around the axis shown in (c). The theorem tells us that the changing number of full turns under deformation is controlled by the shape of the axis (c).

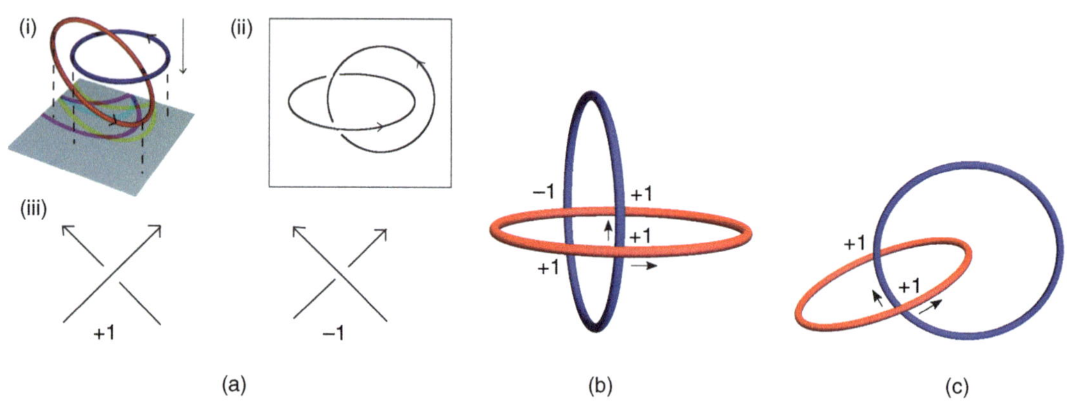

Figure 9.2 Visual interpretations of the linking *Lk* of a pair of curves. (a) In (i)→ (ii), the curves are projected onto a plane. Where the curves intersect, the underlying strand is cut. (iii) the crossings are assigned a value ±1 depending on whether the overlaying strand is running left to right or vice versa. (b) A link projection with four crossings whose sum is +2. (c) A projection of the same curves as in (b), with only *two* crossings. Its sum is still +2.

(see, e.g., Frankel (2011)). The integral in equation (9.3) represents the *average* linking calculated by projection over all projections or viewpoints. This might seem an unnecessary calculation when first encountered; and indeed, for the evaluation of Lk alone, it is. One can calculate the quantity much more quickly than a naive discretization of the double integral in equation (9.3) by various methods (see Qu and James (2021) for an overview of the various possibilities). This chapter will give two reasons the averaged calculation in equation (9.3) is still relevant from a practical perspective. The first is Călugăreanu's theorem from equation (9.1), and the second concerns magnetic field topology.

Călugăreanu asked, "What happens when the ribbon width ϵ tends to zero and the integrand in equation (9.3) develops a (regularizable) singularity?" The answer, in short, is that Lk, given by equation (9.3), can be decomposed into the writhe Wr and twist Tw:

$$Wr = \frac{1}{4\pi}\oint_x\oint_x \mathbf{T}_\mathbf{x}(s)\times\mathbf{T}_\mathbf{x}(t)\cdot\frac{\mathbf{x}(s)-\mathbf{x}(t)}{\|\mathbf{x}(s)-\mathbf{x}(t)\|^3}\,\mathrm{d}s\,\mathrm{d}t, \quad (9.4)$$

$$Tw = \frac{1}{2\pi}\oint_x \mathbf{T}_x(s)\cdot\mathbf{V}(s)\times\frac{\mathrm{d}\mathbf{V}(s)}{\mathrm{d}s}\,\mathrm{d}s. \quad (9.5)$$

One can see that the writhe integral is essentially the self-linking; in fact, this was the name given to the quantity by Călugăreanu and also Pohl and White, who gave a clearer exposition of the theorem after its original definition (Pohl, 1968; White, 1969). Fuller, whose more widely accessible articles have influenced the biological

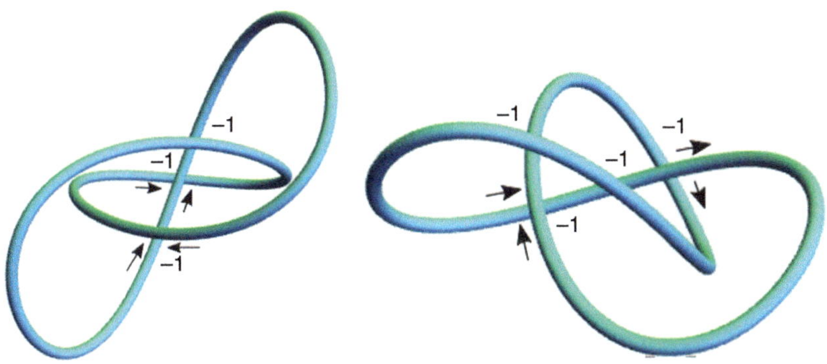

Figure 9.3 Viewpoints and directions of a trefoil knot, the curve **x** shown in Fig. 9.1. The signed crossing calculations are shown to differ.

literature, coined the term *writhe*. The interpretation of equation (9.5) is the average signed (self) crossing of all projections of the curve **x**, as indicated in Fig. 9.3. The critical difference here, in comparison to the projected linking of two curves, is that the sum of signed crossings differs for differing projections, so averaging over all projections is necessary, and Wr is real-valued rather integer-valued (in general); see, e.g., Berger and Prior (2006).

The twisting equation (9.5) is a far more straightforward quantity and represents the rotation of the vector **V** around the curve **x** (represented in the formula by its tangent vector $\mathbf{T}_x$).

So we can now be a little more precise when discussing Fig. 9.1. The linking arises from the observation of crossings along the direction:

$$\mathbf{m} = \frac{\mathbf{x}(s) - \mathbf{y}(t)}{|\mathbf{x}(s) - \mathbf{y}(t)|} = \frac{\mathbf{x}(s) - \mathbf{x}(t) - \epsilon\mathbf{V}(t)}{|\mathbf{x}(s) - \mathbf{x}(t) - \epsilon\mathbf{V}(t)|}, \tag{9.6}$$

Since ϵ is small, a significant component of the variation of the vector **m** comes from the vector

$$\mathbf{m}_x = \frac{\mathbf{x}(s) - \mathbf{x}(t)}{|\mathbf{x}(s) - \mathbf{x}(t)|}, \tag{9.7}$$

which is to say a significant amount of the linking of the two curves in Fig. 9.1 arises from the knotting of the curve **x**. The remarkable fact about the theorem is that Tw exactly quantifies the difference between the two (Lk and Wr). A common biophysical application of this theorem is for the linking Lk to be used as a fixed model constraint on the number of DNA sequence repeats. The constraint is then applied to an energy model of the DNA backbone, which often has twisting Tw as an elastic energy component (Klenin et al., 1989; Fain et al., 1997; Bouchiat and Mézard, 1998; Marko and Neukirch, 2012; Lillian et al., 2011; Lam and Zhen, 2018). The writhe can then be used to constrain the allowed global (axial) shape of the molecule. In DNA models, the writhe measures supercoiling.

Linking has also long played a role in solar physics applications through magnetic helicity, as most readers of this book are no doubt aware. Magnetic helicity is most commonly introduced as the following quantity:

$$H = \int_{\Omega} \mathbf{A} \cdot \mathbf{B}\, \mathrm{d}V, \tag{9.8}$$

where $\mathbf{B} \cdot \mathbf{n} = 0$ on $\partial\Omega$ and **A** is a vector potential satisfying $\mathbf{B} = \nabla \times \mathbf{A}$. As noted by Moreau (1961) and Moffatt (1969), use of the Biot-Savart gauge for **A**

$$\mathbf{A} = \frac{1}{4\pi}\int_{\Omega} \mathbf{B}(\mathbf{y}) \times \frac{\mathbf{x} - \mathbf{y}}{|\mathbf{x} - \mathbf{y}|^3}\mathrm{d}^3 y \tag{9.9}$$

reveals the magnetic helicity to be

$$H = \frac{1}{4\pi}\int_{\mathbb{R}^3}\int_{\mathbb{R}^3} \mathbf{B}(\mathbf{x}) \cdot \mathbf{B}(\mathbf{y}) \times \frac{\mathbf{x} - \mathbf{y}}{|\mathbf{x} - \mathbf{y}|^3}\, \mathrm{d}^3 x\, \mathrm{d}^3 y. \tag{9.10}$$

Since $\mathbf{B} = |\mathbf{B}|\mathbf{T}$, this is just the linking weighted by field strength. As the reader of this book is no doubt aware, magnetic helicity is considered a fundamental quantity in astrophysical contexts as it is (approximately) a conserved quantity (Berger, 1984; Pariat et al., 2015; Ji et al., 1995; Pipin et al., 2013). An intriguing study by Laing et al. (2015) indicated that the writhe component of helicity is conserved under antiparallel reconnection (under changing topology), a result they relate to applications in both solar physics and biophysics. This is a neat example of results from the two subject areas providing mutual insight.

The critical point about these discussions is that they only strictly apply to closed (or ergodic) curve pairs. The theorem in equation (9.1), as derived by Călugăreanu is not valid if the curves **x**, **y** are rectifiable and have endpoints; we refer to these as *open curves* in what follows. In DNA modeling and solar physics, a significant proportion of the potential (likely the majority) of applications of topological entanglement involve open-ended curves. Much of this chapter will detail how the two fields have developed the topological concepts discussed so far to deal with open-curve topology, particularly highlighting possible developments that transpose results from one field to another to provide novel insight.

9.2. OPEN LINKING AND WRITHE

The linking integral in equation (9.3) is a well-defined quantity for open ribbons, an example of which is shown in Fig. 9.4a. However, it is no longer an integer, nor is it invariant under isotopy, even if the ends of the curves are prevented from rotating (Sierzega et al., 2021). One can envisage why in terms of link projections. As in the closed ribbon case, if we project the link along a direction $\mathbf{n}$ and count crossings, we can define a planar linking. $Lk(\mathbf{x}, \mathbf{y})$ is again the average over all such projections, but as not all viewpoints see the same crossing number, Lk is generally not an integer. The lack of invariance can be understood by noting that the number of crossings of specific projections can change if the ribbon's *interior* deforms to pass under/above one of its endpoints in projection, as indicated in Fig. 9.4b. This interior deformation distinction is important. We should expect that if the ribbon's ends are prevented from rotating (deformation is constrained to the domain's interior), some measure of its entanglement should be preserved. This conservation clearly cannot be imposed by the quantity Lk.

With regard to magnetic fields, the analogous observation is that the Biot-Savart gauge in equation (9.9), used to yield a definition of magnetic helicity as the field strength weighted linking, is not a valid vector potential if the magnetic field has a nontrivial normal boundary component. That is to say, $\nabla \times \mathbf{A} \neq \mathbf{B}$ if $\mathbf{A}$ is given by equation (9.9) because of a surface integral that depends on $\mathbf{B} \cdot \mathbf{n}$, with $\mathbf{n}$ the surface normal (see Cantarella et al. (2001a)). This coincides with magnetic fields, which have open-ended field lines and whose linking is not of integral value and not a topological invariant to ideal (isotopic) flows. This hints at the notion that *valid* vector potentials satisfy a physical principle: their associated helicity measures provide measures of the field's topology that are preserved under an ideal flow on the domain's interior (or interior isotopy).

In both cases, there was a need to develop quantities that could enforce invariance for open-ended fields and extend the notion of writhing.

9.2.1. Fuller and Closures

Fuller (1978) had an idea that would have significant consequences in solar physics, although his considerations were toward biophysical applications of topology. He considered open-ended DNA structures, typically tethered to something – whether a cell membrane, protein, or experimental apparatus – to measure its mechanical response. In such scenarios, the total entanglement of the molecule is still conserved (or controlled at its ends). He suggested creating an artificial closure of the ribbon structure, which would interfere as little as possible with the structure and yield a well-defined invariant Lk for the closed structure. This is illustrated in Fig. 9.5a.

This method has proven widely popular in DNA and biophysical modeling (see, e.g., Bouchiat and Mézard (1998), Fain et al. (1997), Rossetto and Maggs (2003), Maffeo et al. (2010), Medalion et al. (2010), Chou et al. (2014), Lepage et al. (2015)) and elastic tube applications (Van der Heijden and Thompson, 2000; van der Heijden et al., 2007; Patil et al., 2020). It is complicated by the

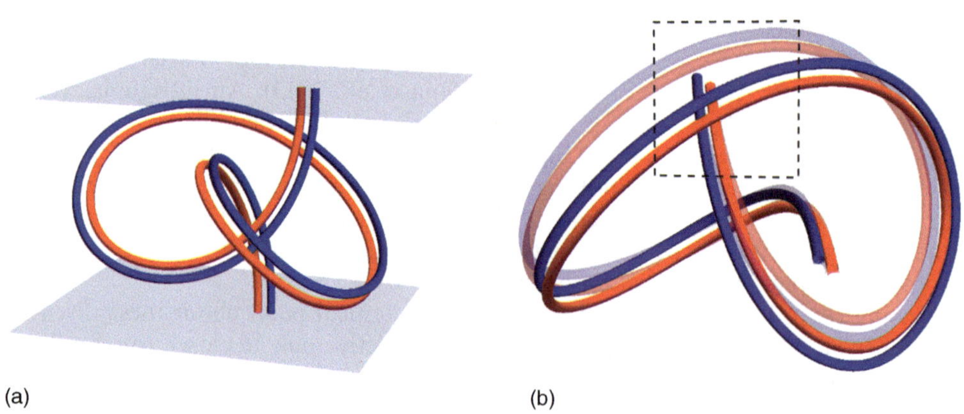

Figure 9.4 Open ribbons. (a) An open-ended ribbon whose endpoints stay fixed in the bounding planes but whose interior may deform smoothly. (b) The ribbon from a point of view included in the evaluation of *Lk* but not the net-winding *L*. One can calculate the planar linking of this projection by assigning an orientation to the curves using the crossing rules indicated in Fig. 9.2. The value changes in the subsequent deformation of the interior of the ribbon (shown opaquely), as we see a set of crossings is lost in the opaque figure as a section of the interior passes over the ribbon's end.

fact that the "closure" generally contributes to both the writhing and the linking of the composite, and hence clear geometrical insight cannot always be achieved from the calculated quantities (Neukirch and Starostin, 2008; Prior and Neukirch, 2016). Alternative approaches have been formulated that approximate the writhing value (Fuller, 1978; Bouchiat and Mézard, 1998; Fain et al., 1997; Rossetto and Maggs, 2003; Samuel et al., 2006; van der Heijden et al., 2007). However, these generally do not characterize the linking/writhing (Neukirch and Starostin, 2008; Prior and Neukirch, 2016) and hence cannot be used consistently to constrain the topology of the ribbon. In short, a huge amount of intellectual effort has been expended in creating imaginary structures, but no consistent and satisfactory method has been found utilizing this approach. One aim of this chapter is to show how developments in solar physics yielded a consistent route to avoid this closure methodology. One of the authors has recently made an attempt to translate this back to the biophysical community (Sierzega et al., 2021); time will judge its success! For now, we summarize the journey to that point.

Closures and Relative Helicity

Despite the previously stated pessimism about the limitations of the closure method in a biological context, its idea inspired Berger and Field (1984) to provide a solution to the problem of the topological definition of helicity for magnetic fields with a nonvanishing boundary component. We take a magnetic field $\mathbf{B}$, defined on a domain V, with some nontrivial normal boundary component $\mathbf{B} \cdot \mathbf{n}$ on ∂V, and extend the field at the boundaries into a larger domain V', such that the extended field has no normal component on V'. This means the extended field topology can be described in terms of the field strength weighted linking. As shown in Fig. 9.5b, this is equivalent to "closing off" tubes of magnetic flux to form closed tubes. As the biophysics community has found, this closure significantly affects the value of the helicity measure. However, as Berger and Field (1984) showed, if we define the relative helicity

$$H(\mathbf{B}) - H(\mathbf{B}'), \tag{9.11}$$

where $\mathbf{B}'$ is some other "reference" field whose normal boundary component on V is the same as that of $\mathbf{B}$, then this difference is independent of the closure used (that the relative helicity can always be defined as equation (9.11) was shown in Prior and Yeates (2014)). In subsequent work, Berger used this relative framework to develop a consistent framework for defining the entanglement of magnetic field lines, *without* the need to consider a closure or reference field (Berger, 1988; Berger and Prior, 2006). We will now describe this approach as it transpired to be crucial in solving the problems in a biophysical context.

9.2.2. Net Winding and the Polar Writhe

If we consider curves that can be bound between two planes, as shown in Fig. 9.6a, we can define the total entanglement by splitting the domain into a set of planes of constant z value. This is achieved by tracking the total rotation of an angle Θ made by the vector $\mathbf{r}$ joining the curves in these planes (and dividing by 2π):

$$\begin{aligned}\mathcal{L}(\mathbf{x},\mathbf{y}) &:= \frac{1}{2\pi}\int_0^h \frac{\mathrm{d}}{\mathrm{d}z}\Theta(\mathbf{x}(z),\mathbf{y}(z))\,\mathrm{d}z \\ &= \frac{1}{2\pi}(\Theta(h)-\Theta(0)) + \mathcal{N},\end{aligned}$$

where N is an integer number of full windings of the vector $\mathbf{r}$. This quantity is invariant to isotopic motions, which vanish on the end boundaries (Berger and Prior, 2006) – more specifically, those that leave the difference $(\Theta(h) - \Theta(0))$ unchanged (this allows for solid body rotations). This appears to be the sought-after invariant.

To extend this measure to curves whose height function is multivalued, we split the curve via its *turning points*, for which $\mathrm{d}x_z/\mathrm{d}z$. We mark each curve as rising or falling with

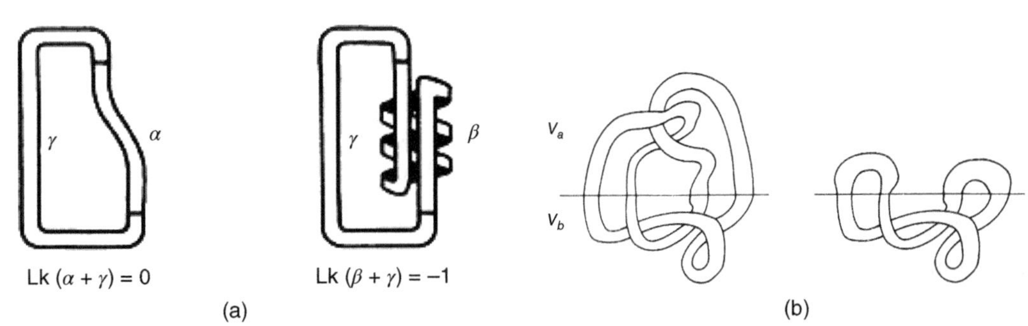

Figure 9.5 Example closure figures. (a) Relative linking. Fuller, 1978, (b) Relative helicity. Berger and Field, 1984/Cambridge University Press.

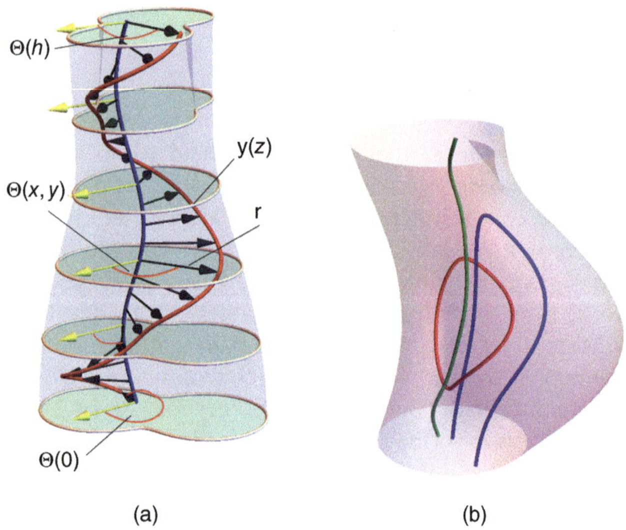

Figure 9.6 Illustrations of winding. (a) A pair of open-ended curves $(\mathbf{x}, \mathbf{y})$ embedded in a domain. Planes of constant z value foliate the domain, and a vector $\mathbf{r}$ joining the curves in each plane is shown. The angle $\Theta(\mathbf{x}, \mathbf{y})$ that this makes with a fixed direction is used to quantify the curves' entanglement. Its total rotation is unchanged if the difference $\Theta(h) - \Theta(0)$ is fixed. (b) Examples of curves to which the quantity $\mathcal{L}$ can be applied.

the following indicator function:

$$\sigma(\mathbf{x}_i) = \begin{cases} 1 & \text{if } \mathrm{d}x_z/\mathrm{d}z > 0, \\ -1 & \text{if } \mathrm{d}x_z/\mathrm{d}z < 0. \end{cases}$$

Then the following sum is invariant under isotopy

$$\mathcal{L}(\mathbf{x}, \mathbf{y}) := \sum_{i=1}^{n+1}\sum_{j=1}^{m+1} \frac{\sigma(\mathbf{x}_i)\sigma(\mathbf{y}_j)}{2\pi} \int_{z_{ij}^{min}}^{z_{ij}^{max}} \frac{\mathrm{d}\Theta\left(\mathbf{x}_i(z), \mathbf{y}_j(z)\right)}{\mathrm{d}z} \mathrm{d}z$$

(Berger and Prior, 2006) *if* the isotopy vanishes at the end boundary. This is valid for combinations of curves that span the planes, are anchored on both planes, and/or are closed in the domain (as shown in Fig. 9.6b).

To see how this relates to the Gauss linking number in equation (9.3), we can define the map

$$\mathbf{m}_z(z) = \frac{\mathbf{y}(z) - \mathbf{x}(z)}{|\mathbf{y}(z) - \mathbf{x}(z)|} = \frac{\mathbf{r}}{|\mathbf{r}|}, \tag{9.12}$$

which is very similar to the map $\mathbf{m}(s, t)$ (equation (9.6)), except it restricts to points on the curve that have the *same* z coordinate (the vector $\mathbf{r}$). It was shown in Berger and Prior (2006) that equation (9.12) is equal to the average of the crossings calculated by projection for all directions $\mathbf{n}$ that lie in the plane: that is, $\mathbf{n} = (\cos(\Theta), \sin(\Theta), 0), \forall\, \Theta \in [0, 2\pi]$. When $\mathbf{m}_z = \mathbf{n}$, one would see a crossing of the curves. This is obviously a far less degenerate calculation than equation (9.3). One can think of this as restricting the set of observers who inform on the linking of the curves, the crucial point being that it does not consider viewpoints that see sections of the interior pass over the ends of the curves, as shown in Fig. 9.4b. This is because all viewpoints (directions of projection) are parallel to the end planes, and we can only see the endpoints crossing in these planes. Thus, if the endpoints do not rotate in these planes, there is no change in their crossing with the rest of the ribbon structure. Critically, if the curves $\mathbf{x}$ and $\mathbf{y}$ are closed curves, the net winding $\mathcal{L}$ is equal to the linking Lk (although the calculation is not the same); it is in some sense a more general invariant than Lk as it can apply a topological constraint in all scenarios where Lk does but also in open-ended cases where Lk is not invariant.

In Prior and Yeates (2014), it was shown that the following choice of vector potential,

$$\mathbf{A}^w = \frac{1}{2\pi}\int_{S_z} \mathbf{B}(\mathbf{y}) \times \frac{\mathbf{x} - \mathbf{y}}{|\mathbf{x} - \mathbf{y}|^2} \mathrm{d}^2 y, \tag{9.13}$$

where S_z is a surface with constant z value, gives a helicity measure in the form of the flux-weighted winding of the curves in the field

$$\begin{aligned} H^w(\mathbf{B}) &= \int_\Omega \mathbf{A}^w \cdot \mathbf{B}\, \mathrm{d}V \\ &= \frac{1}{2\pi}\int_0^h \int_{S_z}\int_{S_z} \frac{\mathrm{d}}{\mathrm{d}z}\Theta(\mathbf{x}, \mathbf{y}) B_z(\mathbf{x}) B_z(\mathbf{y}) \mathrm{d}^2 x\, \mathrm{d}^2 y\, \mathrm{d}z, \end{aligned} \tag{9.14}$$

where h is the domain height ($z \in [0, h]$). Just like $L = Lk$ for closed curves, the helicity given by this definition is equal to the helicity defined in equation (9.10) if the field has no normal boundary component (this derives simply from the fact that the helicity is gauge-invariant in this case).

For the remainder of this chapter, we focus on the fact that the net winding $\mathcal{L}$ in equation (9.12) yields a sensible writhing definition that is equal to $Lk - Tw$, thus extending the Călugăreanu theorem to open-ended ribbons.

The Polar Writhe

Since we have a topological invariant for open ribbons, the net winding $\mathcal{L}(\mathbf{x}, \mathbf{y})$, and the twist is perfectly well defined for open ribbons, we should be able to define a notion of writhing that satisfies a Călugăreanu relation that is a quantity equal to $\mathcal{L} - Tw$.

The polar writhe, introduced in Berger and Prior (2006), is that measure. It is the sum of a local component (W_{pl}) and a nonlocal component (W_{pnl}):

$$W_p = W_{pl} + W_{pnl}. \tag{9.15}$$

The nonlocal component W_{pnl} is just the net winding of the curve with itself:

$$W_{pnl}(\mathbf{x}) = \sum_{i=0}^{n}\sum_{j=0}^{n}\mathcal{L}(\mathbf{x}_i, \mathbf{x}_j). \tag{9.16}$$

Here the curve $\mathbf{x}$ has been split into sections moving up or down in z, as with the net winding. The contributions from identical sections $(\mathbf{x}_i, \mathbf{x}_i)$ are formally set to zero in this sum and instead represented by the local polar writhe W_{pl}:

$$W_{pl}(\mathbf{x}) = \frac{1}{2\pi}\sum_{i=1}^{n}\int_{z_i^{min}}^{z_i^{max}} \frac{\hat{z}\cdot\mathbf{T}_i \times \frac{\mathrm{d}\mathbf{T}_i}{\mathrm{d}z}}{1+|\hat{z}\cdot\mathbf{T}_i|}\,\mathrm{d}z. \tag{9.17}$$

The numerator represents the rotation of the unit tangent vector around the vertical direction. It is positive if the curve coils in a right-handed fashion and negative if left-handed. The denominator indicates that (1) the rotation is given more weight if the curve is rotating relatively tightly, and (2) the modulus sign means the orientation would be the same for a curve whether it points up or down: that is, it only depends on the local helical chirality of the curve. The reason for the name *polar writhe* is that this rotation is measured around the $\hat{z}$ direction, which on a unit sphere of directions is the north pole, and that W_{pl} can be interpreted as an area on the unit sphere bound by the curve $\mathbf{T}$ and the pole (see Berger and Prior (2006)).

9.2.3. Helical-to-Loop Transition and Plectoneme Formation

A clear example that highlights the geometrical meaning of the two components W_{pl} and W_{pnl} is the curve transitioning from a locally helically coiled curve (configuration (i)) to a looped curve (configuration (ii)). This development is shown in Fig. 9.7, which is an equilibrium configuration of a twisted ribbon model under an increasing applied torque (Nizette and Goriely, 1999) (such models are commonly used to describe DNA micromanipulation experiments (Bouchiat and Mézard, 1998; Fain et al., 1997)). The development of its W_p values as the torque is increased in 200 evenly spaced steps is charted on the plot. The polar writhe W_p steadily increases as the curve becomes increasingly helically coiled at its center and then increasingly looped. For curves 1–138, all the contribution to W_p is local due to the helical kinking of the curve around its center (W_{pnl}=0). After this, the loop begins to form, and the curve develops nonlocal writhing ($W_{pnl} > 0$). There is a dramatic interconversion of local and nonlocal writhing while the sum changes smoothly through this transition. This indicates that the helical coiling becomes distorted to form a nonlocally wound loop. This is the first step in plectoneme formation, a process by which DNA molecules coil up to occupy a significantly reduced area (Travers and Muskhelishvili, 2015).

One can think of nonlocal writhing as self-knotting. An example of a far more significantly knotted curve is shown in Fig. 9.7b. It is taken from end-constrained DNA simulations performed in Sierzega et al. (2021). From our intuition of writhing as self-crossing, we should expect the nonlocal writhing of this curve to have a magnitude of around 8 due to the eight crossings marked on the curve. In fact, for this curve, $W_p = 8.88$, $W_{pnl} = 8.55$, and $W_{pl} = 0.33$ (all to three significant figures).

Use of Polar Writhe

In a biophysics/elastic tube context, polar writhe has been used to evaluate the topological development of computational models re-creating DNA micromanipulation experiments (Sierzega et al., 2021). In particular, to show the error of alternative proposed methods, see Fig. 9.8a. It has also been used to model the unknotting of DNA (Prior and Neukirch, 2016), in automated

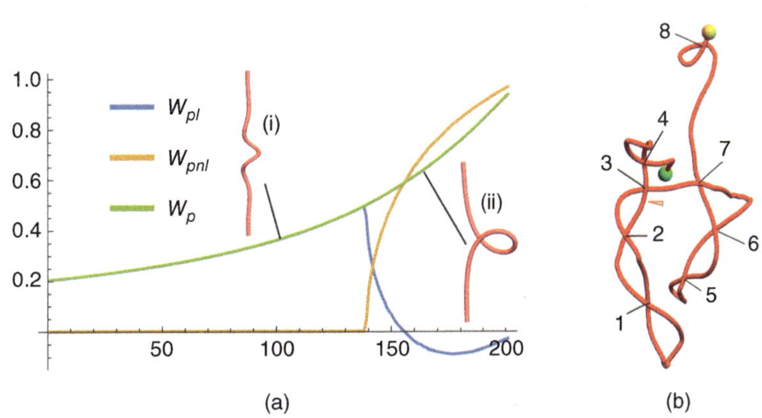

Figure 9.7 Polar writhe calculations of plectoneme geometries. (a) W_p values of a loop forming curve deformation. (b) A curve with significant plectoneme structure. The individual loops are marked.

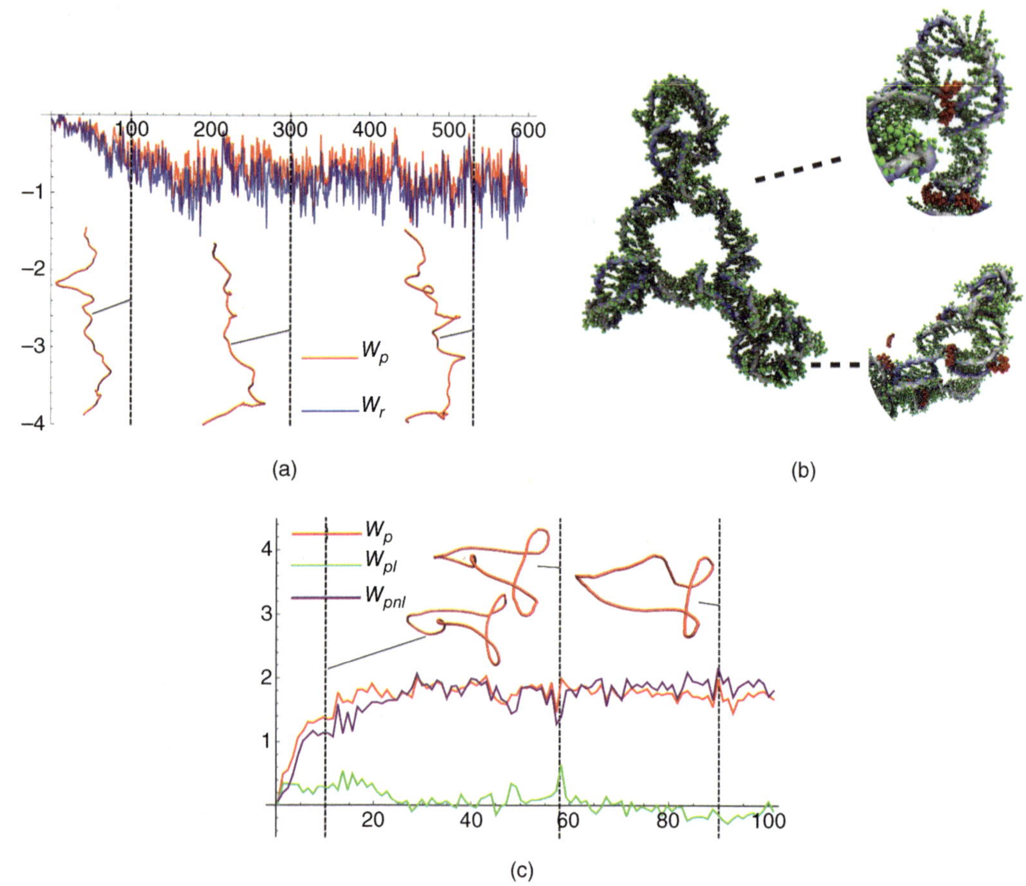

Figure 9.8 Example DNA applications of the polar writhe W_{pnl}. (a) W_p (blue) and an alternative writhe measure (red) for the evolution of an under-twisted and end-constrained DNA molecule. (b) A still from an all-atomistic simulation of DNA loop formation by Spermine bridging. (c) Writhe values for the simulation. The middle vertical dashed line corresponds to a spike in W_{pl}, which was found to herald the formation of a loop. This spike guided when and where the loops in the molecules developed (they are hard to locate manually). Sierzega et al., 2021/Springer Nature/Licensed under CC BY 4.0.

algorithms for aneurysm detection (Lauric et al., 2010), to locate stable and topologically valid elastic knots without self-contact (Moulton et al., 2018), and to detect loop formation in DNA minicircles due to Spermine bridging (Sierzega et al., 2021) (as shown Figs. 9.8b and (c)). In addition, as discussed in some detail in the supplement of Sierzega et al. (2021), various closure- and difference-based writhe measures used successfully in DNA studies are simplified formulae for the polar-writhe measure, which can be used when confined to a subset of permissible curve geometries. Later, we will also discuss some significant biophysical applications of the classical writhe Wr.

In a solar physics context, the nonlocal polar writhe W_{pnl} has been used to estimate the sign of active region field helicity by tracking filament rotation (Liu et al., 2008; Pevtsov et al., 2014; Cobo and Puschmann, 2012; Xu et al., 2020). Evidence for this being a suitable proxy was provided in simulations of eruptive flux ropes (Török et al., 2010, 2014; Guo et al., 2017). In contrast, a survey of general force-free sigmoidal geometries (Prior and Berger, 2012) suggests that some caution should be exercised in making this interpretation. It was also used in Prior and Yeates (2018) to identify differing regions of field line geometry and their mixing (kinked and not kinked) in experimental plasma data.

9.2.4. Writhing as a Point of View

The majority of applications of writhing so far have concerned measuring the writhing of the axis of a ribbon structure or magnetic flux rope. In these cases, it forms part of the invariant sum of writhing plus twisting. For open ribbons, polar writhe is the "correct" quantity as it adds to the twist to give an invariant quantity: the net winding $\mathcal{L}$. However, many applications, such as protein geometry, concern the classification of the geometry of individual curves where linking is not defined.

In this case, both Wr, given by equation (9.4), and the polar writhe W_p are valid measures of the entanglement of the curve. There is no topological relationship to make one "better" than the other. The difference between them is the set of viewpoints over which their net crossings are averaged. Wr considers all projections; polar writhe can be considered directional because it only considers projections for directions orthogonal to the z direction. When applied to the plectonemed curve in Fig. 9.7b, they both give approximately 8, as almost all viewpoints agree on the number of crossings. However, they can often differ significantly, especially if the endpoints of the curve lie at the core of the curve. In such cases, it is not immediately obvious which measure is clearer. Indeed, we shall present a set of methods to extract information from them both. Again, some of these methods will be guided by results from the solar physics literature.

As a final point, before we consider writhing applied to protein structures, it is worth mentioning the use of the writhe Wr in studies of DNA and tangle topology. Studies show strong links between the biological function of DNA and its topology. For example, in Marenduzzo et al. (2013), the topological complexity of DNA as measured by its knot type plays a key role in the rate of DNA ejection for bacteriophages. In Bednar et al. (1994), it was demonstrated, using cryo-electron microscopy, that a decrease of DNA-DNA repulsion by increasing concentrations of counterions causes a higher fraction of the linking number deficit to be partitioned into writhe. In Arsuaga et al. (2005), it was shown that DNA molecules from bacteriophage P4 were highly knotted. Both experimental and computational modeling techniques indicated a tendency toward high-writhe configurations. In particular, it was demonstrated that DNA knots cannot be obtained by confinement alone but must include writhe bias in the conformation sampling. This is just a small subset of the research on this topic. For the interested reader, a far more comprehensive review of DNA topology from a mathematical perspective was conducted by Sumners (2020).

9.3. BOUNDING THE WRITHE OF PROTEINS

9.3.1. Introduction to Protein Knotting

Protein folding is another field in which the study of local and global entanglement is key. Proteins play a huge role in virtually every biological process, from DNA replication (Salas, 1991) to fighting disease (Samrat et al., 2020). To achieve their functionality, proteins must fold from a long chain of amino acids to attain a specific three-dimensional structure (Anfinsen, 1972; Dobson, 2003). We can describe this fold as a discrete three-dimensional curve, the backbone curve, with each point representing the center ($c\alpha$ molecule) of the amino acid. These structures are commonly visualized as a ribbon diagram, as shown in Fig. 9.9, allowing the user to clearly pick out specific recognizable helical substructures. The specific sequence of amino acids that forms this chain is called the *primary* structure of the protein, and it is always identifiable. It has been shown (e.g., Otaki et al. (2010)) that this primary sequence dictates the so-called secondary structure of the protein. The secondary structural elements can be thought of as the building blocks of the protein curve. There are three main types: α-helices, β-strands, and linkers, again shown in Fig. 9.9. The α-helix and β strand structures are rigid and (relatively) unchanging, while the linker sections show much more significant flexibility and are responsible for the folding of the structure (Kabsch and Sander, 1983). The tertiary structure of a protein then is the relative position and orientation of these secondary structure elements: how they are packed and tangled together.

One potential way to study and compare the native structure of proteins (compare the three-dimensional geometry of their discrete backbone curve) would be to use techniques from the rich field of knot theory.

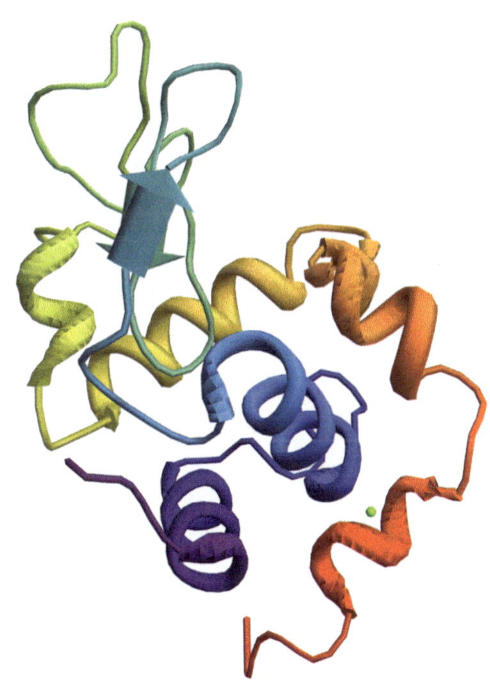

Figure 9.9 The ribbon diagram for lysozyme (Protein Data Bank [PDB]: 1DPX). The α-helical sections are represented by helices, β-sheets by flat arrows (seen toward the top of the diagram in green), and linkers as thin tubes connecting the other two secondary structures. Adapted from Weiss et al., 2000.

It has been shown, however (Mansfield, 1994), (Lua and Grosberg, 2006), that nontrivially knotted proteins account for only about 1% of known proteins. Alongside this fact, there is also the issue that knotted proteins are not knots in the strict mathematical sense. This is simply because proteins are not closed curves; their ends are distinct. For this reason, we opt to use writhing measures to quantify and attempt to classify entanglement in proteins, allowing a much richer way of distinguishing different structures. In doing so, we can better study the structure of the 99% of proteins that aren't knotted but are definitely entangled.

Peter Rogen and colleagues (Røgen and Fain, 2003; Harder et al., 2012; Grønbæk et al., 2020) used higher-order writhe integrals to create classification criteria for protein structures. These integrals are based on the Gauss formula from equation (9.4), but with more than two points for each "crossing." In a similar vein, Zhi et al. (2010) used the Gauss formula but applied it to subsets of the protein to generate a metric for protein similarity. Panagiotou and Plaxco (2020) use the writhe Wr (again using the Gauss formula) to show some promising but preliminary relationships between aspects of structural complexity and folding rate. It should be pointed out that these are far from the only structure classification methods. Part of the complexity of the topic is that there is no definitive classification; many aspects of protein functionality can identify differing proteins as in one way similar – far more than can be summarized here. We direct the reader to, for example, Fox et al. (2015) for a contemporary review of the topic. Related applications of writhing in biophysical contexts include the work of Katritch et al. (1996), where the relationship between the writhe and length of ideally compact knots was investigated; we discuss this work in relation to our results shortly. Orlandini et al. (1994) demonstrated the writhe of random walks scaled like the square of their length; we also compare this work in our results shortly.

The preliminary results we present here simply show how techniques developed in solar physics–inclined research can provide additional insight into the complex universe of protein geometry, including via the use of polar writhe.

9.3.2. A Note on Downsampling Proteins

When studying the writhe of proteins, it is important to note that secondary structures can make a significant, yet often unwanted, contribution to the overall writhe of the curve. Consider the two curves shown in Fig. 9.10, for example. The curve in panel Fig. 9.10a representing the protein backbone has a Wr of about 1.3, whereas the curve in panel Fig. 9.10b given by downsampling the backbone to just its secondary structures has a writhe of 0.003. Here we downsample by taking the start point ($c\alpha$ molecule) of each identified secondary structure section and creating a discrete curve from this subset of points (with the endpoints of the protein added on) to reduce the protein to its secondary structural elements. This is a significant difference; by concentrating only on this downsampled curve, we see that the underlying protein structure is essentially unknotted. If we consider the protein backbone in full detail, though, we see the contribution to the writhe from the helical section on the right of Fig. 9.10a. Although this downsampling may look similar to the Călugăreanu theorem introduced earlier, in our study, we are just concerned with the underlying curve shown in Fig. 9.10b, not the decomposition into writhe and twist of the ribbon (Fig. 9.10c). This is because when making predictions about tertiary structure – that is, how the individual secondary structure blocks fold – the secondary structural elements are

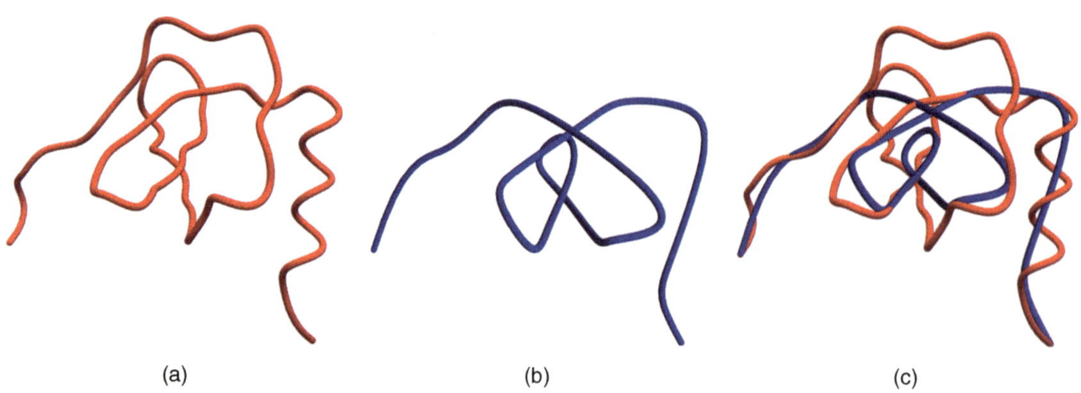

Figure 9.10 A visual example of the contribution of secondary structure to the writhe calculation for the protein CC Chemokine CCL14 (PDB: 2Q8R) (Blain et al., 2007). (a) The backbone curve for 2Q8R, with an α-helical section clearly visible on the right-hand side. (b) A downsampled backbone curve for 2Q8R, representing the underlying topology. (c) The two structures overlaid. Adapted from Blain et al., 2007.

essentially treated as rigid bodies (as in Prior et al. (2020), where this model was used to correctly predict protein structures from experimental scattering data). It is clear then that this downsampling is a vital step in using writhe as a comparative measure.

9.3.3. Borrowing from Solar Physics, Again

Since we have already seen that knotted proteins are rare, it makes sense to ask: Is there a limit on how much writhing we can expect to see for a protein? To approach an answer to this question, we will borrow some key results from the study of magnetic helicity. In particular, we will use the helicity bound derived in Cantarella et al. (2001b), which adapts a bound on the magnetic helicity of a vector field to a bound on the writhing of a knotted tube. This is done by assuming the field (1) is tubular and (2) has no twisting. It is shown that for K, a smooth knotted tube with length L and an embedded tubular neighborhood of radius R, the writhe $Wr(K)$ is bounded by

$$|Wr(K)| < \frac{1}{4}\left(\frac{L}{R}\right)^{\frac{4}{3}}. \tag{9.18}$$

Section-Number Length L_n

Before applying this bound to protein backbone curves, we should consider a few things. First, what is the length of a protein, which is represented here as a discrete curve? One could, for example, imagine the protein backbone completely untangled and straightened and then take the length of this straight-line segment to be L. Since it has been shown in Chakraborty et al. (2013) that the average distance between neighboring amino acid residues is roughly 3.8Å, this choice of L would look like the number of amino acids multiplied by 3.8. It is widely understood that the length of a protein's primary sequence is closely related to its function, so this would seem a sensible choice. We, however, propose something perhaps more fundamental to coincide with the downsampling discussed previously. To use this bound, then, we will define the section-number length L_n of a protein. L_n is simply the number of the protein's secondary structure elements. In the case of the radius R, there isn't an immediately obvious choice here.

Applying the Bound to the Protein Data Bank

We took a set of over 2,000 proteins from the PDB ((Berman et al., 2000)), chosen to have good resolution (>2Å), between 30–300 residues, and redundancies removed by using representative structures at 70% sequence identity (one often finds multiple examples of the same structure in the PDB). We then computed the writhe via the algorithm presented in Klenin and Langowski (2000). Figure 9.11 shows a scatter plot of

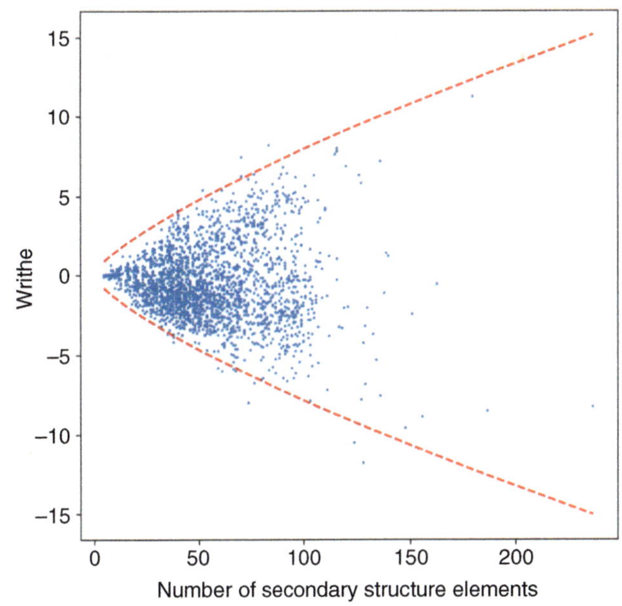

Figure 9.11 In blue, a scatter plot of writhe against the number of secondary structure elements for a sample of roughly 2,000 proteins taken from the PDB. In red, a plot of the curve described by the bound in equation (9.18, with L the number of secondary structure elements and $R = \frac{1}{4}L^{\frac{7}{16}}$. Adapted from Berman et al., 2000.

the writhe of these downsampled proteins as a function of L_n, the number of their secondary structure elements, and the dashed curve described by equation (9.18, with $L = L_n$ and $R = L^{\frac{7}{16}}$. This is close to $L^{1/2}$ growth. Overall, this means the writhe bound of the proteins grows as a function of the length-to-width ratio L_R, as $L_R^{3/4}$. One can see that this bound is a surprisingly good fit, given the amount of structure we ignore, especially for small lengths (<20). Even for longer lengths, most proteins stay within the bound or just over it. This gives rise to some interesting questions we propose for future research in this preliminary study:

- What does the scaling for R represent? Is there some biological explanation for it? Since the curves we study are just the protein's backbone, this radius could represent the space occupied by the remaining atoms we ignore when we use the backbone curve. Indeed, Orlandini et al. (1994) found that the absolute value of the writhe of self-avoiding walks grows with $\sqrt{L}$; our radial scaling is nearly $L_n^{1/2}$, so it represents the self-avoidance of the curve in some sense. However, we found that simply surrounding the curves with a tube of this radius yields overlaps, representing a more general sense of self-avoidance that we should seek to characterize in the future. It is also worth comparing the sublinear $L_R^{3/4}$ length-to-width scaling to the results of Katritch et al. (1996). In that study, it was shown that there is a linear correlation between topological complexity (as described by the average crossing number and the writhe) and the length-to-diameter ratio of a knotted tube of uniform diameter. We speculate that this arises from the fact that unlike the uniformly flexible ideal knots considered in that study, longer proteins often have significant length alpha-helical sections, which behave like highly rigid subsections of the curves, thus limiting their potential complexity.

- We must also question our choice for L: L_n, the number of subsections independent of their individual length. Although we downsampled our proteins according to their secondary structure elements, the curves do not have edges of equal unit length. How does this bound calculated for these fundamental unit-length, unit-radius curves fit so well for our downsampled proteins with arbitrary edge lengths?
- Some proteins are at or close to their bounds, while some have almost no writhe. These curves are not planar but have a roughly equal amount of positive and negative crossing. Is there some functional reason for the structure to exhibit these two extremes of knotting?
- The writhe bound derived in Cantarella et al. (2001b) actually uses the Biot-Savart gauge from equation (9.9). As discussed previously, this is only valid for closed curves, and proteins are open-ended. One plausible reason for some of the proteins to exceed this bound may be contributions to Wr from the endpoints. One might then seek to quantify this by developing a bound using the winding gauge helicity in equation (9.13).

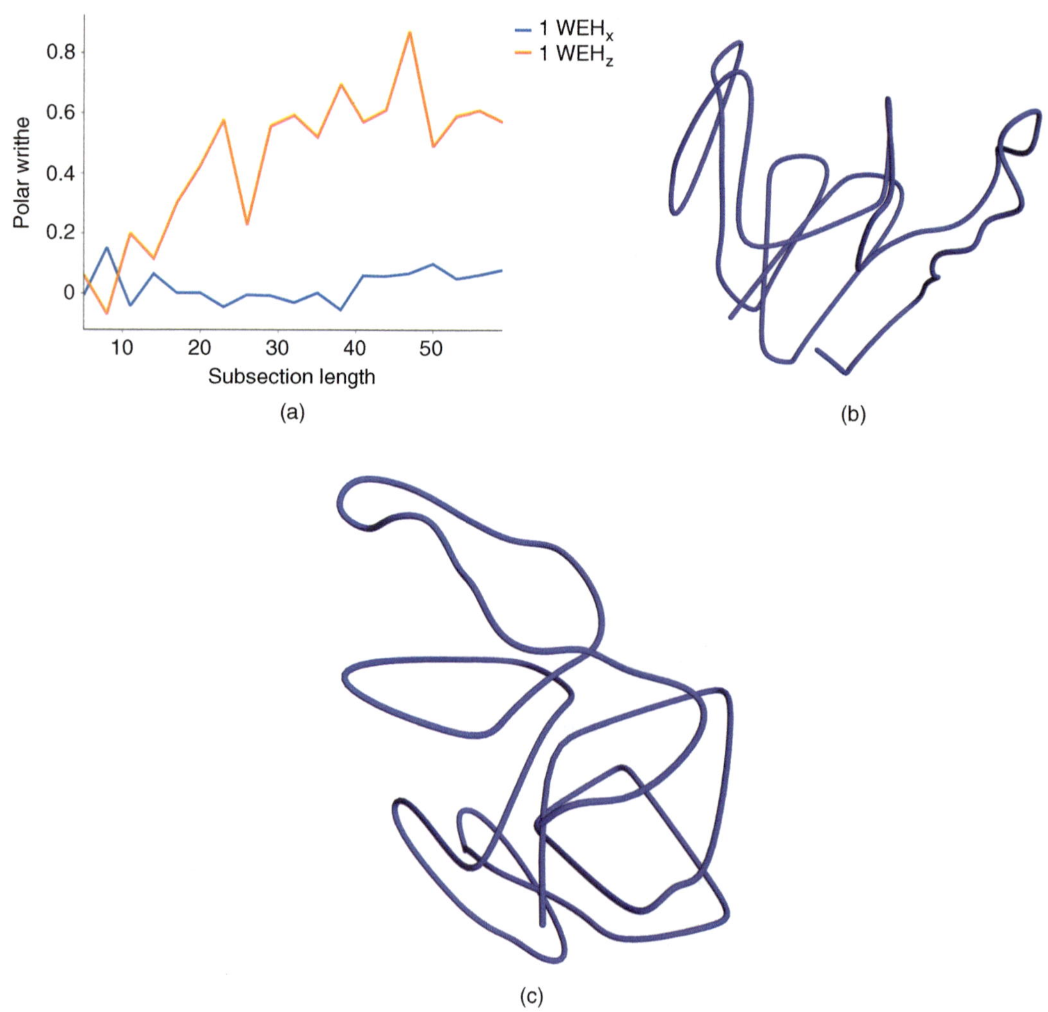

Figure 9.12 (a) A plot of the local polar writhe for the hypothetical protein TT1887 from Thermus thermophilus HB8 (PDB: 1WEH) against its length. In blue: the local polar writhe for the curve oriented along the x-axis. In orange: the local polar writhe for the curve oriented along the z-axis. Adapted from Kukimoto-Niino et al., 2004. (b) and (c) The downsampled protein curve, oriented along the x and z axes, respectively.

9.3.4. Applying Polar Writhe to Proteins

As discussed previously, the two writhe measures Wr and polar writhe W_p differ in that polar writhe only measures self-winding of the curve about the z direction; put another way, it only considers crossings counting from a subset of observers who contribute to the Wr measure of self-entanglement. A natural question is whether any proteins have a systematic directional bias toward their entanglement. Polar writhe is a perfect tool to test this. The z-directional bias stems from the net-winding definition, as we slice the domain into planes normal to $\hat{z}$ and count crossings in these planes. We could just as easily have done this for the $\hat{x}$ or $\hat{y}$ direction, simply replacing z and $\hat{z}$ with, say, x and $\hat{x}$ in the formulae in equations (9.17) and (9.16). One would expect the measures defined along these three directions to differ significantly if there is some directional bias to a protein backbone curve.

Local Helical Coiling

The polar writhe W_p provides an extra layer of information through its local–nonlocal decomposition. We remind the reader that the local contribution W_{pl} measures the local helical coiling rate of the curve, while the nonlocal component is self-knotting. To keep things simple in this preliminary study, we only focus on the local component so we are looking for a directional bias in helical coiling.

Some Examples

To test this idea, we define the "direction" of a protein curve as the vector between its first and last point; call this vector $\mathbf{x}_d$. We can then use this direction as a means to orient our proteins. Figure 9.12b,c shows an example, where the vector $\mathbf{x}_d$ of the (downsampled) backbone curve of the protein TT1887 from Thermus thermophilus HB8 is oriented along the x and z axes for Fig. 9.12b and Fig. 9.12c,

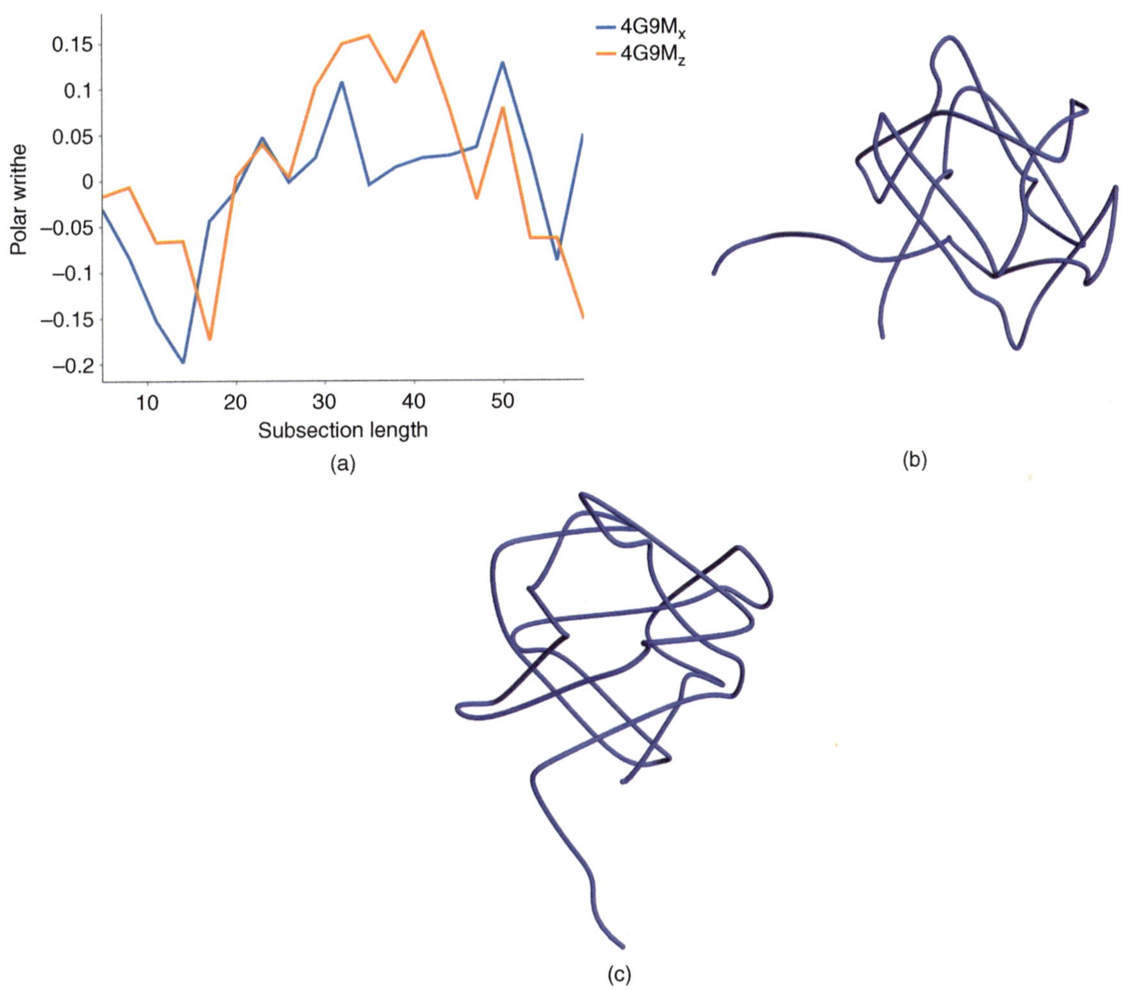

Figure 9.13 (a) A plot of the local polar writhe for Rhizoctonia solani agglutinin (PDB: 4G9M) against its length. In blue: the local polar writhe for the curve oriented along the x-axis. In orange: the local polar writhe for the curve oriented along the z-axis. Adapted from Skamnaki et al., 2013. (b) and (c) The downsampled protein curve, whose endpoint orientation vector $\mathbf{x}_d$ is pointed along the x and z axes, respectively.

respectively. Figure 9.12a shows a plot of W_{pl} for subsets of these two curves $\{\mathbf{x}_i\}_{i=1}^{j}$ for $j = 5, 6 \ldots L_n$ as a function of their length. We see that the local polar writhe calculated with a z-orientation, in orange, dominates that of the x-oriented protein, in blue. This suggests that the protein does exhibit some sense of helical directionality. This coiling is on a significantly larger scale than an alpha helix structure. On the other hand, Fig. 9.13 is an example of a protein showing no directional preference. This can be seen in Fig. 9.13a, where the local polar writhe for the x-oriented protein, in blue, follows a pattern fairly similar to that for the z-oriented protein, in orange.

Future Questions

It remains to determine what this directionality could represent for proteins and how to exploit it for better structural prediction. To this aim, it may be profitable to offer questions that should be addressed in the future:

- Can one correlate any protein functionalities with proteins that exhibit a clear directional writhing bias (or none at all)?
- Are smaller or larger proteins more/less likely to have a direction bias?
- What information is available in the nonlocal polar writhe component W_{pnl}? A bias would somehow represent directional knotting.

REFERENCES

Anfinsen, C.B. (1972). The formation and stabilization of protein structure. *Biochemical Journal*, *128*(4), 737.

Arsuaga, J., Vazquez, M., McGuirk, P., Trigueros, S., Sumners, D.W., & Roca, J. (2005). DNA knots reveal a chiral organization of Dna in phage capsids. *Proceedings of the National Academy of Sciences*, *102*(26), 9165–9169.

Bednar, J., Furrer, P., Stasiak, A., Dubochet, J., Egelman, E.H., & Bates, A.D. (1994). The twist, writhe and overall shape of supercoiled DNA change during counterion-induced transition from a loosely to a tightly interwound superhelix: possible implications for DNA structure in vivo. *Journal of Molecular Biology*, *235*(3), 825–847.

Berger, M.A. (1988). An energy formula for nonlinear force-free magnetic fields. *Astronomy and Astrophysics*, *201*, 355–361.

Berger, M.A. (1984). Rigorous new limits on magnetic helicity dissipation in the solar corona. *Geophysical & Astrophysical Fluid Dynamics*, *30*(1–2), 79–104.

Berger, M.A., & Field, G.B. (1984). The topological properties of magnetic helicity. *Journal of Fluid Mechanics*, *147*, 133–148.

Berger, M.A., & Prior, C. (2006). The writhe of open and closed curves. *Journal of Physics A: Mathematical and General*, *39*(26), 8321.

Berman, H.M., Westbrook, J., Feng, Z., Gilliland, G., Bhat, T.N., Weissig, H., et al. (2000). The protein data bank. *Nucleic Acids Research*, *28*(1), 235–242.

Blain, K.Y., Kwiatkowski, W., Zhao, Q., La Fleur, D., Naik, C., Chun, T.W., et al. (2007). Structural and functional characterization of cc chemokine ccl14. *Biochemistry*, *46*(35), 10008–10015.

Bouchiat, C., & M'ezard, M. (1998). Elasticity model of a supercoiled DNA molecule. *Physical Review Letters*, *80*, 1556–1559. https://doi.org/10.1103/PhysRevLett.80.1556

Călugăreanu, G. (1961). Sur les classes d'isotopie des noeuds tridimensionnels et leurs invariants. *Czechoslovak Mathematical Journal*, *11*(4), 588–625.

Călugăreanu, G. (1959). L'int'egrale de gauss et l'analyse des nœuds tridimensionnels. *Rev. Math. pures appl*, *4*(5).

Cantarella, J., DeTurck, D., & Gluck, H. (2001a). The biot–savart operator for application to knot theory, fluid dynamics, and plasma physics. *Journal of Mathematical Physics*, *42*(2), 876–905.

Cantarella, J., DeTurck, D., & Gluck, H. (2001b). Upper bounds for the writhing of knots and the helicity of vector fields. *AMS IP Studies in Advanced Mathematics*, *24*, 1–22.

Chakraborty, S., Venkatramani, R., Rao, B., Asgeirsson, B., & Dandekar, A. (2013). Protein structure quality assessment based on the distance profiles of consecutive backbone cα atoms. *F1000Research*, 2.

Chou, F.C., Lipfert, J., & Das, R. (2014). Blind predictions of DNA and RNA tweezers experiments with force and torque. *PLoS Computational Biology*, *10*(8).

Cobo, B.R., & Puschmann, K.G. (2012). Twist, writhe, and helicity in the inner penumbra of a sunspot. *The Astrophysical Journal*, *745*(2), 141.

Dobson, C.M. (2003). Protein folding and misfolding. *Nature*, *426*(6968), 884–890.

Fain, B., Rudnick, J., & Östlund, S. (1997). Conformations of linear DNA. *Phys. Rev. E*, *55*, 7364–7368. https://doi.org/10.1103/PhysRevE.55.7364

Fox, N.K., Brenner, S.E., & Chandonia, J.M. (2015). The value of protein structure classification information-surveying the scientific literature. *Proteins: Structure, Function, and Bioinformatics*, *83*(11), 2025–2038.

Frankel, T. (2011). *The geometry of physics: an introduction*. Cambridge University Press.

Fuller, F.B. (1971). The writhing number of a space curve. *Proceedings of the National Academy of Sciences*, *68*(4), 815–819.

Fuller, F.B. (1978). Decomposition of the linking number of a closed ribbon: a problem from molecular biology. *Proceedings of the National Academy of Sciences*, *75*(8), 3557–3561.

Grønbæk, C., Hamelryck, T., & Røgen, P. (2020). Gisa: using Gauss integrals to identify rare conformations in protein structures. *PeerJ*, *8*, e9159.

Guo, Y., Pariat, E., Valori, G., Anfinogentov, S., Chen, F., Georgoulis, M.K., et al. (2017). Magnetic helicity estimations in models and observations of the solar magnetic field. iii. twist number method. *The Astrophysical Journal*, *840*(1), 40.

Harder, T., Borg, M., Boomsma, W., Røgen, P., & Hamelryck, T. (2012). Fast large-scale clustering of protein structures using Gauss integrals. *Bioinformatics*, *28*(4), 510–515.

Ji, H., Prager, S.C., & Sarff, J.S. (1995). Conservation of magnetic helicity during plasma relaxation. *Physical Review Letters*, *74*(15), 2945.

Kabsch, W., & Sander, C. (1983). Dictionary of protein secondary structure: pattern recognition of hydrogen-bonded and geometrical features. *Biopolymers: Original Research on Biomolecules*, *22*(12), 2577–2637.

Katritch, V., Bednar, J., Michoud, D., Scharein, R.G., Dubochet, J., & Stasiak, A. (1996). Geometry and physics of knots. *Nature*, *384*(6605), 142–145. https://doi.org/10.1038/384142a0

Klenin, K., & Langowski, J. (2000). Computation of writhe in modeling of supercoiled DNA. *Biopolymers*, *54*(5), 307–317.

Klenin, K.V., Vologodskii, A.V., Anshelevich, V.V., Klishko, V.Yu, Dykhne, A.M., & Frank-Kamenetskii, M.D. (1989). Variance of writhe for wormlike DNA rings with excluded volume. *Journal of Biomolecular Structure and Dynamics*, *6*(4), 707–714.

Kukimoto-Niino, M., Murayama, K., Kato-Murayama, M., Idaka, M., Bessho, Y., Tatsuguchi, A., et al. (2004). Crystal structures of possible lysine decarboxylases from thermus thermophilus hb8. *Protein Science*, *13*(11), 3038–3042.

Laing, C.E., Ricca, R.L., & Sumners, D.W. (2015). Conservation of writhe helicity under antiparallel reconnection. *Scientific Reports*, *5*(1), 1–6.

Lam, P.M., & Zhen, Y. (2018). Twisting, supercoiling and stretching in protein bound DNA. *Physica A: Statistical Mechanics and its Applications*, *496*, 200–208.

Lauric, A., Miller, E., Frisken, S., & Malek, A.M. (2010). Automated detection of intracranial aneurysms based on parent vessel 3d analysis. *Medical Image Analysis*, *14*(2), 149–159.

Lepage, T., K'epes, F., & Junier, I. (2015). Thermodynamics of long supercoiled molecules: insights from highly efficient Monte Carlo simulations. *Biophysical Journal*, *109*(1), 135–143.

Lillian, T.D., Taranova, M., Wereszczynski, J., Andricioaei, I., & Perkins, N.C. (2011). A multiscale dynamic model of DNA supercoil relaxation by topoisomerase ib. *Biophysical Journal*, *100*(8), 2016–2023.

Liu, R., Gilbert, H.R., Alexander, D., & Su, Y. (2008). The effect of magnetic reconnection and writhing in a partial filament eruption. *The Astrophysical Journal*, *680*(2), 1508.

Lua, R.C., & Grosberg, A.Y. (2006). Statistics of knots, geometry of conformations, and evolution of proteins. *PLOS Computational Biology*, *2*(5), 1–8.

Maffeo, C., Sch"opflin, R., Brutzer, H., Stehr, R., Aksimentiev, A., Wedemann, G., & Seidel, R. (2010). DNA–DNA interactions in tight supercoils are described by a small effective charge density. *Physical Review Letters*, *105*(15), 158101.

Mansfield, M.L. (1994). Are there knots in proteins? *Nature Structural Biology*, *1*, 213–214.

Marenduzzo, D., Micheletti, C., Orlandini, E., & Sumners, D.W. (2013). Topological friction strongly affects viral DNA ejection. *Proceedings of the National Academy of Sciences*, *110*(50), 20081–20086. https://doi.org/10.1073/pnas.1306601110

Marko, J.F., & Neukirch, S. (2012). Competition between curls and plectonemes near the buckling transition of stretched supercoiled DNA. *Physical Review E*, *85*(1), 011908.

Medalion, S., Rappaport, S.M., & Rabin, Y. (2010). Coupling of twist and writhe in short DNA loops. *The Journal of Chemical Physics*, *132*(4), 01B616.

Moffatt, H.K. (1969). The degree of knottedness of tangled vortex lines. *Journal of Fluid Mechanics*, *35*(1), 117–129.

Moffatt, H.K., & Ricca, R.L. (1995). Helicity and the Călugăreanu invariant. In L.H. Kaufmann (Ed.) *Knots and applications* (pp. 251–269). World Scientific.

Moreau, J.J. (1961). Constantes d'un îlot tourbillonnaire en fluide parfait barotrope. *Comptes rendus hebdomadaires des séances de l'Académie des sciences*, *252*, 2810–2812.

Moulton, D.E., Grandgeorge, P., & Neukirch, S. (2018). Stable elastic knots with no self-contact. *Journal of the Mechanics and Physics of Solids*, *116*, 33–53.

Neukirch, S., & Starostin, E.L. (2008). Writhe formulas and antipodal points in plectonemic DNA configurations. *Physical Review E*, *78*, 041912.

Nizette, M., & Goriely, A. (1999). Towards a classification of Euler–Kirchhoff filaments. *Journal of Mathematical Physics*, *40*(6), 2830–2866.

Orlandini, E., Tesi, M.C., Whittington, S.G., Sumners, D.W., & van Rensburg, E.J.J. (1994). The writhe of a self-avoiding walk. *Journal of Physics A: Mathematical and General*, *27*(10), L333–L338.

Otaki, J.M., Tsutsumi, M., Gotoh, T., & Yamamoto, H. (2010). Secondary structure characterization based on amino acid composition and availability in proteins. *Journal of Chemical Information and Modeling*, *50*(4), 690–700.

Panagiotou, E., & Plaxco, K.W. (2020). A topological study of protein folding kinetics. *Topology and Geometry of Biopolymers, AMS Contemporary Mathematics Series*, *746*, 223–233.

Pariat, E., Valori, G., D'emoulin, P., & Dalmasse, K. (2015). Testing magnetic helicity conservation in a solar-like active event. *Astronomy & Astrophysics*, *580*, A128.

Patil, V.P., Sandt, J.D., Kolle, M., & Dunkel, J. (2020). Topological mechanics of knots and tangles. *Science*, *367*(6473), 71–75.

Pevtsov, A.A., Berger, M.A., Nindos, A., Norton, A.A., & van Driel-Gesztelyi, L. (2014). Magnetic helicity, tilt, and twist. *Space Science Reviews*, *186*(1-4), 285–324.

Pipin, V.V., Sokoloff, D.D., Zhang, H., & Kuzanyan, K.M. (2013). Helicity conservation in nonlinear mean-field solar dynamo. *The Astrophysical Journal*, *768*(1), 46.

Pohl, W.F. (1968). Some integral formulas for space curves and their generalization. *American Journal of Mathematics*, *90*(4), 1321–1345.

Prior, C., & Berger, M.A. (2012). On the shape of force-free field lines in the solar corona. *Solar Physics*, *278*(2), 323–345.

Prior, C., & Yeates, A.R. (2014). On the helicity of open magnetic fields. *The Astrophysical Journal*, *787*(2), 100.

Prior, C., & Yeates, A.R. (2018). Quantifying reconnective activity in braided vector fields. *Physical Review E*, *98*(1), 013204.

Prior, C.B., & Neukirch, S. (2016). The extended polar writhe: a tool for open curves mechanics. *Journal of Physics A: Mathematical and Theoretical*, *49*(21), 215201.

Prior, C.B., Davies, O.R., Bruce, D., & Pohl, E. (2020). Obtaining tertiary protein structures by the ab initio interpretation of small angle x-ray scattering data. *Journal of Chemical Theory and Computation*, *16*(3), 1985–2001.

Qu, A., & James, D.L. (2021). Fast linking numbers for topology verification of loopy structures. *ACM Transactions on Graphics*, *40*, 106.

Røgen, P., & Fain, B. (2003). Automatic classification of protein structure by using Gauss integrals. *Proceedings of the National Academy of Sciences*, *100*(1), 119–124.

Rossetto, V., & Maggs, A.C. (2003). Writhing geometry of open DNA. *The Journal of Chemical Physics*, *118*(21), 9864–9874.

Salas, M. (1991). Protein-priming of DNA replication. *Annual Review of Biochemistry*, *60*(1), 39–71.

Samrat, S.K., Tharappel, A.M., Li, Z., & Li, H. (2020). Prospect of SARS-COV-2 spike protein: Potential role in vaccine and therapeutic development. *Virus Research*, *288*, 198141.

Samuel, J., Sinha, S., & Ghosh, A. (2006). DNA elasticity: topology of self-avoidance. *Journal of Physics: Condensed Matter*, *18*(14), S253.

Sierzega, Z., Wereszczynski, J., & Prior, C. (2021). Wasp: a software package for correctly characterizing the topological development of ribbon structures. *Scientific Reports*, *11*(1), 1–15.

Skamnaki, V.T., Peumans, W.J., Kantsadi, A.L., Cubeta, M.A., Plas, K., Pakala, S., et al. (2013). Structural analysis of the rhizoctonia solani agglutinin reveals a domain-swapping dimeric assembly. *The FEBS Journal*, *280*(8), 1750–1763.

Sumners, D.W. (2020). The role of knot theory in DNA research. In C. McCory (Ed.), *Geometry and Topology* (pp. 297–318), CRC Press.

T"or"ok, T., Kliem, B., & Berger, M.A. (2014). The evolution of writhe in kink-unstable flux ropes and erupting filaments. *Plasma Physics and Controlled Fusion*, *56*(6), 064012.

T"or"ok, T., Berger, M.A., & Kliem, B. (2010). The writhe of helical structures in the solar corona. *Astronomy & Astrophysics*, *516*, A49.

Travers, A., & Muskhelishvili, G. (2015). DNA structure and function. *The FEBS Journal*, *282*(12), 2279–2295.

van der Heijden, G., Peletier, M., & Planqu'e, R. (2007). On end rotation for open rods undergoing large deformations. *Quarterly of Applied Mathematics*, *65*(2), 385–402.

Van der Heijden, G.H.M., & Thompson, J.M.T. (2000). Helical and localised buckling in twisted rods: a unified analysis of the symmetric case. *Nonlinear Dynamics*, *21*(1), 71–99.

Weiss, M.S., Palm, G.J., & Hilgenfeld, R. (2000). Crystallization, structure solution and refinement of hen egg-white lysozyme at pH 8.0 in the presence of MPD. *Acta Crystallographica Section D: Biological Crystallography*, *56*(8), 952–958.

White, J.H. (1969). Self-linking and the Gauss integral in higher dimensions. *American Journal of Mathematics*, *91*(3), 693–728.

Xu, Y., Zhu, J., & Guo, Y. (2020). Three-dimensional reconstruction of filament axes and their writhe numbers. *The Astrophysical Journal*, *892*(1), 54.

Zhi, D., Shatsky, M., & Brenner, S.E. (2010). Alignment-free local structural search by writhe decomposition. *Bioinformatics*, *26*(9), 1176–1184.

10

Kinetic Helicity in the Earth's Atmosphere

Otto Chkhetiani and Michael Kurgansky

ABSTRACT

This chapter gives an overview of the helicity of the velocity field (*kinetic* helicity, to distinguish it from the *magnetic* helicity used in magnetohydrodynamics, astrophysics, and solar physics; or simply *helicity* in this chapter) and the role this concept plays in modern research in atmospheric physics and atmospheric turbulence. General information on helicity is presented, including some historical comments and a brief description of helicity's fundamental role in fluid dynamics. Different applications of the helicity concept are discussed in the analysis of various dynamic atmospheric processes, including applications to intense atmospheric vortices such as tropical cyclones, tornadoes, dust devils, and Ekman boundary layer dynamics. The chapter emphasizes helicity balance conditions and the important role of helicity fluxes in their maintenance. Fundamentals of helical turbulence theory are briefly discussed, followed by the helical properties of atmospheric turbulence within the atmospheric boundary layer. In particular, the chapter describes the pioneering efforts of the research team from the Obukhov Institute of Atmospheric Physics (Moscow) to measure turbulent helicity in natural atmospheric conditions.

10.1. INTRODUCTION

This chapter addresses the notion of *kinetic helicity* (or simply *helicity*) and its applications to atmospheric dynamics and atmospheric turbulence. Helicity – or, more precisely, the helicity bulk density – is defined by the dot product $h = \mathbf{v} \cdot \boldsymbol{\omega}$ of the three-dimensional velocity $\mathbf{v}$ and vorticity $\boldsymbol{\omega} = \boldsymbol{\nabla} \times \mathbf{v}$ vectors. It is a pseudoscalar quantity; that is, its sign depends on the chosen frame of reference, either right-handed or left-handed, and this sign changes to the opposite under mirror reflection. Helicity is positive when air parcels move along a helical line following the right-hand rule. On the other hand, the air motion, according to the left-hand rule, corresponds to negative helicity.

A. M. Obukhov Institute of Atmospheric Physics, Russian Academy of Sciences, Moscow, Russia

There is a certain hierarchy of atmospheric motions with respect to vorticity: (1) irrotational (potential) flows; (2) rotational flows (vortices), both two-dimensional and three-dimensional but with zero helicity in the latter case; and (3) helical vortex flows, which are inherently three-dimensional, generally have a complex topology of the vorticity field, and, in a certain sense, are at the top of this hierarchical ladder. In meteorology and atmospheric physics, the interest of many researchers in the concept of helicity is explained primarily by its importance for the study of atmospheric vortex flows, which are usually, but not always, of convective origin and have a complex structure, sometimes topologically nontrivial. These include tornadoes and dust devils (e.g., Tippett et al., 2014; Smith et al., 2014; Clark et al., 2012, 2013), tropical mesoscale vortices (Shenming et al., 2011) and tropical cyclones (e.g., Levina et al., 2009; Molinari and Vollaro, 2010; Yu and Yu, 2011; Levina, 2019; Chen et al., 2021), and turbulence (e.g., Chkhetiani and Golbraikh, 2012;

Helicities in Geophysics, Astrophysics, and Beyond, Geophysical Monograph 283, First Edition.
Edited by Kirill Kuzanyan, Nobumitsu Yokoi, Manolis K. Georgoulis, and Rodion Stepanov.

DOI:10.1002/9781119841715.ch10

Yu et al., 2013; Rorai et al., 2013; Chkhetiani et al., 2018; Vazaeva et al., 2017, 2021). Less attention has been paid to the helicity of large-scale extratropical atmospheric motions, explained by atmospheric baroclinicity and Earth's rotation (Etling, 1985; Kurgansky, 1990; Bluestein, 1992).

This chapter has the following structure. Section 10.2 provides an overview of helicity, including some historical comments, and presents helicity's fundamental role in fluid dynamics. The different applications of helicity to the analysis of various dynamic atmospheric processes are discussed in section 10.3. The fundamentals of the theory of helical turbulence are briefly discussed in section 10.4, followed by the helical properties of atmospheric turbulence within the atmospheric boundary layer. In particular, the pioneering effort of the research team of the Obukhov Institute of Atmospheric Physics (Moscow) to measure turbulent helicity under natural atmospheric conditions is described. Finally, section 10.5 presents concluding remarks and formulates problems that remain in the theory of helical turbulence.

10.2. GENERAL INFORMATION ON HELICITY

In basic fluid dynamics, the conservation law of helicity for barotropic flows of nonviscous fluids in the presence of only potential external forces was discovered by Moreau (1961) and Moffatt (1969) two centuries after L. Euler presented his famous equations for fluid motion in 1757. Moffatt (1969) clarified the topological meaning of helicity as a measure of linkage (knottedness) of vortex lines. This helicity conservation law and its interpretation are closely related to demands of the famous circulation theorem by Lord Kelvin (1868), who, as noted in Moffatt (1969), was fully aware of the possibility of the existence of linked and knotted vortex lines. These ideas go back to the classic works of C.F. Gauss on electromagnetism.

Moreau (1961) and Moffatt (1969) had more immediate predecessors, among them H. Ertel, who, in a joint article with C.-G. Rossby (Ertel and Rossby, 1949), derived the local conservation law for ideal barotropic flows of a compressible fluid:

$$\frac{\mathrm{D}}{\mathrm{D}t}\left\{\frac{1}{\rho}\boldsymbol{\omega}\cdot(\mathbf{v}-\nabla W)\right\}=0,\; W=\int_0^t \mathcal{L}dt. \qquad (10.1)$$

Here, $\mathrm{D}/\mathrm{D}t$ denotes the total (material) derivative in time. Under the integral sign along the fluid particle trajectory, such that $\mathcal{L}=\mathrm{D}W/\mathrm{D}t$, stands the fluid dynamical Lagrange function, with ρ as the fluid density, p the pressure, and Φ the gravity potential,

$$\mathcal{L}=\frac{\mathbf{v}^2}{2}-\left(\Phi+\int\frac{dp}{\rho}\right).$$

The law of conservation of total (integral) helicity follows immediately from equation (10.1) if we multiply it by ρ and integrate over the domain occupied by the fluid motion, given that the vortex lines do not penetrate the boundary of the domain.

Helicity is not a Galilean invariant quantity; see, for example, Bluestein (1992). There is an analogy with neutrino physics (note how Moffatt (1969) proposed the term *helicity* by analogy with the neutrino), where the helicity sign of the neutrino depends on the reference frame in which a neutrino with a nonzero rest mass is observed: that is, moving slower or faster than the neutrino (Goldhaber and Goldhaber, 2011). The same situation is possible in atmospheric physics. In tornado research, helicity is usually considered in the reference frame associated with the paternal storm within which a tornado originates. Moving on to another Galilean frame of reference, in some situations, for example, the right rotation of the wind with height (so-called in meteorology because an observer who stands facing the wind discovers that the wind at a higher level blows from the right side: e.g., Koschmieder (1933)), corresponding to positive helicity, changes (locally) to left rotation, corresponding to negative helicity.

The concept of helicity proved to be the most powerful and demanded for an explanation of the generation of the magnetic field in an electrically conductive fluid due to dynamo action (Moffatt, 1978; Krause and Raedler, 1980). Although nonzero helicity is not necessary for dynamo theory (Gilbert et al., 1988), it greatly simplifies the analysis. Based on observational data, Kuzanyan et al. (2000) calculated the helicity distribution in the solar atmosphere, where the helicity determines the alpha effect responsible for generating the magnetic field (Komm et al., 2014). The rotation of the Sun plays an important role, and it is no coincidence that noticeable (measurable) magnetic fields are found on rotating astronomical objects (Moffatt, 1978). Moiseev et al. (1983, 1988) made a bold attempt to transfer the ideas and methods of dynamo theory to atmospheric physics, but it is not trivial that the application of the theory to atmospheric physics presents more conceptual difficulties, in particular as relates to the conservation of angular momentum.

In flows with high values of $h=\mathbf{v}\cdot\boldsymbol{\omega}$, the vectors $\mathbf{v}$ and $\boldsymbol{\omega}$ are almost collinear, so the Lamb vector $\boldsymbol{\omega}\times\mathbf{v}$ in the Euler equations taken in the Gromeka–Lamb form (Landau and Lifshitz, 1987) is almost zero, and the cascade of energy to smaller scales is blocked (Speziale, 1989). An alternative explanation (Chakraborty, 2007) is that the complex (nontrivial) topology of the vorticity field effectively hampers the stretching of the vortex lines (tubes) in the bounded space, whereas stretching is a primary mechanism of energy transfer along the spectrum. The idea is that a knotted vortex line is forced to effectively occupy a larger volume during stretching, which is difficult.

In an incompressible fluid with uniform density $\rho = 1$, the integrals of helicity $\mathcal{H} = \int \mathbf{v} \cdot \boldsymbol{\omega} d\tau$, of kinetic energy $\mathcal{E} = \frac{1}{2} \int \mathbf{v}^2 d\tau$, and of enstrophy $\mathcal{F} = \frac{1}{2} \int \boldsymbol{\omega}^2 d\tau$, where $d\tau$ is the volume element, are related by the Schwartz inequality $\mathcal{H}^2 \leq 4\mathcal{E}\mathcal{F}$ (Moffatt, 1969). When $\mathcal{E}, \mathcal{H} = \text{const}$, then $\mathcal{F} \geq \frac{1}{4}\mathcal{H}^2/\mathcal{E}$, and the inverse energy cascade to large scales is impossible. However, the joint cascade of energy and helicity to small scales is not restricted (Kraichnan, 1973). The inequality becomes the true equality for pure helical motions, known as Beltrami flows, in which $\boldsymbol{\omega} \times \mathbf{v} = 0$. In atmospheric physics, the concept of Beltrami flow is used with some success to describe the structure of vortices (Kanak and Lilly, 1996). However, it is unclear whether this concept is just a useful idealization or corresponds to a motion that occurs. For example, it has been shown (Kurgansky, 2013a) that a steady adiabatic Beltrami flow of a fluid with three independent thermodynamic functions of the state (moist atmospheric air and/or salt seawater) is dynamically impossible because its existence would contradict the Ertel vorticity theorem (Ertel, 1942). In Nambu mechanics, which represents a generalization of the Hamiltonian formalism in fluid dynamics (Nambu, 1973; Névir and Blender, 1993), the helicity defined as $\mathcal{H} = \frac{1}{2} \int \mathbf{v} \cdot \boldsymbol{\omega} d\tau$ has the same level of importance as the energy $\mathcal{E}$. Note that with this helicity normalization, which is the most natural for the Nambu theory, the previously written Schwartz inequality acquires a simpler canonical form $\mathcal{H}^2 \leq \mathcal{E}\mathcal{F}$. Later, the Nambu formalism was extended to more complex systems of fluid dynamical equations (e.g., Gassmann and Herzog, 2008; Névir and Sommer, 2009; Salazar and Kurgansky, 2010). The main idea is that the original symmetry properties must be preserved for any approximation of the equations of motion (Salmon, 2005, 2007).

Using the Legendre transform, the equation of helicity balance in a Boussinesq fluid can be written in two different forms. The first form uses the potential vorticity (Hide, 1989), and the second uses the vertical component of vorticity (Lilly, 1986; Wu et al., 1992). As an analogue of temperature in the general case of a compressible fluid, the geopotential taken with a minus sign appears in the Boussinesq approximation, which considerably facilitates mathematical analysis. The Legendre transform of thermodynamic variables in a compressible fluid provides no such benefit; see also Kurgansky (2006, 2017). Existing attempts to incorporate the concept of helicity into the vortex theory are limited by the Boussinesq approximation and the case of an incompressible fluid with uniform density. An extension to compressible fluid flows awaits a solution.

The definition of helicity does not include the unit of mass. Therefore, the helicity balance equation can be derived without invoking the mass continuity equation. In barotropic fluid flows, helicity is a conserved quantity when viscosity and nonpotential external forces are absent (Moreau, 1961; Moffatt, 1969). In ideal baroclinic fluid flows, nonuniform nonzero values of the potential vorticity determine the source of helicity (Hide, 1989; Kurgansky, 1989). A relationship between the helicity bulk density $\mathbf{v} \cdot \boldsymbol{\omega} \equiv \mathbf{v} \cdot \nabla \times \mathbf{v}$ and a pseudoscalar quantity $P = \boldsymbol{\omega} \cdot \nabla \ln \rho$ was established in Hide (1989) for a generally rotating compressible fluid with angular velocity $\boldsymbol{\Omega}$. In the absence of general rotation, P coincides with the potential vorticity for a density-stratified incompressible fluid. That is, Hide (1989) established the relationship

$$\frac{\partial}{\partial t}(\mathbf{v} \cdot \boldsymbol{\omega}) = -2\frac{p}{\rho}P + \Psi - \nabla \cdot \mathbf{J}, \tag{10.2}$$

which is reproduced here in symbolic notation. In equation (10.2),

$$\Psi = 2\left[\boldsymbol{\omega} \cdot \mathbf{F} - (2\boldsymbol{\Omega} \cdot \mathbf{v})(\nabla \cdot \mathbf{v})\right]; \tag{10.3}$$

$\mathbf{F}$ denotes the nonpotential forces, including the frictional force and ponderomotive Ampere force (the latter for electrically conductive fluids); vector $\mathbf{J}$ is diagnostically expressed in terms of fluid dynamic variables. For a viscous fluid, in the absence of nonpotential external body forces, $\mathbf{F} = -\nu(\nabla \times \boldsymbol{\omega})$, where ν is the kinematic viscosity. The quantity $\boldsymbol{\omega} \cdot (\nabla \times \boldsymbol{\omega})$ was called "superhelicity" in Hide (1989), and this name seems to have stuck. (However, as R. Hide said later, and as his colleague at the university pointed out to him, it would be linguistically more correct to speak of *hyperhelicity* since the word *helicity* is of Greek origin and the prefix *super* is of Latin origin.) It is worth mentioning that $-2\nu\boldsymbol{\omega} \cdot (\nabla \times \boldsymbol{\omega})$ describes the time rate of dissipation (destruction) of helicity due to viscosity (Moffatt, 1969); and superhelicity is not a conserved quantity for inviscid barotropic fluid flows, although helicity is. The expression in equation (10.3) includes an additional source of helicity associated with the compressibility of the fluid and its general rotation. However, the equation for helicity balance was rewritten in a general invariant form (Kurgansky, 1989), which includes the Ertel potential vorticity $\Pi_s = (\boldsymbol{\omega}_a \cdot \nabla s)/\rho$ for a stratified compressible fluid in a rotating reference frame:

$$\begin{aligned}\frac{\partial}{\partial t}\{(\boldsymbol{\omega} + 4\boldsymbol{\Omega}) \cdot \mathbf{v}\} = -\nabla \cdot \{(\boldsymbol{\omega} + 4\boldsymbol{\Omega})\mathcal{B} + (\omega_a \times \mathbf{v}) \\ \times \mathbf{v} - T\nabla s \times \mathbf{v} - \mathbf{F} \times \mathbf{v}\} \\ + 2T\rho\Pi_s + 2\omega_a \cdot \mathbf{F}.\end{aligned} \tag{10.4}$$

Here, s is the specific entropy; $\mathcal{B} = (\mathbf{v}^2/2) + w + \Phi$ is the Bernoulli function, with w as the specific enthalpy and

Φ the gravity potential; T is the absolute temperature in Kelvin; and $\boldsymbol{\omega}_a = \nabla \times \mathbf{v} + 2\boldsymbol{\Omega}$ is the absolute vorticity. At the same time, it was necessary to generalize the concept of helicity to take into account the overall rotation of the fluid:

$$\mathbf{v} \cdot \boldsymbol{\nabla} \times \mathbf{v} \Rightarrow \mathbf{v} \cdot (\boldsymbol{\nabla} \times \mathbf{v} + 4\boldsymbol{\Omega}) \equiv \mathbf{v} \cdot (\boldsymbol{\omega}_a + 2\boldsymbol{\Omega}). \quad (10.5)$$

The appearance of a multiplier 4 in equation (10.5) is explained by the quadratic nonlinearity of helicity with respect to the velocity field. Opinions differ about what quantity should be considered the absolute helicity. Sometimes (e.g. Névir and Sommer, 2009), the scalar product of the absolute velocity and absolute vorticity, $\mathbf{v}_a \cdot \boldsymbol{\omega}_a \equiv (\mathbf{v} + \boldsymbol{\Omega} \times \mathbf{x}) \cdot (\boldsymbol{\omega} + 2\boldsymbol{\Omega})$, is taken as the absolute helicity, but the presence of the position vector $\mathbf{x}$ in equations written in Euler variables is not always convenient. Sometimes the absolute helicity is taken as a scalar product of the relative velocity and absolute vorticity, $\mathbf{v} \cdot \boldsymbol{\omega}_a = \mathbf{v} \cdot (\boldsymbol{\omega} + 2\boldsymbol{\Omega})$ (Pichler and Schaffhauser, 1998). However, this definition does not fully meet the symmetry requirements, and it is not possible to construct the Nambu fluid dynamical brackets (cf. Névir and Sommer, 2009; Salazar and Kurgansky, 2010). From this perspective, the definition in equation (10.5) has advantages.

10.3. HELICITY IN DYNAMIC ATMOSPHERIC PROCESSES

In the baroclinic atmosphere, when the potential vorticity is identically zero, $\Pi_s \equiv 0$, it follows from equation (10.4) that helicity is conserved in an ideal adiabatic case (Kurgansky, 2006). Equality to zero, $\Pi_s \equiv 0$, is characteristic only of meso- and small-scale atmospheric motions. These flows were extensively studied in the literature, starting from pioneering papers by A. Eliassen (Eliassen and Kleinschmidt (1957), Eliassen (1987) and references therein). Also, based on his helicity balance analysis from equation (10.2), Hide (1989) argued that vanishing potential vorticity is generally valid for both natural and laboratory convection. For large-scale extratropical atmospheric processes, the vector $\boldsymbol{\omega}_a$ "permeates" the isentropic surfaces at a nonzero angle. Therefore, nonzero values Π_s and thus nonzero helicity sources always exist. For large-scale, 2½-dimensional processes (cf. Kurgansky, 2002), helicity quantifies the non-self-similarity of the velocity field at different height levels: for example, when a change in wind direction with height occurs due to the thermal wind effect. Helicity vanishes in barotropic or equivalent-barotropic atmospheric motions. The significant change of wind direction with height, and the accompanying helicity concentration, occur in the Ekman boundary layer. This is the famous Ekman spiral, with right rotation of the wind with height (a veering wind that turns clockwise with height) and positive helicity in the main bulk of the Ekman boundary layer in the Northern Hemisphere, and left wind rotation (a backing wind that turns counterclockwise with height) and negative helicity in the Southern Hemisphere.

Etling (1985) and Lilly (1986) introduced the concept of helicity into atmospheric physics. In the decade that followed, only a few articles devoted to atmospheric helicity were published. Then the number of articles began to increase steadily, especially those dedicated to rotating supercells and tornadoes and applications of storm-relative environmental helicity (Droegemeier et al., 1993) and/or storm-relative helicity (Markowski et al., 1998). These helicities, measured in the frame of reference that moves with the parental storm, are primary predictors of tornadogenesis; see, for example, Doswell and Schultz (2006). To avoid misunderstanding, we note that in fluid dynamics, the term *relative helicity* usually refers to $\chi = (\mathbf{v} \cdot \boldsymbol{\omega}) / (|\mathbf{v}| \cdot |\boldsymbol{\omega}|)$: that is, to the cosine of the angle between the vectors $\mathbf{v}$ and $\boldsymbol{\omega} = (\boldsymbol{\nabla}) \times \mathbf{v}$. Helicity is useful and in great demand in mesometeorology and micrometeorology. We note in this regard that Han et al. (2006) introduced "shearing wind helicity" and "thermal wind helicity," which differ from the commonly used helicity. According to Han et al. (2006), these newly introduced helicities are useful in analyzing circulation systems and have been applied to the diagnosis of Hurricane Andrew (1992). We agree that helicity is a useful diagnostic for identifying certain types of long-lived storms. The more data is available and evoked to characterize these storms, the better.

In general, helicity is of indisputable importance for diagnosing various atmospheric processes, but does it have predictive value? Levina and Montgomery (2010, 2011) (see also Levina (2019)) showed that (integral) helicity is an important indicator of the development of large-scale instability leading to tropical cyclogenesis. Using output data from the Eta regional atmospheric model, Lavrova et al. (2010) found that the calculated helicity can be one of the earliest predictors of the genesis of Mediterranean cyclones. The development of this research direction is of undoubted interest, in particular as it refers to the relationship between anomalous "atmospheric helicity" and the further development of weather systems (cf. Vazaeva et al., 2021).

Hide (1989), Kurgansky (1989, 1993, 2002), Zhemin and Rongsheng (1994), Chkhetiani (2001), Ponomarev et al. (2003), Ponomarev and Chkhetiani (2005), Deusebio and Lindborg (2014), Chkhetiani et al. (2018) and, most recently, Vazaeva et al. (2021) investigated different dynamic aspects of helicity in the Ekman boundary layer. In particular, Zhemin and Rongsheng (1994) took into account the nonlinearity of the equations of motion and revealed the cyclone/anticyclone asymmetry in helicity values in the Ekman boundary layer. The influence of baroclinicity (thermal wind) was also considered by Zhemin and Rongsheng (1994); but unlike Kurgansky (1989, 1993, 2002), they did not analyze the helicity balance in the Ekman boundary layer, which, given the additional effects indicated in Zhemin and Rongsheng (1994), is an interesting problem for future studies (cf. also Deusebio and Lindborg, 2014; Kurgansky, 2017; Chkhetiani et al., 2018).

Helicity density has a dimension of acceleration, and its proper reference value on Earth is the acceleration due to gravity $g = 9.81\,\mathrm{m\,s^{-2}}$. Helicity quantifies the three-dimensional organization (e.g., Levich and Tzvetkov (1985), as well as Table 10.1, adapted from Cieszelski (1999), who summarized estimates in Etling (1985) and Zhemin and Rongsheng (1994); these were further extended and refined in Vazaeva et al. (2021)) and provides stability and durability to atmospheric vortex structures that possess helicity (Lilly, 1986).

In addition to the vortices indicated in Table 10.1, high values of $|h|$ are characteristic of, for example, horizontal vortex structures in the planetary boundary layer: rolls and "cloud streets" (Etling, 1985). For vertical columnar vortices in approximate cyclostrophic equilibrium, the most convenient estimate of the helicity bulk density is $|h| \sim VW/R$, where V is the characteristic azimuthal velocity, W the characteristic vertical velocity, and R the vortex core radius. More precisely, the contribution of the vertical velocity and vorticity components to helicity is estimated as $(2V/R)\,W\pi R^2 H$, where H is the vortex height. The contribution to the helicity of the radial components of the velocity and vorticity is usually negligible. However, the contribution of the azimuthal components turns out to be approximately equal to the contribution of the vertical components calculated earlier, so the total helicity of the vortex is estimated as $\mathcal{H} = 4\pi VWRH$; this, in contrast to the kinetic energy of a vortex, can be finite, as, for example, occurs for a combined Rankine vortex. This is one more virtue of the helicity concept. It is possible to write equivalently that $\mathcal{H} = 2 \cdot 2\pi RV \cdot WH$ (cf. Kurgansky, 2013b), which is consistent with Moffat's (1978) result for the value of total helicity in the case of entangled toroidal and poloidal circulations. For mesoscale tropical cyclones, the contribution to helicity of toroidal (azimuthal) components of $\mathbf{v}$ and $\boldsymbol{\omega}$ is much greater than the contribution of poloidal components, associated with so-called secondary, or meridional, circulation; the former contribution has an order of magnitude of $10^{-1}\,\mathrm{m\,s^{-2}}$, while the latter's is $\sim 10^{-3}\,\mathrm{m\,s^{-2}}$ (cf. Cieszelski, 1999; Kurgansky, 2017). For large-scale quasigeostrophic atmospheric processes, as the scale estimation shows by taking into account that these motions are quasigeostrophic, the contribution of the horizontal components of $\mathbf{v}$ and $\boldsymbol{\omega}$ to the helicity is approximately five orders of magnitude greater than that of the vertical components (cf. Kurgansky, 2017).

Table 10.1 Scale and helicity bulk density of different atmospheric vortex structures (Cieszelski (1999); cf. Kurgansky (2017) and Vazaeva et al. (2021)).

Vortex structure	Height, m	Radius, m	Helicity, m/s^2
Tropical hurricane	10^4	10^5	10^{-1}
Storm	10^4	10^3	10^{-2}
Tornado	10^3	10^2	10^1
Ekman layer	10^3	–	10^{-1}
Rotating thermal	10^3	10^2	10^{-2}
Dust devil	$10^2 - 10^3$	10^1	10^1

Representing helicity $h = \mathbf{v} \cdot \boldsymbol{\omega}$ as a sum of three contributions corresponding to three (locally) orthogonal directions in space, when each is equal to the product of the corresponding components of velocity and vorticity, is not unique. Equally useful is the alternate representation (Hide, 1975), also as a sum of the three terms but arranged differently. For each term, a direction in space is indicated, and only those components of velocity in the plane orthogonal to this direction are considered. This way, in practically important cases, the main contributors to helicity are grouped in a single term but are not dispersed in different contributions, as occurs with the first, more traditional representation. For example, in the Ekman boundary layer, they are grouped into a term that corresponds to the vertical direction and horizontal velocity components (and their vertical derivatives). For vertical convective vortices, such as tornadoes and dust devils, they are grouped into a term corresponding to the radial direction and azimuthal and vertical velocity (and their radial derivatives).

In the atmosphere, helicity generation and destruction regions do not usually coincide but are generally separated in space. Therefore, helicity fluxes are important to maintain the helicity balance. A scheme of helicity balance in intense atmospheric vortices is presented in Fig. 10.1. It is considered a steady axisymmetric vortex. Most of the helicity production G due to vertical convection in the presence of overall rotation (shown schematically by vertically oriented helical lines) occurs above a horizontal surface A, which marks the top of a surface boundary layer B. The helicity destruction D, due to small-scale turbulent viscosity, is limited to B. Under real circumstances, it is optimal to take A at a height of several meters to a few tens of meters above ground level.

Kurgansky (2008) proposed using vertical helicity flux in small and mesoscale vortices as a measure of their

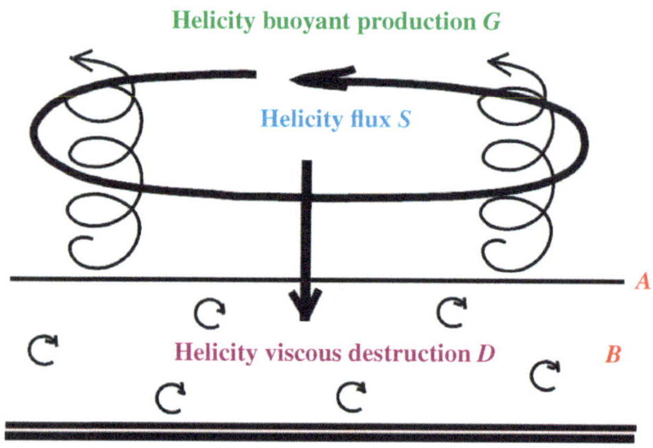

Figure 10.1 A schematic of helicity balance in intense atmospheric vortices. The vortex flow is outlined by a thick curved arrow. Helicity flux S is shown by a downward pointing arrow. Small-scale turbulence within the boundary layer is presented by small curved arrows. A thick double horizontal line shows the Earth's surface.

intensity. For a steady axisymmetric vortex in polar cylindrical coordinates under the neglect of the small radial velocity component u, this downward directed helicity flux through the upper level A of the turbulent boundary layer is (Kurgansky, 2017)

$$S = \iint_A \left(2v^2 - w^2\right) \omega_z \mathrm{d}\sigma. \tag{10.6}$$

Here, v is the azimuthal velocity component, w the vertical velocity component, ω_z the vertical vorticity, and $\mathrm{d}\sigma$ the surface area element; it has been used that $rv\, w^2\big|_{r=0}^{r=\infty} = 0$. Here we come to an important point. As already follows from Moffatt (1969) and confirmed by the analysis of equation (10.4), the helicity flux has both advective (velocity-driven) and non-advective (along the vortex lines) components. As a rule, the latter component – for example, in intense atmospheric vortices – significantly exceeds the former, as accounted for in equation (10.6); see also Kurgansky (2017). The effect of Earth's rotation with the angular velocity $\mathbf{\Omega}$ has been neglected. If needed, accounting for it would result in the additional term $\iint_A (2\mathbf{\Omega} \cdot \mathbf{k})\, v^2 \mathrm{d}\sigma$ on the right side of equation (10.6), where $\mathbf{k}$ is the unit vertical vector.

In atmospheric vortices on the small and meso scales, as a rule, $u < (\ll)\, w < (\ll)\, v$. For example, based on direct measurements from aircraft in waterspouts in Leverson et al. (1977), the characteristic values are as follows: $v = 25\,\mathrm{m\,s^{-1}}$, $w = 8\,\mathrm{m\,s^{-1}}$ and $u = 1\,\mathrm{m\,s^{-1}}$. Therefore, instead of equation (10.6), the approximate formula can be used (see Kurgansky (2008, 2017)):

$$S \cong \iint_A 2v^2 \omega_z \mathrm{d}\sigma \tag{10.7}$$

or, alternatively, $S \cong (8\pi/3)\int_0^\infty v^3 \mathrm{d}r$, if only $v^3 r \to 0$ with $r \to \infty$, where r is the radial coordinate. Glebova et al. (2009) applied the method of helicity fluxes to the Man-Yi (2007) and Wilma (2005) tropical cyclones, and Farahani et al. (2017) did so for tropical storm Gonu (2007).

For the combined Rankine vortex with uniform vertical flow inside the vortex core and neglecting the Coriolis force, the vertical helicity flux in a vortex is estimated by the value $S = 2\pi V\left(V^2 - W^2\right) R$ (see equation (10.6)), where V is the maximum azimuthal velocity, W the vertical velocity, and R the vortex core radius. And if $V \gg W$, then $S \approx 2\pi V^3 R$; see equation (10.7) and Kurgansky (2013b). The natural time scale for vortices is the time of their complete revolution around the axis, $2\pi R/V$. For example, the use of data in Lorenz (2013) allows us to conclude that dry-convective small-scale dust-devil vortices, regardless of their size, have a life duration on average of about 60 revolutions. In large vortices, this revolution takes longer; therefore, in absolute units of time, they live longer. The ratio $\mathcal{H}/S$, where $\mathcal{H} = 4\pi VWRH$ is the total helicity of the vortex flow and H is the height of the vortex (mentioned earlier), determines the characteristic time of the helicity transformation (helicity metabolism) in the vortex flow. This time is short for dust devils and equates to only a few complete revolutions of the vortex around its axis. If we put $H/R = 10$ (Lorenz, 2013; Jackson, 2020) and accept that $V/W = 3$, the estimate shows that the time of the complete renewal of helicity in the vortex is ~6 complete revolutions of the dust devil around its axis; that is, during a typical dust-devil vortex lifetime, the vortex helicity has time to be completely renewed about 10 times. From this perspective, one can say that helicity is a kind of constantly renewing corset (frame) that energy fills. Compared to energy dissipation, the claim of a shorter helicity destruction time is generally consistent with the conclusions of Levshin and Chkhetiani (2013), who showed that helicity in homogeneous turbulence decays faster than energy. The same is true for large-scale extratropical atmospheric motions. The rate of destruction of helicity within the Ekman boundary layer is equal to the downward flux of helicity from the free atmosphere. This flux is determined (per unit area of the Earth's surface) by the product of the Coriolis parameter and the square of the geostrophic wind speed in the free atmosphere (Kurgansky, 1989, 1993, 2002); also Deusebio and Lindborg (2014). Therefore, the helicity transformation time in large-scale extratropical atmospheric processes is given by the reciprocal of the Coriolis parameter (~10^4 s); and it is less than the characteristic duration of large-scale cyclones and anticyclones by a factor of 10.

Until now, the main emphasis has been on observations of helicity but not on its dynamics. The previous arguments about helicity balance and its "metabolism" apply to a quasisteady mature stage of atmospheric vortices. The issue of helicity generation (genesis) is more complex and nontrivial. Helicity is conserved and therefore cannot be generated in barotropic flows of inviscid fluids in the presence of potential forces (Moreau, 1961; Moffatt, 1969). So, the genesis of helicity requires the violation of the earlier conditions; that is, it requires the joint action of atmospheric baroclinicity, which often takes the form of vertical convection leading to vertical flows, and the general rotation of the flow system, either directly or indirectly related to the Earth's rotation. In the latter case, the sense of Earth's rotation is "transmitted" through the sequence of emerging circulation systems of progressively decreasing spatial scale (cold atmospheric fronts, mesocyclones, etc.) in which tornadoes are embedded, almost all of them (with rare exceptions) cyclonic vortices. There are indications that surface friction may also contribute to the genesis of helicity. The small-scale dynamics of helicity are important for tornadoes. In particular, these are Hasimoto solitons (Hasimoto, 1972) propagating along strong vortex filaments (thin tornadoes); Aref and Flinchem (1984) have provided photographs of a tornado near Braman, Oklahoma (11 May 1978) showing localized large-amplitude helical twist in the vortex core (a *soliton*). Holm and Stechmann (2004) showed that the nonlinear Schrödinger equation in Hasimoto (1972) is replaced by the integrable complex modified Korteweg–de Vries equation having soliton solutions when the Hamiltonian in the underlying theory is not the kinetic energy but helicity. The classical case of kinetic energy and the case of helicity discussed in Holm and Stechmann (2004) exhaust all Hamiltonians that have fluid dynamical significance.

10.4. HELICAL TURBULENCE IN THE EARTH'S ATMOSPHERE

Moving from the discussion of general principles in section 10.2 to a more detailed consideration of helical turbulence, particularly in the Earth's atmosphere, we first mention successful attempts to numerically reproduce cascades of energy and helicity directed differently, but taking into account the general rotation of the fluid (Mininni and Pouquet, 2010a,b; Pouquet and Mininni, 2010). However, Biferale et al. (2012, 2013) and Sahoo et al. (2017) showed that without general rotation but by allowing interactions between helical modes of the same sign (homochiral modes), the inverse cascade of energy is possible as in two-dimensional turbulence. Gledzer and Chkhetiani (2015), Chkhetiani and Gledzer (2017), also using the turbulent cascade model with conservation of helicity (Gledzer, 1973), show that to reproduce the inverse energy cascade from small-scale disturbances, a certain level of "helical noise" is necessary for large-scale modes, which is generated by external forcing. These conclusions are also supported by an analysis of second-order moments in the quasinormal approximation. Similar results were also obtained in Stepanov et al. (2015). Another line of research uses the formulation of three-dimensional turbulent fields in terms of helical modes introduced in Knorr et al. (1990) and Waleffe (1992), as applied in Rathmann and Ditlevsen (2017) to investigate details of joined energy and helicity cascades.

An important research line is related to direct numerical simulation (DNS) studies of helicity, described only briefly here. In particular, they include DNS studies of turbulence decay by Holm and Kerr (2002). Their diagnostics showed that the turbulence initial value problem evolves through an early stage, with still-negligible dissipation when vortex sheets form from smooth initial conditions and then coil into helical vortex tubes. Once the helical vortex tubes have formed, viscous dissipation grows rapidly in the DNS, and the energy spectrum gradually approaches the classical $-5/3$ law characteristic of a turbulent energy cascade; but before it is achieved, the energy spectrum remains steeper than $-5/3$; see Fig. 6 in Holm and Kerr (2002). Once a clear $-5/3$ spectrum appears, the energy decay becomes self-similar. A numerical study of the joint cascade of energy and helicity by three-dimensional DNS was carried out by Chen et al. (2003) and verified their theoretical conclusions that the Kolmogorov microscale of turbulence is the minimum scale of helical motions for a joint cascade of energy and helicity. DNS studies of helical turbulent flows continue and are actively developing today. It is beyond the scope of this review to list all relevant papers; see, for example, Biferale et al. (2019).

Chkhetiani (1996) obtained an important relation between the two-point triple correlation of the velocity and the rate of helicity transfer over the spectrum and, therefore, of its viscous destruction at small scales. Later it was called the 2/15 law for helicity, which corresponds to the 4/5 law of A.N. Kolmogorov for the velocity field in three-dimensional nonhelical turbulence and similar to the 4/3 law of A.M. Yaglom for a passive scalar (Chkhetiani, 2008); see also (Golitsyn, 2012, § 6.1.6). The 2/15 law allows us to obtain the spectra of helicity: both the $-5/3$ law, when the helicity behaves as a passive scalar, and the more hypothetical $-4/3$ law, when the helicity cascade is decisive.

Two cases are possible for the turbulent cascade of helicity. The first case is the joint cascade of energy and helicity toward small scales. The spectral slope of helicity coincides in this case with the slope of the energy spectrum and is equal to $-5/3$:

$$\hat{E}(k) = c_E \varepsilon^{2/3} k^{-5/3}, \quad \hat{H}(k) = c_H \eta\, \varepsilon^{-1/3} k^{-5/3}. \qquad (10.8)$$

The second case is the direct cascade of helicity and an inverse cascade of energy (Brissaud et al. 1973):

$$\hat{E}(k) = c_E^* \eta^{2/3} k^{-7/3}, \hat{H}(k) = c_H^* \eta^{2/3} k^{-4/3}. \qquad (10.9)$$

Here, $\hat{E}(k)$ and $\hat{H}(k)$ are the spectral densities of energy $\mathcal{E} = \frac{1}{2}\langle \mathbf{v}^2 \rangle$ and helicity $\mathcal{H} = \langle \mathbf{v} \cdot \boldsymbol{\omega} \rangle$, respectively; $\varepsilon = \nu\langle (\nabla \times \mathbf{v})^2 \rangle$ is the specific kinetic energy dissipation rate; $\eta = \nu \langle \boldsymbol{\omega} \cdot \nabla \times \boldsymbol{\omega} \rangle$ is the time rate of helicity viscous destruction; k is the wavenumber; ν the kinematic viscosity; and angular brackets denote ensemble averaging. Additionally, we note that in this case, the Obukhov–Corrsin spectrum of temperature fluctuations (passive scalar) and the cospectrum of the temperature T and vertical velocity U_3 fluctuations also have different slopes (Moiseev and Chkhetiani, 1996)

$$\hat{E}_{TT}(k) \sim N\eta^{-1/3} k^{-4/3}, \hat{E}_{TU_3}(k) \sim N\eta^{-1/6} k^{-7/6} \qquad (10.10)$$

compared with −5/3 in the usual case. Here, $N = \chi \langle (\nabla T)^2 \rangle$ is the average rate of dissipation of temperature inhomogeneities. Let us also note that the analogue of the Bolgiano–Obukhov scale $L_* = \eta^{-5} N^3 (g\beta)^6$ arises in comparison with $L_{BO} = \varepsilon^{5/4} (g\beta)^{-3/2} N^{-3/4}$, where $\beta = 1/T_0$ with T_0 as the constant reference temperature. For a long time, the existence of the helical cascade in equation (10.9) was considered hypothetical. In particular, the conditions for the realization of inverse energy cascades were considered difficult to fulfill in real conditions, although certain spectra observed in the atmosphere and in laboratory experiments had spectral slopes close to those in equation (10.9) (Chkhetiani et al., 2006). Only relatively recently have such regimes been reproduced in numerical experiments using helicity pumping at different scales (Biferale et al., 2012, 2013; Sahoo et al., 2017; Gledzer and Chkhetiani, 2015).

Helicity in turbulence manifests itself in the appearance in the tensor of two-point triple correlations of the velocity field of those components, which are identically equal to zero in its absence (Chkhetiani et al. 1996), and it holds that

$$\langle \delta \mathbf{v}_l(x|r) \cdot [\mathbf{v}_t(x+r) \times \mathbf{v}_t(x)] \rangle = \frac{2}{15} \eta r^2. \qquad (10.11)$$

The indicated dependence (the 2/15 law) is similar to the Kolmogorov 4/5 law for longitudinal velocity correlations (cf. Monin and Yaglom, 1971, 1975)

$$\left\langle \left(\delta \mathbf{v}_l(x|r) \right)^3 \right\rangle = -\frac{4}{5} \varepsilon r \qquad (10.12)$$

and is the exact result following from the Navier–Stokes equations. Here, the velocity field is divided into longitudinal and transverse components respective to the vector r connecting the observation points, $\mathbf{v}_l = (\mathbf{v}\mathbf{r})\mathbf{r}/r^2$, $\mathbf{v}_t = \mathbf{v} - \mathbf{v}_l$, $\delta \mathbf{v}_l(\mathbf{x}|\mathbf{r}) = (v_l(\mathbf{x}+\mathbf{r}) - v_l(\mathbf{x}))\,\mathbf{r}/r$. Denoting $r_i = r\delta_{ix}$, we obtain for the two-point triple tensor of velocity and vorticity correlations the following representation (Chkhetiani, 2008):

$$T_{i,j,k} = \langle v_i(\mathbf{x}) v_j(\mathbf{x}) \omega_k(\mathbf{x}+\mathbf{r}) \rangle = \frac{\eta r}{15} \delta_{ij} \delta_{kx} - \frac{\eta r}{10} (\delta_{ix} \delta_{jk} + \delta_{ik} \delta_{jx}). \qquad (10.13)$$

Understanding of the presence of helicity in atmospheric turbulence came later. Thus, in the work of Novikov (1972), where the statistical properties of the vorticity flux tensor

$$\langle v_i \omega_j \rangle = \frac{\varepsilon_{ikl}}{4\pi} \int_V \frac{r_l}{r^3} \langle \omega_k(\mathbf{x}') \omega_j(\mathbf{x}) \rangle \mathrm{d}^3\mathbf{x}' \qquad (10.14)$$

were discussed, the diagonal elements of this tensor, which are directly related to helicity, were not considered. According to Novikov (1972), only four nondiagonal elements of equation (10.14) – $\langle u'\omega_2' \rangle$, $\langle v'\omega_1' \rangle$, $\langle w'\omega_1' \rangle$, and $\langle u'\omega_3' \rangle$, where $\mathbf{v} = (u, v, w)$ – should be different from zero in atmospheric boundary layer (ABL) turbulence. The first measurements of the covariance between the velocity and vorticity components and the corresponding cospectra and spectra were carried out at heights of 1.2 and 5 m at the Tsimlyansk test site of the Institute of Atmospheric Physics, Russian Academy of Sciences (Koprov et al., 1988), using an acoustic circulation-meter (the so-called circulimeter) first designed and manufactured by Bovsheverov et al. (1971). The observed differences from zero of the covariance, associated with the circulation in the vertical plane parallel to the mean flow directed along the x-axis and corresponding to $\langle v'\omega_2' \rangle$, were not consistent with the conclusions of Novikov (1972). In addition, a strong correlation was found between the velocity circulation and the temperature field.

To measure helicity, measurements of all three components of the velocity and vorticity are required. Vorticity measurement techniques can be conditionally divided into direct and indirect. Direct measurements with hot-wire arrays were carried out in Wyngaard (1969), Antonia et al. (1988), Vukoslavcevic and Wallace (1996) and Kholmyansky et al. (2001). This technique provides the ability to carry out measurements on scales close to the dissipation range, but it poorly matches the conditions of a real ABL. Indirect methods are associated with measuring circulation along a certain contour and, accordingly, the determination of the average vorticity over the area encircled by the contour, based on the application of the Stokes theorem (Bovsheverov et al., 1971; Koprov et al., 1988, 1994, 2005). Ideologically close techniques for measuring atmospheric circulation were also proposed in Jordan (1980), Ohtou et al. (1983) and Mitsuta and Asai (1984). Such techniques were often used to determine the characteristics of oceanic turbulence (Rossby, 1975; Gaynor et al., 1977; Thwaites et al., 1995; Sanford et al., 1999). The procedure for calculating the velocity circulation was discussed in detail by Longuet-Higgins (1982). In Koprov et al. (1994, 2005), synchronized measurements of velocity and circulation were made using one- and two-component acoustic circulimeters with a contour diameter of 1 m and an acoustic anemometer. All three sensors were placed on top of a 46-m-high mast, with the anemometer in the center and the circulimeters on both sides at a distance of 0.8 m. The spectra of helicity components with a slope close to −5/3 were observed (Fig. 10.2); see Koprov et al. (2005). An integral estimate of helicity from the observed spectrum gives the values 0.02–0.03 ms^{-2}. For the components of the two-point tensor triple product of the velocity and vorticity components $\langle v_i(\mathbf{x})\, v_j(\mathbf{x})\, \omega_k(\mathbf{x}+\mathbf{r})\rangle$ in the same frequency range as for the helicity components, a spectrum slope close to −1 is observed (Fig. 10.3).

In a field campaign in Tsimlyansk, Russia, in August 2012, Koprov et al. (2015) used four three-component acoustic anemometers in the vertices of a rectangular tetrahedron with a side measuring 5 m and its base 5.5 m above the ground (Fig. 10.4). Two different methods of calculating helicity were applied. First the velocity circulations along the contours encircling the tetrahedron faces were calculated; then the Stokes theorem was applied to calculate the corresponding average vorticities ω_x, ω_y, and ω_z, and thus the helicity. Second, an alternative finite-difference method for approximating ω_x, ω_y, and ω_z at point A (see Fig. 10.5) was applied to calculate helicity:

$$H_A = u_A(\omega_x)_A + v_A(\omega_y)_A + w_A(\omega_z)_A$$
$$\approx \frac{w_A u_B - u_A w_B}{L_{AB}} + \frac{v_A u_D - u_A v_D}{L_{AD}} + \frac{w_A v_C - v_A w_C}{L_{AC}}. \qquad (10.15)$$

The three components (u, v, w) of velocity are measured at points A, B, C, and D; L_{AB}, L_{AD}, and L_{AC} are the lengths of the tetrahedron edges (see Fig. 10.5). Both methods give similar results.

Measured values for turbulent helicity are shown in Table 10.2. Primes denote deviations from the values obtained by running mean averaging with a 30-second rectangular window; θ' is the temperature fluctuation. Hereafter, an overbar denotes averaging over the entire measurement session. More details are given in Chkhetiani et al. (2018).

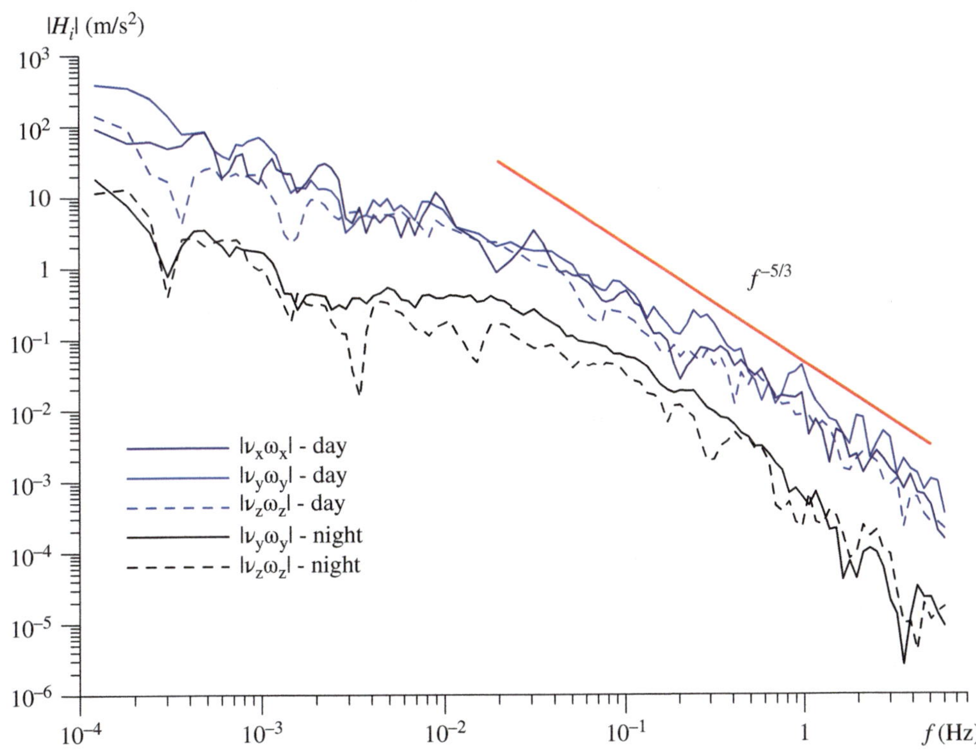

Figure 10.2 Frequency spectra of contributions to helicity calculated based on two-hour records: nighttime of 11 September 2004 with onset at 22:17 local time, and daytime of 12 September 2004 with onset at 13:41 local time. Koprov et al., 2005 / Springer Nature.

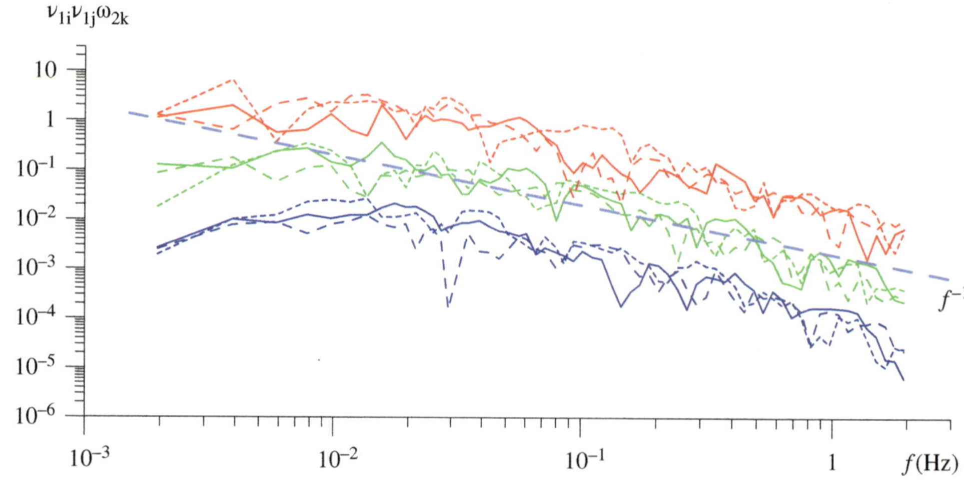

Figure 10.3 Triple correlation frequency spectra calculated from two-hour records of the velocity and circulation components from 28 August 2004 with onset at 11:38 local time (red line) and 16:41 local time (green line) and from 09 November 2004 with onset at 22:17 local time (blue line). The solid line shows the $\langle v_y v_z \omega_y \rangle$ spectrum, the long dashed line the $\langle v_x v_z \omega_y \rangle$ spectrum, and the short dashed line the $\langle v_x v_y \omega_y \rangle$ spectrum.

Figure 10.4 Photo of a tetrahedron composed of four sonic anemometers to measure helicity in Koprov et al. (2015); see also Chkhetiani et al. (2018). Upper-left corner: an enlarged image of the uppermost anemometer at a height of 10 m assembled with a cup anemometer and a wind vane (temperature, pressure, and humidity sensors). Photo by O. Chkhetiani.

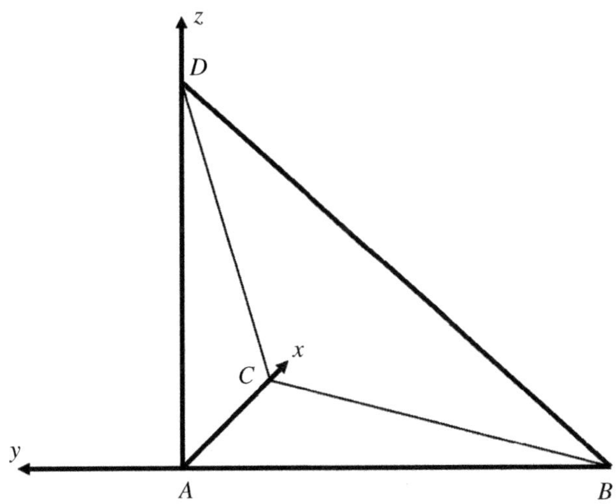

Figure 10.5 Schematic of the rectangular tetrahedron *ABCD* used for helicity measurements; the (x, y) axes are directed north and west, respectively. Three anemometers are positioned at the vertices of a rectangular isosceles triangle *ABC* with sides *AB* and *AC* equal to 5 m. The fourth anemometer at point *D* is located 5 m higher, just above point *A*. Chkhetiani et al. 2018 / Springer Nature.

We emphasize that for the negative mean helicity, the instantaneous helicity values can be of any sign, negative values dominate only in a statistical sense, and the corresponding standard deviation is significantly higher than the mean value (see Fig. 10.6).

Figure 10.7 shows the histogram of turbulent helicity H' values presented in Fig. 10.6. The measurement session lasted 14,113 seconds; therefore, dividing the N values in Fig. 10.7 by 14,113, we infer the probability density function of H' values. It can be compared with Fig. 7 in Deusebio and Lindborg (2014) for an idealized Ekman layer in the Northern Hemisphere, where, in contrast, positive turbulent helicity values dominate.

We now compare the experimental results by Koprov et al. (2015) with theoretical estimates following the idealized conceptual scheme by Deusebio and Lindborg (2014). More details are given in Chkhetiani et al. (2018). Here we describe only the main stages of reasoning. In the Northern Hemisphere, positive helicity is injected from the free atmosphere into the Ekman boundary layer with the time rate $f\mathbf{u}_g^2$, where $\mathbf{u}_g$ is the vector of geostrophic wind. Furthermore, helicity is cascaded as an effective passive scalar quantity toward smaller scales and finally destroyed by viscosity. Then it follows that the helicity destruction rate in the lower part of the Ekman boundary layer is

$$\eta = 2f\mathbf{u}_g^2/h_E, \tag{10.16}$$

where $h_E = \sqrt{2K/f}$ is the Ekman boundary layer thickness, K the small-scale eddy viscosity, and f the Coriolis parameter, and the factor 2 is derived from more detailed considerations of helicity and superhelicity of the classical Ekman spiral solution in Kurgansky (1989) (more in Chkhetiani et al. (2018)). Turbulent helicity $H' \equiv \overline{\mathbf{u}' \cdot \boldsymbol{\omega}'} = H'(l)$ on the given spatial scale l that belongs to an inertial interval can be estimated from Brissaud et al. (1973):

$$H^t(l) = C_H \eta \varepsilon^{-1/3} l^{2/3}, \tag{10.17}$$

which corresponds to the joint cascade of energy and helicity to small scales; compare to equation (10.9), and see also Chen et al. (2003). In equation (10.17), C_H has been estimated either as 2.3 (André and Lesieur, 1977) or 2.5 (Avinash et al., 2006). Taking $h_E = 400\,\text{m}$, $f = 10^{-4}\text{s}^{-1}$, $|u_g| = 10\ \text{m s}^{-1}$, $\varepsilon = 10^{-3}\,\text{m}^2\text{s}^{-3}$, $C_H = 2.4$, and $l = 10\,\text{m}$ (compare to Koprov et al. (2005)), we obtain $H^t \approx 5.5 \times 10^{-3}\ \text{m s}^{-2}$: that is, several times smaller than the daytime H^t values measured in Koprov et al. (2005).

In fact, a lower estimate was obtained for H^t, which ignores other possible sources of turbulent helicity in the bottom part of the planetary boundary layer. Chkhetiani et al. (2018) discuss these and other possible complicating factors. The analysis of meteorological conditions during the field campaign in Koprov et al. (2015) suggests that the joint action of the Ekman spiral flow and sea-breeze circulation explains the resulting negative sign of H' and its magnitude; see Table 10.2 and Chkhetiani et al. (2018). This hypothesis was tested using the balance model of H' in the planetary boundary layer (Chkhetiani et al., 2018).

It is also mentioned in Chkhetiani et al. (2018) that the scale $\lambda_D \sim \varepsilon^{2/7}\nu^{3/7}\left(2fu_g^2/h_E\right)^{-3/7}$ obtained by equating equation (10.16) and $\eta = 2\nu\overline{\boldsymbol{\omega}' \cdot \nabla \times \boldsymbol{\omega}'}$ corresponds to the Ditlevsen scale (Ditlevsen and Giuliani (2001a,b); see also

Table 10.2 Values of co-variances obtained experimentally in Koprov et al. (2015), see also Chkhetiani et al. (2018). The *x*-axis is directed north, the *y*-axis west, and the *z*-axis up. Values in the middle row correspond to the measurement session from 12:38 to 16:30 local time (UTC + 3 h) on 8 August 2012. The numbers in the lower row correspond to the measurement session from 22:46 to 00:34 local time, the end of the session on 9 August 2012.

	$\overline{u'\omega'_x}$, ms^{-2}	$\overline{v'\omega'_y}$, ms^{-2}	$\overline{w'\omega'_z}$, ms^{-2}	$\overline{w'\theta'}$, mKs^{-1}	$\overline{\omega'_z\theta'}$, Ks^{-1}
Day	–0.0170	–0.0276	0.0143	0.1545	0.0456
Night	–0.0138	–0.0010	0.0085	–0.0172	–0.0123

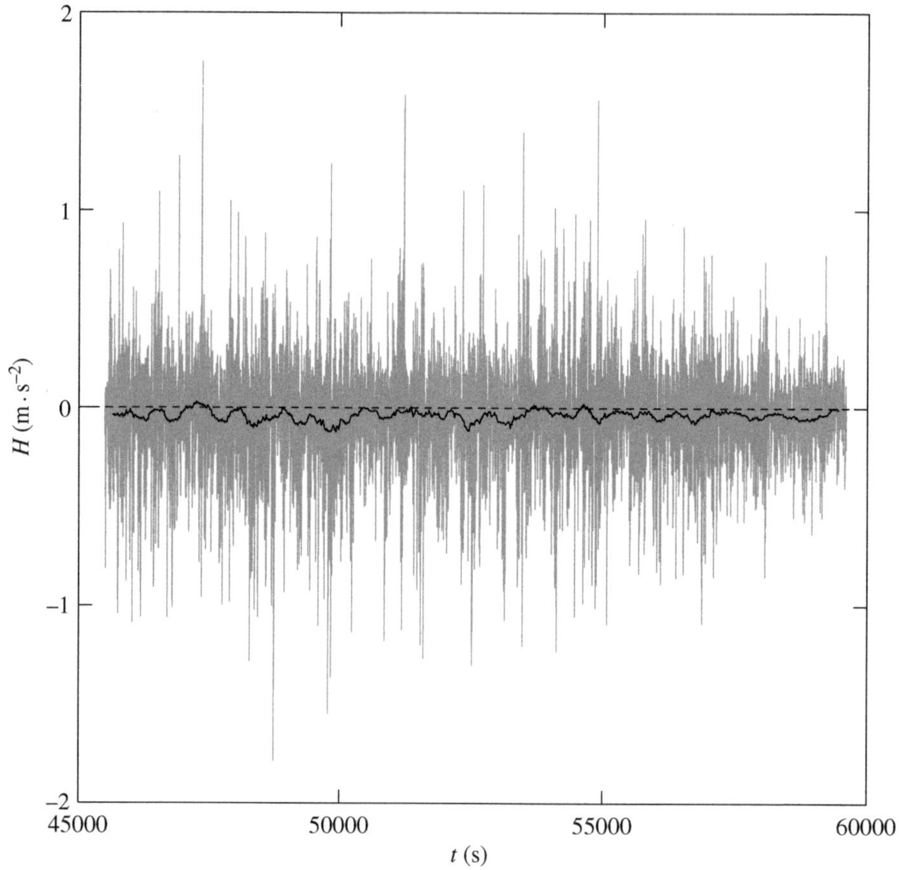

Figure 10.6 Turbulent helicity H' values (in ms^{-2}) measured during the day in Koprov et al. (2015) and averaged over 1 sec (thin grey line); see Table 10.2 and Chkhetiani et al. (2018). The abscissa shows the elapsed time (in seconds) since the beginning of the day. A solid line shows 5-min averages of H'.

Ditlevsen (2010)). Here, ν is the air molecular viscosity. For the parameter values pertinent to the measurement site in Tsymlyansk, $\lambda_D \sim 1.1 \times 10^{-1}$m is two orders of magnitude greater than the Kolmogorov microscale of turbulence, which is the minimum scale of helical motions for a joint cascade of energy and helicity (Chen et al. (2003); see also, Briard and Gomez (2017)).

For a thermally stratified boundary layer, the turbulent helicity is (Chkhetiani et al., 2018)

$$H' = 2K\tau\left(\frac{d^2U}{dz^2}\frac{dV}{dz} - \frac{d^2V}{dz^2}\frac{dU}{dz}\right) + f\frac{d\,\overline{w'^2}}{dz}\tau + 2\alpha g\overline{\omega_z'\theta'}\,\tau, \tag{10.18}$$

where $\alpha > 0$ is the coefficient of thermal expansion of air and τ the turbulent relaxation time. There are three different terms in the expression for H'. The first term corresponds mainly to $\overline{u'\omega_x'} + \overline{v'\omega_y'}$, and the expression in parentheses represents the superhelicity S of the mean flow. In the absence of the other two terms in equation (10.18), this term denotes $H' < 0$ for $S < 0$. The second term, generally positive in the Northern Hemisphere, is two or three orders of magnitude smaller than the first term. The third term is due to thermal stratification and corresponds to the three contributions to helicity, including $\overline{w'\omega_z'}$. As the data in Table 10.2 show, it is positive during the day and negative at night.

For the third term in equation (10.18), we have in a steady, statistically equilibrium turbulent regime

$$\tau^{-1}\overline{\omega_z'\theta'} \approx w'\left(\overline{\boldsymbol{\omega}}\cdot\nabla\theta'\right) + \overline{w'\left(\boldsymbol{\omega}'\cdot\nabla\theta'\right)} \tag{10.19}$$

It should be noted that the right-hand side of equation (10.19) is nothing but a mutual correlation of the velocity fluctuations and the turbulent Ertel potential vorticity. In Fig. 10.8, we show the time dependence of two fluctuating quantities, $P \equiv \omega_z'\theta'$ and $Q \equiv \left(w'\left(\overline{\boldsymbol{\omega}}\cdot\nabla\theta'\right) + w'\left(\boldsymbol{\omega}'\cdot\nabla\theta'\right)\right)$, for the daytime

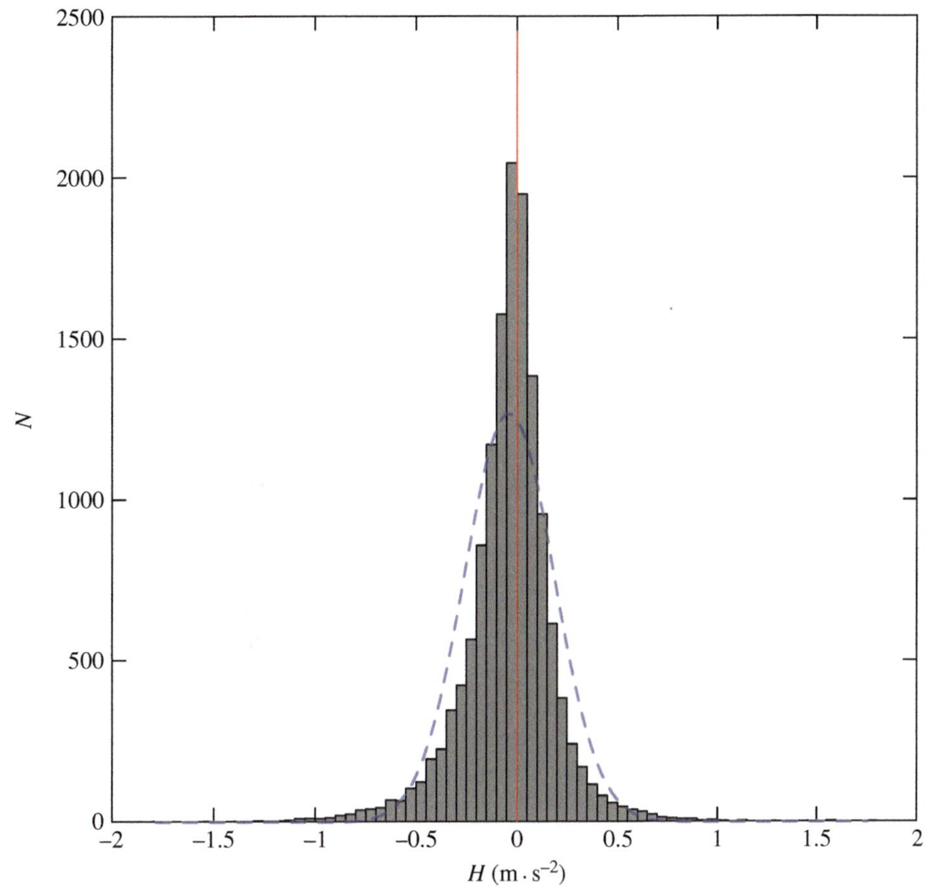

Figure 10.7 Histogram of instantaneous turbulent helicity values measured in Tsimlyansk, southern Russia, on 8 August 2012 at 12:38–16:30 local time (see the text for details). The horizontal axis shows turbulent helicity values (in ms^{-2}); the width of each bin equals 0.05 ms^{-2}. On the vertical axis, the numbers show the number N of helicity values falling in the corresponding bin. The sum of all N values equals 14,113 (see the text). A vertical red line shows zero helicity values $H = 0$. For comparison, a Gaussian distribution with the same dispersion is shown by a blue dashed line. Adapted from Koprov et al., 2015.

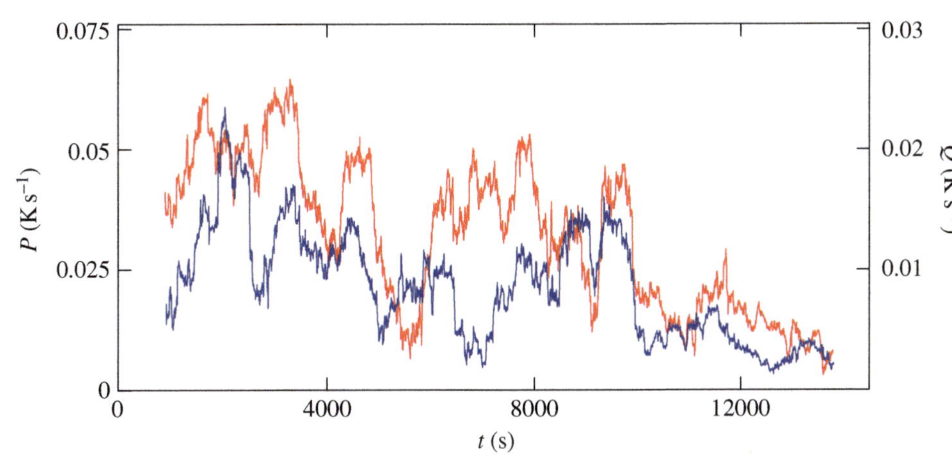

Figure 10.8 Time dependence of P (in Ks^{-1}, red line) and Q (in Ks^{-2}, blue line) values for the daytime measurements in Koprov et al. (2015); see also Table 10.2 and the text for details. On the horizontal axis, zero corresponds to 12:48 local time (Chkhetiani et al., 2018).

measurements in Koprov et al. (2015); see Table 10.2. A 10-min average is used to plot P and Q, and the correlation between these quantities is evident in Fig. 10.8. Analogous results are obtained for the nighttime measurements in Koprov et al. (2015).

In 2014, measurements were carried out using a similar scheme with a basic setup scale of 0.7 m at heights of 3.5, 13.1, and 25 meters (Koprov et al., 2018). Differences from zero of all three diagonal elements of the vorticity flux tensor, the correlation of the vorticity with the temperature field, and spectral dependencies for helicity were found close to those following from the concept of the constancy of the helicity flux toward small scales in the inertial interval.

10.5. CONCLUDING REMARKS AND OUTLOOK

The role and significance of helicity in atmospheric physics will be fully appreciated in the future. Among other things, this chapter discusses the relationship between helicity and potential vorticity, which is especially nontrivial in a compressible fluid. There is a certain complementarity of the concepts of potential vorticity and helicity. The first – historically, at least – was primarily needed to clarify the principles of the dynamics of large-scale processes. At the same time, the concept of helicity for such motions is not very significant (meaningful). The situation changes as the transition to meso- and then small-scale atmospheric motions occurs. This is not to say that the concept of potential vorticity is completely losing its meaning. In a sense, it becomes less constructive since the "invertibility principle" (Hoskins et al., 1985) – that is, the possibility of restoring all other (balanced) hydrodynamic fields from the known field of potential vorticity – gradually loses its applicability. For large-scale motions, this possibility is guaranteed by the quasigeostrophic theory and its improvements. For meso- and small-scale motions, considerations of symmetry (axial, etc.) are indispensable. At the same time, the helicity concept acquires more weight, and its alignment occurs in the conceptual sense with the potential vorticity, which makes its connection with the latter more important.

Much remains to be done in the future in the study of helical turbulence, which is the central theme of the second part of this chapter. We can highlight three aspects related to problems that remain to be solved. First, the sign of helicity is a tricky question since unlike energy or enstrophy, helicity is not a positive definite quantity. In particular, the results presented in this chapter show that the sign of turbulent helicity inferred in direct field measurements within the atmospheric boundary layer is highly dependent on the local physical and meteorological conditions at the observational site and on its geographical position. Therefore, it is highly desirable to conduct further research in this direction – particularly helicity field measurements in the Southern Hemisphere. Second, there is some hope that with the invocation of the helicity concept, we will be able to get closer to understanding the enigmatic problem of intermittence of atmospheric (geophysical) turbulence, but much future work is needed in this direction. Third – which is also important and somewhat related to the two previous aspects – is the issue of the inverse cascade of energy toward larger scales in three-dimensional helical turbulence, which could help explain the genesis and maintenance of meso- and small-scale vortex structures and further clarify the role and even the need for general rotation in this process.

ACKNOWLEDGMENTS

The authors deeply honor the memory of Raymond Hide, Semen Moiseev, and Boris Koprov, who laid the foundation for many of the ideas underlying this review and made significant contributions to the study of helicity. We thank Peter Ditlevsen and an anonymous reviewer for their valuable comments and suggestions that improved this chapter.

REFERENCES

André, J. C., & Lesieur, M. (1977). Influence of helicity on the evolution of isotropic turbulence at high Reynolds number. *Journal of Fluid Mechanics*, *81*(01), 187. https://doi.org/10.1017/s0022112077001979

Antonia, R. A., Browne, L. W. B., & Shah, D. A. (1988). Characteristics of vorticity fluctuations in a turbulent wake. *Journal of Fluid Mechanics*, *189*, 349–365. https://doi.org/10.1017/s0022112088001053

Aref, H., & Flinchem, E. P. (1984). Dynamics of a vortex filament in a shear flow. *Journal of Fluid Mechanics*, *148*, 477–497. https://doi.org/10.1017/S0022112084002457

Avinash, V., Verma, M. K., & Chandra, A. V. (2006). Field-theoretic calculation of kinetic helicity flux. *Pramana*, *66*(2), 447–453. https://doi.org/10.1007/bf02704397

Biferale, L., Musacchio, S., & Toschi, F. (2012). Inverse energy cascade in three-dimensional isotropic turbulence. *Physical Review Letters*, *108*(16), 164501. https://doi.org/10.1103/physrevlett.108.164501

Biferale, L., Musacchio, S., & Toschi, F. (2013). Split energy–helicity cascades in three-dimensional homogeneous and isotropic turbulence. *Journal of Fluid Mechanics*, *730*, 309–327. https://doi.org/10.1017/jfm.2013.349

Biferale, L., Gustavsson, K., & Scatamacchia, R. (2019). Helicoidal particles in turbulent flows with multi-scale helical injection. *Journal of Fluid Mechanics*, *869*, 646–673. https://doi.org/10.1017/jfm.2019.237

Bluestein, H. (1992). *Synoptic-dynamic meteorology in midlatitudes: principles of kinematics and dynamics* (Vol. 1). Oxford University Press.

Bovsheverov, V. M., Gurvich, A. S., Kochetkov, A. N., & Lomadze, S. O. (1971). Measurement of the frequency spectrum of small-scale circulation of velocity in a turbulent flow. *Izvestiya, Atmospheric and Oceanic Physics*, *7*(4), 245–248.

Briard, A., & Gomez, T. (2017). Dynamics of helicity in homogeneous skew-isotropic turbulence. *Journal of Fluid Mechanics*, *821*, 539–581. https://doi.org/10.1017/jfm.2017.260

Brissaud, A., Frisch, U., Leorat, J., Lesieur, M., & Mazure, A. (1973). Helicity cascades in fully developed isotropic turbulence. *Phys Fluids*, *16*(8), 1366–1367. https://doi.org/10.1063/1.1694520

Chakraborty, S. (2007). Signatures of two-dimensionalisation of 3d turbulence in the presence of rotation. *Europhysics Letters (EPL)*, *79*(1), 14002. https://doi.org/10.1209/0295-5075/79/14002

Chen, N., Tang, J., Zhang, J. A., Ma, L., & Yu, H. (2021). On the distribution of helicity in the tropical cyclone boundary layer from dropsonde composites. *Atmospheric Research*, *249*, 105298. https://doi.org/10.1016/j.atmosres.2020.105298

Chen, Q., Chen, S., & Eyink, G. L. (2003). The joint cascade of energy and helicity in three-dimensional turbulence. *Physics of Fluids*, *15*(2), 361–374. https://doi.org/10.1063/1.1533070

Chkhetiani, O. G. (1996). On the third moments in helical turbulence. *Journal of Experimental and Theoretical Physics Letters*, *63*(10), 808–812. https://doi.org/10.1134/1.567095

Chkhetiani, O. G. (2001). On the helical structure of the Ekman boundary layer. *Izv Atmos Ocean Phys*, *37*(5), 369–375.

Chkhetiani, O. G. (2008). On the local structure of helical turbulence. *Doklady Physics*, *53*(10), 513–516. https://doi.org/10.1134/s1028335808100030

Chkhetiani, O. G., & Gledzer, E. B. (2017). Helical turbulence with small-scale energy and helicity sources and external intermediate scale noises as the origin of large scale generation. *Physica A: Statistical Mechanics and its Applications*, *486*, 416–433. https://doi.org/10.1016/j.physa.2017.05.027

Chkhetiani, O. G., & Golbraikh, E. (2012). Turbulent field helicity fluctuations and mean helicity appearance. *International Journal of Non-Linear Mechanics*, *47*(3), 113–117. https://doi.org/10.1016/j.ijnonlinmec.2011.12.001

Chkhetiani, O. G., Eidelman, A., & Golbraikh, E. (2006). Large- and small-scale turbulent spectra in MHD and atmospheric flows. *Nonlinear Processes in Geophysics*, *13*(6), 613–620. https://doi.org/10.5194/npg-13-613-2006

Chkhetiani, O. G., Kurgansky, M. V., & Vazaeva, N. V. (2018). Turbulent helicity in the atmospheric boundary layer. *Boundary-Layer Meteorology*, *168*, 361–385. https://doi.org/10.1007/s10546-018-0356-4

Cieszelski, R. (1999). Studies on turbulence parametrizations for flows with helicity. *Izvestiya, Atmospheric and Oceanic Physics*, *35*(2), 157–170.

Clark, A. J., Kain, J. S., Marsh, P. T., Correia, J., Xue, M., & Kong, F. (2012). Forecasting tornado pathlengths using a three-dimensional object identification algorithm applied to convection-allowing forecasts. *Weather and Forecasting*, *27*(5), 1090–1113. https://doi.org/10.1175/waf-d-11-00147.1

Clark, A. J., Gao, J., Marsh, P. T., Smith, T., Kain, J. S., Correia, J., et al. (2013). Tornado pathlength forecasts from 2010 to 2011 using ensemble updraft helicity. *Weather and Forecasting*, *28*(2), 387–407. https://doi.org/10.1175/waf-d-12-00038.1

Deusebio, E., & Lindborg, E. (2014). Helicity in the Ekman boundary layer. *Journal of Fluid Mechanics*, *755*, 654–671. https://doi.org/10.1017/jfm.2014.307

Ditlevsen, P. D. (2010). *Turbulence and shell models*. Cambridge University Press. https://doi.org/10.1017/cbo9780511919251

Ditlevsen, P. D., & Giuliani, P. (2001). Dissipation in helical turbulence. *Physics of Fluids*, *13*(11), 3508–3509. https://doi.org/10.1063/1.1404138

Ditlevsen, P. D., & Giuliani, P. (2001). Cascades in helical turbulence. *Physical Review E*, *63*(3), 036304. https://doi.org/10.1103/physreve.63.036304

Doswell, C. A., & Schultz, D. M. (2006). On the use of indices and parameters in forecasting severe storms. *Electronic Journal of Severe Storms Meteorology*, *1*(3), 1–22.

Droegemeier, K. K., Lazarus, S. M., & Davies-Jones, R. (1993). The influence of helicity on numerically simulated convective storms. *Monthly Weather Review*, *121*(7), 2005–2029. https://doi.org/10.1175/1520-0493(1993)121<2005:tiohon>2.0.co;2

Eliassen, A. (1987). Entropy coordinates in atmospheric dynamics. *Meteorologische Zeitschrift*, *37*(1), 1–11.

Eliassen, A., & Kleinschmidt, E. (1957). Dynamic meteorology. In J. Bartels (Ed.), *Geophysik II / Geophysics II. Handbuch der physik / Encyclopedia of physics* (Vol. 10 / 48). Berlin, Heidelberg: Springer. https://doi.org/10.1007/978-3-642-45881-1_1

Ertel, H. (1942). Ein neuer hydrodynamischer. *Wirbelsatz. Meteorol Z*, *59*, 277–281.

Ertel, H., & Rossby, C.-G. (1949). A new conservation-theorem of hydrodynamics. *Geofisica Pura e Applicata*, *14*(3-4), 189–193. https://doi.org/10.1007/bf01981973

Etling, D. (1985). Some aspects of helicity in atmospheric flows. *Beiträge zur Physik der Atmosphäre*, *58*(1), 88–100.

Farahani, M. M., Khansalari, S., & Azadi, M. (2017). Evaluation of helicity generation in the tropical storm Gonu. *Meteorology and Atmospheric Physics*, *129*(3), 333–344. https://doi.org/10.1007/s00703-016-0465-x

Gassmann, A., & Herzog, H.-J. (2008). Towards a consistent numerical compressible non-hydrostatic model using generalized Hamiltonian tools. *Quarterly Journal of the Royal Meteorological Society*, *134*(635), 1597–1613. https://doi.org/10.1002/qj.297

Gaynor, J. E., Hall, F. F., Edinger, J. G., & Ochs, G. R. (1977). Measurement of vorticity in the surface layer using an acoustic echo sounder array. *Remote Sensing of Environment*, *6*(2), 127–137. https://doi.org/10.1016/0034-4257(77)90010-4

Gilbert, A. D., Frisch, U., & Pouquet, A. (1988). Helicity is unnecessary for alpha effect dynamos, but it helps. *Geophysical & Astrophysical Fluid Dynamics*, *42*(1–2), 151–161. https://doi.org/10.1080/03091928808208861

Glebova, E. S., Levina, G. V., Naumov, A. D., & Trosnikov, I. V. (2009). The helical feature calculation for the velocity field of a developing tropical cyclone. *Russian Meteorology and Hydrology*, *34*(9), 572–580. https://doi.org/10.3103/s1068373909090027

Gledzer, E. B. (1973). System of hydrodynamic type admitting two quadratic integrals of motion. *Dokl Akad Nauk SSSR*, *209*(5), 1046–1048.

Gledzer, E. B., & Chkhetiani, O. G. (2015). Inverse energy cascade in developed turbulence at the breaking of the symmetry of helical modes. *JETP Letters*, *102*(7), 465–472. https://doi.org/10.1134/s0021364015190066

Goldhaber, A. S., & Goldhaber, M. (2011). The neutrino's elusive helicity reversal. *Physics Today*, *64*(5), 40–43. https://doi.org/10.1063/1.3592004

Golitsyn, G. S. (2012). *The statistics and dynamics of natural processes and phenomena: methods, instrumentation, and results* (in Russian). Moscow: Krasand.

Han, Y., Wu, R., & Fang, J. (2006). Shearing wind helicity and thermal wind helicity. *Advances in Atmospheric Sciences*, *23*(4), 504–512. https://doi.org/10.1007/s00376-006-0504-5

Hasimoto, H. (1972). A soliton on a vortex filament. *Journal of Fluid Mechanics*, *51*, 477–485. https://doi.org/10.1017/S0022112072002307

Hide, R. (1975). A note on helicity. *Geophysical Fluid Dynamics*, *7*(1), 157–161. https://doi.org/10.1080/03091927508242617

Hide, R. (1989). Superhelicity, helicity and potential vorticity. *Geophysical & Astrophysical Fluid Dynamics*, *48*(1-3), 69–79. https://doi.org/10.1080/03091928908219526

Holm, D. D., & Kerr, R. (2002). Transient vortex events in the initial value problem for turbulence. *Physical Review Letters*, *88*(24), 244501. https://doi.org/10.1103/PhysRevLett.88.244501

Holm, D. D., & Stechmann, S. N. (2004). Hasimoto transformation and vortex soliton motion driven by fluid helicity. arXiv:nlin/0409040v1 [nlin.SI], 1–15.

Hoskins, B. J., McIntyre, M. E., & Robertson, A. W. (1985). On the use and significance of isentropic potential vorticity maps. *Quarterly Journal of the Royal Meteorological Society*, *111*(470), 877–946. https://doi.org/10.1002/qj.49711147002

Jackson, B. (2020). On the relationship between dust devil radii and heights. *Icarus*, *338*, 113523. https://doi.org/10.1016/j.icarus.2019.113523

Jordan, A. R. (1980). Pressure and vorticity transients from summer storms and aircraft. *Journal of Applied Meteorology*, *19*(10), 1223–1233. https://doi.org/10.1175/1520-0450(1980)019<1223:pavtfs>2.0.co;2

Kanak, K. M., & Lilly, D. K. (1996). The linear stability and structure of convection in a circular mean shear. *Journal of the Atmospheric Sciences*, *53*(18), 2578–2593. https://doi.org/10.1175/1520-0469(1996)053<2578:tlsaso>2.0.co;2

Kholmyansky, M., Tsinober, A., & Yorish, S. (2001). Velocity derivatives in the atmospheric surface layer at $Re_{\lambda}=104$. *Physics of Fluids*, *13*(1), 311–314. https://doi.org/10.1063/1.1328358

Knorr, G., Lynov, J. P., & Pécseli, H. L. (1990). Self-organization in three-dimensional hydrodynamic turbulence. *Zeitschrift für Naturforschung*, *45 a*, 1059–1073.

Komm, R., Gosain, S., & Pevtsov, A. A. (2014). Hemispheric distribution of subsurface kinetic helicity and its variation with magnetic activity. *Solar Physics*, *289*(7), 2399–2418. https://doi.org/10.1007/s11207-014-0477-y

Koprov, B. M., Azizyan, G. V., & Kalugin, V. V. (1988). Spectra of velocity circulation in the surface layer of the atmosphere. *Boundary-Layer Meteorology*, *42*(1-2), 137–143. https://doi.org/10.1007/bf00119879

Koprov, B. M., Kalugin, V. V., & Tieme, N. S. (1994). Turbulent vorticity flux. *Izvestiya, Atmospheric and Oceanic Physics*, *30*(1), 10–14.

Koprov, B. M., Koprov, V. M., Ponomarev, V. M., & Chkhetiani, O. G. (2005). Experimental studies of turbulent helicity and its spectrum in the atmospheric boundary layer. *Doklady Physics*, *50*(8), 419–422. https://doi.org/10.1134/1.2039983

Koprov, B. M., Koprov, V. M., Kurgansky, M. V., & Chkhetiani, O. G. (2015). Helicity and potential vorticity in surface turbulence. *Izvestiya, Atmospheric and Oceanic Physics*, *51*(6), 565–575. https://doi.org/10.1134/s0001433815060092

Koprov, B. M., Koprov, V. M., Solenaya, O. A., Chkhetiani, O. G., & Shishov, E. A. (2018). Technique and results of measurements of turbulent helicity in a stratified surface layer. *Izvestiya, Atmospheric and Oceanic Physics*, *54*(5), 446–455. https://doi.org/10.1134/s0001433818050067

Koschmieder, H. (1933). *Dynamischer meteorologie (Dynamic meterorology)*. Leipzig: Akademische Verlagsgesellschaft.

Kraichnan, R. H. (1973). Helical turbulence and absolute equilibrium. *Journal of Fluid Mechanics*, *59*(4), 745–752. https://doi.org/10.1017/s0022112073001837

Krause, F., & Raedler, K.-H. (1980). *Mean-field magnetohydrodynamics and dynamo theory*. Berlin: Academie. https://doi.org/10.1016/c2013-0-03269-0

Kurgansky, M. V. (1989). Relationship between helicity and potential vorticity in a compressible rotating fluid. *Izvestiya, Atmospheric and Oceanic Physics*, *25*(12), 979–981.

Kurgansky, M. V. (1990). *Adiabatic invariants in problems of atmospheric dynamics* (doctoral dissertation). Moscow: Hydrometeorological Science Research Center.

Kurgansky, M. V. (1993). *Introduction to large-scale atmospheric dynamics (adiabatic invariants and their use)* (in Russian). Saint Petersburg: Gidrometeoizdat.

Kurgansky, M. V. (2002). *Adiabatic invariants in large-scale atmospheric dynamics*. London-New York: Taylor and Francis. https://doi.org/10.1201/b12540

Kurgansky, M. V. (2006). Helicity production and maintenance in a baroclinic atmosphere. *Meteorologische Zeitschrift*, *15*(4), 409–416. https://doi.org/10.1127/0941-2948/2006/0148

Kurgansky, M. V. (2008). Vertical helicity flux in atmospheric vortices as a measure of their intensity. *Izvestiya, Atmospheric and Oceanic Physics*, *44*(1), 64–71. https://doi.org/10.1134/s0001433808010076

Kurgansky, M. V. (2013a). On helical vortex motions of moist air. *Izvestiya, Atmospheric and Oceanic Physics*, *49*(5), 479–484. https://doi.org/10.1134/s000143381305006x

Kurgansky, M. V. (2013b). Simple models of helical baroclinic vortices. *Procedia IUTAM*, 7. https://doi.org/10.1016/j.piutam.2013.03.023

Kurgansky, M. V. (2017). Helicity in dynamic atmospheric processes. *Izvestiya, Atmospheric and Oceanic Physics*, *53*(2), 127–141. https://doi.org/10.1134/s0001433817020074

Kuzanyan, K., Zhang, H., & Bao, S. (2000). Probing signatures of the alpha-effect in the solar convection zone. *Solar Physics*, *191*, 231–246.

Landau, L. D., & Lifshitz, E. M. (1987). *Fluid mechanics*. Oxford: Pergamon Press.

Lavrova, A. A., Glebova, E. S., Trosnikov, I. V., & Kaznacheeva, V. D. (2010). Modeling the evolution of the family of Mediterranean cyclones using the regional model of the atmosphere. *Russian Meteorology and Hydrology*, *35*(6), 363–370. https://doi.org/10.3103/s1068373910060014

Leverson, V. H., Sinclair, P. C., & Golden, J. H. (1977). Waterspout wind, temperature and pressure structure deduced from aircraft measurements. *Monthly Weather Review*, *105*(6), 725–733. https://doi.org/10.1175/1520-0493(1977)105<0725:wwtaps>2.0.co;2

Levich, E., & Tzvetkov, E. (1985). Helical inverse cascade in three-dimensional turbulence as a fundamental dominant mechanism in mesoscale atmospheric phenomena. *Physics Reports*, *128*(1), 1–37. https://doi.org/10.1016/0370-1573(85)90036-5

Levina, G., Glebova, E., Naumov, A., & Trosnikov, I. (2009). Application of helical characteristics of the velocity field to evaluate the intensity of tropical cyclones. In A. Talamelli, J. Peinke, M. Oberlack (Eds.), *Progress in turbulence III* (Vol. 131, pp. 259–262). Berlin Heidelberg: Springer. https://doi.org/10.1007/978-3-642-02225-8_63

Levina, G. V. (2019). A realization of the turbulent vortex dynamo in the atmosphere: based on the 21st century knowledge. *Journal of Physics: Conference Series*, *1336*(1), 012007. https://doi.org/10.1088/1742-6596/1336/1/012007

Levina, G. V., & Montgomery, M. T. (2010). A first examination of the helical nature of tropical cyclogenesis. *Doklady Earth Sciences*, *434*(1), 1285–1289. https://doi.org/10.1134/s1028334x1009031x

Levina, G. V., & Montgomery, M. T. (2011). Helical scenario of tropical cyclone genesis and intensification. *Journal of Physics: Conference Series*, *318*(7), 072012. https://doi.org/10.1088/1742-6596/318/7/072012

Levshin, A. O., & Chkhetiani, O. G. (2013). Decay of helicity in homogeneous turbulence. *JETP Letters*, *98*(10), 598–602. https://doi.org/10.1134/s0021364013230070

Lilly, D. K. (1986). The structure, energetics and propagation of rotating convective storms. Part II: Helicity and storm stabilization. *Journal of the Atmospheric Sciences*, *43*(2), 126–140. https://doi.org/10.1175/1520-0469(1986)043<0126:tseapo>2.0.co;2

Longuet-Higgins, M. (1982). On triangular tomography. *Dynamics of Atmospheres and Oceans*, *7*(1), 33–46. https://doi.org/10.1016/0377-0265(82)90004-5

Lorenz, R. (2013). The longevity and aspect ratio of dust devils: Effects on detection efficiencies and comparison of landed and orbital imaging at Mars. *Icarus*, *226*(1), 964–970. https://doi.org/10.1016/j.icarus.2013.06.031

Markowski, P. M., Straka, J. M., Rasmussen, E. N., & Blanchard, D. O. (1998). Variability of storm-relative helicity during vortex. *Monthly Weather Review*, *126*(11), 2959–2971. https://doi.org/10.1175/1520-0493(1998)126<2959:vosrhd>2.0.co;2

Mininni, P. D., & Pouquet, A. (2010a). Rotating helical turbulence. i. global evolution and spectral behavior. *Physics of Fluids*, *22*(3), 035105. https://doi.org/10.1063/1.3358466

Mininni, P. D., & Pouquet, A. (2010b). Rotating helical turbulence. ii. intermittency, scale invariance, and structures. *Physics of Fluids*, *22*(3), 035106. https://doi.org/10.1063/1.3358471

Mitsuta, Y., & Asai, H. (1984). A sonic anemometer for the measurement of vorticity and its transport in the surface layer. *Experiments in Fluids*, *2*(3), 150–152. https://doi.org/10.1007/bf00296432

Moffatt, H. K. (1969). The degree of knottedness of tangled vortex lines. *Journal of Fluid Mechanics*, *35*(1), 117–129. https://doi.org/10.1017/s0022112069000991

Moffatt, H. K. (1978). *Magnetic field generation in electrically conducting fluids*. Cambridge University Press.

Moiseev, S. S., & Chkhetiani, O. G. (1996). Helical scaling in turbulence. *Journal of Experimental and Theoretical Physics*, *83*(1), 192–198.

Moiseev, S. S., Sagdeev, R. Z., Tur, A. V., Khomenko, G. A., & Yanovskii, V. V. (1983). Theory of the origin of large-scale structures in hydrodynamic turbulence. *Soviet Physics JETP*, *58*(6), 1149–1153.

Moiseev, S. S., Rutkevich, P. B., Tur, A. V., & Yanovskii, V. V. (1988). Vortex dynamo in a convective medium with helical turbulence. *Soviet Physics JETP*, *67*(2), 294–299.

Molinari, J., & Vollaro, D. (2010). Distribution of helicity, cape, and shear in tropical cyclones. *Journal of the Atmospheric Sciences*, *67*(1), 274–284. https://doi.org/10.1175/2009jas3090.1

Monin, A. S., & Yaglom, A. M. (1971, 1975). *Statistical fluid mechanics*. Cambridge, MA: The MIT Press.

Moreau, J.-J. (1961). Constantes d'un ilot tourbillonnaire en fluide parfait barotrope. *Comptes Rendus de l'Académie des Sciences - Series IIA - Earth and Planetary Science*, *252*, 2810–2812.

Nambu, Y. (1973). Generalized Hamiltonian dynamics. *Physical Review D*, *7*(8), 2405–2412. https://doi.org/10.1103/physrevd.7.2405

Névir, P., & Blender, R. (1993). A Nambu representation of incompressible hydrodynamics using helicity and enstrophy. *Journal of Physics A: Mathematical and General*, *26*(22), L1189–L1193. https://doi.org/10.1088/0305-4470/26/22/010

Névir, P., & Sommer, M. (2009). Energy–vorticity theory of ideal fluid mechanics. *Journal of the Atmospheric Sciences*, *66*(7), 2073–2084. https://doi.org/10.1175/2008jas2897.1

Novikov, E. A. (1972). Vorticity transport. *Izvestiya, Atmospheric and Oceanic Physics*, *8*(7), 432–434.

Ohtou, A., Maitani, T., & Seo, T. (1983). Direct measurement of vorticity and its transport in the surface layer over a paddy field. *Boundary-Layer Meteorology*, *27*(2), 197–207. https://doi.org/10.1007/bf00239615

Pichler, H., & Schaffhauser, A. (1998). The synoptic meaning of helicity. *Meteorology and Atmospheric Physics*, *66*, 23–34. https://doi.org/10.1007/bf01030446

Ponomarev, V. M., & Chkhetiani, O. G. (2005). Semiempirical model of the atmospheric boundary layer with parametrization of turbulent helicity effect. *Izv Atmos Ocean Phys*, *41*(4), 418–432.

Ponomarev, V. M., Khapaev, A. A., & Chkhetiani, O. G. (2003). Role of helicity in the formation of secondary structures in the Ekman boundary layer. *Izv Atmos Ocean Phys*, *39*(4), 391–400.

Pouquet, A., & Mininni, P. D. (2010). The interplay between helicity and rotation in turbulence: implications for scaling laws and small-scale dynamics. *Philosophical Transactions of the Royal Society A: Mathematical, Physical and Engineering Sciences*, *368*(1916), 1635–1662. https://doi.org/10.1098/rsta.2009.0284

Rathmann, N. M., & Ditlevsen, P. D. (2017). Pseudo-invariants contributing to inverse energy cascades in three-dimensional turbulence. *Physical Review Fluids*, *2*(5), 054607. https://doi.org/10.1103/PhysRevFluids.2.054607

Rorai, C., Rosenberg, D., Pouquet, A., & Mininni, P. D. (2013). Helicity dynamics in stratified turbulence in the absence of forcing. *Physical Review E*, *87*(6), 063007. https://doi.org/10.1103/physreve.87.063007

Rossby, T. (1975). An oceanic vorticity meter. *Journal of Marine Research*, *82*(2), 213–222.

Sahoo, G., Alexakis, A., & Biferale, L. (2017). Discontinuous transition from direct to inverse cascade in three-dimensional turbulence. *Physical Review Letters*, *118*(16), 164501. https://doi.org/10.1103/physrevlett.118.164501

Salazar, R., & Kurgansky, M. V. (2010). Nambu brackets in fluid mechanics and magnetohydrodynamics. *Journal of Physics A: Mathematical and Theoretical*, *43*(30), 305501. https://doi.org/10.1088/1751-8113/43/30/305501

Salmon, R. (2005). A general method for conserving quantities related to potential vorticity in numerical models. *Nonlinearity*, *18*(5), R1–R16. https://doi.org/10.1088/0951-7715/18/5/r01

Salmon, R. (2007). A general method for conserving energy and potential enstrophy in shallow-water models. *Journal of the Atmospheric Sciences*, *64*(2), 515–531. https://doi.org/10.1175/jas3837.1

Sanford, T. B., Carlson, J. A., Dunlap, J. H., Prater, M. D., & Lien, R.-C. (1999). An electromagnetic vorticity and velocity sensor for observing finescale kinetic fluctuations in the ocean. *Journal of Atmospheric and Oceanic Technology*, *16*(11), 1647–1667. https://doi.org/10.1175/1520-0426(1999)016<1647:aevavs>2.0.co;2

Shenming, F., Wanli, L., Jianhua, S., & Rudi, X. (2011). A budget analysis of a long-lived tropical mesoscale vortex over hainan in october 2010. *Meteorology and Atmospheric Physics*, *114*(1-2), 51–65. https://doi.org/10.1007/s00703-011-0156-6

Smith, T. M., Gao, J., Calhoun, K. M., Stensrud, D. J., Manross, K. L., Ortega, K. L., et al. (2014). Examination of a real-time 3dvar analysis system in the hazardous weather testbed. *Weather and Forecasting*, *29*(1), 63–77. https://doi.org/10.1175/waf-d-13-00044.1

Speziale, C. G. (1989). On helicity fluctuations and the energy cascade in turbulence. In S. L. Koh & C. G. Speziale (Eds.), *A symposium dedicated to A. Cemal Eringen, June 20–22 1988 Berkley California (Lect. Notes Eng. 39)* (pp. 50–57). Berlin Heidelberg: Springer. https://doi.org/10.1007/978-3-642-83695-4_5

Stepanov, R., Golbraikh, E., Frick, P., & Shestakov, A. (2015). Hindered energy cascade in highly helical isotropic turbulence. *Physical Review Letters*, *115*(23), 234501. https://doi.org/10.1103/physrevlett.115.234501

Thwaites, F. T., Williams, A. J., Terray, E. A., & Trowbridge, J. H. (1995). A family of acoustic vorticity meters to measure ocean boundary layer shear. In *Proceedings of the IEEE Fifth Working Conference on Current Measurement* (pp. 193–198). IEEE. https://doi.org/10.1109/ccm.1995.516173

Tippett, M. K., Sobel, A. H., Camargo, S. J., & Allen, J. T. (2014). An empirical relation between U.S. tornado activity and monthly environmental parameters. *Journal of Climate*, *27*(8), 2983–2999. https://doi.org/10.1175/jcli-d-13-00345.1

Vazaeva, N. V., Chkhetiani, O. G., Kouznetsov, R. D., Kallistratova, M. A., Kramar, V. F., Lyulyukin, V. S., & Kuznetsov, D. D. (2017). Estimating helicity in the atmospheric boundary layer from acoustic sounding data. *Izvestiya, Atmospheric and Oceanic Physics*, *53*(2), 174–186. https://doi.org/10.1134/s0001433817020104

Vazaeva, N. V., Chkhetiani, O. G., Kurgansky, M. V., & Kallistratova, M. A. (2021). Helicity and turbulence in the atmospheric boundary layer. *Izvestiya, Atmospheric and Oceanic Physics*, *57*(1), 29–46. https://doi.org/10.1134/s0001433821010126

Vukoslavcevic, P., & Wallace, J. M. (1996). A 12-sensor hot-wire probe to measure the velocity and vorticity vectors in turbulent flow. *Measurement Science and Technology*, *7*(10), 1451–1461. https://doi.org/10.1088/0957-0233/7/10/016

Waleffe, F. (1992). The nature of triad interactions in homogeneous turbulence. *Physics of Fluids A*, *4*, 350–363. https://doi.org/10.1063/1.858309

Wu, W.-S., Lilly, D. K., & Kerr, R. M. (1992). Helicity and thermal convection with shear. *Journal of the Atmospheric Sciences*, *49*(19), 1800–1809. https://doi.org/10.1175/1520-0469(1992)049<1800:hatcws>2.0.co;2

Wyngaard, J. C. (1969). Spatial resolution of the vorticity meter and other hot-wire arrays. *Journal of Physics E: Scientific Instruments*, *2*(11), 983–987. https://doi.org/10.1088/0022-3735/2/11/320

Yu, C., Hong, R., Xiao, Z., & Chen, S. (2013). Subgrid-scale eddy viscosity model for helical turbulence. *Physics of Fluids*, *25*(9), 095101. https://doi.org/10.1063/1.4819765

Yu, Z., & Yu, H. (2011). Relation of the second type thermal helicity to precipitation of landfalling typhoons: A case study of typhoon Talim. *Acta Meteorologica Sinica*, *25*(2), 224–237. https://doi.org/10.1007/s13351-011-0029-4

Zhemin, T., & Rongsheng, W. (1994). Helicity dynamics of atmospheric flow. *Advances in Atmospheric Sciences*, *11*(2), 175–188. https://doi.org/10.1007/bf02666544

Part III

Theoretical and Numerical Helicity Modeling

11

Effects of P-Noninvariance, Particles, and Dynamos

Victor B. Semikoz[1] and Dmitry D. Sokoloff[1,2]

ABSTRACT

The origin of magnetic fields in various celestial bodies is primarily associated with the dynamo mechanism: a transformation of kinetic energy into magnetic energy in an electrically conductive medium. As drivers for dynamo action, one considers differential rotation and various forms of mirror asymmetric flows in a given object's convective or turbulent envelope. Differential rotation is insufficient to generate a dynamo, and mirror asymmetry is a weak but crucial additional source of generation. Mirror asymmetry in solar, stellar, galactic, and other celestial bodies is associated with Coriolis or magnetic force action: that is, we remain here in the domain of classical physics and do not include any relativistic and/or quantum effects. However, modern physics suggests another way to produce mirror asymmetry for astrophysical plasma: elementary particles can be mirror asymmetric by themselves. Correspondingly, one can expect dynamo generation in the plasma of the early universe based on the intrinsic mirror asymmetry of particles. This kind of dynamo excitation appears interesting and attractive; however, it is much less investigated than more conventional dynamos because the results of the dynamo action are not seen directly in astronomical observations. Nevertheless, many important results concerning this cosmological dynamo are presented in the scientific literature, and we review some of them here.

11.1. INTRODUCTION

The origin of magnetic fields in various celestial bodies is primarily associated with the dynamo mechanism: a transformation of kinetic energy into magnetic energy in an electrically conductive medium Parker (1979a,b). As drivers for dynamo action, one considers differential rotation and various forms of mirror asymmetric flows in a given object's convective or turbulent envelope. Differential rotation is insufficient to generate a dynamo, and mirror asymmetry is a weak but crucial additional source of generation. Mirror asymmetry in solar, stellar, galactic, and other celestial bodies is associated with Coriolis or magnetic force action: that is, we remain here in the domain of classical physics and do not include any relativistic and/or quantum effects. It is interesting, however, that modern physics suggests another way to produce mirror asymmetry for astrophysical plasma: the difference of right and left fermion densities is mirror-asymmetric due to the corresponding particle helicity. Correspondingly, one can expect dynamo generation in the magnetized plasma of the early universe based on the mirror asymmetry of these particle densities (see, e.g., the derivation in equation 11.7). This kind of dynamo excitation appears interesting and attractive; however, it is much less investigated than more conventional dynamos because the results of the dynamo action are not seen directly in astronomical observations. Nevertheless, many important results concerning this cosmological dynamo are presented in the scientific literature, and we review some of them here.

[1] *Pushkov Institute of Terrestrial Magnetism, Ionosphere and Radiowave Propagation (IZMIRAN), Russian Academy of Sciences, Moscow, Russia*

[2] *Department of Physics, Moscow State University, Moscow, Russia*

Helicities in Geophysics, Astrophysics, and Beyond, Geophysical Monograph 283, First Edition.
Edited by Kirill Kuzanyan, Nobumitsu Yokoi, Manolis K. Georgoulis, and Rodion Stepanov.

DOI:10.1002/9781119841715.ch11

11.2. SEED MAGNETIC FIELDS

To get a finite value of the magnetic field, one needs to suppose the presence of a nonzero initial magnetic field $B(t_0) \neq 0$ in the Faraday induction equation (see equation (11.1)). For the Sun, such a seed field can be a relic field in our Galaxy that presumably originated from a primordial magnetic field (PMF) in the early universe.

An astrophysical origin of the intergalactic magnetic field (IGMF) is also possible: for example, magnetohydrodynamic plasma penetration of intergalactic space through supernovae explosions or active galactic nucleus (AGN) jets. Nevertheless, the cosmological scenario we consider here seems preferable, especially after first estimates of the lower bound on the PMF Neronov and Vovk (2010).

It is well-known that in the standard model of particle physics (the SM model by Weinberg-Salam-Glashow for electroweak interactions), the progenitor of a Maxwellian electromagnetic field A_μ exists before the electroweak phase transition (EWPT) or in the symmetric phase at high temperatures $T > T_{EWPT} \simeq 100 - 160$ GeV. Such a progenitor is the hypercharge field Y_μ, which is also massless as a photon in a vacuum. The corresponding hypermagnetic field $\mathbf{B}_Y = \nabla \times \mathbf{Y}$ arises in the Big Bang simultaneously with ultrarelativistic hot matter after the preceding inflation stage, so it is a seed field for the Maxwellian PMF $\mathbf{B} = \nabla \times \mathbf{A}$. Such a scenario for the origin of magnetic fields of celestial bodies is supported by the recent estimate of a lower bound on the PMF strength $B^{now}_{PMF} > 10^{-16}$ G at the spatial scale $\lambda_B > 1$ Mpc Neronov and Vovk (2010).

The upper bound on the PMF strength has long been known from observations of the Faraday rotation measure of the polarization plane for electromagnetic (EM) waves,

$$\mathrm{RM} = 812 \int_0^L n_e(\mathrm{cm}^{-3}) B_{\parallel}(\mu\mathrm{G}) dl(\mathrm{kpc}) \ \mathrm{rad/m}^2.$$

Here, $B_{\parallel} = (\mathbf{B} \cdot \mathbf{k})/k$ is the field projection on the EM wave direction to a radio telescope, $\mathbf{k} = (0,0,k)$, and $\phi = \mathrm{RM}\ \lambda^2$ is the rotation angle in the (x, y) -polarization plane. Such a bound $B_{PMF} < 10^{-9}$ G Parker (1979b) is confirmed by cosmic microwave background (CMB) polarization measurements – in particular, the Plank experiment, $B_{PMF} <$ 3 nG at the scale $\lambda_B > 1$ Mpc Semikoz et al. (2009).

The PMF amplitude decreases as the universe cools, $B_{PMF} \sim T^2$, following the decrease in temperature $T = T_{now}(1 + z)$ from the start of the EWPT with the red shift $z \simeq 10^{15}$ to the moment of galaxy creation, $z \simeq 10$. Then the PMF feeds the galactic magnetic field up to $B_{gal} \sim 10^{-6}$ G at present ($z = 0$) through either a small-scale dynamo or a mean-field dynamo.

Magnetic field evolution after the EWPT is shown in Fig. 11.1 Semikoz et al. (2009). The solid line represents the Maxwellian magnetic field evolved from the hypermagnetic field starting at $z_{EW} \simeq 10^{15}$ as frozen-in plasma. The dotted line represents the large-scale (1 pc at the epoch of galaxy formation) component of the magnetic field, which becomes causal at the moment after the

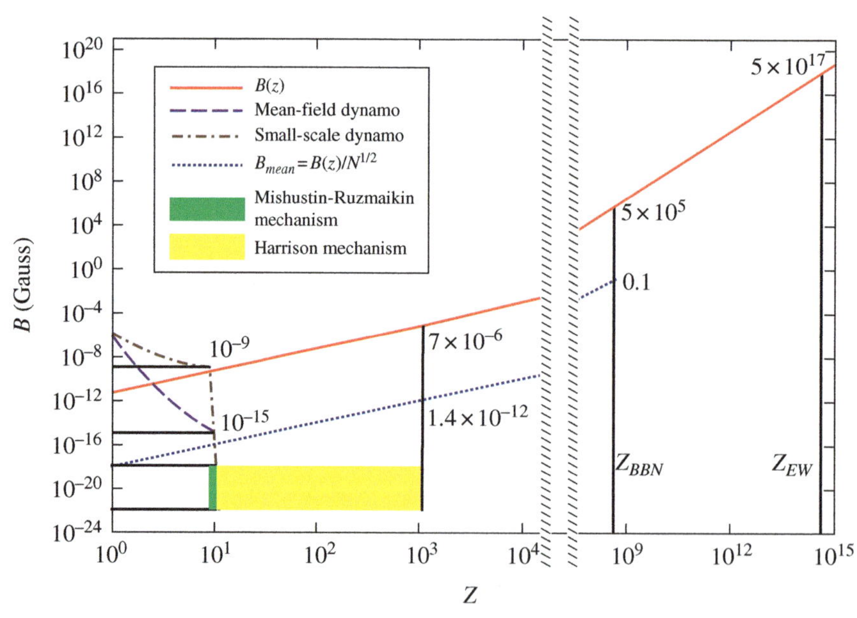

Figure 11.1 Evolution of primordial magnetic fields in the universe after the EWPT ($z \lesssim 10^{15}$) for the two types of dynamo amplification, starting from the moment of galaxy creation $z \sim 10$ Semikoz et al., 2009 / American Physical Society.

EWPT. Note that the mean field $B_{mean} = BN^{-1/2}$ inferred as the statistical mean field with $N = (L_{gal}/l)^3$ and small scale l is causal at Big Bang nucleosynthesis (BBN) time, but not earlier.

11.3. MIRROR ASYMMETRY AND HELICITY

In electrodynamics, spatial parity is conserved at the quantum level in particle physics and in a continuous medium like a cosmic plasma. In particular, in the standard magnetohydrodynamics (MHD) for the mean-field dynamo Krause and Raedler (1980) in the Faraday induction equation,

$$\frac{\partial \mathbf{B}}{\partial t} = \nabla \times (\mathbf{v} \times \mathbf{B}) + \beta \nabla^2 \mathbf{B} + \nabla \times (\alpha \mathbf{B}), \tag{11.1}$$

the helicity parameter $\alpha = -\tau_{corr}\langle \mathbf{u} \cdot (\nabla \times \mathbf{u})\rangle/3$ is a pseudoscalar that changes sign under spatial inversion $\alpha \overset{\mathrm{P}}{\to} -\alpha$. Such a change provides conservation of the P-parity in the Faraday equation. Indeed, **B** is a pseudovector: that is, does not change sign under spatial inversion $\mathbf{B} \overset{\mathrm{P}}{\to} \mathbf{B}$, whereas the fluid velocity **v** is a vector that does change sign under spatial inversion, $\mathbf{v} \overset{\mathrm{P}}{\to} -\mathbf{v}$. As a result, all the terms of the previous equation preserve positive spatial parity, $\eta_{\mathrm{P}} = +1$.

For the azimuthal-symmetric mean field $\mathbf{B}(r, \theta, t)$, such as in the Sun, the simplest choice for the pseudoscalar $\alpha \simeq \alpha_0 \sin\theta$ reflects the property of *mirror asymmetry* for the helicity parameter α that changes its sign based on a transition from the northern hemisphere to the southern hemisphere: that is, with a shift of solar latitude θ, $\theta \to \theta + \pi$.

11.4. MEAN-FIELD DYNAMOS IN RELATIVISTIC PLASMAS

In an isotropic and uniform plasma of the early universe, one can neglect random small-scale velocities at large magnetic field scales $\lambda_B \lesssim l_H$, $\mathbf{u} \to 0$. Then the Krause-Rädler approach becomes inapplicable. Simultaneously, in the isotropic universe, the regular (large-scale) component of the total fluid velocity $\mathbf{v} = \mathbf{V} + \mathbf{u}$ is absent, $\mathbf{V} = 0$, or the advection (dynamo) term $\nabla \times (\mathbf{v} \times \mathbf{B})$ in equation (11.1) vanishes completely. In such a case, the coefficient of the turbulent diffusion $\beta = \beta_T$ changes to the small ohmic diffusion, $\beta_T \Longrightarrow \sigma_{cond}^{-1}$, where $\sigma_{cond} \simeq T/\alpha_{em}$ is the plasma conductivity. The induction equation (11.1) reduces to

$$\frac{\partial \mathbf{B}}{\partial t} = \left(\frac{1}{\sigma_{cond}}\right) \nabla^2 \mathbf{B} + \nabla \times (\alpha \mathbf{B}), \tag{11.2}$$

where α is helicity arising in the anomalous MHD (AMHD),

$$\alpha = \frac{2\alpha_{em}\mu_5}{\pi\sigma_{cond}}. \tag{11.3}$$

Here, $\mu_5 = (\mu_R - \mu_L)/2$ equals the difference of chemical potentials μ_R (and μ_L) for the right- and left-handed charged leptons in a chiral (massless, $m_e \to 0$) medium.

The parameter μ_5 changes sign under spatial inversion, $\mu_R \leftrightarrow \mu_L$: that is, $\mu_5 \overset{\mathrm{P}}{\to} -\mu_5$. As a result, the helicity parameter (11.3) is a pseudoscalar in anomalous MHD as well, $\alpha \overset{\mathrm{P}}{\to} -\alpha$, which does not change the positive P-parity for all terms in the Faraday equation (11.2), $\eta_{\mathrm{P}} = +1$.

Note that in a neutron star (NS), given its preferable rigid rotation, $\partial_\theta \Omega = \partial_r \Omega = 0$, or $\Omega = const$. Thus, in both of the previous cases, the induction equation (11.2) with α from equation (11.3) is appropriate.

The conservation of P-parity (negative $\eta_{\mathrm{P}} = -1$ for vector terms) is valid for the Maxwell equation in both the standard MHD and the AMHD:

$$\nabla \times \mathbf{B} = \sigma_{cond}[\mathbf{E} + \mathbf{v} \times \mathbf{B} + \alpha \mathbf{B}]. \tag{11.4}$$

Here, one neglects the displacement current $\partial_t \mathbf{E}$ for low MHD frequencies $\omega \ll \sigma_{cond}$, and $\mathbf{j}_{Ohm} = \sigma_{cond}[\mathbf{E} + \mathbf{v} \times \mathbf{B}]$ is the standard ohmic current in a moving medium. In both cases (MHD and AMHD), the additional vector current $\mathbf{j} \sim \alpha \mathbf{B}$ is parallel to the magnetic field and does not contribute the Lorenz force $\mathbf{F} \sim \mathbf{j} \times \mathbf{B}$ in the Navier-Stokes equation.

11.4.1. Origin and Evolution of Alpha in the AMHD

Let us explain the origin of the α parameter from equation (11.3) in AMHD. The additional vector current in the Maxwell equation (11.4), the chiral magnetic effect (CME) $\mathbf{J}_{\mathrm{CME}} = (\sigma_{cond}\alpha)\mathbf{B} = e^2\mu_5\mathbf{B}/2\pi^2$, is widely applied in quantum chromodynamics (QCD) plasma Kharzeev et al. (2016) and relativistic astrophysics Son and Surówka (2009), Sigl and Leite (2016), Masada et al. (2018). Such a current is given by the pseudovector current $\mathbf{j} \sim \mathbf{B}$ calculated first by Vilenkin (1980) for massless left- and right-handed electrons and positrons at the main (i.e., nondegenerate over spin) Landau level that move parallel (anti-parallel) to **B** (see, e.g., Fig. 1 in Tashiro et al. (2012)).

The divergence of the *sum* of these four right and left currents,

$$j_R^\mu(\mathbf{x}, t) = \overline{\psi}_R \gamma^\mu \psi_R = \overline{\psi}_e(\mathbf{x}, t)\gamma^\mu(1 + \gamma^5)\psi_e(\mathbf{x}, t)/2$$

and

$$j_L^\mu(\mathbf{x}, t) = \overline{\psi}_L \gamma^\mu \psi_L = \overline{\psi}_e(\mathbf{x}, t)\gamma^\mu(1 - \gamma^5)\psi_e(\mathbf{x}, t)/2,$$

does not depend on the matrix γ^5 and equals zero, $\partial_\mu(j_R^\mu(\mathbf{x}, t) + j_L^\mu(\mathbf{x}, t)) = \partial_\mu j^\mu(\mathbf{x}, t) = 0$. This corresponds to the well-known conservation law for the vector current $j^\mu(\mathbf{x}, t) = \overline{\psi}_e(\mathbf{x}, t)\gamma^\mu\psi_e(\mathbf{x}, t)$:

$$\partial_\mu j^\mu(\mathbf{x}, t) = 0. \tag{11.5}$$

Here, $\psi_e(\mathbf{x}, t)$ are the wave functions for electrons (positrons) obeying the Dirac equation, $\overline{\psi}_e(\mathbf{x}, t) = \psi_e^+(\mathbf{x}, t)\gamma^0$, γ^μ are the Dirac 4×4 matrices, $\mu = 0, 1, 2, 3$, and $\gamma^5 = -i\gamma^0\gamma^1\gamma^2\gamma^3$. Under statistical averaging, the conservation law in equation (11.5) corresponds to the continuity equation in a continuous medium, $\partial_t n_e + \nabla \cdot (n_e \mathbf{v}) = 0$, or the charge conservation after integration over the volume $V^{-1} \int d^3x(\ldots)$, $Q = const$.

On the other hand, the *difference* of the right and left currents, $j_5^\mu(\mathbf{x}, t) = j_R^\mu(\mathbf{x}, t) - j_L^\mu(\mathbf{x}, t) = \overline{\psi}_e\gamma^\mu\gamma^5\psi_e$, is the pseudovector current that is not conserved, $\partial_\mu j_5^\mu(\mathbf{x}, t) \neq 0$. In a vacuum, this current $j_5^\mu(\mathbf{x}, t)$ is not conserved for massive fermions; and in the external electromagnetic field $F_{\mu\nu}$, such nonconservation is known as the Adler-Bell-Jackiw anomaly Adler (1969), Bell and Jackiw (1969) and is valid even in the absence of a medium:

$$\frac{\partial j_5^\mu}{\partial x^\mu} = 2im_e\overline{\psi}_e\gamma^5\psi_e + \frac{e^2}{8\pi^2}F_{\mu\nu}\tilde{F}^{\mu\nu}. \tag{11.6}$$

Here, $\tilde{F}_{\mu\nu}$ is the dual tensor for the electromagnetic tensor, $\tilde{F}_{\mu\nu} = -\varepsilon_{\mu\nu\rho\sigma}F^{\mu\nu}$, resulting in their product, $F^{\mu\nu}\tilde{F}_{\mu\nu} = 4(\mathbf{E} \cdot \mathbf{B})$. In the chiral limit $m_e \to 0$, one neglects the averaged massive term in the anomaly from equation (11.6), $2im_e\langle\overline{\psi}_e\gamma_5\psi_e\rangle \to 0$ (see the calculations in Dvornikov and Semikoz (2018), Copinger et al. (2018)).

In standard MHD, the dynamics of small-scale vortices $\omega = \nabla \times \mathbf{u}$ and random velocities $\mathbf{u}$ entering the averaged helicity parameter $\alpha = -\tau_{corr}\langle\mathbf{u} \cdot \omega\rangle/3$ is a separate, complicated problem. On the other hand, in AMHD, the evolution of the helicity parameter in equation (11.3) is precisely described under a simple statistical averaging Adler anomaly in equation (11.6), also known as the axial Ward identity in quantum electrodynamics (QED). Note that equation (11.6) is statistically averaged in a chiral plasma ($m_e \to 0$) and integrated over the volume $V^{-1} \int d^3x(\ldots)$, leading to the important conservation law in AMHD:

$$\frac{\mathrm{d}}{\mathrm{d}t}\left[n_R(t) - n_L(t) + \frac{\alpha_{em}}{\pi}h(t)\right] = 0. \tag{11.7}$$

Here, $n_{R,L}(t) = V^{-1} \int d^3x\langle\psi_{R,L}^+(\mathbf{x}, t)\psi_{R,L}(\mathbf{x}, t)\rangle$ are the densities of right and left leptons (anti-leptons), and $\langle\ldots\rangle$ is the symbol of the statistical averaging in a given volume V.

Thus, equation (11.7) is a generalization of the conservation law for an ideal plasma ($\sigma_{cond} \to \infty$) in the standard MHD, $\mathrm{d}h(t)/\mathrm{d}t = 0$, where $h(t) = V^{-1} \int d^3x(\mathbf{A} \cdot \mathbf{B})$ is the magnetic helicity density. We recall that at a finite conductivity $\sigma_{cond} \neq \infty$, the evolution equation for magnetic helicity in the standard MHD is

$$\begin{aligned}\frac{\mathrm{d}h(t)}{\mathrm{d}t} &= -\frac{2}{V}\int d^3x(\mathbf{E} \cdot \mathbf{B}) \\ &= -\frac{2}{V\sigma_{cond}}\int d^3x\,[(\nabla \times \mathbf{B}) \cdot \mathbf{B}].\end{aligned} \tag{11.8}$$

In a hot chiral plasma in the early universe, $T \gg m_e$, the densities $n_{R,L}(t)$ in equation (11.7) are determined by the temperature T and corresponding chemical potentials, $n_{R,L} = \mu_{R,L}T^2/6$. In the degenerate ultrarelativistic electron gas in a neutron star (NS), these densities are $n_{R,L} = \mu_e^2\mu_{R,L}/6\pi^2$, where the sum of chemical potentials $\mu_R + \mu_L = 2\mu_e$ gives the standard electron density in the NS for the ultrarelativistic gas with the chemical potential μ_e, $\mu_e \gg m_e$, $n_R + n_L = n_e = \mu_e^3/3\pi^2$.

Note that the *difference of densities* entering the conservation law in equation (11.7), $n_R(t) - n_L(t) = (\mu_R(t) - \mu_L(t))T^2/6 = \mu_5(t)T^2/3$ in a hot plasma, or $n_R(t) - n_L(t) = (\mu_R(t) - \mu_L(t))\mu_e^2/6\pi^2 = \mu_5(t)\mu_e^2/3\pi^2$ in the NS electron gas, is a *pseudoscalar* since $\mu_5 \overset{\mathrm{P}}{\to} -\mu_5$. It changes much faster than the temperature T in the expanding universe or the slowly changing electron chemical potential μ_e in a cooling supernova as a NS progenitor. This allows us to obtain important confirmation of the evolution of the helicity parameter $\alpha(t) \sim \mu_5(t)$ in a chiral medium based on the conservation law in equation (11.7). Relying on that law, one can predict how the magnetic field instability term $\nabla \times \alpha\mathbf{B}$ evolves in the Faraday equation (11.2). Such a term grows with a decrease in the positive magnetic helicity $h(t) > 0$, $\mathrm{d}h/\mathrm{d}t < 0$ and decreases in the case $\mathrm{d}h/\mathrm{d}t > 0$. For the negative case $h(t) < 0$, the predictions are the opposite.

Based on the mass correction due to Coulomb collisions in a hot plasma – the rate of the spin-flip $e_R \leftrightarrow e_L$ in the elastic *ee*-scattering, $\Gamma_f \sim (m_e/T)^2$, which leads to the vanishing chiral imbalance, $\mu_5 = (\mu_R - \mu_L)/2 \to 0$ – one obtains from the conservation law in equation (11.7) the governing equation for α:

$$\frac{\mathrm{d}\alpha}{\mathrm{d}t} = -\left(\frac{6\alpha_{em}^2}{\pi^2T^2\sigma_{cond}}\right)\frac{\mathrm{d}h}{\mathrm{d}t} - \Gamma_f\alpha. \tag{11.9}$$

This kinetic equation should be solved self-consistently with AMHD equations for the Fourier spectra $\rho_B(k, t)$ and $h(k, t)$ that define the magnetic field energy density $\rho_B(t) = B^2(t)/2 = \int dk\rho_B(k, t)$ and the helicity density $h(t) = \int dkh(k, t)$ in equation (11.9).

11.4.2. The Influence of Turbulent Motion the CME in the Early Universe

This section is based on the paper Dvornikov and Semikoz (2017). For a relativistic plasma with the equation of state $P = \rho/3$ or $P + \rho = 4\rho/3$, the complete system of MHD equations in the single fluid approximation is

$$\partial_t\rho + \frac{4}{3}\nabla \cdot (\rho\mathbf{v}) = 0, \tag{11.10}$$

$$\frac{4}{3}\partial_t(\rho\mathbf{v}) - \frac{4}{3}\rho\mathbf{v} \times (\nabla \times \mathbf{v}) = (\mathbf{J} \times \mathbf{B}) - \nabla P + \frac{4}{3}\rho\nu\nabla^2\mathbf{v}, \tag{11.11}$$

$$\mathbf{J} = (\nabla \times \mathbf{B}), \tag{11.12}$$

$$\partial_t \mathbf{B} = -\nabla \times \mathbf{E}, \tag{11.13}$$

where ρ is the energy density of the fluid and P is the pressure, $\mathbf{E}$ is the electric field, and ν is the kinematic viscosity. Using the previous equation and Ohm's law with the anomalous current $\mathbf{J}_{\text{CME}}$ directed along the magnetic field, we arrive at the induction equation in the form

$$\partial_t \mathbf{B} = \nabla \times (\mathbf{v} \times \mathbf{B}) + \eta_m \nabla^2 \mathbf{B} + \alpha \nabla \times \mathbf{B}. \tag{11.14}$$

Here, α is the helicity parameter in equation (11.3), $\eta_m = \sigma_{\text{cond}}^{-1}$ is the magnetic diffusion coefficient for the hot plasma conductivity, and $\sigma_c \simeq 100$.

Using parametrization (Sigl (2002), Dvornikov and Semikoz (2017))

$$\frac{\partial \mathbf{v}}{\partial t} = \frac{1}{P + \rho} (\mathbf{J} \times \mathbf{B}), \tag{11.15}$$

we avoid solving the Navier-Stokes equation. Equation (11.15) is further simplified

$$\mathbf{v} = \frac{\tau_d}{P + \rho} (\mathbf{J} \times \mathbf{B}), \tag{11.16}$$

where $\tau_d = l_{\text{free}} \approx 1/\alpha_{\text{em}}^2 T$ is the drag time estimated as the free path $l_{\text{free}} = (n\sigma)^{-1} \approx 1/\alpha_{\text{em}}^2 T$, where $n \sim T^3$ is the hot plasma density and $\sigma \sim \alpha_{\text{em}}^2/T^2$ is the Coulomb cross-section. We assume that the drag time is much greater than the period of the Larmor rotation. This means the Lorentz force can accelerate the charged fluid until it interacts with other particles in the background.

Note that Sigl (2002), Dvornikov and Semikoz (2017) dropped all the gradients in the Navier-Stokes equation referring to matter variables, including the pressure and kinematic viscosity ($\sim \nu$) terms as well as the nonlinear velocity term. Thus, only the Lorentz force term is retained in equation (11.15), $\mathbf{F}_{\text{L}} \sim (\mathbf{J} \times \mathbf{B})$.

After substituting the current $\mathbf{J} = \nabla \times \mathbf{B}$ into the velocity from equation (11.16), the advection (dynamo) term $\nabla \times \mathbf{v} \times \mathbf{B}$ in the Faraday equation (11.14) is odd over the random magnetic field $\mathbf{B}$, or $\nabla \times \mathbf{v} \times \mathbf{B} \sim B^3$. On the other hand, for binary $\mathbf{B}^* \partial_t \mathbf{B} \ldots$ and $\mathbf{A}^* \partial_t \mathbf{B} \ldots$ in the kinetic equations for the Fourier spectra of the magnetic energy density $\mathcal{E}_B \sim B^2$ and the magnetic helicity density $\mathcal{H}_B \sim AB$, even products $\sim B^4$ account for the velocity from equation (11.16). This allows us to apply the Wick theorem when averaging such kinetic equations over random (small-scale) magnetic fields, leading to the corresponding turbulent corrections illustrated in Fig. 11.2.

The numerical solution of these kinetic equations is shown in Fig. 11.2 for the fully helical PMF. One can see that the stronger the initial magnetic field, the more noticeable the difference between the turbulent and nonturbulent cases. The chiral anomaly parameter $\mu_5(t)$ is supported by matter turbulence at a higher level starting from the EWPT time; then, at a few hundred MeV, it reduces more smoothly and drops a bit earlier than accounting for the CME effect only. This can be explained by the inverse cascade with increased large-scale contributions in spectra when the role of turbulent motion ceases.

The evolution of PMF characteristics through the CME mechanism in a hot plasma in the early universe that is isotropic and uniform is described by self-consistent kinetic equations for the PMF isotropic spectra $\tilde{\mathcal{E}}_{\text{B}}(\tilde{k}, \eta)$, $\tilde{\mathcal{H}}_{\text{B}}(\tilde{k}, \eta)$, and the CME parameter $\tilde{\mu}_5(\eta)$. The effect of turbulent motions for the CME mechanism in a hot plasma can be essential for a strong initial PMF $B_0 \sim 0.1 T_{EW}^2 \simeq 10^{23}$ G (see Fig. 11.2a,c,e). Such a field does not influence the Friedman law for the universe expansion since $B_0^2/2 \ll \rho = \pi^2 g^* T_{EW}^4/30$ and obeys the BBN limit $B \lesssim 10^{11}$ G during universe cooling down to $T_{BBN} \simeq 0.1$ MeV.

11.4.3. Nascent Neutron Stars, Mirror Asymmetry, and Strong Magnetic Fields

This section is based on Dvornikov et al. (2020), which considers another example of applying the effects under discussion to generate strong magnetic fields in magnetars, $B \sim 10^{15}$ G. It is well-known that a small imbalance between the densities of right-handed and left-handed electrons, $n_5 = n_{\text{R}} - n_{\text{L}} > 0$, $n_5 \ll n_e$, where $n_e = n_{\text{R}} + n_{\text{L}}$,

$$\begin{aligned} n_5(\mathbf{x}, t) &= \left[\frac{\mu_{\text{R}}^3}{6\pi^2} - \frac{\mu_{\text{L}}^3}{6\pi^2} \right] + \frac{T^2}{3} (\mu_{\text{R}} - \mu_{\text{L}}) \\ &= \mu_5(\mathbf{x}, t) \frac{\mu_e^2}{\pi^2} \left[1 + \frac{2\pi^2}{3} \left(\frac{T}{\mu_e} \right)^2 \right], \end{aligned} \tag{11.17}$$

arises in the environment of a contracting protostar due to the direct Urca process. (The wording was first suggested by Gamov using Odessa's (Ukraine) jargon: *Urca* = "fast robbery." That allowed him to explain fast exhaustion of the proton content in a NS via the capture of electron neutrinos in the process $p + e_L \to n + \nu_{e_L}$ and following a contraction of the NS). The chemical potential imbalance $\mu_5(t) = (\mu_{\text{R}} - \mu_{\text{L}})/2$ driven by a huge temperature gradient at the narrow shock front enters the anomalous current $\mathbf{J}_{\text{CME}} = e^2 \mu_5 \mathbf{B}/2\pi^2$ in the Maxwell equation. Such a temperature gradient near the stalled shock is considered by Janka (2001).

The governing equation for μ_5 related with n_5 in equation (11.17) reads as a new CME source:

$$\frac{\partial \mu_5}{\partial t} \sim \mathbf{B} \cdot \nabla T; \tag{11.18}$$

see the details in Dvornikov et al. (2020).

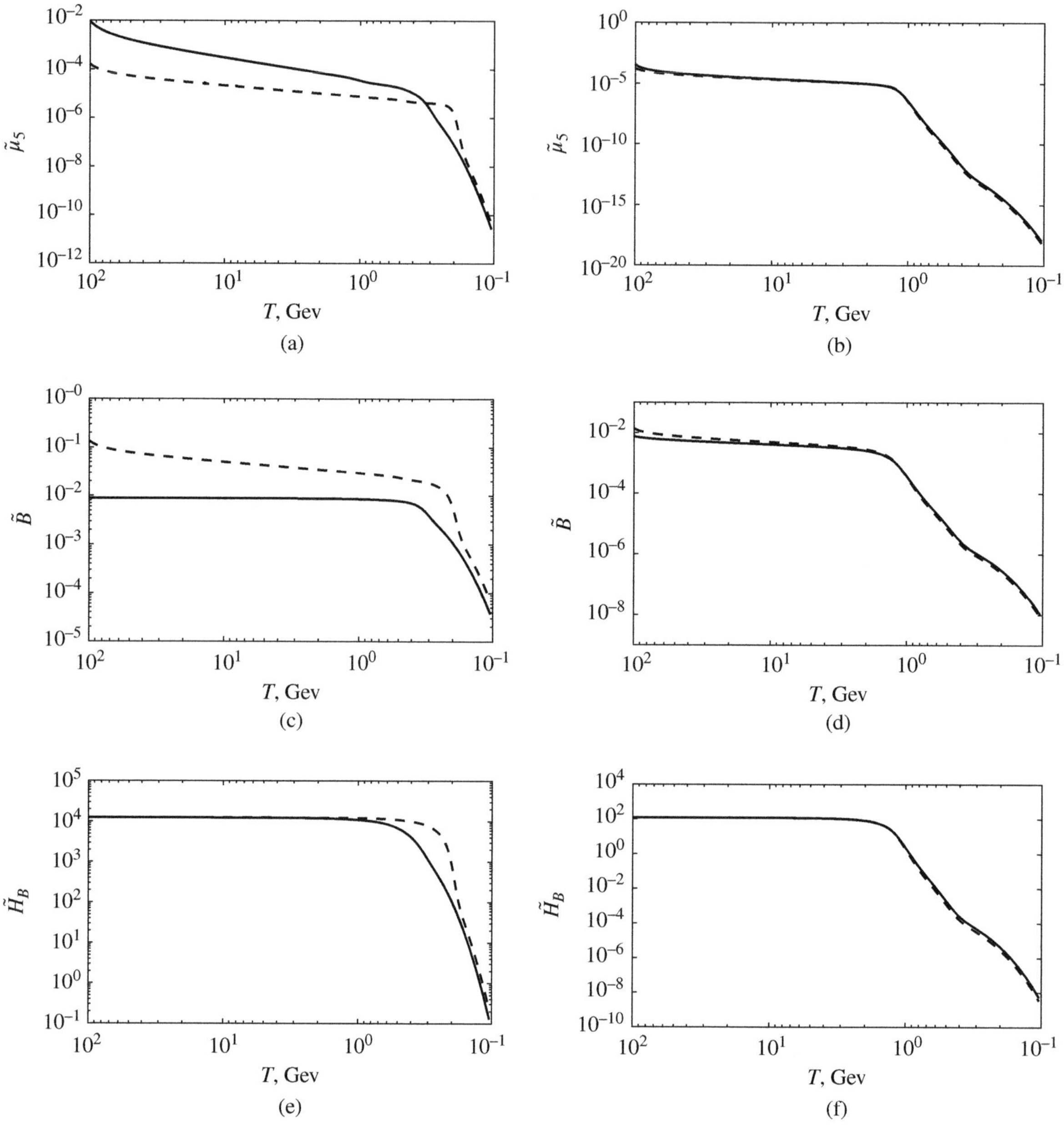

Figure 11.2 The evolution of the chiral imbalance (a, b), the magnetic energy density (c, d), and the helicity density (e, f) in plasma of the early universe; $\tilde{B}_0 = 10^{-1}$ (a, c, e) and $\tilde{B}_0 = 10^{-2}$ (b, d, f). Solid lines show both turbulence effects and CME, whereas dashed lines show CME effects only.

Ohm's law in a nascent NS is modified due to the appearance of a new thermoelectric term in the electric field $\mathbf{E} \sim \nabla T$ (see the details in Dvornikov et al. (2020)). Then the corresponding product $\mathbf{B} \cdot \mathbf{E} \sim \mathbf{B}\nabla T \sim \mathrm{d}h/\mathrm{d}t$ is proportional to the helicity change in equation (11.7) being connected with $\mathrm{d}n_5/\mathrm{d}t \sim \mathrm{d}\mu_5/\mathrm{d}t$ via equations (11.17) and (11.18). Therefore, the helicity evolution $\mathrm{d}\alpha/\mathrm{d}t \sim \mathrm{d}\mu_5/\mathrm{d}t$ is driven by a huge temperature gradient at the narrow shock front $\mathrm{d}T/\mathrm{d}r|_{r=R_s} \simeq -\Delta T/(\Delta r)_{\text{front}}$; see equations (11.17) and (11.18). Such a temperature gradient is seen near the stalled shock at a distance $r = R_s$ from the NS center in Fig. 2 in Janka (2001).

We consider the axisymmetric magnetic field $\mathbf{B}(r, \theta, t)$ in a supernova as a NS progenitor. In a thin layer at $r = R_s$, we use the low-mode approximation from Nefedov and Sokoloff (2010) for all AMHD functions $B_\varphi = B(\theta, t)$, $A_\varphi = A(\theta, t)$, including the helicity parameter $\alpha(\theta, t) \sim \mu_5$

governed by the CME through a shock with a temperature jump dT/dr at $r = R_s$, taken from Janka (2001).

The complete system of dynamic equations for the 3D dynamo model with an axisymmetric magnetic field in a thin layer at $r \approx R_s$ includes the following three AMHD equations:

1. For $\alpha(\theta, t) \approx \alpha_1(t) \sin 2\theta + \alpha_2(t) \sin 4\theta + \dots$ as the helicity parameter originated by the CME.
2. For the azimuthal potential $A_\varphi(\theta, t) \approx a_1(t) \sin\theta + a_2(t) \sin 3\theta + \dots$.
3. For the toroidal magnetic field $B_\varphi(\theta, t) \approx b_1(t) \sin 2\theta + b_2(t) \sin 4\theta + \dots$.

Thus, one can see in Fig. 11.3 the growth of both magnetic field components: poloidal ($a_{1,2}$) and toroidal ($b_{1,2}$), on three orders of magnitude from a natural initial value 10^{10}G at a distance $r > 150$km ahead of the narrow shock front to the strong field $B = 10^{13}$G behind the shock $r < R_s$. Such amplification is sufficient to get through the following matter inflow onto the NS surface the value $B \approx 2.25 \times 10^{15}$G in a future magnetar at the NS radius 10 km.

The CME outcome is associated with the growing helicity $\alpha \sim \mu_5$ (Fig. 11.4). Note the downward-directed spikes of the toroidal components $b_{1,2}$ (Fig. 11.3a,b), which indicate sign changes in the evolution of the toroidal magnetic field. Toroidal components become negative exactly at the moments when spikes appear in Fig. 11.3a,b.

Fig. 11.3 demonstrates the growth of both magnetic field components, poloidal $B_p \sim a_{1,2}$ and toroidal $B \sim b_{1,2}$, on three orders of magnitude from an initial value $\sim 10^{10}$ G at the distance $r \gtrsim 150$ km ahead the narrow shock front up to the strong field $B \simeq 10^{13}$ G behind shock $r \lesssim R_s$. Such amplification is sufficient to get through the following matter inflow onto the neutron star surface the great value $B \sim 2.25 \times 10^{15}$ G in a future magnetar at the neutron star radius 10 km.

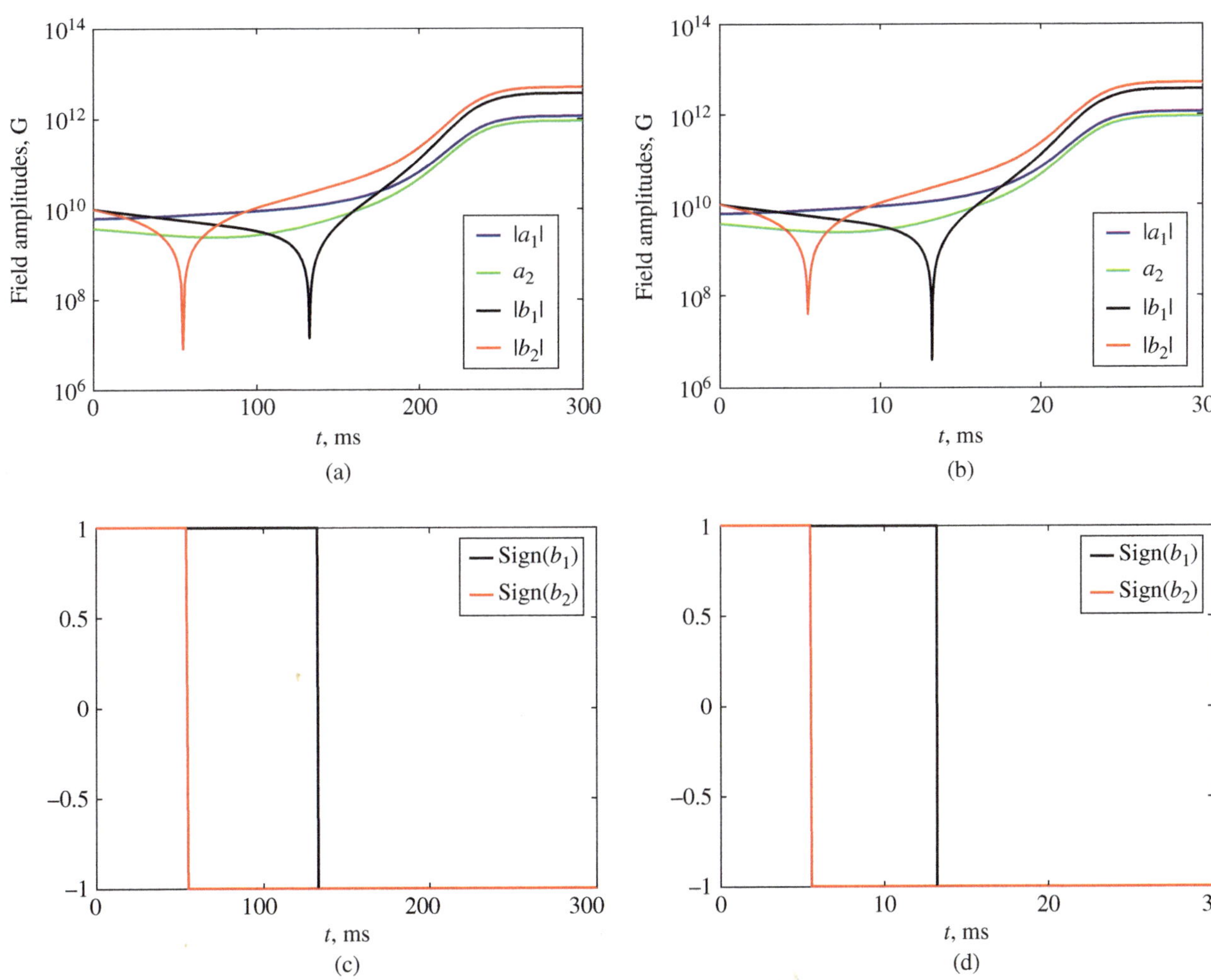

Figure 11.3 The evolution of the Fourier modes (a, b) for the toroidal field (given in Gauss) and signs of $b_{1,2}$ (c, d) behind the shock at $r \lesssim R_s = 150$ km. (a, c) $(\Delta r)_{\text{front}} = 10^{-10}$ cm; (b, d) $(\Delta r)_{\text{front}} = 10^{-11}$ cm.

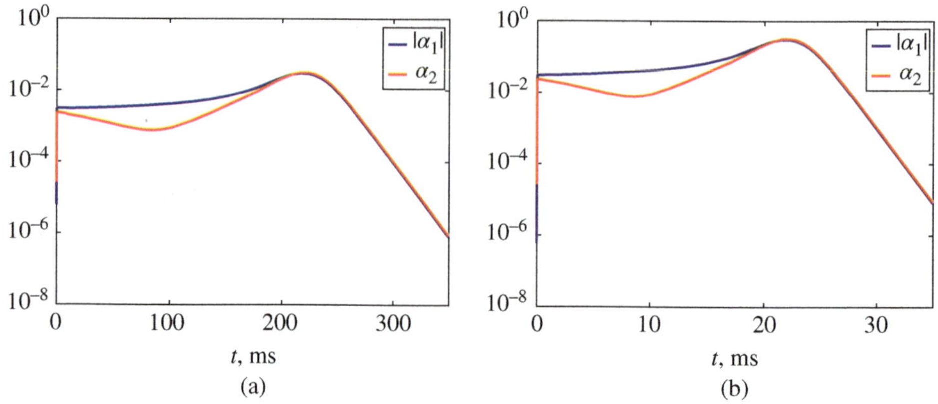

Figure 11.4 The evolution of the Fourier modes $\alpha_{1,2}$ behind the shock at $r \lesssim R_s = 150$ km, (a) $(\Delta r)_{\text{front}} = 10^{-10}$ cm, (b) $(\Delta r)_{\text{front}} = 10^{-11}$ cm.

11.5. DYNAMO DRIVEN BY NEUTRINO ASYMMETRY IN THE EARLY UNIVERSE

The section is based on Semikoz and Sokoloff (2004). The unusual helicity parameter α in the AMHD induction equation (11.2) arises in a hot plasma of the early universe due to electroweak $\nu_a e$ interaction Semikoz and Sokoloff (2004):

$$\alpha = \frac{\ln 2}{4\sqrt{2}\pi^2}\left(\frac{G_F T}{\lambda^{(\nu)}_{\text{fluid}}}\right)\sum_a c^{(A)}_{\nu_a}\left(\frac{n_{\nu_a} - n_{\bar{\nu}_a}}{n_{\nu_a}}\right), \qquad a = e, \mu, \tau. \tag{11.19}$$

Here, $G_F \simeq 10^{-5}/m_p^2$ is the Fermi constant; m_p is the proton mass; $\lambda^{(\nu)}_{\text{fluid}}$ is the neutrino free path in plasma; $\lambda^{(\nu)}_{\text{fluid}} \ll \Lambda_B$, where Λ_B is the PMF spatial scale; and $\Lambda_B \ll l_H$; $c^{(A)}_{\nu_a} = \mp 0.5$ is the axial coupling constant for $\nu_a e$ interactions in standard model (upper sign for electron neutrinos).

Since the neutrino asymmetry in hot plasma, $n_{\nu_a} - n_{\bar{\nu}_a} \approx \mu_{\nu_a} T^2/6$, depends on the scalar chemical potential μ_{ν_a}, $\mu_{\nu_a} \xrightarrow{\text{P}} \mu_{\nu_a}$, or α in equation (11.19) is also a scalar, $\alpha \xrightarrow{\text{P}} \alpha$, the instability term $\nabla \times \alpha\mathbf{B}$ is a vector, $\nabla \times \alpha\mathbf{B} \xrightarrow{\text{P}} -\nabla \times \alpha\mathbf{B}$. This property leads to a violation of P- parity in the Faraday equation as a whole:

$$\partial_t \mathbf{B} = \eta\nabla^2\mathbf{B} + \nabla \times \alpha\mathbf{B}, \qquad \eta = \sigma^{-1}_{cond}, \tag{11.20}$$

given the transformation $\mathbf{B} \xrightarrow{\text{P}} \mathbf{B}$. This CME case in AMHD differs from the standard MHD, where the helicity parameter α is a pseudoscalar, $\alpha \xrightarrow{\text{P}} -\alpha$, and the pseudovector term $\nabla \times \alpha\mathbf{B}$ remains a pseudovector as it should be in QED plasma; see equation (11.3).

Whereas the P-parity is not conserved in the induction equation (11.20), CP conservation in the same MHD induction equation (11.20) is valid due to the CP conservation for weak interactions in the SM of particle physics.

Indeed, the helicity parameter α is CP odd in equation (11.19), $\alpha \xrightarrow{\text{CP}} -\alpha$, due to the transformation

$$(n_{\nu_a} - n_{\bar{\nu}_a}) \xrightarrow{\text{CP}} -(n_{\nu_a} - n_{\bar{\nu}_a}). \tag{11.21}$$

This occurs when a neutrino transforms to an antineutrino under the charge conjugation $\nu_a \xrightarrow{\text{C}} \bar{\nu}_a$ and simultaneously the particle helicity (spin projection on neutrino velocity, $(\sigma \cdot \mathbf{p})/\text{p})$ changes, $\nu_L \leftrightarrow \bar{\nu}_R$. For Dirac neutrinos, both transformations are important, because $\nu^{(D)}$ is left-handed, $\nu^{(D)} = \nu_L$, while $\bar{\nu}^{(D)}$ is right-handed, $\bar{\nu}^{(D)} = \bar{\nu}_R$.

Thus, the transformation in equation (11.21) provides CP conservation for the Faraday equation as a whole with parity $\eta_{\text{CP}} = -1$ for all terms. The charge conjugation of the electromagnetic field in QED, $CA_\mu C^{-1} \to -A_\mu$, means the **B**-field that is a pseudovector under space inversion, $\mathbf{B} \xrightarrow{\text{P}} \mathbf{B}$, is CP -odd, $\mathbf{B} \xrightarrow{\text{CP}} -\mathbf{B}$: that is, $\eta_{\text{CP}} = -1$. The same transformation is valid for the instability term, $\nabla \times \alpha\mathbf{B} \xrightarrow{\text{CP}} -\nabla \times \alpha\mathbf{B}$. This happens because the vector $\alpha\mathbf{B}$ transforms as the vector potential $\mathbf{A}$, $\mathbf{A} \xrightarrow{\text{CP}} \mathbf{A}$, or $\alpha\mathbf{B} \xrightarrow{\text{CP}} \alpha\mathbf{B}$ due to the helicity property $\alpha \xrightarrow{\text{CP}} -\alpha$ provided by the rule in equation (11.21). As a result, this instability term is CP-odd, $\eta_{\text{CP}} = -1$.

In the absence of an advection (dynamo) term, the induction equation has the α^2-dynamo solution Semikoz and Sokoloff (2004)

$$B(t) = B_0 \exp\left[\int_{t_0}^{t}\left(\frac{\alpha^2(t')}{4\eta(t')}\right)dt'\right]. \tag{11.22}$$

11.6. CONCLUSIONS AND DISCUSSION

In recent decades, a new branch of MHD applying to dense ultrarelativistic (chiral) plasmas and originating from quantum effects in particle physics has been developing extensively. In this chapter, we discussed different anomalous magnetohydrodynamics (AMHD) effects. One of them is the chiral magnetic effect (CME) given by the vector current $\mathbf{J}_{\mathrm{CME}} = \sigma_{cond}\alpha\mathbf{B}$ and the new helicity parameter $\alpha \sim \mu_5$ in equation (11.3), which is a pseudoscalar similar to the Krause-Rädler result for α in standard MHD. We discussed in detail the origin of this quantum effect and AMHD consequences such as the new conservation law in equation (11.7) for combined helicity (particle chirality plus magnetic helicity) and the corresponding evolution of $\alpha(t)$ in equation (11.9), or the chiral imbalance $\mu_5(t)$.

We considered applications including the influence of the turbulent motion of the CME in the early universe (Dvornikov and Semikoz (2017)) and the generation of strong magnetic fields in a nascent neutron star through the CME driven by shock in its supernova progenitor (Dvornikov et al. (2020)).

The different helicity parameter α in equation (11.19), which is a scalar, unlike the pseudoscalar α in CME and standard MHD, amplifies a seed field via an α^2 dynamo in equation (11.22) due to electroweak $\nu_a e$ interactions in the presence of nonzero neutrino asymmetry $(n_{\nu_a} - n_{\bar{\nu}_a}) \neq 0$ Semikoz and Sokoloff (2004). While P-parity is not conserved in the induction equation, the α parameter is CP-odd due to the condition in equation (11.21), resulting in the conservation of negative $\eta_{\mathrm{CP}} = -1$ in the entire Faraday equation – which is as it should be for electroweak interactions in a vacuum (particle physics) or a medium.

We did not cover interesting application of AMHD such as leptogenesis before the electroweak phase transition, $T > T_{EWPT} \simeq 100$ GeV, and the corresponding generation of the baryon asymmetry of the universe (BAU), $\eta_B = (n_B - n_{\bar{B}})/s \simeq 10^{-10}$, driven by hypermagnetic field $\mathbf{B}_Y = \nabla \times \mathbf{Y}$ and its helicity density $h_Y(t) = V^{-1} \int d^3x \mathbf{Y} \cdot \mathbf{B}_Y$ (see, e.g., Semikoz et al. (2016)).

We also considered some ideas concerning particle effects in MHD and related topics. There are various other important issues in the field, including the evolution of turbulent flows in the conditions of the early universe (e.g., Brandenburg et al. (2017), Schober et al. (2019)).

REFERENCES

Adler, S. L. (1969). Axial-vector vertex in spinor electrodynamics. *Physical Review*, *177*(5), 2426–2438. https://doi.org/10.1103/PhysRev.177.2426

Bell, J. S., & Jackiw, R. (1969). A PCAC puzzle: $\pi^0 \to \gamma\gamma$ in the σ-model. *Nuovo Cimento A Serie*, *60*(1), 47–61. https://doi.org/10.1007/BF02823296

Brandenburg, A., Kahniashvili, T., Mandal, S., Pol, A. R., Tevzadze, A. G., & Vachaspati, T. (2017). Evolution of hydromagnetic turbulence from the electroweak phase transition. *Physical Review D*, *96*(12), 123528. https://doi.org/10.1103/PhysRevD.96.123528

Copinger, P., Fukushima, K., & Pu, S. (2018). Axial ward identity and the Schwinger mechanism: applications to the real-time chiral magnetic effect and condensates. *Physical Review Letters*, *121*(26), 261602. https://doi.org/10.1103/PhysRevLett.121.261602

Dvornikov, M., & Semikoz, V. B. (2017). Influence of the turbulent motion on the chiral magnetic effect in the early universe. *Physical Review D*, *95*(4), 043538. https://doi.org/10.1103/PhysRevD.95.043538

Dvornikov, M., & Semikoz, V. B. (2018). Magnetic helicity evolution in a neutron star accounting for the Adler-Bell-Jackiw anomaly. *Journal of Cosmology and Astroparticle Physics*, *2018*(8), 021. https://doi.org/10.1088/1475-7516/2018/08/021

Dvornikov, M., Semikoz, V. B., & Sokoloff, D. D. (2020). Generation of strong magnetic fields in a nascent neutron star accounting for the chiral magnetic effect. *Physical Review D*, *101*(8), 083009. https://doi.org/10.1103/PhysRevD.101.083009

Janka, H. Th. (2001). Conditions for shock revival by neutrino heating in core-collapse supernovae. *Astronomy & Astrophysics*, *368*, 527–560. https://doi.org/10.1051/0004-6361:20010012

Kharzeev, D. E., Liao, J., Voloshin, S. A., & Wang, G. (2016). Chiral magnetic and vortical effects in high-energy nuclear collisions-A status report. *Progress in Particle and Nuclear Physics*, *88*, 1–28. https://doi.org/10.1016/j.ppnp.2016.01.001

Krause, F., & Raedler, K. H. (1980). *Mean-field magnetohydrodynamics and dynamo theory*. Oxford University Press.

Masada, Y., Kotake, K., Takiwaki, T., & Yamamoto, N. (2018). Chiral magnetohydrodynamic turbulence in core-collapse supernovae. *Physical Review D*, *98*(8), 083018. https://doi.org/10.1103/PhysRevD.98.083018

Nefedov, S. N., & Sokoloff, D. D. (2010). Parker's dynamo as specific behavior of a dynamical system. *Astronomy Reports*, *54*(3), 247–253. https://doi.org/10.1134/S1063772910030066

Neronov, A., & Vovk, I. (2010). Evidence for strong extragalactic magnetic fields from fermi observations of TeV blazars. *Science*, *328*(5974), 73. https://doi.org/10.1126/science.1184192

Parker, E. N. (1979a). *Cosmical magnetic fields. Their origin and their activity.* Clarendon.

Parker, E. N. (1979b). *Cosmical magnetic fields. Their origin and their activity.* Gordon and Breach.

Schober, J., Brandenburg, A., Rogachevskii, I., & Kleeorin, N. (2019). Energetics of turbulence generated by chiral MHD dynamos. *Geophysical & Astrophysical Fluid Dynamics*, *113*(1–2), 107–130. https://doi.org/10.1080/03091929.2018.1515313

Semikoz, V. B., & Sokoloff, D. D. (2004). Large-scale magnetic field generation by α effect driven by collective neutrino-plasma interaction. *Physical Review Letters*, *92*(13), 131301. https://doi.org/10.1103/PhysRevLett.92.131301

Semikoz, V. B., Sokoloff, D. D., & Valle, J. W. F. (2009). Is the baryon asymmetry of the Universe related to galactic magnetic fields? *Physical Review D*, *80*(8), 083510. https://doi.org/10.1103/PhysRevD.80.083510

Semikoz, V. B., Smirnov, A. Yu., & Sokoloff, D. D. (2016). Generation of hypermagnetic helicity and leptogenesis in the early Universe. *Physical Review D*, *93*(10), 103003. https://doi.org/10.1103/PhysRevD.93.103003

Sigl, G. (2002). Cosmological magnetic fields from primordial helical seeds. *Physical Review D*, *66*(12), 123002. https://doi.org/10.1103/PhysRevD.66.123002

Sigl, G., & Leite, N. (2016). Chiral magnetic effect in protoneutron stars and magnetic field spectral evolution. *Journal of Cosmology and Astroparticle Physics*, *2016*(1), 025. https://doi.org/10.1088/1475-7516/2016/01/025

Son, D. T., & Surówka, P. (2009). Hydrodynamics with triangle anomalies. *Physical Review Letters*, *103*(19), 191601. https://doi.org/10.1103/PhysRevLett.103.191601

Tashiro, H., Vachaspati, T., & Vilenkin, A. (2012). Chiral effects and cosmic magnetic fields. *Physical Review D*, *86*(10), 105033. https://doi.org/10.1103/PhysRevD.86.105033

Vilenkin, A. (1980). Equilibrium parity-violating current in a magnetic field. *Physical Review D*, *22*(12), 3080–3084. https://doi.org/10.1103/PhysRevD.22.3080

12

Stability of Plasmas Through Magnetic Helicity

Simon Candelaresi[1] and Fabio Del Sordo[2,3]

ABSTRACT

Magnetic helicity and, more broadly, magnetic field line topology constrain plasma dynamics. Helically interlocked magnetic rings are in a nontrivial topological state. It is harder to bring them into a topologically trivial state than two rings that are not linked. This restriction causes helical plasmas to exhibit increased stability in laboratory devices, the Sun, and the intergalactic medium. Here we discuss how a magnetic field stabilizes the plasma and prevents it from being disrupted by magnetic helicity. We present observational results, numerical experiments, and analytical results that illustrate how helical magnetic fields contribute strongly to the long-term stability of some plasmas. We discuss several cases, such as the solar corona, toroidal fusion devices, and the galactic and extragalactic medium, with a special emphasis on extragalactic bubbles.

12.1. INTRODUCTION

The physics of magnetized fluids and helical magnetic fields plays a special role in magnetohydrodynamics (MHD). It is important both in amplifying magnetic fields (e.g., Brandenburg and Subramanian, 2005; Bhat et al., 2016) and for characterizing the saturation stage of magnetic field evolution in MHD instabilities (e.g., Chatterjee et al., 2011; Bonanno et al., 2012). In this chapter, we discuss the importance of the law of conservation of helicity (e.g., Bekenstein, 1987) for the evolution and stability of plasmas. Large-scale magnetic helicity is known to stabilize plasmas in different contexts. One remarkable example is rising flux tubes below the solar photosphere. Their rise is aided by an internal twist in the rising and erupting plasma patches (Leka et al., 1996). Without a twisted magnetic field, rising flux tubes are more prone to disruption as they rise through the convective zone. With a topologically trivial and nonhelical internal field, coronal mass ejections (CMEs) would disintegrate before reaching the solar photosphere.

Two formalisms tell us how magnetic helicity acts to stabilize the system. One is the near-perfect conservation of helicity in most astrophysical systems with high magnetic Reynolds numbers. The other is the realizability condition (Arnold, 1986), which poses a lower bound on magnetic energy in the presence of helicity.

Helicity is more than a stabilizer. As magnetic fields are continuously wound and braided, they may be subject to the MHD kink instability (e.g., Bonanno, 2013), generated by a strong twist of the magnetic field lines. Fields in the solar atmosphere can become very twisted (e.g., Mei, Z. X. et al. (2018)). However, based on theoretical evidence, (Aschwanden, 2019) argued that for active regions, the induced twist may be too small to lead to a kink instability.

Here we will explain how magnetic helicity plays an important role in plasma stability, particularly in astrophysical magnetic fields, from small scales in fusion

[1]*School of Mathematics and Statistics, University of Glasgow, Glasgow, UK*

[2]*Institute of Space Sciences, Barcelona, Spain*

[3]*Catania Astrophysical Observatory, National Institute for Astrophysics, Catania, Italy*

Helicities in Geophysics, Astrophysics, and Beyond, Geophysical Monograph 283, First Edition.
Edited by Kirill Kuzanyan, Nobumitsu Yokoi, Manolis K. Georgoulis, and Rodion Stepanov.

DOI:10.1002/9781119841715.ch12

devices to kiloparsec-sized plasmas in the intergalactic medium.

In this chapter, we will begin in section 12.2 by illustrating how conservation of magnetic helicity works. In section 12.3, we will discuss the final equilibrium state that plasmas should reach and the constraints that play a role in reaching this equilibrium (section 12.4). In section 12.5, we will see some examples of plasmas that reach stability due to large-scale magnetic helicity; we will concentrate on plasmas on the galactic and intergalactic scales in section 12.6. In section 12.7, we will move to a specific example, intergalactic cavities, for which a mechanism of stabilization through large-scale magnetic helicity has been recently proposed. Finally, we will draw some conclusions in section 12.8.

12.2. CONSERVATION OF HELICITY

The conservation of magnetic helicity in regimes with high magnetic Reynolds numbers can be easily derived from the MHD equations for a resistive, viscous compressible gas (e.g., Biskamp (2003)):

$$\frac{\partial \boldsymbol{A}}{\partial t} = \boldsymbol{u} \times \boldsymbol{B} - \eta \boldsymbol{J}, \tag{12.1}$$

$$\frac{\mathrm{D}\boldsymbol{u}}{\mathrm{D}t} = -c_{\mathrm{s}}^{2}\boldsymbol{\nabla} \ln \rho + \boldsymbol{J} \times \boldsymbol{B}/\rho + \boldsymbol{F}_{\mathrm{visc}}, \tag{12.2}$$

$$\frac{\mathrm{D}\ln \rho}{\mathrm{D}t} = -\boldsymbol{\nabla} \cdot \boldsymbol{u}, \tag{12.3}$$

with magnetic vector potential $\boldsymbol{A}$, velocity $\boldsymbol{u}$, magnetic field $\boldsymbol{B} = \boldsymbol{\nabla} \times \boldsymbol{A}$, magnetic resistivity η, isothermal speed of sound c_{s}, density ρ, current density $\boldsymbol{J} = \boldsymbol{\nabla} \times \boldsymbol{B}$, viscous forces $\boldsymbol{F}_{\mathrm{visc}}$, and Lagrangian time derivative $\mathrm{D}/\mathrm{D}t = \partial/\partial t + \boldsymbol{u} \cdot \boldsymbol{\nabla}$. Here, the viscous forces are given as $\boldsymbol{F}_{\mathrm{visc}} = \rho^{-1}\boldsymbol{\nabla} \cdot 2\nu\rho\boldsymbol{S}$, with the kinematic viscosity ν and traceless rate of strain tensor $\boldsymbol{S}_{ij} = \frac{1}{2}(u_{i,j} + u_{j,i}) - \frac{1}{3}\delta_{ij}\boldsymbol{\nabla} \cdot \boldsymbol{u}$. Because it is an isothermal gas, we have $p = c_{\mathrm{s}}^{2}\rho$ for the pressure. As an alternative to the induction equation in vector potential form, we could write it directly using the magnetic field $\boldsymbol{B}$:

$$\frac{\partial \boldsymbol{B}}{\partial t} = \boldsymbol{\nabla} \times (\boldsymbol{u} \times \boldsymbol{B}) + \eta \nabla^{2}\boldsymbol{B}. \tag{12.4}$$

Here, we use the induction equation (12.1) expressed using the magnetic vector potential $\boldsymbol{A}$ instead of the magnetic field $\boldsymbol{B}$. Doing so has several advantages. First, we can compute the magnetic helicity density directly

$$h_{\mathrm{m}} = \boldsymbol{A} \cdot \boldsymbol{B} \tag{12.5}$$

in all space for all time without needing to find the inverse curl. Second, the solenoidal condition of $\boldsymbol{\nabla} \cdot \boldsymbol{B} = 0$ is trivially fulfilled. This is particularly useful in numerical calculations as it ensures not incurring nonzero divergence $\boldsymbol{B}$, which could lead to spurious results (see, e.g., Brandenburg and Scannapieco, 2020). Of course, we are left with the gauge choice since the vector potential $\boldsymbol{A}' = \boldsymbol{A} + \boldsymbol{\nabla}\phi$, with the differentiable scalar field ϕ, leads to the same vector field $\boldsymbol{B} = \boldsymbol{\nabla} \times \boldsymbol{A}'$. The implicit gauge in our induction equation is the Weyl gauge with vanishing electrostatic potential.

The helicity density depends on space and time. For conservation of helicity, we are interested in the total helicity: that is, the volume integrated magnetic helicity density

$$H_{\mathrm{m}} = \int_{V} \boldsymbol{A} \cdot \boldsymbol{B} \, \mathrm{d}V. \tag{12.6}$$

We will assume that the volume we are interested in will remain static. Later, we will contrast the magnetic helicity with the magnetic energy

$$E_{\mathrm{m}} = \frac{1}{2}\int_{V} \boldsymbol{B} \cdot \boldsymbol{B} \, \mathrm{d}V. \tag{12.7}$$

From the MHD equations, we can now derive the rate of change in time of the magnetic helicity as

$$\partial_t H_{\mathrm{m}} = \int_{V} \partial_t \boldsymbol{A} \cdot \boldsymbol{B} + \boldsymbol{A} \cdot \partial_t \boldsymbol{B}. \tag{12.8}$$

Using the induction equation for $\boldsymbol{A}$ from equations (12.1) and $\boldsymbol{B}$ (12.4) and $\boldsymbol{\nabla} \times \boldsymbol{B} = \boldsymbol{J}$, we find

$$\begin{aligned} \partial_t H_{\mathrm{m}} &= \int_{V} (\boldsymbol{u} \times \boldsymbol{B} + \eta \nabla^{2}\boldsymbol{A}) \cdot \boldsymbol{B} \, \mathrm{d}V \\ &+ \int_{V} \boldsymbol{A} \cdot (\boldsymbol{\nabla} \times (\boldsymbol{u} \times \boldsymbol{B}) + \eta \nabla^{2}\boldsymbol{B}) \, \mathrm{d}V \\ &= -2\eta \int_{V} \boldsymbol{J} \cdot \boldsymbol{B} \, \mathrm{d}V + \int_{\partial V} \boldsymbol{A} \times (\boldsymbol{B} \times \boldsymbol{u} - \eta \boldsymbol{J}) \cdot \mathrm{d}\boldsymbol{S}. \end{aligned} \tag{12.9}$$

The first term is the resistive helicity dissipation/generation. For turbulent dissipative systems, the electric current density and magnetic field are, on average, aligned. This makes this term *in practice* a dissipation term. The last two terms are surface terms. They can potentially be helicity injection or helicity loss in a finite domain. For instance, this is the case for domains located in the solar atmosphere or bordering the solar photosphere. On the photosphere, this helicity injection can be substantial (e.g., Linan et al. (2018), Gupta, M. et al. (2021)). In these cases, the fields are finite, and there is a possible helicity flux through the surface, even for small diffusivities, as the diffusivity does not enter the surface terms. The same concept applies on large scales, such as galaxies, which may experience helicity loss through the halo and then to the extragalactic medium. This helicity flux may be relevant for galactic dynamos (Ntormousi et al., 2020) and may play a key role in galactic magnetic fields reaching observed values (Del Sordo et al., 2013; Rincon, 2021).

For systems with periodic boundaries or closed boundaries, or where the boundaries are so distant that the quantities can be considered zero, the surface terms vanish. In that case, we can focus on the resistive dissipation term. This dissipation is small for most astrophysical settings, as η is so small that structures of the size in question diffuse on much longer time scales than typical dynamical time scales (e.g., Berger (1984)). So, we can assume that magnetic helicity is conserved in such systems.

In the limit of vanishing magnetic diffusion ($\eta = 0$), the induction equation (12.1) simplifies to

$$\frac{\partial \boldsymbol{A}}{\partial t} = \boldsymbol{u} \times \boldsymbol{B}; \tag{12.10}$$

or, expressed in terms of the magnetic field $\boldsymbol{B}$, we can write

$$\frac{\partial \boldsymbol{B}}{\partial t} = \nabla \times (\boldsymbol{u} \times \boldsymbol{B}). \tag{12.11}$$

This equation describes a magnetic field that is Lie-transported under the velocity field $\boldsymbol{u}$ (Candelaresi et al., 2014). We then say that the field is "frozen in" the fluid (Alfvén, 1942; Batchelor, 1950; Priest and Forbes, 2000). Simply put, a gas/fluid compression perpendicular to $\boldsymbol{B}$ increases the local magnetic field strength by the compression factor. We can also show that every open advected surface S preserves its magnetic flux

$$\Phi = \int_S \boldsymbol{B} \cdot \mathrm{d}S. \tag{12.12}$$

This can be expressed in terms of differential two-forms (e.g., Arnold and Khesin (2013), Candelaresi et al. (2014)): $\beta = \boldsymbol{B}_x \mathrm{d}y \wedge \mathrm{d}z + \boldsymbol{B}_y \mathrm{d}z \wedge \mathrm{d}x + \boldsymbol{B}_z \mathrm{d}x \wedge \mathrm{d}y$

$$\frac{\mathrm{d}}{\mathrm{d}t}\beta = \mathscr{L}_{\boldsymbol{u}}(\beta), \tag{12.13}$$

where $\mathscr{L}_{\boldsymbol{u}}(\beta)$ is the Lie-transport of β under the velocity $\boldsymbol{u}$.

As the magnetic field is frozen in, magnetic field lines obtain a physical meaning because they cannot be broken up and reconnected, which manifests the restrictions in the field's dynamics. Together with the magnetic field, the magnetic helicity density is Lie-transported in the ideal limit. Because the helicity is a density, every advected volume V preserves its helicity. Consequently, the entire domain (which is also a volume) conserves its magnetic helicity:

$$\frac{\mathrm{d}}{\mathrm{d}t} H_{\mathrm{m}} = 0. \tag{12.14}$$

So, we have exact conservation of magnetic helicity for $\eta = 0$.

The discussion is similar for the limit case, $\lim_{\eta \to 0}$. We can perform the same calculations and arrive at the same result:

$$\lim_{\eta \to 0} \frac{\mathrm{d}}{\mathrm{d}t} H_{\mathrm{m}} = 0. \tag{12.15}$$

12.3. RELAXATION EQUILIBRIUM STATE

What consequences does the conservation of helicity have for the system's dynamics, particularly relaxing/decaying magnetic fields? During any stage of its evolution – including its end state – the system must go through states of the same magnetic helicity. This excludes a large class of fields, such as those with connectivity with a different magnetic helicity.

So, similar to the conservation of energy in a nondissipative system like a pendulum, we can make predictions about what behavior is excluded and, to some extent, make qualitative predictions about the geometry of the magnetic field lines during their evolution.

We know that a field in a nonequilibrium state will try to reach a minimum-energy state. This is true for any relaxing magnetic fields in a closed or infinitely large MHD system. Without any constraints, this lowest energy would simply be zero. Woltjer (1958) showed that under conservation of magnetic helicity, the lowest magnetic energy state is the force-free state with $\nabla \times \boldsymbol{B} = \alpha \boldsymbol{B}$, where α is the force-free parameter. This state cannot always be reached during dynamical time evolution; it depends on the geometry and topology of the magnetic field lines. Magnetic helicity imposes restrictions, and other topological quantities may also affect the obtainable equilibrium state. For instance, if the minimum energy state can only be reached by reconnecting magnetic field lines, it won't be accessible under ideal evolution.

Since in the ideal ($\eta = 0$) case the field evolves under a Lie transport, every advected subvolume of the domain conserves its helicity. In particular, every finite neighborhood of the magnetic field lines conserves its helicity content. This can be easily imagined for closed field lines of finite length. We then have infinitesimally thin magnetic closed flux tubes (potentially knotted or braided) that conserve helicity. This is true for all neighboring flux tubes that may contain a different amount of helicity. For ergodic field lines that fill a finite space in three dimensions, the interpretation becomes more interesting. In that case, we have constant helicity not just on a subvolume with measure 0, but potentially with finite measure and finite volume (e.g., Arnold and Khesin (2013)).

Under ideal conditions with no breaking and coalescing of magnetic field lines (no reconnection), Taylor (1974) suggested that helicity is conserved for every subvolume bounded by the magnetic field lines. That means not only the total helicity but also subhelicities are conserved. The minimum energy state (relaxed state) is then a nonlinear force-free field where the force-free parameter α depends on the field line:

$$\nabla \times \boldsymbol{B} = \alpha(a, b)\boldsymbol{B}. \tag{12.16}$$

For ergodic field lines, we have finite volumes with the same parameter α, as we are dealing with the same field lines. However, under real conditions with magnetic reconnection, Taylor (1974) further argued that only the total magnetic helicity is conserved, as subvolumes bounded by magnetic field lines do not make any physical sense. In that case, the minimum energy state is a linear force-free state of the form

$$\nabla \times \boldsymbol{B} = \alpha \boldsymbol{B}, \tag{12.17}$$

where α is a constant in the domain.

12.4. FIELD RELAXATION CONSTRAINTS

From the discussion by Woltjer and Taylor, we cannot deduce if the given magnetic field will ever reach such an equilibrium. Topological constraints may inhibit the plasma from approaching a linear or nonlinear force-free state.

One such constraint is given by the realizability condition (Arnold, 1986). It defines a lower limit for the magnetic energy in the presence of magnetic helicity as

$$2E(k) \geq k|H(k)|, \tag{12.18}$$

where k is the inverse wavelength. This can be integrated over the parameter k, which results in a global constraint. In an experiment or numerical simulation, this effect can be easily observed by comparing the magnetic helicity with the magnetic energy. After some initial free relaxation with a drop in energy, the energy reaches a near-constant value in low-resistivity environments. Independent of the magnetic resistivity, the ratio of the two quantities will reach a near constant.

An analogous exact relation exists for a purely hydrodynamical case that follows the compressible viscous Navier-Stokes equations. But instead of relating the kinetic energy to the kinetic helicity, the relation is between the kinetic helicity and the enstrophy, which is the volume integral of the vorticity squared. It can be easily derived that the enstrophy is bound from below by the kinetic helicity (Candelaresi et al., 2021). Although there is no existing exact relation, the same authors found experimentally a similar realizability condition that limits the kinetic energy from below by the presence of unsigned kinetic helicity.

A naïve interpretation of the realizability condition can be drawn from a simple thought experiment. Take two closed magnetic flux tubes with no internal twist that are linked once. This configuration is helical. A minimum energy state cannot be reached without reconnection. However, we can conceive an experiment with three linked flux tubes, and by defining the correct signs of the magnetic field in each tube, we can make this configuration nonhelical. This is a slightly troubled situation, as the naïve interpretation of realizability breaks down. From numerical experiments by Del Sordo et al. (2010), we know that the actual linking plays only a minor role, while helicity poses the real restrictions, as expressed in the realizability condition.

Taylor relaxation cannot always be realized. Due to additional constraints from existing topological quantities, the proposed equilibrium state cannot always be reached without breaking the field's topology. For instance, Yeates et al. (2010) showed that for a field containing finite field line helicity, the equilibrium state was not the Taylor state.

12.5. PLASMA STABILITY

12.5.1. Solar Eruptions

Analyzing soft X-ray images of solar active regions, Canfield et al. (1999) showed that sigmoidal structures are more likely to result in solar eruptions than nonsigmoidal structures. Sigmoids result from the twisting of magnetic flux tubes, where an increase in twist leads to kinking. These flux tubes are generated below the photosphere, where they obtain their internal twist from solar rotation (e.g., Leka et al. (1996)), although flux tubes that have emerged can be twisted through photospheric foot points where the tubes are anchored (e.g., Candelaresi et al. (2016)). As they rise through the convection zone, some are disrupted and do not pinch through the photosphere as an intact flux tube. However, helical flux tubes survive longer and may result in later CMEs (Gibson et al., 2002).

While twist leads to stability below the photosphere, it can also lead to a kink instability if it is sufficiently large (e.g., Finkelstein and Weil (1978), Craig and Sneyd (1990), Rust and Kumar (1996), Ebrahimi and Karami (2016), Vemareddy et al. (2017)). Such large flux-tube twists can be generated through photospheric motions (Canfield et al., 1999).

At the same time, a twist induced into a more complex photospheric magnetic loop structure can potentially move the flux tubes such that reconnection may occur. Even on a relatively small scale, such as that of sunspots, helicity can be induced into flux tubes via sunspot rotation (e.g., Wang et al. (2016)). These reconnection events lead to high electric current density and high particle acceleration. Such events may not be CMEs but rather bright flares that can be observed in white light or extreme ultraviolet.

12.5.2. Toroidal Fusion Plasma Fields

A magnetically confined toroidal fusion device produces controlled fusion reactions in hot plasma (e.g.,

Wesson and Campbell (1997)). To confine the plasma in the shape of a torus, strong magnetic fields are used (e.g., Taylor (1993), Jardin et al. (2015)). A helical magnetic field in fusion plasma can easily maintain its energy: this is evident from the realizability condition in equation (12.18). Such a helical field can be generated by two currents. One is poloidal: that is, a ring current around the torus' main axis. The other component is a current along the main axis, which can be artificially injected.

From Taylor (1974) and Taylor (1986), we know that without any energy input, the magnetic field relaxes into a force-free state with $\nabla \times \boldsymbol{B} = \alpha \boldsymbol{B}$. This is the minimum energy state under conservation of helicity. Therefore, we can deduce the minimum energy input required to keep the plasma in a stable configuration, allowing us to maintain it in a stationary state.

12.5.3. Numerical Experimental Evidence

To understand the mechanism behind the stabilizing effect of magnetic helicity in plasmas, it is useful to conduct a series of experiments. Real-world experiments do not currently have the power to measure the mechanisms in sufficient detail. But we need to know about the magnetic field relaxation mechanism, the nature of reconnection events, and the energy conversion sites (magnetic to kinetic/thermal).

A solution is to model the plasma using appropriate equations. For a nonrelativistic plasma, as we assume the solar, galactic, and fusion plasmas to be, we can use the MHD equations (12.1)–(12.3).

From a purely geometrical interpretation (Moffatt, 1969), one can argue that linking (helicity) stabilizes by preventing/restricting flux tubes from reconnecting arbitrarily. An early numerical experiment was devised by Del Sordo et al. (2010). They inserted three magnetic flux rings into a box (initial condition). Two outer rings were linked to the inner ring, resulting in a linked configuration. By changing the sign of the magnetic flux of one of the outer rings, they changed the system from one with net magnetic helicity to no magnetic helicity. A control configuration with three unlinked flux tubes was also used. With the nonhelical field relaxing similarly to the trivial nonlinked field, it was shown that the helicity, rather than the actual linkage, restrict the field relaxation.

Similar experiments were carried out by Candelaresi and Brandenburg (2011), who studied the relaxation of helical and nonhelical knots. One of their results was that topologically nontrivial and nonhelical knots show a kind of intermediate stability between topologically trivial fields and helical fields. For nonhelical braids, Yeates et al. (2010) observed similar enhanced stability compared to the trivial case. They argued that further topological invariants should be considered. The consequences would be far-ranging and have implications particularly for the stability of toroidal fusion plasma fields. Prior and Yeates (2016a,b) introduced a technique for generating tubular magnetic fields with arbitrary axial geometry and internal topology to investigate the behavior of flux ropes whose field lines have more complex entangled configurations. One of the main applications of this technique is the study of flux ropes in the solar corona. They concluded that magnetic field lines with sufficient complex entanglement can suppress large-scale morphological changes, and magnetic energy can be reduced through reconnection and expansion of the ropes.

12.6. GALACTIC AND INTERGALACTIC MEDIA

12.6.1. Observations

Much of the discussion in the literature about helicity in plasmas focuses on the Sun and, to a lesser extent, fusion plasmas. To explain the observed large-scale fields in galaxies, magnetic helicity has been considered to model the magnetic field amplification through the dynamo effect. Since the parameters are unsuitable for direct numerical simulations, many authors (Kleeorin et al., 2000; Sur et al., 2007; Shukurov et al., 2008; Brandenburg et al., 2009; Chamandy et al., 2014; Chamandy, 2016) have used the mean-field approach in which the given fields are separated into fluctuating and mean parts, assuming there is scale separation.

While magnetic fields in galaxies are less studied than those in stars and planets, the field in the intergalactic medium is even less well explored. From a number of X-ray observations, we know that some galaxies in clusters like Cygnus A (Carilli et al., 1994) emit material through jets that form bubble-like structures above the galactic plane. The X-ray signature is an effect of the internal magnetic field, which can have a strength of $10\,\mu G$–$40\,\mu G$ (Carilli and Taylor, 2002). Further observations by Taylor et al. (2002), Churazov et al. (2001), Bîrzan et al. (2004), McNamara and Nulsen (2007), and Montmerle (2011) in different galaxy clusters, including the Virgo cluster with the galaxy M87, suggest that these lobes, or bubbles, have a higher temperature and lower density than the surrounding intergalactic medium by a factor of about three. With a size of roughly 5 kpc, they are smaller, although not significantly, than the galaxies from which they emanate. Like other galaxies, the Milky Way displays bubbles rising above and below its midplane. They are called Fermi bubbles and emanate multiwavelength radiation from radio waves to γ-ray waves (Cheng et al., 2015). Possible mechanisms leading to these bubbles are galactic fountain flows (Bregman, 1980), the ejection of hot material through jets, and compression of the intergalactic medium (Churazov et al.,

2001). Energy input from an active galactic nucleus can provide the necessary power to heat the bubbles.

Plasma in the bubbles is magnetized and relatively hot. Because the bubbles are hot and underdense, they rise from the galactic plane much like a hot air balloon, but without a membrane. This leads to a velocity shear at the bubble's boundary and should lead to Kelvin-Helmholtz instability. However, from measurements of their total lifetime (e.g., Bîrzan et al. (2004)), we know that they can survive for 10 Myr–100 Myr). This is much longer than estimated from the instability analysis. So, apart from their origin, it is puzzling how they survive so long.

12.6.2. Kelvin-Helmholtz Instability

In pure hydrodynamics, a shearing flow with velocity $\boldsymbol{u} = f(x)\boldsymbol{e}_y$ can be unstable, depending on the shearing flow profile $f(x)$ (typically its steepness) and the density profile of the gas (Chandrasekhar, 1961). In its simplest form, we can consider two phases. One phase consists of a fluid or gas with density ρ_1 and constant (in space) velocity $\boldsymbol{u}_1$. The other phase has density ρ_2 and parallel velocity $\boldsymbol{u}_2 \neq \boldsymbol{u}_1$. According to the Navier-Stokes equations, such a system will not change over time. Through the Kelvin-Helmholtz instability, however, a small perturbation will be amplified exponentially, depending on the wavelength of the perturbation.

For a magnetized gas or plasma, we must consider MHD. The gas and magnetic field are coupled through the Lorentz force. Through magnetic tension building up through the fluid flow from the onsetting instability, one could imagine a stabilizing effect from the Lorentz force. Chandrasekhar (1961) studied the effect of a homogeneous external magnetic field on the Kelvin-Helmholtz instability. There are two simple cases. The first is a field perpendicular to the shearing flow ($\boldsymbol{B}_0 \perp \boldsymbol{u}$). This case does not have any effect on the instability. However, a parallel magnetic field stabilizes the system. The suppressed wavelengths depend on the strength of the field. At a field strength of

$$\boldsymbol{B}^2 \geq 2\pi(u_1 - u_2)^2(\rho_1\rho_2)/(\rho_1 + \rho_2), \qquad (12.19)$$

all modes are suppressed.

So, an external parallel magnetic field is a viable stabilizer for buoyant intergalactic bubbles. Whether such a field exists or is strong enough is a matter to be determined by observers. We will discuss this case in the next section and compare it to our proposed stabilizing mechanism.

12.7. HELICAL INTERGALACTIC BUBBLES

From helical galactic dynamo models (e.g., Kleeorin et al. (2000), Sur et al. (2007)), we can assume that the generated magnetic field is also helical. If these bubbles originate from the galactic interior, where they obtain their magnetic field, we can also assume that this field is helical. If the bubbles are helical, we can hypothesize that they have intrinsic stability.

It is hard to make any precise predictions about the geometry and topology of this internal field. Therefore, we will discuss two different internal helical magnetic fields and their effects on the stability of intergalactic bubbles. For a more in-depth discussion and details, see Candelaresi and Del Sordo (2020).

12.7.1. Numerical Experiment

To test whether a helical magnetic field can stabilize intergalactic bubbles such that they survive the observed 10–100 Myr, we performed a number of MHD simulations (Candelaresi and Del Sordo, 2020). The setup consists of an underdense hot bubble surrounded by a colder stratified medium (Fig. 12.1). Buoyancy is generated by a homogeneous gravitational acceleration g. Since the galactic disk is larger than the bubble, a constant gravitational pull is a justified approximation and simplifies the numerical calculations.

In the direct numerical simulations, we solve the full resistive, viscous, and compressible MHD equations with the energy equation, which includes the temperature. The density and temperature contrast between bubble and medium is a factor of three, which is approximately what has been observed. The numerical code used, PencilCode (Brandenburg and Dobler, 2002; Brandenburg, 2020) (https://github.com/pencil-code), solves for the

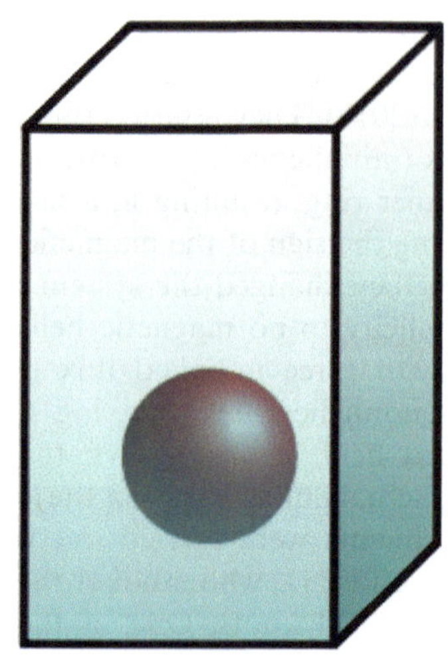

Figure 12.1 Schematic representation of the initial condition of the numerical experiment. The hot underdense bubble with homogeneous temperature and density is embedded in a stratified cold medium of greater density. The galactic disk is at the lower boundary and exerts a constant gravitational force.

magnetic vector potential $\boldsymbol{A}$ instead of the magnetic field $\boldsymbol{B} = \nabla \times \boldsymbol{A}$. This ensures that the solenoidal condition is fulfilled and facilitates the computation of the magnetic helicity. The equations we solve are as follows:

$$\frac{\partial \boldsymbol{A}}{\partial t} = \boldsymbol{u} \times \boldsymbol{B} + \eta \nabla^2 \boldsymbol{A}, \tag{12.20}$$

$$\frac{\mathrm{D}\boldsymbol{u}}{\mathrm{D}t} = -c_s^2 \nabla \left(\frac{\ln T}{\gamma} + \ln \rho \right) + \frac{\boldsymbol{J} \times \boldsymbol{B}}{\rho} - \boldsymbol{g} + \boldsymbol{F}_{\mathrm{visc}}, \tag{12.21}$$

$$\frac{\mathrm{D} \ln \rho}{\mathrm{D}t} = -\nabla \cdot \boldsymbol{u}, \tag{12.22}$$

$$\frac{\partial \ln T}{\partial t} = -\boldsymbol{u} \cdot \nabla \ln T - (\gamma - 1) \nabla \cdot \boldsymbol{u} + \frac{1}{\rho c_v T} \left(\nabla \cdot (K \nabla T) + \eta \boldsymbol{J}^2 + 2\rho\nu \boldsymbol{S} \otimes \boldsymbol{S} + \zeta \rho (\nabla \cdot \boldsymbol{u})^2 \right), \tag{12.23}$$

with the magnetic vector potential $\boldsymbol{A}$, magnetic field $\boldsymbol{B} = \nabla \times \boldsymbol{A}$, fluid velocity $\boldsymbol{u}$, constant magnetic resistivity (diffusivity) η, advective derivative $\mathrm{D}/\mathrm{D}t = \partial/\partial t + \boldsymbol{u} \cdot \nabla$, speed of sound $c_s = \gamma p/\rho$, adiabatic index $\gamma = c_p/c_v$, heat capacities c_p and c_v at constant pressure and volume, temperature T, density ρ, electric current density $\boldsymbol{J} = \nabla \times \boldsymbol{B}$, gravitational acceleration $\boldsymbol{g}$, viscous force $\boldsymbol{F}_{\mathrm{visc}}$, heat conductivity K, and bulk viscosity ζ. The viscous force is given as $\boldsymbol{F}_{\mathrm{visc}} = \rho^{-1} \nabla \cdot 2\nu\rho \boldsymbol{S}$, with the traceless rate of strain tensor $S_{ij} = \frac{1}{2}(u_{i,j} + u_{j,i}) - \frac{1}{3}\delta_{ij} \nabla \cdot \boldsymbol{u}$. The equation of state used here is for an ideal monatomic gas; it appears implicitly in our equations, as we eliminated the pressure p. Here the gas is monatomic with $\gamma = 5/3$.

12.7.2. Helicity as a Stabilizer

We insert a helical magnetic field into our numerical bubble. We use different setups to exclude effects from the field's geometry. One consists of an ABC-type field of the form

$$\boldsymbol{A} = f(r) A_0 \begin{pmatrix} \cos((y - y_c)k) + \sin((z - z_c)k) \\ \cos((z - z_c)k) + \sin((x - x_c)k) \\ \cos((x - x_c)k) + \sin((y - y_c)k) \end{pmatrix}, \tag{12.24}$$

where the wave number k can be used to regulate the helicity relative to the magnetic energy, which is regulated by the amplitude A_0. Here the subscript c stands for a cavity. The function $f(r)$ depends only on the radius from the bubble's center and ensures that the field goes to zero smoothly at the boundary without forming a current layer.

The second helical configuration is the spheromak, which consists of twisted magnetic fields in a toroidal shape. Those fields are embedded into each other, forming a highly helical field (Fig. 12.2). One advantage of this configuration is that it smoothly approaches zero field strength at the bubble's boundary. Similar to the ABC case, we can adjust the magnetic energy and magnetic helicity independently. For a more detailed formulation, see Candelaresi and Del Sordo (2020).

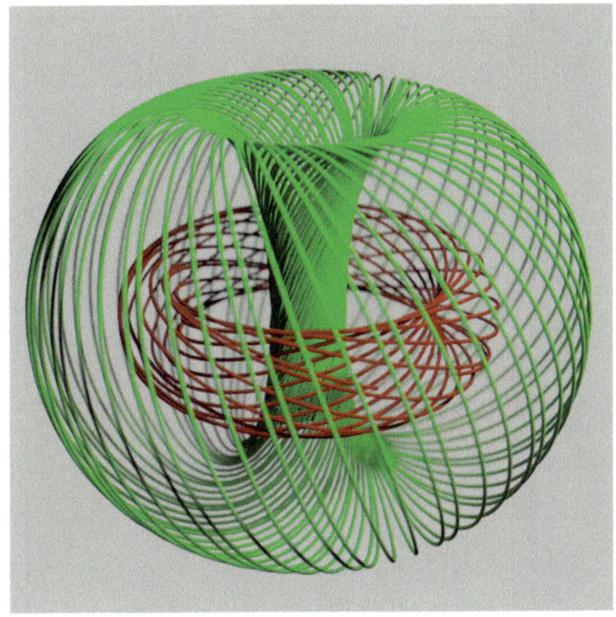

Figure 12.2 Helical spheromak configuration showing two of the magnetic field lines. The original color image was monochrome. Candelaresi and Del Sordo, 2020/IOP Publishing/Licensed under CC BY 4.0.

In both cases, we keep the magnetic energy constant while adjusting the helicity such that the high-helicity case has four times the helicity of the low-helicity case. As control cases, we also run simulations with no magnetic field and two with only an external field. This gives us seven different simulations.

The stability of the bubble is measured using a coherence measure d_{mean}, which is simply the mean distance of the points within the bubble. Whether a point belongs to the bubble is determined by the temperature. By definition, all points above a temperature threshold belong to the bubble. So, a large value of d_{mean} tells us that the bubble has undergone significant disruption.

From our calculations (Fig. 12.3), we see that a helical internal magnetic field stabilizes the field. This is independent of the type of helical field used (ABC versus spheromak). At a minimum level of helicity, we reach stability. However, if we want the same stability from an external field, we require much greater total magnetic energy. We estimate that a helical field with a maximum strength on the order of 10-5 G can stabilize the bubble over a time scale of about 250 Myr. For comparison, these simulations also show that in a purely hydrodynamical case, the bubbles can be stable for about 80 Myr.

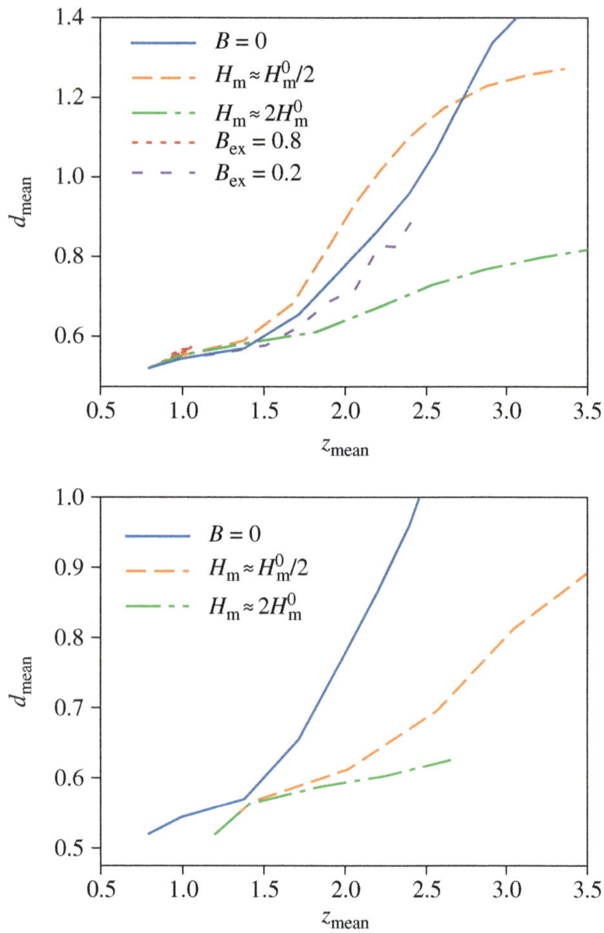

Figure 12.3 Coherence measure d_{mean} of the intergalactic bubbles depending on the mean height. The upper panel shows the control case with no magnetic field ($\boldsymbol{B} = 0$), with the ABC field with two different helicity values and the same energy, and the case of an external magnetic field with strength close to the predicted value for suppressing instability on all wavelengths. The red dotted line, corresponding to $\boldsymbol{B}_{ex} = 0.8$, is very short because the strong field prevents the bubble from rising. The lower panel shows the case for the spheromak configuration. The original color figures were monochrome. Candelaresi and Del Sordo, 2020/IOP Publishing/Licensed under CC BY 4.0.

12.8. CONCLUSIONS

We have discussed the role of helical magnetic fields in stabilizing plasmas. First, we have seen how magnetic helicity is a conserved quantity in ideal MHD and in the high-conductivity limit – that is, when magnetic diffusivity is very small. This is the case for various astrophysical plasmas on scales with very different lengths.

Conservation of helicity is well illustrated by considering interlocked magnetic tubes. When their configuration is characterized by finite magnetic helicity, it results in a much more stable system than in nonhelical cases. On the other hand, the topological linkage of such tubes plays no role in stabilizing the plasma. Therefore, conservation of magnetic helicity imposes strong constraints on the relaxation of magnetic fields whenever helicity is present.

We have discussed conservation of helicity in the stability of structures in the solar corona, such as CMEs, as well as in fusion plasma fields. We see similar behavior on much larger scales, such as in galactic and extragalactic environments. In particular, some extragalactic structures moving at high velocity through the extragalactic medium are observed to be surprisingly stable toward disruption by the Kelvin-Helmoltz instability. Magnetic helicity may therefore play an important role in such a disruption.

In the last section of the chapter, we examined the possibility that a helical magnetic field may stabilize extragalactic bubbles even at low magnetic energy densities. These bubbles are observed rising from the midplane of our galaxy and other galaxies into the intergalactic medium. We justify the presence of helicity in these bubbles by the fact that they are inflated by an active galactic nucleus or from jets coming from the galactic center, which yield helical fields.

We tested two different types of helical magnetic fields: ABC flow and a spheromak field. This field initially sits inside the bubble rising through buoyancy in an otherwise stably stratified medium. We quantified the disruption of the bubble and observed that this bubble is stabilized by the helicity toward the Kelvin-Helmholtz instability in the nonlinear regime if the magnetic field is sufficiently helical. We estimate that a helical field with a maximum strength on the order of 10-5 G can stabilize the bubble over a time scale of at least 250 Myr, which is more than three times longer than the stability of a nonmagnetized bubble. This remarkable example illustrates the role of magnetic helicity in stabilizing structures in plasmas. Moreover, it also suggests that helical magnetic fields are present in astrophysical magnetized plasmas wherever stable configurations are observed and at several scales, from stars to extragalactic structures.

ACKNOWLEDGMENTS

Fabio Del Sordo acknowledges support from Marie Curie Action of the European Union (grant agreement 101030103) and the program Unidad de Excelencia María de Maeztu CEX2020-001058-M, and he thanks Scuola Normale Superiore in Pisa for hospitality during 2020–2021.

REFERENCES

Alfvén, H. (1942). Existence of electromagnetic-hydrodynamic waves. *Nature*, *150*, 405–406.

Arnold, V., & Khesin, B. (2013). *Topological methods in hydrodynamics*. New York: Springer.

Arnold, V. I. (1986). The asymptotic Hopf invariant and its applications. *Selecta Mathematica, New Series*, *5*, 327–345.

Aschwanden, M. J. (2019). Helical twisting number and braiding linkage number of solar coronal loops. *The Astrophysical Journal*, *874*(2), 131.

Batchelor, G. K. (1950). On the spontaneous magnetic field in a conducting liquid in turbulent motion. *Proceedings of the Royal Society of London. Series A, Mathematical and Physical Sciences*, *201*, 405–416.

Bekenstein, J. D. (1987). Helicity conservation laws for fluids and plasmas. *The Astrophysical Journal*, *319*, 207.

Berger, M. A. (1984). Rigorous new limits on magnetic helicity dissipation in the solar corona. *Geophysical and Astrophysical Fluid Dynamics*, *30*, 79–104.

Bhat, P., Subramanian, K., & Brandenburg, A. (2016). A unified large/small-scale dynamo in helical turbulence. *Monthly Notices of the Royal Astronomical Society*, *461*(1), 240–247.

Bîrzan, L., Rafferty, D. A., McNamara, B. R., Wise, M. W., & Nulsen, P. E. J. (2004). A systematic study of radio-induced x-ray cavities in clusters, groups, and galaxies. *The Astrophysical Journal*, *607*(2), 800–809.

Biskamp, D. (2003). *Magnetohydrodynamic turbulence*. Cambridge University Press.

Bonanno, A. (2013). Solar dynamo and toroidal field instabilities. *Solar Physics*, *287*(1–2), 185–196.

Bonanno, A., Brandenburg, A., Del Sordo, F., & Mitra, D. (2012). Breakdown of chiral symmetry during saturation of the Tayler instability. *Physical Review E*, *86*(1), 016313.

Brandenburg, A. (2020). Scientific usage of the pencil code. https://doi.org/10.5281/zenodo.3947506

Brandenburg, A., Candelaresi, S., & Chatterjee, P. (2009). Small-scale magnetic helicity losses from a mean-field dynamo. *Monthly Notices of the Royal Astronomical Society*, *398*, 1414–1422.

Brandenburg, A., & Dobler, W. (2002). Hydromagnetic turbulence in computer simulations. *Computer Physics Communications*, *147*(1–2), 471–475.

Brandenburg, A., & Scannapieco, E. (2020). Magnetic helicity dissipation and production in an ideal MHD code. *The Astrophysical Journal*, *889*(1), 55.

Brandenburg, A., & Subramanian, K. (2005). Astrophysical magnetic fields and nonlinear dynamo theory. *Physics Reports*, *417*, 1–209.

Bregman, J. N. (1980). The galactic fountain of high-velocity clouds. *The Astrophysical Journal*, *236*, 577–591.

Candelaresi, S., & Brandenburg, A. (2011). Decay of helical and nonhelical magnetic knots. *Physical Review E*, *84*(1), 016406.

Candelaresi, S., & Del Sordo, F. (2020). Stabilizing effect of magnetic helicity on magnetic cavities in the intergalactic medium. *The Astrophysical Journal*, *896*(1), 86.

Candelaresi, S., Hornig, G., Podger, B., & Pontin, D. I. (2021). Topological constraints in the reconnection of vortex braids. *Physics of Fluids*, *33*(5), 056101.

Candelaresi, S., Pontin, D. I., & Hornig, G. (2014). Mimetic methods for Lagrangian relaxation of magnetic fields. *SIAM Journal on Scientific Computing*, *36*(6), B952–B968.

Candelaresi, S., Pontin, D. I., & Hornig, G. (2016). Effects of field-line topology on energy propagation in the corona. *The Astrophysical Journal*, *832*, 150.

Canfield, R. C., Hudson, H. S., & McKenzie, D. E. (1999). Sigmoidal morphology and eruptive solar activity. *Geophysical Research Letters*, *26*, 627–630.

Carilli, C. L., Perley, R. A., & Harris, D. E. (1994). Observations of interaction between cluster gas and the radio lobes of Cygnus A. *Monthly Notices of the Royal Astronomical Society*, *270*(1), 173–177.

Carilli, C. L., & Taylor, G. B. (2002). Cluster magnetic fields. *Annual Review of Astronomy and Astrophysics*, *40*(1), 319–348.

Chamandy, L. (2016). An analytical dynamo solution for large-scale magnetic fields of galaxies. *Monthly Notices of the Royal Astronomical Society*, *462*(4), 4402–4415.

Chamandy, L., Shukurov, A., Subramanian, K., & Stoker, K. (2014). Non-linear galactic dynamos: a toolbox. *Monthly Notices of the Royal Astronomical Society*, *443*(3), 1867–1880.

Chandrasekhar, S. (1961). *Hydrodynamic and hydromagnetic stability*. Oxford University Press.

Chatterjee, P., Mitra, D., Brandenburg, A., & Rheinhardt, M. (2011). Spontaneous chiral symmetry breaking by hydromagnetic buoyancy. *Physical Review E*, *84*(2), 025403.

Cheng, K. S., Chernyshov, D. O., Dogiel, V. A., & Ko, C. M. (2015). Multi-wavelength emission from the Fermi bubble. II. Secondary electrons and the hadronic model of the bubble. *The Astrophysical Journal*, *799*(1), 112.

Churazov, E., Brüggen, M., Kaiser, C. R., Böhringer, H., & Forman, W. (2001). Evolution of buoyant bubbles in M87. *The Astrophysical Journal*, *554*(1), 261–273.

Craig, I. J. D., & Sneyd, A. D. (1990). Nonlinear development of the kink instability in coronal flux tubes. *The Astrophysical Journal*, *357*, 653–661.

Del Sordo, F., Candelaresi, S., & Brandenburg, A. (2010). Magnetic-field decay of three interlocked flux rings with zero linking number. *Physical Review E*, *81*(3), 036401.

Del Sordo, F., Guerrero, G., & Brandenburg, A. (2013). Turbulent dynamos with advective magnetic helicity flux. *Monthly Notices of the Royal Astronomical Society*, *429*(2), 1686–1694.

Ebrahimi, Z., & Karami, K. (2016). Resonant absorption of kink magnetohydrodynamic waves by a magnetic twist in coronal loops. *Monthly Notices of the Royal Astronomical Society*, *462*(1), 1002–1011.

Finkelstein, D., & Weil, D. (1978). Magnetohydrodynamic kinks in astrophysics. *International Journal of Theoretical Physics*, *17*(3), 201–217.

Gibson, S. E., Fletcher, L., De Zanna, G., Pike, C. D., Mason, H. E., Mandrini, C. H., et al. (2002). The structure and evolution of a sigmoidal active region. *The Astrophysical Journal*, *574*(2), 1021.

Gupta, M., Thalmann, J. K., & Veronig, A. M. (2021). Magnetic helicity and energy budget around large confined and eruptive solar flares. *Astronomy & Astrophysics*, *653*, A69.

Jardin, S. C., Ferraro, N., & Krebs, I. (2015). Self-organized stationary states of tokamaks. *Physical Review Letters*, *115*, 215001.

Kleeorin, N., Moss, D., Rogachevskii, I., & Sokoloff, D. (2000). Helicity balance and steady-state strength of the dynamo generated galactic magnetic field. *Astronomy & Astrophysics, 361*, L5–L8.

Leka, K. D., Canfield, R. C., McClymont, A. N., & van Driel-Gesztelyi, L. (1996). Evidence for current-carrying emerging flux. *The Astrophysical Journal, 462*, 547.

Linan, L., Pariat, E., Moraitis, K., Valori, G., & Leake, J. (2018). Time variations of the nonpotential and volume-threading magnetic helicities. *The Astrophysical Journal, 865*(1), 52.

McNamara, B., & Nulsen, P. (2007). Heating hot atmospheres with active galactic nuclei. *Annual Review of Astronomy and Astrophysics, 45*, 117–175.

Mei, Z. X., Keppens, R., Roussev, I. I., & Lin, J. (2018). Parametric study on kink instabilities of twisted magnetic flux ropes in the solar atmosphere. *Astronomy & Astrophysics, 609*, A2.

Moffatt, H. K. (1969). The degree of knottedness of tangled vortex lines. *Journal of Fluid Mechanics, 35*, 117–129.

Montmerle, T. (2011). Galactic and extragalactic hot bubbles: Feedback from massive stars. *EAS Publications Series, 51*, 299–310).

Ntormousi, E., Tassis, K., Del Sordo, F., Fragkoudi, F., & Pakmor, R. (2020). A dynamo amplifying the magnetic field of a Milky-Way-like galaxy. *Astronomy & Astrophysics, 641*, A165.

Priest, E. R., & Forbes, T. G. (2000). *Magnetic reconnection: MHD theory and applications*. Cambridge University Press.

Prior, C., & Yeates, A. R. (2016a). Twisted versus braided magnetic flux ropes in coronal geometry. I. Construction and relaxation. *Astronomy & Astrophysics, 587*, A125.

Prior, C., & Yeates, A. R. (2016b). Twisted versus braided magnetic flux ropes in coronal geometry. II. Comparative behaviour. *Astronomy & Astrophysics, 591*, A16.

Rincon, F. (2021). Helical turbulent nonlinear dynamo at large magnetic Reynolds numbers. arXiv:2108.12037.

Rust, D. M., & Kumar, A. (1996). Evidence for helically kinked magnetic flux ropes in solar eruptions. *The Astrophysical Journal Letters, 464*, L199.

Shukurov, A., Sokoloff, D., Subramanian, K., & Brandenburg, A. (2008). Galactic dynamo and helicity losses through fountain flow. *Astronomy & Astrophysics, 448*, L33–L36.

Sur, S., Shukurov, A., & Subramanian, K. (2007). Galactic dynamos supported by magnetic helicity fluxes. *Monthly Notices of the Royal Astronomical Society, 377*, 874–882.

Taylor, G. B., Fabian, A. C., & Allen, S. W. (2002). Magnetic fields in the Centaurus cluster. *Monthly Notices of the Royal Astronomical Society, 334*, 769–776.

Taylor, J. B. (1974). Relaxation of toroidal plasma and generation of reverse magnetic fields. *Physical Review Letters, 33*, 1139–1141.

Taylor, J. B. (1986). Relaxation and magnetic reconnection in plasmas. *Reviews of Modern Physics, 58*, 741–763.

Taylor, J. B. (1993). Filamentation, current profiles, and transport in a tokamak. *Physics of Fluids B, 5*(12), 4378–4383.

Vemareddy, P., Gopalswamy, N., & Ravindra, B. (2017). Prominence eruption initiated by helical kink instability of an embedded flux rope. *The Astrophysical Journal, 850*, 38.

Wang, R., Liu, Y. D., Wiegelmann, T., Cheng, X., Hu, H., & Yang, Z. (2016). Relationship between sunspot rotation and a major solar eruption on 12 July 2012. *Solar Physics*, 1–13.

Wesson, J., & Campbell, D. (1997). *Tokamaks*. Oxford Engineering Science Series. Clarendon Press.

Woltjer, L. (1958). A theorem on force-free magnetic fields. *Proceedings of the National Academy of Sciences of the United States of America, 44*, 489–491.

Yeates, A. R., Hornig, G., & Wilmot-Smith, A. L. (2010). Topological constraints on magnetic relaxation. *Physical Review Letters, 105*(8), 085002.

13

Helicity-Conserving Relaxation in Unstable and Merging Twisted Magnetic Flux Ropes

Philippa K. Browning[1], Mykola Gordovskyy[1], and Alan W. Hood[2]

ABSTRACT

Twisted magnetic flux ropes are reservoirs of free magnetic energy. In a highly conducting plasma such as the solar corona, energy release through multiple magnetic reconnections can be modeled as a helicity-conserving relaxation to a minimum energy state. One possible trigger for this relaxation is the ideal kink instability in a twisted flux rope. We show that this provides a good description for confined solar flares and we develop the approach from idealized cylindrical models to realistic models of coronal loops. Using 3D magnetohydrodynamic simulations combined with test-particle simulations of nonthermal electrons and ions, we predict multiple observational signatures of such flares. We then show how interactions and mergers of flux ropes can release free magnetic energy, using relaxation theory to complement simulations of merging-compression formation in spherical tokamaks and heating avalanches in the solar corona.

13.1. INTRODUCTION

Magnetic flux ropes – twisted bundles of magnetic field lines – are ubiquitous structures in laboratory, space, and astrophysical plasmas. An idealized theoretical flux rope comprises a cylindrically symmetric magnetic field with axial (B_z) and azimuthal (B_θ) components (Priest, 2014). But of course, this is only an approximation to actual structures: for example, in magnetically confined fusion devices, twisted magnetic fields are generally bent into tori, while astrophysical flux ropes are 3D and often embedded within magnetic fields of complex topology.

Flux ropes carry electrical currents and hence are non-potential magnetic fields and reservoirs of free magnetic energy. This feature of flux ropes is the main focus of this chapter. We will show how the release of this stored magnetic energy can lead to heating of the plasma and acceleration of charged particles, with application to solar flares and solar coronal heating, as well as the formation of spherical tokamaks in the laboratory. Flux ropes play a very important role in eruptions in the solar corona (Liu, 2020; Patsourakos et al., 2020). Relaxing to a minimum energy state with observed magnetic helicity provides a powerful framework for predicting energy release and interpreting numerical simulations and observed phenomena with wide relevance.

Magnetic helicity is defined as

$$K = \int_V \mathbf{A} \cdot \mathbf{B} dV, \qquad (13.1)$$

where $\mathbf{A}$ is the vector potential corresponding to the magnetic field $\mathbf{B}$. As discussed in (for example) Berger (1999) and elsewhere in this volume, helicity quantifies the interlinkage of magnetic flux or, perhaps more usefully when considering flux ropes, the overall "twistedness" of the magnetic field. It is well known that K is conserved in ideal magnetohydrodynamics (MHD) (the limit of zero resistivity). The helicity of every closed magnetic flux tube

[1]Department of Physics and Astronomy, University of Manchester, Manchester, UK
[2]School of Mathematics and Statistics, University of St Andrews, St Andrews, UK

Helicities in Geophysics, Astrophysics, and Beyond, Geophysical Monograph 283, First Edition.
Edited by Kirill Kuzanyan, Nobumitsu Yokoi, Manolis K. Georgoulis, and Rodion Stepanov.

DOI:10.1002/9781119841715.ch13

is invariant – related to the fact that K is a topological property conserved when magnetic field lines are frozen to the plasma. Building on earlier work by Woltjer (1958), Taylor (1974) proposed that in a "real" plasma with small but finite resistivity, the only surviving invariant from this infinite set is the total helicity, in which the integral in equation (13.1) is taken over the whole volume. Taylor further hypothesized that in a turbulent plasma, subject to multiple small-scale magnetic reconnections, the plasma will relax to a state of minimum magnetic energy while conserving helicity. The minimum energy or relaxed state can be shown to be a linear or constant-α force-free field

$$\nabla \times \mathbf{B} = \alpha \mathbf{B}, \qquad (13.2)$$

where $\alpha = \mu_0 \mathbf{j} \cdot \mathbf{B}/B^2$, the ratio of the parallel current to the magnetic field, is spatially uniform. This a special case of a force-free field $\mathbf{j} \times \mathbf{B} = 0$, which can also be described by equation (13.2) but with α being generally a function of position, though constant along each field line.

Taylor's hypothesis, as it is generally known, was successfully applied to explain the appearance of the eponymous layer of reversed toroidal magnetic field in reverse-field pinch devices (Taylor, 1974) and to many other laboratory plasmas (Taylor, 1984, 2000). Helicity conservation during a relaxation event on a reversed-field pinch has been experimentally demonstrated by Ji et al. (1995). Relaxation theory has also been successfully applied to spheromaks (Bellan, 2000) and spherical tokamaks, while localized relaxation models can be used to predict the fields following tokamak sawteeth crashes and edge localized modes (Gimblett, 2006; Liang et al., 2010). The theory has also been widely applied to the solar corona and other astrophysical phenomena, as discussed in this chapter. It provides a nice example of the synergies between laboratory and astrophysical plasma physics.

For Taylor's hypothesis to apply, magnetic helicity must be "better" conserved than magnetic energy in the highly conducting plasma environment of the solar corona – noting that in ideal MHD, both quantities are conserved. This will be the case if dissipation is mainly concentrated within small-scale structures (current sheets) associated with magnetic reconnection (Browning et al., 2008). More rigorous limits on the relative magnitudes of helicity and energy dissipation in the solar corona are provided by Berger (1986), and observational evidence for Taylor relaxation in the flaring corona is given in Nandy et al. (2003) and Murray et al. (2013).

One of the major unsolved problems in solar physics is to explain the existence of the hot corona, whose temperature ($\approx 10^6$ K) is significantly higher than the photosphere ($6000K$) (e.g., Klimchuk, 2006, Parnell and De Moortel, 2012; De Moortel and Browning, 2015). Coronal heating is associated with the transfer and subsequent dissipation of magnetic energy from the solar interior into the corona through MHD waves or magnetic reconnection. The latter scenario is the combined effect of many small flare-like energy releases known as *nanoflares* (Parker, 1988). Heyvaerts and Priest (1984) first suggested that Taylor's hypothesis could be applied to this problem. They proposed that slow motions of the photospheric foot points of the coronal magnetic energy cause the coronal magnetic field to evolve through a sequence of force-free fields with free magnetic energy. The stressed field then relaxes, conserving magnetic helicity but releasing free magnetic energy to a constant-α state as in equation (13.2). A series of such relaxation events may heat the corona, but for heating to be effective, there must be significant departures from the relaxed state. This raises the important question of how far free energy can build up before a relaxation event occurs: or, what triggers relaxation. One possible answer is that relaxation is triggered by the onset of an ideal instability, such as the kink instability in a twisted flux rope. This will be discussed in the next section.

13.2. KINK INSTABILITY AND RELAXATION IN THE SOLAR CORONA

A flux rope is unstable to the ideal kink instability if the field line twist (the angle rotated by the field lines from one end of the rope to the other) exceeds a critical value. For a periodic cylinder, this is the well-known Kruskal-Shafranov limit of 2π. For solar coronal loops, the coronal magnetic field is anchored in the dense photosphere, and the resulting line-tying of the foot points is stabilizing. Hood and Priest (1979) found the critical twist thus to be somewhat larger, up to 6π, depending on the current profile. The resistive kink instability becomes unstable at a slightly lower threshold, but this grows slowly in the highly conducting corona. Thus the ideal kink is likely to be far more important as a trigger for energy release. There are many observations of the ideal kink instability in the solar corona apparently leading to both eruptive and non-eruptive events (e.g., Srivastava et al., 2010; Yang et al., 2016). The development of the kink has also been widely studied in laboratory astrophysics experiments (e.g., Bellan, 2018).

13.2.1. Relaxation in Kink-Unstable Flux Ropes

As discussed, helicity-conserving relaxation to a minimum energy state can be useful for predicting the energy release in a solar flare or smaller events (nanoflares) associated with coronal heating. To know how much magnetic energy is available for release, we must first know the magnetic field configuration when relaxation starts. Indeed, for significant free energy to be available, the field must depart significantly from a constant-α state, and relaxation can occur only intermittently. Browning and Van der Linden (2003) suggested that one likely trigger for relaxation is the onset of the ideal kink instability in a twisted magnetic flux rope. This was motivated by laboratory experiments in a spheromak showing that the kink instability plays a crucial role in relaxation (Duck et al., 1997) and also by MHD simulations demonstrating the formation of a helical current sheet in the nonlinear phase of the ideal kink instability (Baty and Heyvaerts, 1996; Lionello et al., 1998; Gerrard et al., 2001). Vorticial motions in the photosphere slowly twist the flux rope until the instability threshold is reached; the flux rope then rapidly develops a strong helical distortion on Alfvenic time scales, and a current sheet forms, allowing magnetic reconnection and energy dissipation. The kink instability has been shown to play a significant role in the onset of solar eruptions (e.g., Török et al., 2004; Williams et al., 2005), but here we focus on energy release within flux ropes that do not erupt.

Browning and Van der Linden (2003) applied this concept to a family of nonlinear force-free field profiles in a cylinder with piecewise constant $\alpha(r)$. Considering finite-length flux ropes with line-tied boundary conditions, the energy release was calculated at the threshold for the onset of the ideal kink instability, assuming that the field relaxes to a constant-α field with the same helicity. This is taken to be a single heating event or nanoflare. The energy release varies over a large range, depending on the specific current profile $\alpha(r)$ at instability onset. Importantly, this model predicts a minimum nanoflare size for given loop parameters: taking a loop of length 20 Mm, radius 1 Mm, and longitudinal magnetic field 30 G gives a minimal or elemental nanoflare of about 10^{18} J, with a possibility of events 10–100 times bigger within the same loop.

This model was extended by Bareford et al. (2010, 2011) to determine the distribution of nanoflare energies arising from random photospheric twisting motions. Considering a family of force-free fields having a piecewise constant-α profile, with $\alpha = \alpha_1$ in the inner part of the loop and $\alpha = \alpha_2$ in the outer part, it was found that the kink instability threshold forms a closed curve in (α_1, α_2) parameter space. Slow photospheric motions cause the field profile to "wander" within this space, at some point reaching the instability threshold and then undergoing helicity-conserving relaxation – with a release of stored energy – to a constant-α state. The whole process repeats many times, generating a distribution of nanoflare energies. For a loop of given dimensions and field strength, this produces a range of nanoflare sizes over more than one order of magnitude; in the case of zero-net current loops, a power law energy distribution is found (Bareford et al., 2011). A further key feature of this model is that it predicts an average shear (ratio of horizontal field to vertical field) of about 0.5, consistent with the required limit to achieve an adequate energy flux for observed coronal heating (Parker, 1988). Any successful theory of coronal heating must account for this necessary build-up of free energy and energy dissipation.

To test the hypothesis of helicity-conserving relaxation and determine the actual mechanisms by which relaxation occurs, Browning et al. (2008) and Hood et al. (2009) performed 3D resistive MHD simulations using the LARE3D code (Arber et al., 2000) for cylindrical loops in an initially kink-unstable equilibrium, respectively with and without net current. These simulations revealed that in the nonlinear phase of the kink, a helical current sheet formed (as seen previously); but this subsequently fragmented into multiple current sheets distributed throughout the loop volume, as shown in Fig. 13.1. The splitting of the current sheet is likely to be a 3D analog of the well-known plasmoid instability. Magnetic reconnection develops within these current sheets, leading to turbulent reconnection in which outflows from each current sheet drive further reconnections. As reconnection occurs at multiple locations, the field lines mix considerably, flattening the α profile and causing the field to evolve toward the constant-α state. Magnetic reconnection dissipates magnetic energy, initially leading to a rise in kinetic energy associated with reconnection outflows but eventually heating the plasma as the field settles down to a new (lower-energy) equilibrium. In the zero-net-current case in which the loop is embedded within the untwisted ambient field, reconnection between the loop and its surroundings also causes the loop to expand radially as it relaxes.

These simulations show that magnetic helicity is much better conserved than energy, as expected, although the helicity dissipation is not entirely negligible due to the relatively coarse numerical resolution. The spatial distribution of $\alpha(\mathbf{r})$ in the final state is not obviously constant, reflecting a lot of small-scale structure remaining in the current; however, the magnetic field is well-approximated by a constant-α state with a value of α, which in most cases is close to that predicted through helicity conservation. Furthermore, the energy release is well predicted by relaxation theory. However, the actual energy released in most cases is somewhat lower than predicted by relaxation theory, presumably as a fully relaxed state is not attained when the simulations are ended. This confirms the validity of Taylor's hypothesis for calculating the energy release, justifying the use of a semianalytical approach, which is particularly powerful for investigating large parameter spaces. For example, Bareford et al. (2010, 2011) use relaxation theory to calculate the energy releases for 10^5 different initial unstable equilibria, a task that would not be feasible with 3D numerical simulations.

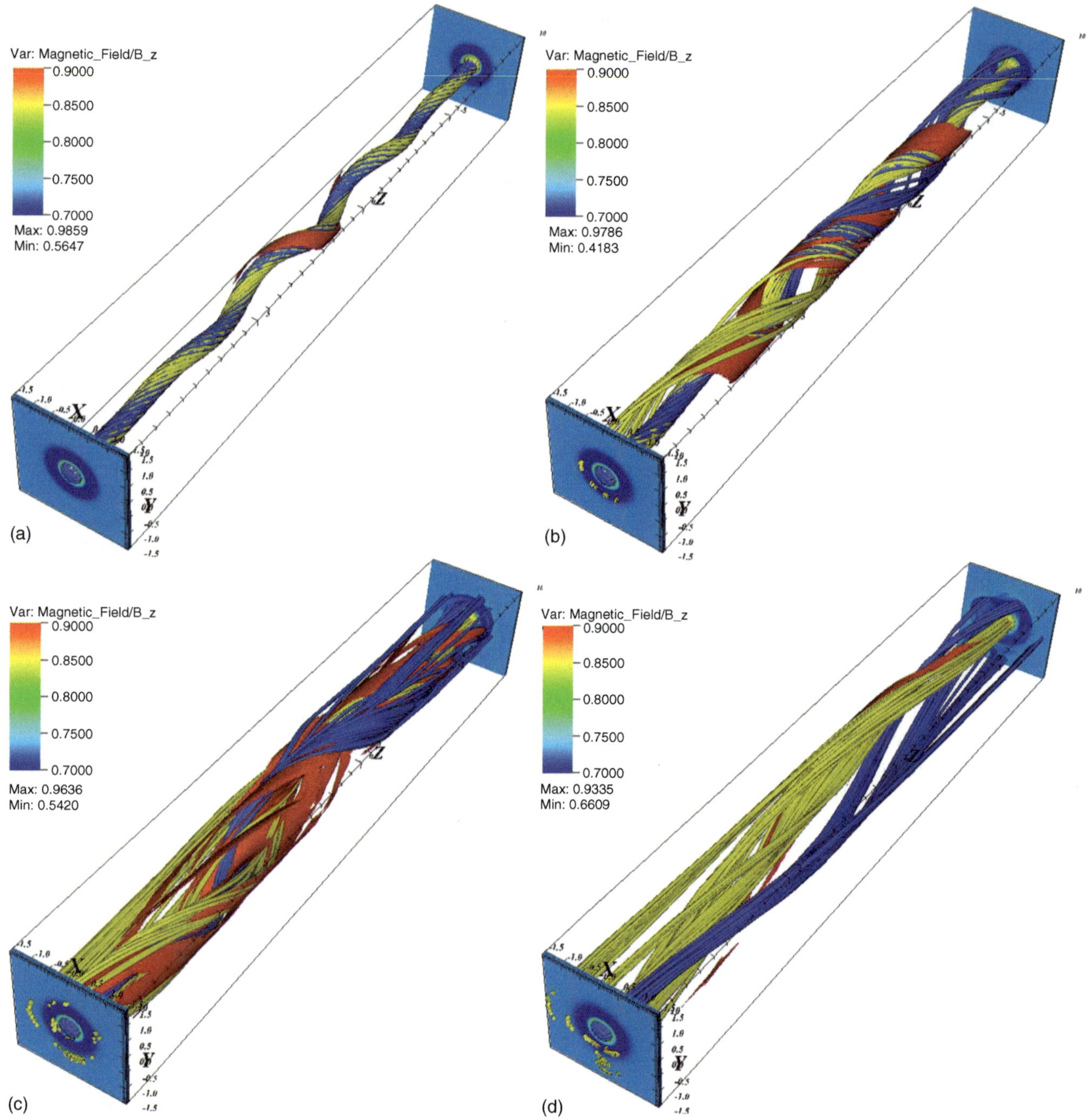

Figure 13.1 Magnetic field evolution in a kink-unstable cylindrical flux rope with zero net current at normalized times (a) $t = 55$, (b) 85, (c) 150, and (d) 295. The field lines are color-coded by their starting locations at one end of the loop, and the dark grey surface denotes a current isosurface. Adapted from Hood et al., 2009.

Simulations with more complex initial states consisting of braided magnetic fields have been shown to relax to a state somewhat different from a Taylor relaxed state, consisting approximately of two very weakly twisted flux ropes with opposite signs of α. This has been interpreted as being due to additional topological constraints not included in Taylor's hypothesis (Yeates, 2015). On the other hand, it should be noted that there is no expectation within the framework of relaxation theory that a minimum energy state should *always* be attained: this should only occur if there is a sufficiently high level of turbulent reconnection to allow most field lines to reconnect. Otherwise, relaxation may not occur at all, or it may cease at some metastable state with energy higher than the minimum. Indeed, the case of two flux ropes with opposite twist is a good example of a state with free magnetic energy but in which the reconnection required to access the minimum energy state (a potential field, in this case) is unlikely to occur since the azimuthal fields at the interface between the flux ropes are in the same direction.

It should also be noted that relaxation theory has to be modified when considering astrophysical plasmas since (unlike laboratory plasmas) there are no conducting walls to define the extent of relaxation. Therefore, we must have *partial relaxation*, in which the field relaxes over a volume of limited extent while the surrounding magnetic field remains undisturbed (Bareford et al., 2013). The extent of the partial relaxation region depends on the turbulent reconnection driving the relaxation. In a kink-unstable loop, this is controlled by the nonlinear amplitude of the unstable mode; typically, the relaxed loop has a radius of around 1.5 times its initial value. We must rely on numerical simulations to determine this parameter; however, the energy release is not highly sensitive to it (Bareford et al., 2013).

13.2.2. Observable Signatures of Energy Release in Flux Ropes

We have considered so far a rather idealized cylindrical geometry, whereas coronal loops have both global curvature and field expansion. Also, it has been assumed that the initial state is unstable to the kink instability, whereas in reality, a loop must be slowly twisted across the kink-instability threshold. We now describe more realistic models, which furthermore can properly determine the plasma heating by incorporating radiation and thermal conduction in the energy equation.

A major question in solar flare physics is to explain the origin of the large numbers of nonthermal electrons and ions, which carry a significant fraction of the released energy. While the electric fields associated with magnetic reconnection can accelerate particles, it is difficult to accelerate sufficiently large numbers of particles within a narrow current sheet, and it has thus been suggested that distributed acceleration throughout a large volume of the flaring region is more feasible (e.g., Cargill et al., 2012). The scenario of a kink-unstable flux rope provides a means to do this due to the fragmented nature of the current sheets.

Using test particles combined with 3D MHD simulations, Gordovskyy and Browning (2011, 2012) showed that 5–10% of protons and electrons could be accelerated out of the thermal distribution, with energies ranging from about 1 keV up to 1 MeV and hard power-law-like spectra with indices from 0.8 to 1.5. The particles gain energy stepwise as they repeatedly encounter current sheets, and the loop quickly fills with energetic particles. Acceleration is predominantly through the parallel electric field in the current sheets; hence, the pitch angle distribution of the energetic particles is mainly parallel.

One important feature of "real" coronal loops compared with cylinders is magnetic convergence – the field lines converge toward the photospheric foot point sources, and the field strength is thus significantly weaker at the looptop than at the foot points. The effects of this on a kink-unstable flux rope have been addressed by Gordovskyy et al. (2013, 2014), as well as the effects on particle energization of Coulomb collisions with the background plasma (Figs. 13.2 and 13.3). There is no useful analytical form for the equilibrium, so the twisted loop is formed numerically by slow rotational motions of the foot points until it becomes kink unstable. At this point, the loop behaves similarly to the earlier cylindrical models: kinking, forming fragmented current sheets, and relaxing to a state of lower magnetic energy. The magnetic convergence somewhat changes the spatial distribution of accelerated electrons, as electrons are more strongly accelerated in the foot-point regions due to the stronger currents associated with field line convergence. Collisions with the background plasma are much more frequent in the dense plasma layers near the foot points (the chromosphere) – and the resultant pitch-angle scattering there enhances the trapping of energetic electrons in the loop since they are mostly scattered out of the loss cone.

The combination of test particles with 3D MHD simulations – the MHDTP approach – provides a powerful basis for forward modeling of both thermal and nonthermal emissions from solar flares (Gordovskyy et al., 2019). To properly model the distribution of energetic particles, the population of test particles must be tracked through the time-evolving electromagnetic fields, interpolating the fields in space and time to the particles' locations (Gordovskyy et al., 2010). Three-dimensional MHD simulations of a confined solar flare in a twisted coronal loop are performed by slowly twisting the foot points of a loop within a dipolar magnetic region embedded in a gravitationally stratified atmosphere until an instability sets in. The subsequent evolution is qualitatively similar to earlier studies of cylindrical loops (Bareford et al., 2016). However, the curved loops usually also undergo an upward "failed eruption" and large-scale reconnection before relaxing to a lower-energy weakly twisted loop as shown in Fig. 13.2 (Bareford et al., 2016, Pinto et al., 2016). The release of stored magnetic energy provides significant plasma heating. However, since the soft X-ray/extreme ultraviolet (EUV) emission from this heated plasma is mainly visible after the reconnection has significantly reduced the field line twist, the initial strong twist of the field lines is not visible in the thermal

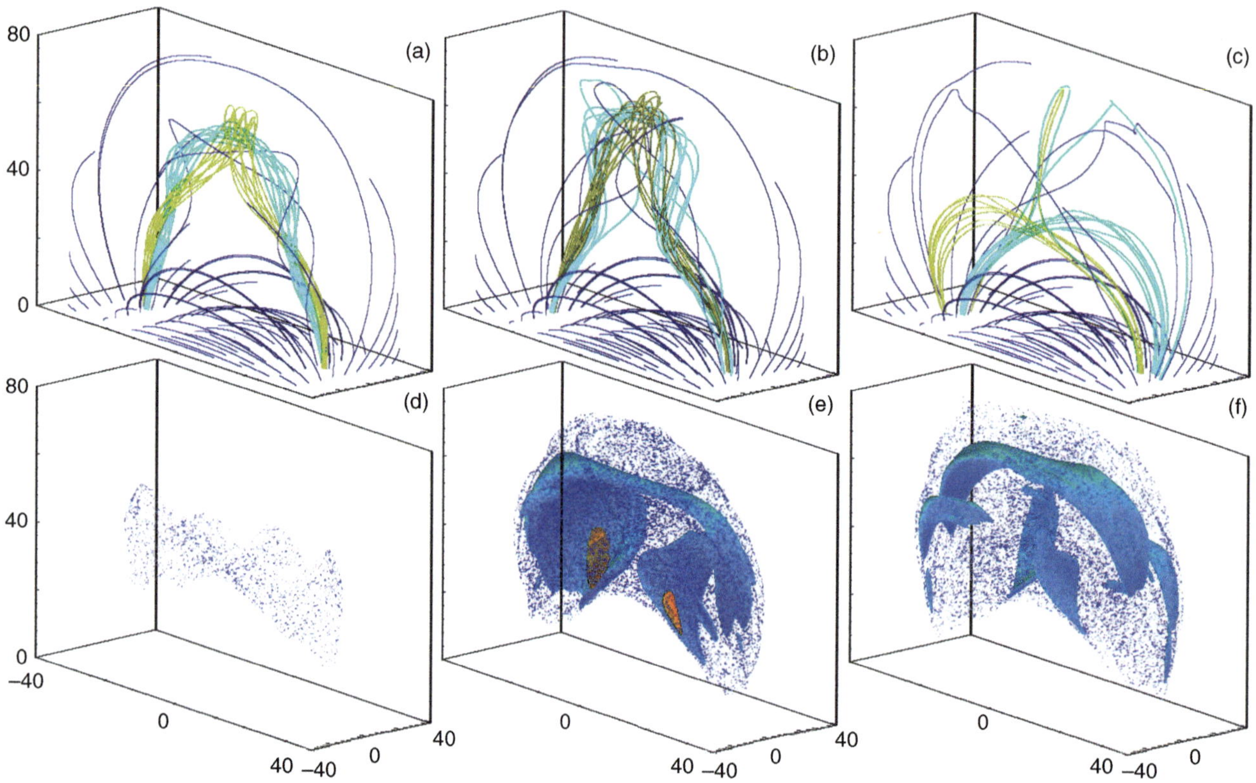

Figure 13.2 Evolution of a reconnecting twisted loop in a gravitationally stratified corona. (a–c) Magnetic field lines; different shades demonstrate the change in connectivity. (d–f) Distribution of the hot plasma, with a dark grey patchy surface corresponding to 2 MK and medium grey and light grey surfaces corresponding to 3 and 4 MK, respectively. (a, d), (b, e), and (c, f) correspond to times of about 15 s, 43 s, and 72 s after the kink, respectively. Axis units are in megameters. Adapted from Gordovskyy et al., 2014; Gordovskyy et al., 2017.

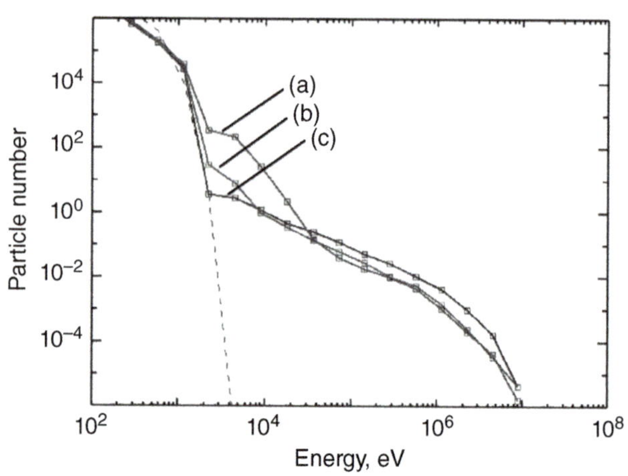

Figure 13.3 Electron energy spectra in the model of a reconnecting twisted loop. Lines (a), (b), and (c) correspond to approximately 15 s, 30 s, and 120 s after the onset of instability. The dashed line denotes the initial Maxwellian distribution. Adapted from Gordovskyy et al., 2014.

emission (Pinto et al. 2015, Pinto et al. 2016). Thus, other observational features must be used to confirm the presence of twisted magnetic fields.

The fragmented current sheets in the kink-unstable flux rope allow effective acceleration of particles throughout the loop volume, leading to strong nonthermal tails in the particle energy distributions (Fig. 13.3). The MHDTP approach in the twisted-loop scenario has allowed forward modeling and analysis of a range of observational signatures in thermal and nonthermal emission, as follows. Hard X-ray spatial locations show foot-point sources growing in time, sometimes looptop or extended loop emissions (Gordovskyy et al., 2014). Predicted soft X-ray and hard X-ray light curves (Pinto et al., 2016) are broadly consistent with observations, showing spiky and transient nonthermal emission, with smoother and more protracted thermal emission. The average line-of-sight velocities and velocity dispersion leading to nonthermal spectral line broadening have been investigated in a relaxing twisted loop (Gordovskyy et al., 2016), showing positive correlations between velocity dispersion and temperature consistent with observations.

Perhaps the most specific signature of energy release within twisted magnetic fields comes from the polarization of microwaves created by gyrosynchrotron emission from mildly relativistic electrons. It has been shown (Gordovskyy et al., 2017) that the characteristic patterns in microwave polarization characteristic of twisted magnetic fields are most likely to be observable in higher microwave frequencies (>30GHz) and for loops seen on the limb; also, the effect is strongest in the initial phases of reconnection and energy release, when the number of nonthermal electrons is greatest.

Studies in more idealized cylindrical geometry have also predicted observational signatures in soft X-rays (Pinto, 2015); Hinode EUV imaging spectrometer (EIS) emission lines (Snow et al., 2017), notably loop expansion and intensity enhancements at the loop edge; and multiple signatures potentially observable with the Daniel K. Inouye Solar Telescope (DKIST) (Snow et al., 2018).

We may summarize the predicted observable signatures of energy release in reconnecting twisted loops and corresponding observational evidence as follows:

- Expanding foot point hard X-ray source. Some observational evidence (e.g., Kontar et al., 2011).
- Weakly twisted filamentary extreme UV and soft X-ray emission, with the apparent twist angle significantly lower than the twist angle expected for the kink instability, 3.5π. Widely observed (e.g., Srivastava et al., 2009) but not necessarily characteristic of this scenario.
- Spiky and transient hard X-ray light curves, with more prolonged soft X-ray emission. A general feature of observed flares but not specifically characteristic of this scenario.
- Correlation of velocity dispersion (nonthermal line-broadening) with temperature and other features in flow patterns. Agrees with observations very well (e.g., Doschek et al., 2008), but this behavior may be typical of plasma heating with turbulent reconnection outflows, not necessarily in twisted fields.
- Microwave polarization patterns. Some observations (e.g., Sharykin et al., 2018), with potential for future observations with high-resolution facilities.
- Actual or failed eruption. (By failed eruption, we mean an initial expansion of the twisted loop after the onset of energy release, followed by the contraction of the loop. Both types of coronal loop eruptions are widely observed (e.g., Leaman et al., 2003, Alexander et al., 2006, Srivastava et al. 2009, Kuridze et al. 2013).)

The foot-point expansion and cross-loop microwave polarization gradient can be considered key observational signatures indicating energy release and particle acceleration in a twisted coronal loop. Note that no single observation could provide conclusive evidence of energy release in twisted loops. Rather, a combination of multiwavelength observations is needed. Many of these signatures are characteristic of energy release in distributed reconnection sites, which can arise in different flaring scenarios and not just in twisted fields; indeed, the methodology developed here could very usefully be applied to configurations beyond the single-twisted-loop scenario. However, energy release within a twisted loop does provide an attractive scenario for confined flares in which a loop transiently brightens but does not erupt, as well as so-called "failed" eruptions, and the modeling described here may provide a guide for the interpretation of observations and suggest observations for new instruments.

13.3. MERGING FLUX ROPES

13.3.1. Flux Rope Merging in Spherical Tokamaks

So far, we have discussed only individual twisted loops. However, fields with multiple twisted flux ropes may contain free magnetic energy, which could be released as the flux ropes merge through magnetic reconnection (Browning et al., 1986). As twisted flux ropes merge into a smaller number of ropes, the typical scale length of the field increases: this may be interpreted as an inverse cascade of energy from small scales to large scales or as a self-organization process. Reconnection and energy release in merging flux ropes are likely to be important in the solar corona and have been modeled (e.g., Linton et al., 2001; Kliem et al., 2014) as well as being observed in many eruptive events (see Liu (2020) and references therein). Furthermore, the merger of flux ropes may play a role in the pre-eruption formation of a large-scale flux rope (Patsourakis et al., 2013, 2020; Kliem et al. 2021). The merger of magnetic flux ropes has also been studied in a number of dedicated laboratory reconnection experiments. However, here we consider reconnection in a spherical tokamak that, due to the high plasma temperature, is in a parameter regime more closely akin to the solar corona than most dedicated experiments. The plasma β (around 10^{-4} - 10^{-2}) is comparable to the solar corona, and the Lundquist number S is also large, up to 10^7 (Browning et al., 2014).

The merger of two flux ropes has been classified as "cohelicity" or "counter-helicity" (Yamada et al., 1990), depending on whether, respectively, the flux-rope helicities are of the same or opposite signs. We focus here mainly on the former case, cohelicity, which is found in the spherical tokamak and seems more likely in the solar corona. In this case, the system initially has a nonzero helicity that constrains relaxation, the final state being a single twisted flux tube. In the counter-helicity case, the initial total helicity is zero, so the field could relax to a potential state, releasing more free magnetic energy.

The spherical tokamak is a fusion device consisting of a toroidal magnetic configuration with a very low aspect ratio: essentially, it is a conventional tokamak with no central hole, instead having a central column carrying the toroidal field current and (often) a solenoid-generating poloidal field. There is considerable interest in developing a start-up scheme not involving the central solenoid, and merging compression, in which two current-carrying plasma tori merge, is an attractive scenario that has been explored in the Mega Ampere Spherical Tokamak (MAST) and other devices (Gryaznevich and Sykes, 2017). Ramping down the current in the poloidal field coils inductively forms two current-carrying flux ropes attracted by the mutual currents, which thus detach from the field coils and move together. The two flux ropes merge into a single twisted flux rope with significant heating, thus forming a hot spherical tokamak plasma (see, e.g., Fig. 1 of Stanier et al. (2014) for a sketch of the field and plasma geometries).

Stanier et al. (2013) performed two-fluid and resistive MHD simulations of the merging-compression process in MAST. The initial state comprises a superposition of two current-carrying flux ropes, which are separately in force-free equilibrium but experience an attractive Lorentz force due to their parallel currents. The flux ropes move together, forming a current sheet and reconnecting, with some oscillations, eventually forming a single flux rope with a spherical-tokamak magnetic field configuration (in low-aspect-ratio simulations). The magnetic reconnection is associated with significant plasma heating, and the temperature profiles in the two-fluid simulations are in good agreement with measured ion and electron temperatures (Browning et al., 2016). Flux-rope merging is also associated with particle acceleration, with observations of high-energy ions and more indirect evidence for electrons, as predicted by modeling the reconnecting fields in MAST (McClements, 2019).

This process can be usefully modeled as a Taylor relaxation to a minimum energy state as in equation (13.2) (where, in the fusion literature, α is denoted by μ). A semi-analytical model has been developed for a plasma contained within a conducting chamber of rectangular cross-section in both infinite-aspect-ratio geometry (Browning et al., 2014) and a tight-aspect-ratio cylindrical configuration (Browning et al., 2016). The initial field is modeled as two adjacent flux ropes of rectangular cross-section; each flux rope is a constant-α field. However, a current sheet (discontinuity in the poloidal field) between the flux ropes means the system has free energy. This is assumed to relax to a final state, which is a constant-α field throughout the region, where the value of α and the peak value of the field are calculated by applying the constraints of helicity and toroidal flux conservation. A typical case is shown in Fig. 13.4. The final state naturally has lower magnetic energy than the initial state, and the loss in magnetic energy is converted into plasma thermal energy. Taking parameters representative of MAST, with a plasma current of 300 kA in each initial flux rope, the average temperature rise is estimated to be 150 eV, consistent with observations. The free magnetic energy – and hence the final temperature – are shown to have approximately quadratic scaling with plasma current.

The relaxation model is potentially useful for making quick predictions of the field configurations and heating across a wide parameter space. It can also be easily adapted to model merging of coronal loops, where different initial configurations, such as multiple loops or loops with opposite twist, could also arise (Browning et al., 2016). For two flux ropes, the energy release is found to be substantially greater if the flux ropes initially have opposite currents (twisted in reverse senses) than if the currents are in the same direction. This is because the former configuration has zero helicity; hence, the relaxed state has lower energy. However, oppositely twisted flux ropes are much less likely to relax and release the free energy since neither the toroidal nor poloidal magnetic field components reverse at the interface between the flux ropes, and there is no obvious way – apart perhaps from a strong external perturbation – in which magnetic reconnection could be initiated. If an initial larger array of flux ropes is considered, the energy release scales almost linearly with the number of flux ropes since the free magnetic energy is associated with the current sheets between flux ropes. This relaxation from many flux ropes to a single rope, releasing magnetic energy, exhibits the natural tendency for an inverse cascade from small to large length scales.

13.3.2. Heating Avalanches in the Solar Corona

It has long been considered that flux-rope interactions and mergers in the solar corona may release free magnetic energy, leading to a large-scale flare or smaller events associated with solar coronal heating. Building on the modeling of individual kink-unstable flux ropes described previously, Tam et al. (2015) used 3D MHD simulations to investigate the dynamics of a pair of adjacent flux ropes representing separately twisted threads within a coronal loop. It was discovered that if an unstable thread is located close to a stable thread, the instability in the

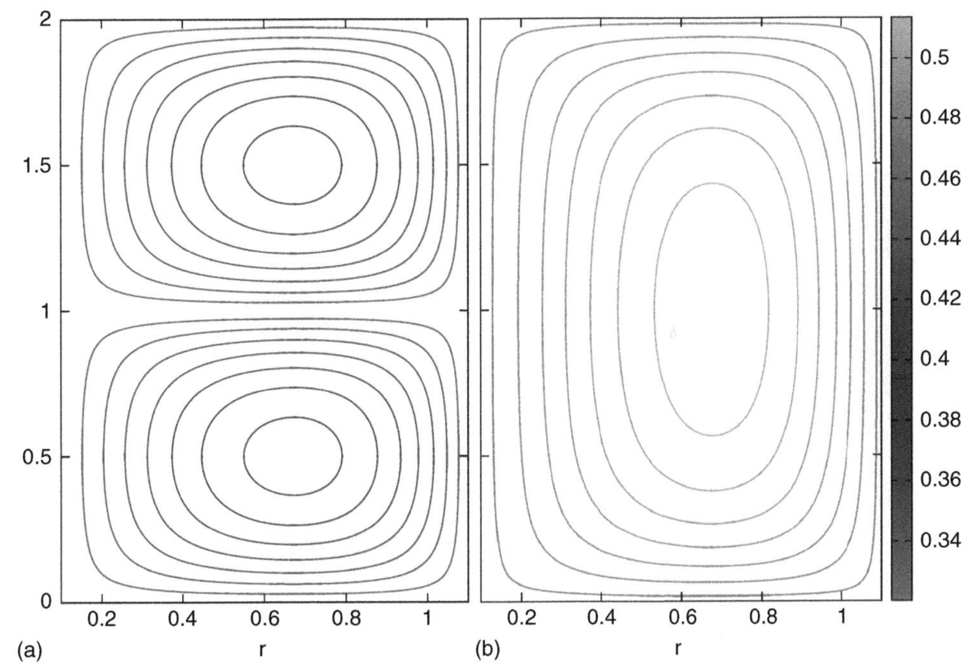

Figure 13.4 Poloidal field lines for the relaxation model of merging flux ropes in a tight-aspect-ratio rectangular container. (a) Each initial flux rope has $\alpha H = 1.25$. (b) The final relaxed state has $\alpha H = 0.849$, where $2H$ is the container height. Browning et al., 2016/IOP Publishing.

former could trigger energy release in both threads. The threads interact through magnetic reconnection, and the final state is a single weakly twisted flux rope. Thus, free magnetic energy from a twisted flux rope can be released even if it is kink-stable. An alternative scenario for interacting flux ropes is considered by Ripperda et al. (2017a, 2017b), with oppositely directed (and hence repelling) current channels in which the tilt instability causes reconnection in its nonlinear phase. Similarly to Gordovskyy et al. (2010) and subsequent work described earlier, the relativistic guiding-center equations are used to model the evolution of nonthermal ions and electrons. Strong particle acceleration is shown to take place, forming particle energy distributions with significant peaks at high energies ("bump on tail" distributions).

These studies lead to the question of whether a single unstable twisted thread could trigger energy release from multiple stable threads: Hood et al. (2016) demonstrated that this could indeed be the case. They used 3D MHD simulations to investigate the evolution of an array of 22 stable threads with only 1 unstable thread, showing that an "avalanche" of heating ensued, with the initially unstable flux rope interacting and merging successively with most of its stable neighbors (see Fig. 13.5). This was the first demonstration using an MHD model of an avalanche process, often postulated in cellular automaton models of coronal heating. It has major implications for solar coronal heating since localized instability can trigger heating in a much larger volume.

This scenario of merging flux ropes can be usefully described as a helicity-conserving relaxation process (Hussain et al., 2017). Two (or more) twisted flux ropes may relax to a minimum energy state of a single constant-α twisted flux rope with the same total helicity as the initial configuration but lower energy. This semi-analytical model was benchmarked against the MHD simulations of Tam et al. (2015), giving good agreement in the values of released energy. The energy release does depend on the radius of the relaxed flux rope, as relaxation will only extend over a limited volume, with the surrounding potential magnetic field remaining undisturbed (Bareford et al., 2013); a simple conservation-of-volume approach gives good agreement with numerical results. The relaxation model can be applied to larger collections of flux ropes, including the specific system modeled by Hood et al. (2016); the relaxation model predicts that the magnetic energy decreases in approximate steps as additional flux ropes are "consumed" in the avalanche. While 3D numerical simulations are computationally expensive, the relaxation approach allows the exploration of wide parameter spaces. It opens up the possibility of developing a cellular-automaton-type model with rules for interactions and energy release developed from relaxation theory.

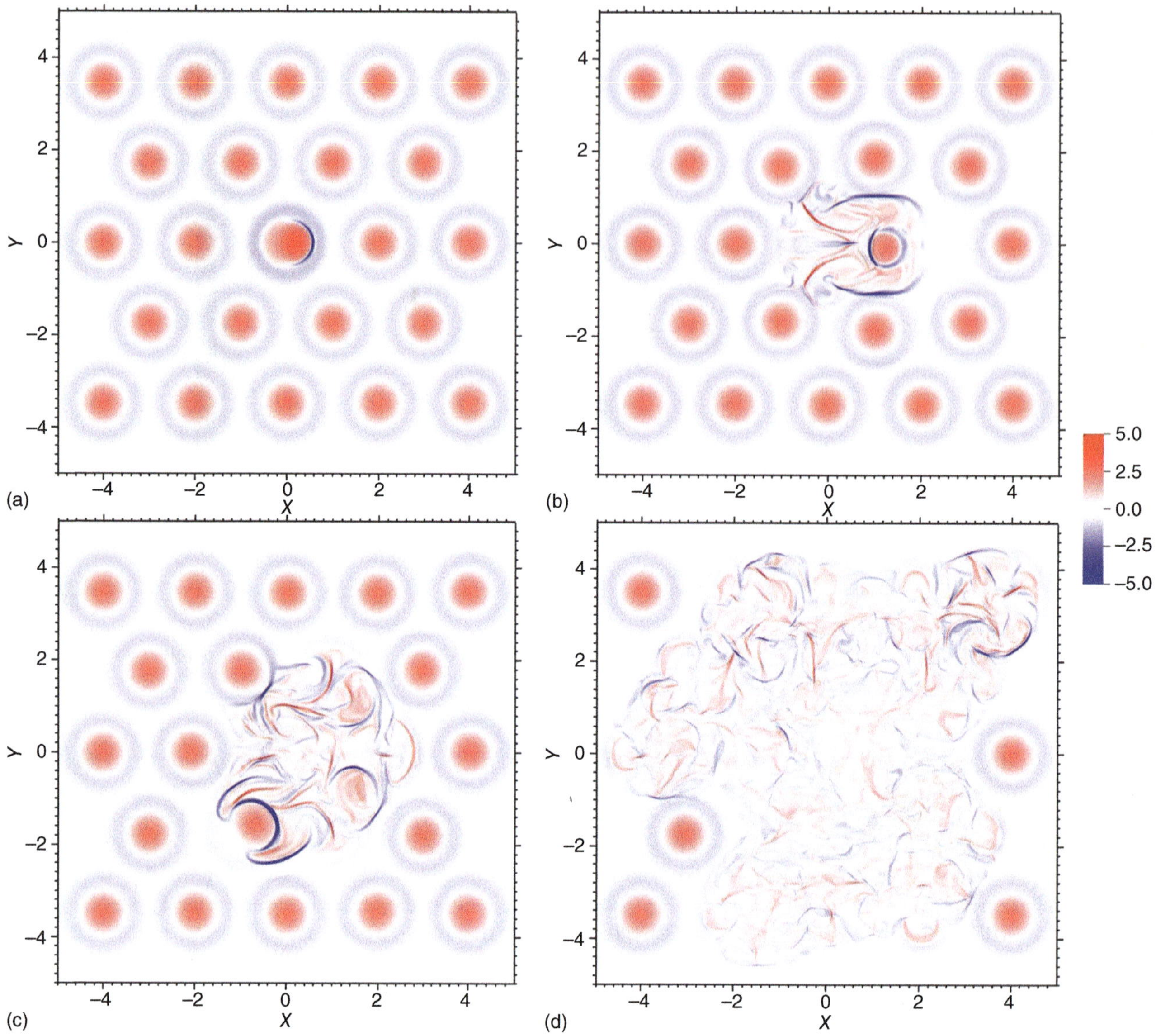

Figure 13.5 Contours of axial current in the midplane of an array of 23 parallel twisted threads, with one initially kink-unstable central thread and all other threads initially stable, at normalized times (a) $t = 75$, (b) 165, (c) 290, and (d) 800. Hood et al., 2016/IOP Publishing.

However, relaxation theory cannot predict whether two twisted threads will or will not interact and relax: the conditions under which this will occur must be determined through MHD simulations. Preliminary work by Hussain (2018) suggests that relaxation is unlikely to occur between oppositely twisted threads – or if it does so, it will occur more slowly and requires stronger triggering. This is because, as discussed, there is no reversal of the poloidal field component at the interfaces between threads in this case, even though the relaxed state is closer to a potential field; thus the free energy is greater than for like-twisted threads. Furthermore, relaxation takes place if threads are sufficiently close, with a separation of less than about 0.25–0.5 radii required for interaction.

The simulations of Tam et al. (2015) and Hood et al. (2016) consider an initial setup with one thread in a kink-unstable state. This was extended by Reid et al. (2018) to show that similar behavior can arise in a field that – more realistically – starts in a stable, untwisted state and is continuously driven by foot-point motions until it becomes unstable. As the driving is maintained, a series of bursts of heating ensue. Indeed, we expect the corona to be generally in a state of "driven relaxation" in which energy input from foot-point motions statistically balances energy dissipation through relaxation. Driven relaxation associated with external helicity injection is also integral to spheromaks (Bellan, 2000) and spherical tokamaks such as the National Spherical Torus Experiment (NSTX), where the current drive is through coaxial helicity injection.

13.4. DISCUSSION AND CONCLUSIONS

Twisted flux ropes store free magnetic energy, which may be released either in individual flux ropes in the nonlinear phase of the ideal kink instability or through the interaction of multiple twisted flux ropes. In the former case, the field relaxes to a state of minimum magnetic energy through internal magnetic reconnections and reconnection with ambient untwisted field lines. In the latter case, multiple flux ropes reduce their energy by reconnecting into a single flux rope. The theory of helicity-conserving relaxation to a minimum energy state – a constant-α force-free field (Taylor, 1974) – provides a powerful tool for modeling such processes and a framework for interpreting observations and simulations. A key strength of this approach is that the final equilibrium state and the energy released during reconnection can often be calculated analytically, allowing exploration of very large parameter spaces unfeasible with numerical simulations.

Numerical MHD simulations of unstable or merging flux ropes provide considerable insight into the process of relaxation and the underlying magnetic reconnection. Furthermore, they allow predictions of observable signatures of energy release, including both thermal and nonthermal emissions (if a test-particle model for nonthermal particles is integrated with the MHD model). Energy release in twisted coronal loops provides a viable scenario for confined solar flares, with the release of stored magnetic energy leading to both plasma heating and particle acceleration. Consideration of the characteristics of a range of observables in both thermal and nonthermal emission is required in order to find clear evidence for energy release in twisted magnetic fields in flares.

A comparison of 3D MHD simulations with the predictions of relaxation theory shows that helicity-conserving relaxation is a useful model, especially for calculating magnetic energy changes. The energy of the final state of the magnetic field may be well approximated by the energy of the constant-α state, even if α is not obviously close to being constant. The reasons are twofold. First, since energy is an integral of the magnetic field, which is itself an integral of the current, it is insensitive to fluctuations in the current. Second, because the relaxed state is an energy minimum, the energy deviation from it depends quadratically on deviations in the current profile. Numerical simulations also provide insight into the mechanisms by which relaxation occurs, whereby multiple reconnections create many interchanges of magnetic field lines and hence mix the α profile.

However, relaxation theory must be adapted when applied to the solar corona and other astrophysical contexts. Relaxation can be only partial, and the extent of the relaxed region has to be postulated using knowledge gained from numerical simulations or other means. For example, a zero-net-current flux rope will undergo some reconnection with the ambient field, and hence the relaxed flux rope is somewhat greater in size than the initial flux rope – but the ambient field at larger distances does not participate in the relaxation, so the relaxation is highly localized. Furthermore, the coronal field is constantly subject to disturbances by photospheric foot-point motions. Thus there is an interplay between relaxation and driving, and a fully relaxed state can never be attained. Similar considerations apply in some laboratory plasmas.

We have shown that helicity-conserving relaxation can be applied to both solar coronal loops and spherical tokamaks and that concepts and methods can be fruitfully shared between these fields. Many other applications have been considered, some also emphasizing the role of the kink instability: for example, Bromberg et al. (2019) consider the kink instability in relativistic astrophysical jets and find that the relaxed state can be described as a Taylor state.

ACKNOWLEDGMENTS

PKB and MG acknowledge financial support from the UK Science and Technology Facilities Council (grant ST/T00035X/1). AWH also acknowledges STFC (grant ST/S000402/1).

REFERENCES

Alexander, D., Liu, R., & Gilbert, H.R. (2006). What is the role of the kink instability in solar coronal eruptions? *Astrophysical Journal Letters*, *569*, L255–L258, https://doi.org/10.1086/379530

Arber, T.D., Longbottom, A.W., Gerrard, C.L., & Milne, A.M. (2001). A staggered grid, Lagrangian-Eulerian Remap code for 3D–MHD simulations. *Journal of Computational Physics*, *171*, 151, https://doi.org/10.1006/jcph.2001.6780

Bareford, M.R., Browning, P.K., & Van der Linden, R.A.M. (2010). A nanoflare distribution generated by repeated relaxations triggered by kink instability. *Astronomy & Astrophysics*, *521*, A70, https://doi.org/10.1051/0004-6361/201014067

Bareford, M.R., Browning, P.K., & Van der Linden, R.A.M. (2011). The flare-energy distributions generated by kink-unstable ensembles of zero-net-current coronal loops. *Solar Physics*, *273*, 93–115, https://doi.org/10.1007/s11207-011-9832-4

Bareford, M.R., Hood, A.W., & Browning, P.K. (2013). Coronal heating by the partial relaxation of twisted loops. *Astronomy & Astrophysics*, *550*, A40, https://doi.org/10.1051/0004-6361/201219725

Bareford, M.R., Gordovskyy, M., Browning, P.K., & Hood, A.W. (2016). Energy release in driven twisted coronal loops. *Solar Physics*, *291*, 187–209, https://doi.org/10.1007/s11207-015-0824-7

Baty, H., & Heyvaerts, J. (1996). Electric current concentration and kink instability in line-tied coronal loops. *Astronomy & Astrophysics*, *308*, 935–950.

Bellan, P.M. (2000). *Spheromaks: a practical application of magnetohydrodynamic dynamos and plasma self-organization*. Imperial College Press.

Bellan, P.M. (2018). Experiments relevant to astrophysical jets. *Journal of Plasma Physics*, *84*, 755840501, https://doi.org/10.1017/S002237781800079X

Berger, M.A. (1984). Rigorous new limits on magnetic helicity dissipation in the solar corona. *Geophysical and Astrophysical Fluid Dynamics*, *30*, 79–104, https://doi.org/10.1080/03091928408210078

Berger, M.A. (1999). An Introduction to magnetic helicity. *Plasma Physics and Controlled Fusion*, *41*, 167B, https://doi.org/10.1088/0741-3335/41/12B/312

Bromberg, O., Singh, C.B., Davelaar, J., & Philippov, A.A. (2019). Kink instability: evolution and energy dissipation in relativistic force-free nonrotating jets. *Astrophysical Journal*, *884*, 39, https://doi.org/10.3847/1538-4357/ab3fa5

Browning, P.K., Sakurai, T., & Priest, E.R. (1986). Coronal heating in closely-packed flux tubes: a Taylor-Heyvaerts relaxation theory. *Astronomy & Astrophysics*, *158*, 217–227.

Browning, P.K., & Van der Linden, R.A.M. (2003). Solar coronal heating by relaxation events. *Astronomy & Astrophysics*, *400*, 355–367, https://doi.org/10.1051/0004-6361:20021887

Browning, P.K., Stanier, A., Ashworth, G., McClements, K.G., & Lukin, V.S. (2014). Self-organization during spherical torus formation by flux rope merging in the Mega Ampere Spherical Tokamak. *Plasma Physics and Controlled Fusion*, *56*, 064009, https://doi.org/10.1088/0741-3335/56/6/064009

Browning, P.K., Cardnell, S., Evans, M., Arese Lucini, F., Lukin, V.S., McClements, K.G., et al. (2016). Two-fluid and magnetohydrodynamic modelling of magnetic reconnection in the MAST spherical tokamak and the solar corona. *Plasma Physics and Controlled Fusion*, *58*, 014041, https://doi.org/10.1088/0741-3335/58/1/014041

Cargill, P.J., Vlahos, L., Baumann, G., Drake, J.F., & Nordlund, A. (2012). Current fragmentation and particle acceleration in solar flares. *Space Science Reviews*, *173*, 223, https://doi.org/10.1007/s11214-012-9888-y

Doschek, G.A., Warren, H.P., Mariska, J.T., Muglach, K., Culhane, J.L., Hara, H., & Watanabe, T. (2008). Flows and nonthermal velocities in solar active regions observed with the EUV imaging spectrometer on Hinode: a tracer of active region sources of heliospheric magnetic fields. *Astrophysical Journal*, *686*, 1362–1371, https://doi.org/10.1086/591724

Duck, R.C., Browning, P.K., Cunningham, G., Gee, S., al-Kharkhy, A., Martin, R., & Rusbridge, M.G. (1997). Structure of the n = 1 mode responsible for relaxation and current drive during sustainment of the SPHEX spheromak. *Plasma Physics and Controlled Fusion*, *39*, 715–736, https://doi.org/10.1088/0741-3335/39/5/004

Gerrard, C.L., Arber, T.D., Hood, A.W., & Van der Linden, R.A.M. (2001). Numerical simulations of kink instability in line-tied coronal loops. *Astronomy & Astrophysics*, *373*, 1089–1098, https://doi.org/10.1051/0004-6361:20010678

Gimblett, C.G., Hastie, R.J., & Helander, P. (2006). Model for current-driven edge-localized modes. *Physical Review Letters*, *96*, 035006, https://doi.org/10.1103/PhysRevLett.96.035006

Gordovskyy, M., Browning, P.K., & Vekstein, G.E. (2010). Particle acceleration in fragmenting periodic reconnecting current sheets in solar flares. *Astrophysical Journal*, *720*, 1603–1611, https://doi.org/10.1088/0004-637X/720/2/1603

Gordovskyy, M., & Browning, P.K. (2011). Particle acceleration by magnetic reconnection in a twisted coronal loop. *Astrophysical Journal*, *729*, 101, https://doi.org/10.1088/0004-637X/729/2/101

Gordovskyy, M., & Browning, P.K. (2012). Magnetic relaxation and particle acceleration in a flaring twisted coronal loop. *Solar Physics*, *277*, 299–316, https://doi.org/10.1007/s11207-011-9900-9

Gordovskyy, M., Kontar, E.P., & Browning, P.K. (2016). Plasma motions and non-thermal line broadening in flaring twisted coronal loops. *Astronomy & Astrophysics*, *589*, A104, https://doi.org/10.1051/0004-6361/201527249

Gordovskyy, M., Browning, P.K., & Kontar, E. (2017). Polarisation of microwave emission from reconnecting twisted coronal loops. *Astronomy & Astrophysics*, *604*, A116, https://doi.org/10.1051/0004-6361/201629334

Gordovskyy, M., Browning, P.K., & Pinto, R.F. (2019). Combining MHD and kinetic modelling of solar flares. *Advances in Space Research*, *63*, 1453–1465, https://doi.org/10.1016/j.asr.2018.09.024

Gryaznevich, M.W., & Sykes, A. (2017). Merging-compression formation of high temperature spherical tokamak plasma. *Nuclear Fusion*, *57*, 072003, https://doi.org/10.1088/1741-4326/aa4ffd

Heyvaerts, J., & Priest, E.R. (1984). Coronal heating by reconnection in DC current systems - A theory based on Taylor's hypothesis. *Astronomy & Astrophysics*, *137*, 63.

Hood, A.W., & Priest, E.R. (1979). Kink instability of solar coronal loops as the cause of solar flares. *Solar Physics*, *64*, 303–321, https://doi.org/10.1007/BF00151441

Hood, A.W., Browning, P.K., Cargill, P.J., & Tam, K.V. (2016). An MHD avalanche in a multi-threaded coronal loop. *Astrophysical Journal*, *817*, 5, https://doi.org/10.3847/0004-637X/817/1/5

Hussain, A.S., Browning, P.K., & Hood, A.W. (2017). A relaxation model of coronal heating in multiple interacting flux ropes. *Astronomy & Astrophysics*, *600*, A5, https://doi.org/10.1051/0004-6361/201629589

Hussain, A.S. (2018). Synergistic effects of neutrons and plasma on materials in fusion reactors and relaxation of merging magnetic flux ropes in fusion and solar plasmas (PhD thesis). University of Manchester.

Ji, H., Prager, S., & Sarff, J. (1995). Conservation of magnetic helicity during plasma relaxation. *Physical Review Letters*, *74*, 2945–2948, https://doi.org/10.1103/PhysRevLett.74.2945

Kliem, B., Török, T., Titov, V.S., Lionello, R., Linker, J.A., Liu, R., et al. (2014). Slow rise and partial eruption of a double-decker filament. II. A double flux rope model. *Astrophysical Journal*, *792*, 107, https://doi.org/10.1088/0004-637X/792/2/107

Kliem, B., Lee, J., Liu, R., White, S.M., Liu, C., & Masuda, S. (2021). Nonequilibrium flux rope formation by confined flares preceding a solar coronal mass ejection. *Astrophysical Journal*, *909*, 91, https://doi.org/10.3847/1538-4357/abda37

Klimchuk, J. (2006). On solving the coronal heating problem. *Solar Physics*, *234*, 41–77, https://doi.org/10.1007/s11207-006-0055-z

Kontar, E.P., Hannah, I.G., & Bian, N.H. (2011). Acceleration, magnetic fluctuations, and cross-field transport of energetic electrons in a solar flare loop. *Astrophysical Journal Letters*, *730*, L22, https://doi.org/10.1088/2041-8205/730/2/L22

Kuridze, D., Mathioudakis, M., Kowalski, A.F., Keys, P.H., Jess, D.B., Balasubramaniam, K.S., & Keenan, F.P. (2013). Failed filament eruption inside a coronal mass ejection in active region 11121. *Astronomy & Astrophysics*, *552*, A55, https://doi.org/10.1051/0004-6361/201220055

Leamon, R.J., Canfield, R.C., Blehm, Z., & Pevtsov, A.A. (2003). What is the role of the kink instability in solar coronal eruptions? *Astrophysical Journal Letters*, *495*, L255–L258.

Liang, Y., Gimblett, C.G., Browning, P.K., Devoy, P., Koslowski, H.R., Jachmich, S., et al. (2010). Multiresonance effect in type-I edge-localized mode control with low n fields on JET. *Physical Review Letters*, *105*, 065001, https://doi.org/10.1103/PhysRevLett.105.065001

Linton, M.G., Dahlburg, R.B., & Antiochos, S.K. (2001). Reconnection of twisted flux tubes as a function of contact angle. *Astrophysical Journal*, *553*, 905–921, https://doi.org/10.1086/320974

Lionello, R., Velli, M., Einaudi, G., & Mikic, Z. (1998). Nonlinear magnetohydrodynamic evolution of line-tied coronal loops. *Astrophysical Journal*, *494*, 840–850, https://doi.org/10.1086/305221

Liu, R. (2020). Magnetic flux ropes in the solar corona; structure and evolution towards eruption. *Research in Astronomy and Astrophysics*, *20*, 165, https://doi.org/10.1088/1674-4527/20/10/165

De Moortel, I., & Browning, P.K. (2015). Recent advances in coronal heating. *Philosophical Transactions of the Royal Society A*, *373*, 20140269–20140269, https://doi.org/10.1098/rsta.2014.0269

McClements, K.G. (2019). Reconnection and fast particle production in tokamak and solar plasmas. *Advances in Space Research*, *63*, https://doi.org/10.1016/j.asr.2018.08.045

Murray, S.A., Bloomfield, D.S., & Gallagher, P.T. (2013). Evidence for partial Taylor relaxation from changes in magnetic geometry and energy during a solar flare. *Astronomy & Astrophysics*, *550*, A119, https://doi.org/10.1051/0004-6361/201219964

Nandy, D., Hahn, M., Canfield, R.C., & Longcope, D.W. (2003). Detection of a Taylor-like plasma relaxation process in the Sun. *Astrophysical Journal Letters*, *597*, L73–L76, https://doi.org/10.1086/379815

Parker, E.N. (1988). Nanoflares and the solar X-ray corona. *Astrophysical Journal*, *330*, 474–479, https://doi.org/10.1086/166485

Parnell, C.E., & De Moortel, I. (2012). A contemporary review of coronal heating. *Philosophical Transactions of the Royal Society A*, *370*, 3217–3240, https://doi.org/10.1098/rsta.2012.0031

Patsourakos, S., Vourlidas, A., & Stenborg, G. (2013). Direct evidence for a fast coronal mass ejection driven by the prior formation and subsequent destabilization of a magnetic flux rope. *Astrophysical Journal*, *764*, 125, https://doi.org/10.1088/0004-637X/764/2/125

Patsourakos, S., Vourlidas, A., Török, T., Kliem, B., et al. (2020). Decoding the pre-eruptive magnetic field configurations of coronal mass ejections. *Space Science Reviews*, *216*, 131, https://doi.org/10.1007/s11214-020-00757-9

Pinto, R.F., Vilmer, N., & Brun, A.S. (2015). Soft X-ray emission in kink-unstable coronal loops. *Astronomy & Astrophysics*, *576*, A37, https://doi.org/10.1051/0004-6361/201323358

Pinto, R.F., Gordovskyy, M., Browning, P.K., & Vilmer, N. (2016). Thermal and non-thermal emission from reconnecting twisted coronal loops. *Astronomy & Astrophysics*, *585*, A159, https://doi.org/10.1051/0004-6361/201526633

Priest, E.R. (2014). *Magnetohydrodynamics of the Sun*. Cambridge University Press.

Reid, J., Hood, A.W., Parnell, C.E., Browning, P.K., & Cargill, P.J. (2018). Coronal energy release by MHD avalanches: continuous driving. *Astronomy & Astrophysics*, *615*, A84, https://doi.org/10.1051/0004-6361/201732399.

Ripperda, B., Porth, O., Xia, C., & Keppens, R. (2017a). Reconnection and particle acceleration in interacting flux ropes – I. Magnetohydrodynamics and test particles in 2.5D. *Monthly Notices of the Royal Astronomical Society*, *467*, 3279–3298, https://doi.org/10.1093/mnras/stx379

Ripperda, B., Porth, O., Xia, C., & Keppens, R. (2017b). Reconnection and particle acceleration in interacting flux ropes – II 3D effects on test particles in magnetically dominated plasmas. *Monthly Notices of the Royal Astronomical Society*, *467*, 3465–3482, https://doi.org/10.1093/mnras/stx1875

Sharykin, I.N., Kuznetsov, A.A., & Myshyakov, I.I. (2018). Probing twisted magnetic field using microwave observations in an M class solar flare on 11 February, 2014. *Solar Physics*, *293*, 34, https://doi.org/10.1007/s11207-017-1237-6

Snow, B., Botha, G.J.J., Regnier, S., Morton, R.J., Verwichte, E., & Young, P.R. (2017). Observational signatures of a kink-unstable coronal flux rope using Hinode/EIS. *Astrophysical Journal*, *842*, 16, https://doi.org/10.3847/1538-4357/aa6d0e

Snow, B., Botha, G.J.J., Scullion, E., McLaughlin, J.A., Young, P.R., & Jaeggli, S.A. (2018). Predictions of DKIST/DL-NIRSP observations for an off-limb kink-unstable coronal loop. *Astrophysical Journal*, *863*, 172, https://doi.org/10.3847/1538-4357/aad3bc

Srivastava, A.K., Zaqarashvili, T.V, Kumar, P., & Khodachenko, M.L. (2010). Observation of kink instability during small B5.0 solar flare on 2007 June 4. *Astrophysical Journal*, *795*, 292, https://doi.org/10.1088/0004-637X/715/1/292

Stanier, A., Browning, P.K., McClements, K., Gordovskyy, M., Gryaznevich, M., & Lukin, V. (2013). Two fluid simulations of driven reconnection in the mega ampere spherical tokamak. *Physics of Plasmas*, *20*, 122302, https://doi.org/10.1063/1.4830104

Tam, K.V., Hood, A.W., Browning, P.K., & Cargill, P.J. (2015). Coronal heating in multiple magnetic threads. *Astronomy & Astrophysics*, *580*, A122, https://doi.org/10.1051/0004-6361/201525995

Tanabe, H., Yamada, T., Watanabe, T., et al. (2017). Recent progress on magnetic reconnection research in the MAST spherical tokamak. *Physics of Plasmas*, *24*, 056108, https://doi.org/10.1063/1.4977922

Taylor, J.B. (1974). Relaxation of toroidal plasma and generation of reverse magnetic field. *Physical Review Letters*, *30*, 1139–1141, https://doi.org/10.1103/PhysRevLett.33.1139

Taylor, J.B. (1984). Relaxation and magnetic reconnection in plasmas. *Reviews of Modern Physics*, *58*, 741–763, https://doi.org/10.1103/RevModPhys.58.741

Taylor, J.B. (2000). Relaxation revisited. *Physics of Plasmas*, *7*, 1623, https://doi.org/10.1063/1.873984

Threlfall, J., Hood, A.W., & Browning, P.K. (2018). Flare particle acceleration in the interaction of twisted coronal flux ropes. *Astronomy & Astrophysics*, *611*, A40, https://doi.org/10.1051/0004-6361/201731915

Török, T., Kliem, B., & Titov, V.S. (2004). Ideal kink instability of a magnetic loop equilibrium. *Astronomy & Astrophysics*, *413*, L27–L30, https://doi.org/10.1051/0004-6361:20031691

Williams, D.R., Török, T., Démoulin, P., van Driel-Gesztelyi, L., & Kliem, B. (2005). Eruption of a kink-unstable filament in NOAA active region 10696. *Astrophysical Journal Letters*, *628*, L163–L166, https://doi.org/10.1086/432910

Woltjer, L. (1958). A theorem on force-free magnetic fields. *Proceedings of the National Academy of Sciences*, *44*, 489–491, https://doi.org/10.1073/pnas.44.6.489

Yamada, M., Ono, Y., Hayakama, A., Katsurai, M., & Perkins, F.W. (1990). Magnetic reconnection of plasma toroids with cohelicity and counterhelicity. *Physical Review Letters*, *65*, 721.

Yang, K., Guo, Y., & Ding, M.D. (2016). Quantifying the topology and evolution of a magnetic flux rope associated with multi-flare activities. *Astrophysical Journal*, *824*, 148, https://doi.org/10.3847/0004-637X/824/2/148

Yeates, A.M., Russell, A.J.B., & Hornig, G. (2015). Physical role of magnetic constraints in localized magnetic relaxation. *Proceedings of the Royal Society A*, *471*, 20150012, https://doi.org/10.1098/rspa.2015.0012

14

Emergence of Magnetic Structure in Supersonic Isothermal Magnetohydrodynamic Turbulence

Jean-Mathieu Teissier[1] and Wolf-Christian Müller[1,2]

ABSTRACT

The inverse transfer of magnetic helicity is a fundamental process that may explain large-scale magnetic structure formation and sustainment. Until very recently, direct numerical simulations (DNS) of the inverse transfer in magnetohydrodynamics (MHD) turbulence have been done in incompressible MHD or at only low Mach numbers. We review the first results obtained through DNS of the isothermal MHD equations at Mach numbers ranging from subsonic to about 10. The spectral exponent of the magnetic helicity spectrum becomes flatter with increasing compressibility. When considering the Alfvén velocity in place of the magnetic field, however, results found in incompressible MHD, including a dynamic balance between shear and twist, can be extended to supersonic MHD. In the global picture of an inverse transfer of magnetic helicity, three phenomena are at work: a local direct transfer mediated by the large-scale velocity field, a local inverse transfer mediated by the intermediate-scale velocity field, and a nonlocal inverse transfer mediated by the small-scale velocity field. The compressive part of the velocity field is geometrically favored in the local direct transfer. It contributes to the nonlocal inverse transfer but plays no role in the local inverse transfer.

14.1. INTRODUCTION

Flows are mostly turbulent, especially at geophysical or astrophysical scales, so intricate nonlinear interactions lead to chaotic behavior. In three-dimensional hydrodynamic turbulence, large-scale kinetic energy cascades to smaller and smaller scales through the breakup of large eddies into smaller ones, which successively experience a similar process until viscous effects dissipate this energy into heat. In this sense, turbulence can be seen as a mechanism "destroying" continuously large-scale coherence: that is, the "structure" of the flow.

Despite this, large-scale magnetic structures embedded in turbulent plasmas are observed throughout the universe. They exist, for example, in and around planets with liquid cores, stars, galaxies, and clusters of galaxies (Brandenburg and Subramanian, 2005). It is thus important to understand the physical mechanisms responsible for generating and sustaining large-scale magnetic structures.

Two key processes in this respect are the dynamo and the inverse transfer of magnetic helicity (ITMH in the following). "Inverse transfer" refers here to a transfer from small to large scales, contrary to the "direct cascade of kinetic energy" in 3D hydrodynamics described previously. In Fourier space, this corresponds to a transfer from large wave numbers to small ones.

Dynamos are thought to be, for example, at the heart of planetary and stellar magnetospheres (Brandenburg and Subramanian, 2005). In a dynamo, a weak seed magnetic field is amplified through the conversion of kinetic to magnetic energy until nonlinear saturation. This process generates and sustains a large-scale magnetic

[1]*Center for Astronomy and Astrophysics, Technical University of Berlin, Berlin, Germany*

[2]*Max-Planck/Princeton Center for Plasma Physics, Princeton, NJ, USA*

Helicities in Geophysics, Astrophysics, and Beyond, Geophysical Monograph 283, First Edition.
Edited by Kirill Kuzanyan, Nobumitsu Yokoi, Manolis K. Georgoulis, and Rodion Stepanov.

DOI:10.1002/9781119841715.ch14

field. While crucial in dynamo theory (Blackman, 2015; Vishniac and Cho, 2001), helical magnetic fields are subject to other important structure formation processes. The magnetic helicity is defined as $\mathcal{H}^M = \langle \boldsymbol{a} \cdot \boldsymbol{b} \rangle$, with $\langle \cdot \rangle$ denoting a volume average over the whole system and $\boldsymbol{a}$ the magnetic vector potential; $\boldsymbol{b} = \nabla \times \boldsymbol{a}$ and is hence a purely magnetic quantity. The theoretical possibility of ITMH was first established through absolute equilibrium statistical models in Frisch et al. (1975) and rapidly confirmed through diverse numerical experiments.

Magnetic helicity is a quadratic ideal invariant of the induction equation (Elsässer, 1956; Woltjer, 1958)

$$\partial_t \boldsymbol{b} = \nabla \times (\boldsymbol{v} \times \boldsymbol{b}). \tag{14.1}$$

The proof of invariance does not require the velocity field $\boldsymbol{v}$ to be incompressible or even an actual solution of, for example, the magnetohydrodynamics equations (Aluie, 2017). Magnetic helicity is gauge invariant in the setting considered in this chapter (a closed periodic domain without a mean magnetic field (Berger, 1997)). In general, unless the volume is closed and the magnetic field component normal to the boundary vanishes, magnetic helicity is not gauge invariant. In such situations, one resorts to the so-called "relative helicity" (Berger and Field, 1984; Finn and Antonsen, 1985).

Magnetic helicity quantifies topological aspects of the magnetic field lines, such as their degree of linkage, twist, writhe, and knottedness (Moffatt, 1969). It is well conserved even when low resistivity is present (for example, in magnetic reconnection), significantly better than the total energy (Berger, 1999). In many geophysical and astrophysical plasmas, where low resistivity is typical, the quasiconservation of magnetic helicity imposes strong constraints on magnetic field dynamics. Apart from constraints on the geodynamo, conservation of magnetic helicity is thought to play a crucial role in, for example, solar flares and coronal mass ejections (Kumar and Rust, 1996; Low, 1994).

Most direct numerical simulations (DNS) of the ITMH to date have been performed in the incompressible case (Alexakis et al., 2006; Linkmann et al., 2017; Meneguzzi et al., 1981; Müller et al., 2012; Pouquet and Patterson, 1978; Stepanov et al., 2015). This choice is mostly made for the sake of conceptual and numerical simplicity:

- On the conceptual side, incompressibility assumes a constant and uniform density, which is equivalent to the strict solenoidality of the velocity field. In that case, a degree of freedom for the velocity field is removed (in Fourier space, the velocity modes are in the plane orthogonal to their wavevector $\boldsymbol{k}$: the available manifold for each mode is not 3D but 2D). The thermal pressure is thus reduced to a passive quantity since one can solve for the vorticity instead of the velocity. Also, quadratic invariants are conserved in each nonlinear Fourier triad. However, in compressible turbulence, the situation is significantly more difficult. Some well-known theoretical results in incompressible turbulence have only recently been extended to the compressible case. For example, the existence of an inertial range for hydrodynamic compressible turbulence has been rigorously proven only in the last decade (Aluie, 2011, 2013; Aluie et al., 2012). Similarly, considerations about, for example, an extension of or a variance from the incompressible Kolmogorov $-\frac{5}{3}$ spectral power law for velocity fluctuations in compressible flows have been derived theoretically and verified through numerical experiments around the same years (Federrath, 2013; Galtier and Banerjee, 2011) (see also Ferrand et al. (2020)), as well as considerations about a compressible analog to Kolmogorov's four-fifths law (Kritsuk et al., 2013). Other results in incompressible magnetohydrodynamics (MHD), including exact relations, await an extension to compressible MHD (e.g., Politano et al. (2003), which derives a von Kármán-Howarth relation in helical MHD flows).
- On the numerical side, incompressible flows with periodic boundary conditions can be dealt with very efficiently by pseudospectral methods (Canuto et al., 1988). Supersonic compressible flows, on the other hand, lead to shocks and discontinuities for which a Fourier representation is not suitable (e.g., because of the Gibbs phenomenon). Reconstructing abruptly varying quantities naively through polynomial interpolation on uniform grids produces strong oscillations. Resolving such cases properly without sacrificing too much accuracy in smooth regions required the advent of higher-order shock-capturing methods able to adjust themselves depending on whether the zone considered presents a shock. The first essentially non-oscillatory (ENO) schemes, of discretization order higher than two and able to adapt to discontinuities, were designed at the end of the 1980s and then extended to the more stable weighted ENO (WENO) schemes in the 1990s (Harten et al., 1987; Jiang and Shu, 1996; Liu et al., 1994; Shu, 2009). High-order time integration techniques avoiding the addition of oscillations as part of the time-stepping procedure were developed about 15 years ago (Ketcheson, 2008). Techniques allowing a numerically more efficient dimension-by-dimension approach were designed a couple of years later (Buchmüller and Helzel, 2014; McCorquodale and Colella, 2011). For very strong shocks, maintaining the positivity of the mass density and the thermal pressure is not a trivial matter and remains an active area of research (Balsara, 2012; Balsara et al., 2016; Wu and Shu, 2019). Thus, viable numerical schemes allowing high-accuracy supersonic DNS with acceptable numerical costs are relatively

recent. Consequently, the few DNS studies using an adiabatic or isothermal equation of state have been performed in the subsonic or transonic case (Balsara and Pouquet, 1999; Brandenburg, 2001; Christensson et al., 2001).

However, incompressibility is of limited applicability in many astrophysical situations, where high Mach numbers (the ratio of the flow speed to the speed of sound) are common. For example, the interstellar medium exhibits root mean square (RMS) turbulent Mach numbers typically ranging from about 0.1 to 10 (Elmegreen and Scalo (2004), section 4.2).

In this chapter, we review some results of the ITMH in incompressible MHD and the first results of the ITMH in compressible supersonic MHD, obtained through fourth-order accurate finite-volume implicit large-eddy simulation: a DNS of the ideal MHD equations including grid-scale numerical dissipation, without physical viscosity and resistivity (Teissier and Müller, 2021a,b). In the presence of a mechanical and an electromotive forcing, they are

$$\partial_t \rho = -\nabla \cdot (\rho \boldsymbol{v}), \tag{14.2}$$

$$\partial_t(\rho \boldsymbol{v}) = -\nabla \cdot \left(\rho \boldsymbol{v}\boldsymbol{v}^T + (\rho c_s^2 + \frac{1}{2}|\boldsymbol{b}|^2)\boldsymbol{I} - \boldsymbol{b}\boldsymbol{b}^T \right) + \rho \boldsymbol{f}_V, \tag{14.3}$$

$$\partial_t \boldsymbol{b} = \nabla \times (\boldsymbol{v} \times \boldsymbol{b}) + \boldsymbol{f}_M, \tag{14.4}$$

with the solenoidality constraint

$$\nabla \cdot \boldsymbol{b} = 0. \tag{14.5}$$

The symbols have their usual meaning: ρ is the mass density, $\boldsymbol{v}$ the velocity, c_s the constant speed of sound in the isothermal case so that the pressure is $p = \rho c_s^2$, $\boldsymbol{b}$ is the magnetic field, and $\boldsymbol{I}$ is the 3×3 identity matrix. The mechanical forcing $\boldsymbol{f}_V$ injects kinetic energy at the largest scales through an acceleration field, and the electromotive driving $\boldsymbol{f}_M$ injects small-scale maximally helical magnetic fluctuations along the positive helical eigenvector in Fourier space (section 14.2.2). Both forcings are described in section 14.4. The magnetic forcing is required to drive a sufficiently strong inverse transfer of magnetic helicity. It might represent a source of small-scale Alfvén wave activity that generates magnetic fluctuations, but in our finite-size system, it serves as a well-defined small-scale source of magnetic helicity to ensure a respective spectral scaling range of sufficient extent.

The previous equations are written in conservative form, as appropriate when using a finite-volume solver. The MHD approximation describes the dynamics of a conducting fluid well, such as planetary fluid metallic cores or astrophysical plasmas at large enough time and space scales (Kivelson, 1995). The isothermal equation of state is a reasonable approximation, for example, in the early homologous phase of a gravitational collapse of a molecular cloud, where the heat generated through the collapse is removed by radiative cooling. During a later phase, the plasma becomes optically thick, and the temperature rises in a nearly adiabatic way (Caroll and Ostlie (2007), section 12.2).

Thus, the isothermal approximation can be considered the next step of complexity when departing from strict incompressibility. It introduces one new variable: a nonconstant density associated with the internal energy density $u = \rho c_s^2 \ln(\rho/\rho_0)$, with ρ_0 the mean density of the system (Kritsuk et al., 2013). The variable density allows for finite $\nabla \cdot \boldsymbol{v}$.

The main aim of the investigations of supersonic isothermal MHD flow (Teissier and Müller, 2021a,b) reviewed in this chapter is to explore similarities and differences between scaling laws in incompressible and supersonic MHD turbulence. The role of the compressive velocity field is also in focus. For this, DNS without a mean magnetic field under periodic boundary conditions is performed so that the magnetic helicity is gauge invariant. For the sake of simplicity, only one sign of magnetic helicity is injected. Although it is not physically motivated, since both signs are expected in nature, the results in this already complex setting should help interpret more realistic configurations in the future.

This chapter is structured as follows. The spectral analysis tools used, which allow disentangling helical components and obtaining information about a locality in Fourier space, are presented in section 14.2. The results of the isothermal supersonic DNS are presented in section 14.5 after a review of some previously established results in incompressible MHD turbulence (section 14.3) and a presentation of the numerical method used (section 14.4). Concluding remarks are given in section 14.6.

14.2. FOURIER DIAGNOSTICS AND SHELL-TO-SHELL AND HELICAL TRANSFERS

14.2.1. Shells, Spectra, and Cospectra

In the chosen setup with periodic boundary conditions in a cubic box of size L, each field $\boldsymbol{u}$ can be represented as a discrete sum of its Fourier harmonics, with Fourier coefficients $\hat{\boldsymbol{u}}_{\boldsymbol{k}}$ at wavevector $\boldsymbol{k}$. The power spectrum $\mathcal{P}(\boldsymbol{u})$ is thus computed through a summation over Fourier shells $S_K = \{\boldsymbol{k} | K \leq |\boldsymbol{k}|/\kappa < K+1\}$, with $K \in \mathbb{N}$ and $\kappa = \frac{2\pi}{L}$ the smallest wave number in the system:

$$\mathcal{P}(\boldsymbol{u})_K = \frac{1}{2} \sum_{\boldsymbol{k} \in S_K} |\hat{\boldsymbol{u}}_{\boldsymbol{k}}|^2 = \frac{1}{2} \langle |\boldsymbol{u}_K|^2 \rangle, \tag{14.6}$$

where the second equality comes from Parseval's theorem, with $\boldsymbol{u}_K$ the inverse Fourier transform of the shell-filtered $\boldsymbol{u}$ at shell K (i.e., keeping only $\hat{\boldsymbol{u}}_{\boldsymbol{k}}$ where $\boldsymbol{k} \in S_K$ and setting the other Fourier coefficients to 0).

Similarly, the cospectrum of two fields $\boldsymbol{u}, \boldsymbol{w}$ is defined by

$$Q(\boldsymbol{u}, \boldsymbol{w})_K = \sum_{\boldsymbol{k} \in S_K} \hat{\boldsymbol{u}}_{\boldsymbol{k}} \cdot \hat{\boldsymbol{w}}^*_{\boldsymbol{k}} = \langle \boldsymbol{u}_K \cdot \boldsymbol{w}_K \rangle, \tag{14.7}$$

where the star in the superscript denotes the complex conjugate. The magnetic helicity spectrum is thus $H^M = Q(\boldsymbol{a}, \boldsymbol{b})$.

14.2.2. Helical Triadic Interactions

The decomposition of a field in helical components is based on the diagonalization of the curl operator in Fourier space $i\boldsymbol{k} \times \cdot$. This operator possesses the eigenvalues $(0, +k, -k)$, with $k = |\boldsymbol{k}|$ associated, respectively, with the unitary eigenvectors $(\hat{\boldsymbol{h}}^0_{\boldsymbol{k}}, \hat{\boldsymbol{h}}^+_{\boldsymbol{k}}, \hat{\boldsymbol{h}}^-_{\boldsymbol{k}})$, where $\hat{\boldsymbol{h}}^0_{\boldsymbol{k}} = \boldsymbol{k}/|\boldsymbol{k}|$ and (Waleffe (1992); see also Brandenburg and Subramanian (2005)):

$$\hat{\boldsymbol{h}}^{\pm}_{\boldsymbol{k}} = \frac{1}{\sqrt{2}} \frac{\boldsymbol{k} \times (\boldsymbol{k} \times \hat{\boldsymbol{e}}) \mp i k(\boldsymbol{k} \times \hat{\boldsymbol{e}})}{k^2 \sqrt{1 - (\boldsymbol{k} \cdot \hat{\boldsymbol{e}}/k)^2}}, \tag{14.8}$$

with $\hat{\boldsymbol{e}}$ an arbitrary unitary vector nonparallel to $\boldsymbol{k}$.

This decomposition can be seen as an extension of the Helmholtz decomposition in compressive (parallel to $\boldsymbol{k}$) and solenoidal (orthogonal to $\boldsymbol{k}$) components. Indeed, the plane orthogonal to $\boldsymbol{k}$ is spanned by the eigenvectors $\hat{\boldsymbol{h}}^{\pm}_{\boldsymbol{k}}$ corresponding to circularly polarized waves with opposite polarity. In configuration space, flow lines corresponding to these eigenvectors form a left- or right-handed helix.

A rigorous mathematical framework for the helical decomposition has been introduced by Constantin and Majda (1988) (where it is called "Beltrami decomposition"). In three-dimensional hydrodynamic turbulence, this decomposition allows us to find helical interactions responsible for a subdominant inverse transfer in the global picture of a direct cascade of energy (Biferale et al., 2012; Waleffe, 1992), without the need for rotation or confinement (quasi-2D flow). This may be relevant for turbulence models in large-eddy simulations (Alexakis, 2017). On a mathematical level, a helical decimation of the Navier-Stokes equations (i.e., their projection on the positive helical eigenvectors) gives some insight regarding, for example, the existence and uniqueness of weak solutions (Biferale and Titi, 2013). Extensions to the MHD case may provide information to disentangle the intertwined dynamics of magnetic and kinetic helicities (Alexakis and Biferale, 2018; Lessinnes et al., 2009; Linkmann et al., 2016; Linkmann and Dallas, 2017; Linkmann et al., 2017).

A magnetic field mode can thus be represented as $\hat{\boldsymbol{b}}_{\boldsymbol{k}} = \sum_{s_k \in \{+,-\}} b^{s_k}_{\boldsymbol{k}} \hat{\boldsymbol{h}}^{s_k}_{\boldsymbol{k}}$, where $s_k = 0$ is not allowed since the magnetic field is solenoidal. In contrast, a velocity field mode can have a compressive component so that $\hat{\boldsymbol{v}}_{\boldsymbol{p}} = \sum_{s_p \in \{+,0,-\}} v^{s_p}_{\boldsymbol{p}} \hat{\boldsymbol{h}}^{s_p}_{\boldsymbol{p}}$.

Using this decomposition, the time variation of helical component $b^{s_k}_{\boldsymbol{k}}$ of the magnetic field is given in incompressible MHD by (Lessinnes et al., 2009)

$$\partial_t b^{s_k}_{\boldsymbol{k}} = s_k k \sum_{\boldsymbol{k}+\boldsymbol{p}+\boldsymbol{q}=0} \sum_{s_p, s_q} v^{s_p *}_{\boldsymbol{p}} b^{s_q *}_{\boldsymbol{q}} g^{s_k, s_p, s_q}_{k,p,q}. \tag{14.9}$$

This relation remains valid in compressible MHD, the only difference being that $s_p \in \{+, 0, -\}$ and no longer only $s_p = \pm$. The geometric factor $g^{s_k,s_p,s_q}_{k,p,q}$ depends on the triad shape and the helical components considered. Its modulus quantifies the importance of the respective triad interaction and is given by (Waleffe, 1992)

$$G^{s_k, s_p \in \{+,-\}, s_q}_{k,p,q} = \frac{|s_k k + s_p p + s_q q|}{2\sqrt{2}} \times \frac{\sqrt{2k^2p^2 + 2p^2q^2 + 2q^2k^2 - k^4 - p^4 - q^4}}{2kpq} \tag{14.10}$$

for the incompressible modes (an extra factor $1/(2\sqrt{2})$ is present, as compared to (Waleffe, 1992), because of the helical eigenvector normalization) and by

$$G^{s_k, s_p = 0, s_q = \pm s_k}_{k,p,q} = \frac{|(q \mp k)(p^2 - k^2 - q^2 \mp 2qk)|}{4kpq} \tag{14.11}$$

when the velocity field component considered is the compressive one (Teissier and Müller, 2021b).

14.2.3. Shell-to-Shell Transfers

Unlike energy, which can be transferred from the kinetic energy reservoir to the magnetic one and vice versa, magnetic helicity is a purely magnetic quantity. The velocity field in the triadic interactions (see equation (14.9)) only plays the role of a mediator. Summing over all the wave numbers in a certain shell K, one can derive the transfer of magnetic helicity from a shell Q to a shell K, mediated by the velocity field shell P (Alexakis et al., 2006; Plunian et al., 2019):

$$\mathcal{T}^{H^M}(Q, P, K) = 2\langle \boldsymbol{b}_K \cdot (\boldsymbol{v}_P \times \boldsymbol{b}_Q) \rangle, \tag{14.12}$$

with the subscripts on $\boldsymbol{b}$ and $\boldsymbol{v}$ denoting the shell at which the fields have been filtered (see equation (14.6)). We recall that the volume averaging $\langle \cdot \rangle$ takes place in configuration space. The derivations of (Alexakis et al., 2006) done in the incompressible case remain valid in compressible MHD because of the magnetic field solenoidality. The interpretation as a transfer rate of magnetic helicity between shells is consistent with the antisymmetric property

$\mathcal{T}^{\mathcal{H}^M}(Q,P,K) = -\mathcal{T}^{\mathcal{H}^M}(K,P,Q)$. Then in the absence of electromotive driving and dissipation, the magnetic helicity in the shell K, $H_K^M = \langle \boldsymbol{a}_K \cdot \boldsymbol{b}_K \rangle$, is governed by

$$\partial_t H_K^M = \sum_Q \sum_P \mathcal{T}^{\mathcal{H}^M}(Q,P,K). \qquad (14.13)$$

To disentangle the role of the different helical components, the $\boldsymbol{v}$ and $\boldsymbol{b}$ filtered fields can first be projected on the helical eigenvectors $(\hat{\boldsymbol{h}}_k^0, \hat{\boldsymbol{h}}_k^+, \hat{\boldsymbol{h}}_k^-)$, giving (Teissier and Müller, 2021b)

$$\mathcal{T}^{\mathcal{H}^M}_{s_K s_P s_Q}(Q,P,K) = 2\langle \boldsymbol{b}_K^{s_K} \cdot (\boldsymbol{v}_P^{s_P} \times \boldsymbol{b}_Q^{s_Q}) \rangle,$$

$$\partial_t H_K^M = \sum_Q \sum_P \sum_{s_K, s_P, s_Q} \mathcal{T}^{\mathcal{H}^M}_{s_K s_P s_Q}(Q,P,K). \qquad (14.14)$$

In this expression, $\boldsymbol{b}_K^{s_K}$ represents the positively (for $s_K = +$) or negatively ($s_K = -$) helical magnetic field (i.e., projected on $\hat{\boldsymbol{h}}^{s_K}$) filtered at shell K. The last summation on the right occurs over 12 terms: $s_K, s_Q \in \{+,-\}$, $s_P \in \{+,0,-\}$. The terms where $s_K = s_Q$, labeled "homochiral," are antisymmetric and can be interpreted as magnetic helicity shell-to-shell transfers, depending on the helical components. However, the terms where $s_K = -s_Q$ must be considered in pairs to exhibit the antisymmetric property: $\mathcal{T}^{\mathcal{H}^M}_{s_K s_P s_Q} + \mathcal{T}^{\mathcal{H}^M}_{s_Q s_P s_K}$. In the considered investigations, these "heterochiral" terms have a relatively low magnitude and are therefore not treated in much detail.

14.3. RESULTS FROM PREVIOUS RESEARCH IN INCOMPRESSIBLE MHD

The first prediction regarding the magnetic helicity spectral exponent was obtained through a dimensional analysis "à la Kolmogorov" (Pouquet et al., 1976). Assuming local transfers and a constant flux of magnetic helicity in an inertial range, a dimensional analysis gives $[\boldsymbol{a}][\boldsymbol{b}]/T = [\boldsymbol{b}]^2[\boldsymbol{v}] \sim const.$ In the incompressible case, $[\boldsymbol{v}] \sim [\boldsymbol{b}]$, so $b_l^3 \sim (H_l^M/l)^{3/2} \sim const.$, where the subscript l means typical values for eddies of size l. This leads to $H_l^M \sim l$, so for the magnetic helicity spectrum, $H_K^M \sim K^{-2}$. The spectral exponent of -2 has been observed in numerical simulations of the eddy-damped quasinormal Markovian (EDQNM) model extended to MHD (Pouquet et al., 1976).

However, later DNS of incompressible MHD and Lagrangian-averaged MHD have shown a significantly steeper spectral exponent of the H^M spectrum, in the range $[-3.7, -3.3]$ (Graham et al., 2011; Mininni and Pouquet, 2009; Müller et al., 2012). This echoes the word of caution announced in Pouquet et al. (1976), warning that the -2 scaling may be altered by nonlocal effects. Indeed, Alexakis et al. (2006) have shown through Fourier shell-to-shell analysis that the ITMH presents both local and nonlocal features.

Based on the EDQNM closure model, a dynamical equilibrium between shearing and twisting effects leads to the balance (Müller and Malapaka (2010), see also Mininni and Pouquet (2009), Graham et al. (2011), Müller et al. (2012))

$$\left(\frac{E^V}{E^M}\right)^g \propto \frac{H^V}{H^J}, \qquad (14.15)$$

with $E^V = \mathcal{P}(\boldsymbol{v})$ and $E^M = \mathcal{P}(\boldsymbol{b})$ the (specific) kinetic and magnetic energy spectra, $H^V = \mathcal{Q}(\boldsymbol{v}, \boldsymbol{\omega})$ and $H^J = \mathcal{Q}(\boldsymbol{b}, \nabla \times \boldsymbol{b}) \approx (\frac{2\pi}{L}K)^2 H^M$ the kinetic and current helicity spectra, and g a constant exponent. This balance has been verified empirically with an exponent $g = 1$ in decaying turbulence experiments with large-scale helical magnetic fields (Graham et al., 2011; Mininni and Pouquet, 2009). For the incompressible investigation of the ITMH (Müller and Malapaka, 2010), the data is also compatible with an exponent $g = 1$, even though later analysis has shown that an exponent $g = 2$ fits the data better (Müller et al., 2012), presumably because of different domains where the individual spectra E^V, E^M, H^V, and H^J present scaling laws. In Graham et al. (2011), this equilibrium has been interpreted as a "partial Alfvénization of the flow," so we refer to equation (14.15) as the *Alfvénic balance.*

Regarding the strength and locality of the ITMH, a shell-to-shell analysis in incompressible MHD (Alexakis et al., 2006) shows that the transfer is more local at early times and nonlocal at later times when magnetic helicity is present at the largest scales. This study also shows local direct transfer of magnetic helicity. Furthermore, using a helical decomposition in the framework of linear stability analysis, Linkmann et al. (2017) have shown that transfers between two like-signed helical magnetic modes (e.g., from b_q^+ to b_k^+) are stronger and more nonlocal when mediated by a like-signed helical velocity mode (i.e., v_p^+), as compared to an oppositely signed one (v_p^-).

These results in incompressible MHD are compared with those in the supersonic isothermal case (Teissier and Müller, 2021a,b) in section 14.5.

14.4. NUMERICAL METHOD

The results in supersonic MHD reviewed here (Teissier and Müller, 2021a,b) have been obtained through DNS of the ideal isothermal MHD equations (14.2)–(14.5). The finite-volume shock-capturing numerical solver used (Verma et al. (2019); see also Teissier (2020)) is fourth-order accurate in space and time, ensures magnetic field solenoidality through the constrained transport method, and resorts to lower-order reconstruction in the vicinity of strong discontinuities (fallback approach) to keep the mass density positive at high Mach numbers.

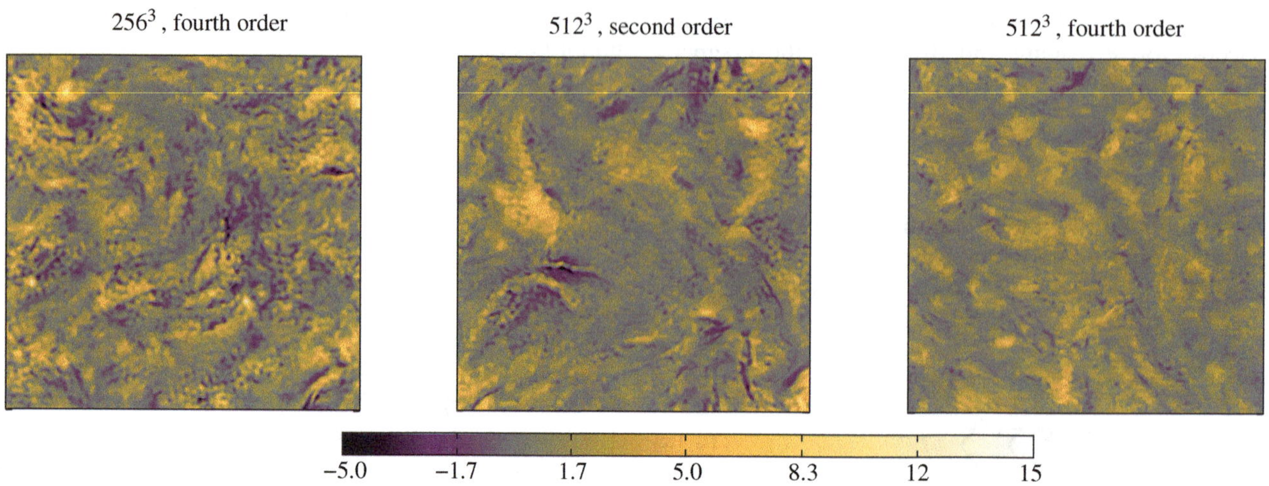

Figure 14.1 Slices of the magnetic helicity density, normalized by its mean in the system, for three runs at an instant when the magnetic helicity integral scale $\mathcal{I}_{HM} \approx L/6$. These runs correspond to a Mach number close to unity in the statistically stationary state and are done at different resolutions and using numerical schemes of different discretization order.

Using a fourth-order numerical code allows results of good accuracy at the moderate 512^3 resolution (see a comparison with a second-order solver in Fig. 14.1). The main results are confirmed at resolution 1024^3.

Initially, the plasma is at rest ($\rho = \rho_0 = 1$ is constant and $\boldsymbol{v} = 0$) in a triply periodic cubic box $[0, L = 1]^3$, and no magnetic field is present. The isothermal speed of sound is taken as $c_s = 0.1$. An Ornstein-Uhlenbeck mechanical forcing $\boldsymbol{F}^{OU}$ injects kinetic energy at the largest scales with a specified degree of compressivity. The forcing is defined in Fourier space, initialized with $\boldsymbol{F}^{OU} = 0$, and evolved in time according to

$$d\boldsymbol{F}_{\boldsymbol{k}}^{OU}(t) = -\boldsymbol{F}_{\boldsymbol{k}}^{OU}(t)\frac{dt}{t_{auto}} + F_0\left(\frac{2\sigma(\boldsymbol{k})^2}{t_{auto}}\right)^{1/2} \underline{\mathcal{P}^{\zeta}} \cdot d\boldsymbol{W}(t), \tag{14.16}$$

with $\sigma(\boldsymbol{k}) = 1$ for the largest scales $\boldsymbol{k} \in S_1 \cup S_2$ and $\sigma(\boldsymbol{k}) = 0$ otherwise. The autocorrelation time is set to an estimate of the turbulent turnover time: $t_{auto} = L/(2c_s\mathcal{M}^*)$, with $\mathcal{M}^*$ the expected stationary state root mean square (RMS) Mach number. A three-dimensional continuous random walk is performed via the Wiener process, $d\boldsymbol{W}(t) = dt\boldsymbol{N}(0, dt)$, with $\boldsymbol{N}(0, dt)$ a 3D Gaussian distribution with zero mean and standard deviation dt. The compressiveness of the forcing is determined by the projector $\mathcal{P}_{ij}^{\zeta}(\boldsymbol{k}) = \zeta\delta_{ij} + (1 - 2\zeta)\frac{k_i k_j}{|k|^2}$. Only the extreme cases $\zeta = 1$ (purely solenoidal forcing) and $\zeta = 0$ (purely compressive forcing) are considered (compare to Federrath (2013)). The field $\boldsymbol{F}^{OU}$ is back-transformed in configuration space and normalized at each timestep so that a constant energy injection rate ϵ_{inj}^{K} is achieved (Teissier and Müller, 2021a). As a consequence, the amplitude F_0 can be chosen arbitrarily. The finite-volume formalism guarantees the conservation of momentum but not of mean velocity. Hence the weak fluctuating mean velocity field is removed at each iteration. Indeed, even though a spatially uniform mean velocity field does not affect hydrodynamic turbulence due to the Galilean invariance of the Navier-Stokes equations, this is generally not the case for the MHD equations, which are not Galilean invariant.

The kinetic energy cascades to smaller and smaller scales. At the smallest scales, numerical dissipation is dominant so that a hydrodynamic statistically stationary state is reached with a desired degree of compressivity and RMS Mach number $\mathcal{M}$, determined by both ϵ_{inj}^{K} and ζ. To cover typical Mach numbers found in the interstellar medium, the following runs were performed:

- With purely solenoidal driving: M01s, M1s, M5s, M7s, M11s, where the number represents the average RMS turbulent Mach number in the statistically stationary state, respectively ($\mathcal{M} \approx 0.12, 1.1, 5.1, 7.0, 11.$)
- With purely compressive driving: M1c, M3c, M5c, M8c, with $\mathcal{M} \approx 0.80, 2.8, 5.1, 7.9$, respectively

The mass density ρ exhibits an increasing number of filamentary structures with increasing $\mathcal{M}$. The compressively driven runs, with a high $\nabla \cdot \boldsymbol{v}$ component, have significantly stronger mass density variations than the solenoidally driven ones. This is made apparent through large regions of very low density and comparison of the probability distribution functions of $\log(\rho/\rho_0)$ (Fig. 14.2; see also Federrath et al. (2010), Federrath (2013)). The compressively driven runs also present pronounced shock fronts, already at Mach numbers of order unity (Teissier and Müller, 2021a).

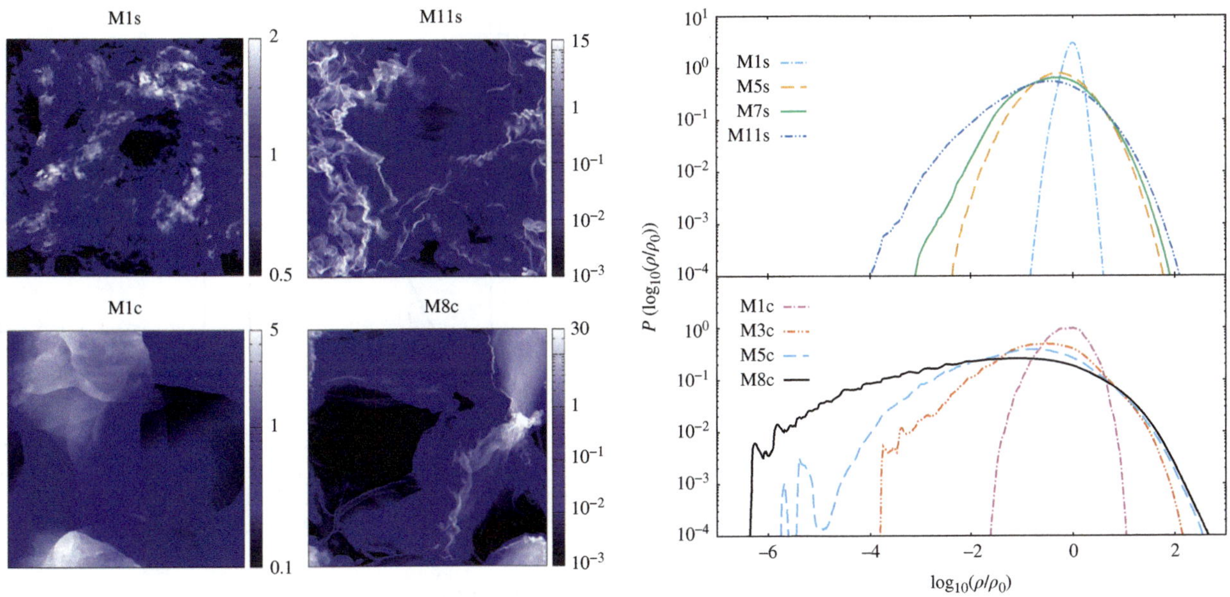

Figure 14.2 Mass density properties during the hydrodynamic statistically stationary state. Left: Mass density slices for the M1s, M11s, M1c, and M8c runs. Right: Probability distribution function of $\log(\rho/\rho_0)$ averaged through roughly one turbulent turnover time.

At a chosen instant in the statistically stationary state, an electromotive forcing at small scales is switched on while the large-scale mechanical forcing is maintained. The electromotive forcing is a delta-correlated (white noise in time), maximally positive helical driving: that is, with all the components along the $\hat{\boldsymbol{h}}_k^+$ eigenvector (see equation (14.8)). Generated in Fourier space for the shells $\bigcup_{48\leq K\leq 52} S_K$, it is back-transformed and normalized in configuration space so that a constant magnetic energy injection rate ϵ_{inj}^M is achieved. For all the runs, $\epsilon_{inj}^M = \epsilon_{inj}^K$, apart from M01s and M1s where it is $4\epsilon_{inj}^K$ and $2\epsilon_{inj}^K$, respectively, to more quickly reach the convergence to a power-law behavior of the magnetic helicity spectrum (Teissier and Müller, 2021a). As verified numerically, the cross-helicity remains low for all the runs, so the dynamics are dominated by the direct cascade of energy and the ITMH.

The simultaneous forcing at two scales may seem odd at first, but it is necessary to observe the effects of the compressive velocity field. Indeed, its effects enter the magnetic field dynamics through a term $\sim \nabla \cdot \boldsymbol{v}$. Due to the direct cascade of kinetic energy, only a large-scale driving can generate a dynamically important compressive velocity field in the range of scales characteristic of the inverse transfer of magnetic helicity. Secondary velocity fluctuations generated by Alfvénic oscillations of the growing magnetic structures are insufficient.

The injected magnetic helicity is subsequently transferred to ever-larger scales. Since there is no large-scale energy sink, the asymptotic state of the system would be a condensate of magnetic helicity at the largest scales, dominating the dynamics. In particular, this finite-size effect would strongly affect the locality of transfers and the spectral exponents (Alexakis et al., 2006; Linkmann and Dallas, 2016). Hence, the results presented here consider an instant in time when an energetically dominant largest-scale magnetic structure has not yet emerged (more precisely, when the magnetic helicity integral scale $\mathcal{I}_{\mathcal{H}^M} = L(\int_K K^{-1} H^M dK)/(\int_K H^M dK)$ is about $L/8$ or $L/6$). Although not ideal, this choice of stopping the runs earlier presents a pragmatic compromise between an asymptotic scaling range polluted by a large-scale sink of magnetic helicity and perfect statistical stationarity. This allows us to observe dependencies of the spectral scaling of magnetic helicity and derived quantities on the Mach number and the forcing mechanism with limited numerical expense.

14.5. RESULTS

Sections 14.5.1 and 14.5.2 present the main results of (Teissier and Müller, 2021a,b), respectively. Additional material and more detailed analysis can be found therein.

14.5.1. Spectra and Alfvénic Balance

The results presented in this subsection are taken at an instant when the magnetic helicity integral scale $\mathcal{I}_{\mathcal{H}^M} \approx L/8$ (see section 14.4) so that an energetically dominant largest-scale magnetic structure has not yet emerged.

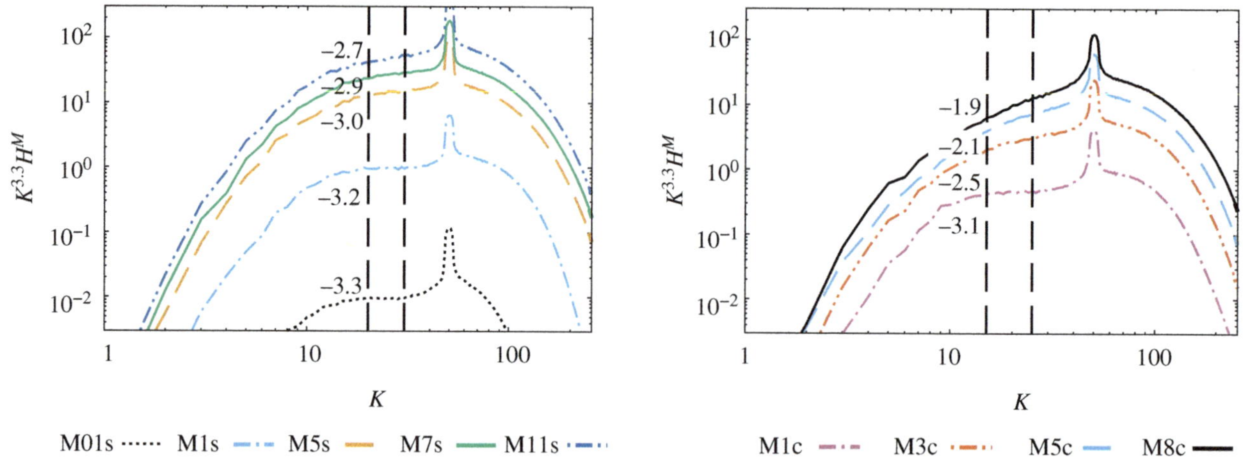

Figure 14.3 Magnetic helicity spectra compensated by $K^{3.3}$ at an instant when $\mathcal{I}_{H^M} \approx L/8$. The numbers indicate the spectral slope measured through a least-squares fit in the domain delimited by the two vertical dashed lines.

At a low Mach number (for the M01s, M1s, and M1c runs), the magnetic helicity spectra exhibit a scaling consistent with results found in incompressible MHD: $H_K^M \sim K^m$ with $m \approx -3.3$ (Müller et al., 2012). For increasing compressibility, the spectra become flatter ($|m|$ decreases); see Fig. 14.3. The scaling depends significantly more on the flow's compressivity (the $\nabla \cdot \boldsymbol{v}$ component) than the RMS Mach number. For example, deviations from the incompressible scaling are greater for the M3c run ($H_K^M \sim K^{-2.5}$) than for the M11s run ($H_K^M \sim K^{-2.7}$).

Although the trend toward flatter spectra is very clear, the exponents' precise numerical values should be taken with appropriate caution since only a particular snapshot is considered. For example, starting to inject magnetic helicity at a different instant in the hydrodynamic statistically stationary state of the M5c run, when the mass density PDF is peaked at lower values, results in a flatter scaling $H_K^M \sim K^{-1.1}$. Furthermore, at a later instant in time, the departure of the most compressible solenoidally driven runs from the incompressible case is smaller (Teissier and Müller, 2021a). The shortness of the inertial range, due to the small spectral distance between the magnetic helicity integral scale and the electromotive forcing scale, should not be disregarded.

During the ITMH, the magnetic field experiences topological changes, which require magnetic reconnection processes. The changes are communicated through Alfvén waves that propagate at the Alfvén speed $|\boldsymbol{v}_A| = |\boldsymbol{b}/\sqrt{\rho}|$ along the affected magnetic field lines. In incompressible MHD, the magnetic field and the Alfvén velocity are the same up to a constant multiplicative factor. On the other hand, in the compressible case, low-density regions are associated with shorter Alfvénic timescales $\propto 1/|\boldsymbol{v}_A|$. This suggests considering the Alfvén velocity in place of the magnetic field. The spectral "Alfvénic helicity" $H^A = \mathcal{Q}(\boldsymbol{v}_A, \nabla \times \boldsymbol{v}_A)$ exhibits for all the runs considered a very similar scaling $H_K^A \sim K^m$ with $m \approx -1.3$ (Fig. 14.4). This scaling is, for all the runs other than M1c, close to that of the corresponding quantity in incompressible MHD: $H^J = \mathcal{Q}(\boldsymbol{b}, \nabla \times \boldsymbol{b})$, the current helicity, which scales as $H_K^J \sim K^2 H_K^M \sim K^{-1.3}$. The fact that the scaling exponent does not change significantly over a wide range of compressibility indicates a systematic scale-dependent correlation between the magnetic field and the density. The M1c run may show different dynamics because the compressive ratio of the velocity field stays high in the inverse transfer region, unlike the higher-Mach-number runs, where extended low-density regions are strongly affected by the magnetic field.

This density-magnetic field correlation suggests extending the Alfvénic balance from equation (14.15) by considering the Alfvén velocity in place of the magnetic field:

$$\left(\frac{E^{V,sol}}{E^{A,sol}}\right)^g \propto \frac{H^V}{H^A}, \tag{14.17}$$

with $E^{A,sol} = \mathcal{P}^{sol}(\boldsymbol{v}_A)$ and $E^{V,sol} = \mathcal{P}^{sol}(\boldsymbol{v})$ the power spectra obtained by considering only the solenoidal components of the fields (orthogonal to $\boldsymbol{k}$ in Fourier space). This balance is well followed for the subsonic and transonic runs M01s, M1s, and M1c with an exponent $g = 2$ (Fig. 14.5), consistent with investigations in incompressible MHD (Müller et al., 2012). For the most compressive supersonic runs M3c, M5c, and M8c, it is well followed with an exponent $g = 1$. The supersonic solenoidally driven runs M5s, M7s, and M11s display an intermediate behavior.

The relation in equation (14.17) may be seen as the most straightforward extension of equation (14.15) for compressible flows: that is, considering only the solenoidal

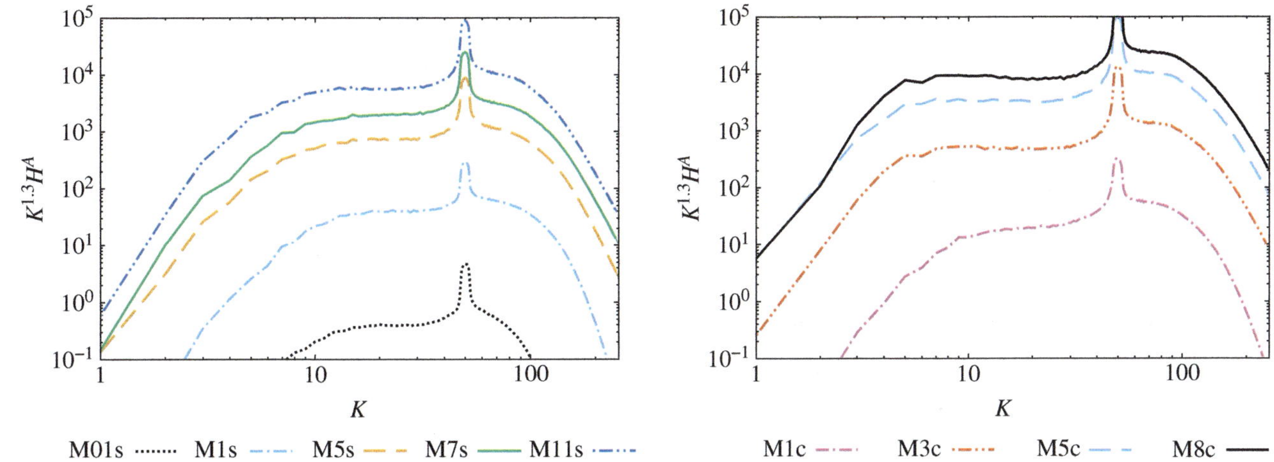

Figure 14.4 Alfvénic helicity spectra compensated with $K^{1.3}$, at the same instant as in Fig. 14.3.

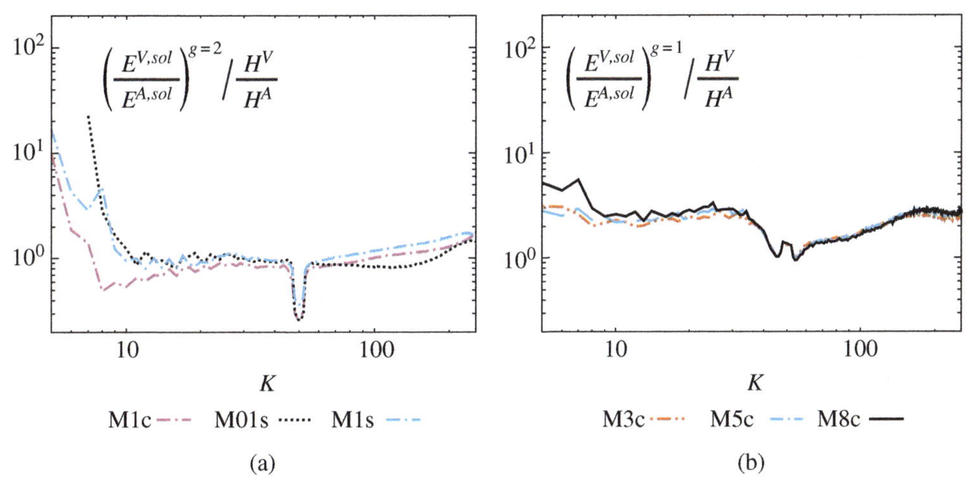

Figure 14.5 Test of the Alfvénic balance with (a) $g = 2$ for low-Mach-number runs and (b) $g = 1$ for the most compressive runs. The same instant as in Fig. 14.3 is considered.

part of the velocity field (the only one that is present in incompressible flows as well). Other possibilities have been tried – for example, considering $\mathcal{P}(\sqrt{\rho}\boldsymbol{v})$, $\frac{1}{2}\mathcal{Q}(\boldsymbol{v}, \rho\boldsymbol{v})$, among others, in place of $E^{V,sol}$ – but the best agreement giving the least spread of the curves has been obtained with equation (14.17). This has also been confirmed at the higher 1024^3 resolution.

The extension of the Alfvénic balance suggests that the shear, not the compressive waves, balances the twisting effects in supersonic isothermal MHD.

14.5.2. Shell-to-Shell Transfers

We review here shell-to-shell transfer results from the most compressive M8c run at an instant when $\mathcal{I}_{\mathcal{H}^M} \approx L/6$. The displayed transfer rates are in units of the magnetic helicity injection rate (or rather, an estimate thereof): $\epsilon_{inj}^{\mathcal{H}^M} = (2\epsilon_{inj}^M h_f)/(2\pi K_f^M/L)$, with $h_f = +1$ the helical fraction of the electromotive driving occurring around $K_f^M = 50$.

The shell-to-shell transfer analysis reveals three features in the global picture of the ITMH (Fig. 14.6): a local direct transfer (LDT), a local inverse transfer (LIT), and a nonlocal inverse transfer (NLIT) from the electromotive forcing scale up to the largest ones. This is consistent with the findings in incompressible MHD (Alexakis et al., 2006), which report both an LDT and an inverse transfer: LIT at early times and NLIT at later times. In that investigation, the electromotive driving takes place around the shell $K = 8$ so that although both LIT and NLIT are present later, they are more difficult to separate.

These three features can be associated with different spatial scales of the mediating velocity field. This is shown in Fig. 14.7, which displays the transfer of

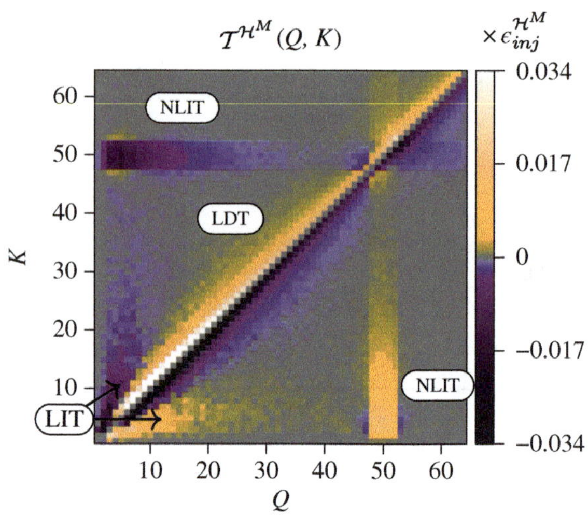

Figure 14.6 Magnetic helicity transfer rates from shell Q to shell K, $\mathcal{T}^{\mathcal{H}^M}(Q,K) = \sum_P \mathcal{T}^{\mathcal{H}^M}(Q,P,K)$ (compare to equation (14.12)) for the M8c run at an instant when $\mathcal{I}_{\mathcal{H}^M} \approx L/6$. The closer to (farther from) the main diagonal, the more local (nonlocal) the transfer. Above the diagonal, a bright orange color means a direct transfer (from large to small scales), and a dark purple color is an inverse transfer, and vice versa under the diagonal. LDT stands for local direct transfer, LIT for local inverse transfer, and NLIT for nonlocal inverse transfer.

magnetic helicity from shell Q to shell K mediated by a range of velocity field shells P. The LDT is associated with a direct cascade of magnetic energy and is mediated by the large-scale velocity field ($1 \le P \lesssim 3$). The NLIT is mediated by the small-scale velocity field ($30 \lesssim P$) and can be interpreted as merging small-scale magnetic fluctuations with a much larger magnetic structure. The LIT corresponds to the merging of similar-sized magnetic structures (within a factor of about two) and is mediated by the intermediate scales of the velocity field ($4 \lesssim P \lesssim 29$).

The helical decomposition also allows us to associate the LDT, LIT, and NLIT with different helical components of the velocity field. For the M8c run, the largest terms are those mediated by the helically like-signed velocity field and its compressive part (Fig. 14.8): $\mathcal{T}^{\mathcal{H}^M}_{+++}$ and $\mathcal{T}^{\mathcal{H}^M}_{+0+}$ (see equation (14.14)). The other helical-transfer terms are relatively small and therefore not shown here (apart from $\mathcal{T}^{\mathcal{H}^M}_{+-+}$, to which we refer in the next point).

Which helical component plays a role for which of the LDT, LIT, and NLIT features can be explained by considering the geometric helical transfer moduli (equations (14.10)–(14.11), plotted in Fig. 14.9). Comparing Figs. 14.8 and 14.9, one observes the following:

1. The LIT is mediated by the like-signed helical velocity field (and, to a smaller extent, by the oppositely signed helical one), but the compressive part of the velocity field plays no role in it. Indeed, for these local transfers $k \approx q \approx p$ mediated by the intermediate scale velocity field, G^{+++} presents a maximum, whereas G^{+-+} is significantly smaller and G^{+0+} vanishes.

2. The NLIT is mediated by both the like-signed helical velocity field and the compressive velocity field, consistently with high values of G^{+++} and G^{+0+} for small velocity scales: $q \ll p \approx k$ and $k \ll p \approx q$. The contribution from the compressive velocity field is more nonlocal (compare Fig. 14.8a with Fig. 14.8b) since G^{+0+} vanishes for $k \approx q$. The oppositely signed helical velocity field plays a negligible role since G^{+-+} vanishes for $p \gg k$ and $p \gg q$.

3. The LDT, which corresponds to the large-scale velocity field $p \ll k \approx q$, is mediated by all helical components. The compressive velocity field is, however, geometrically favored, since (1) on the $q = k \pm p$ lines, $G^{+0+} = 1$ while $G^{+++} = G^{+-+} = 0$; and (2) the $q = k$ line, where G^{+++} and G^{+-+} are high at low p, does not play a role in the magnetic

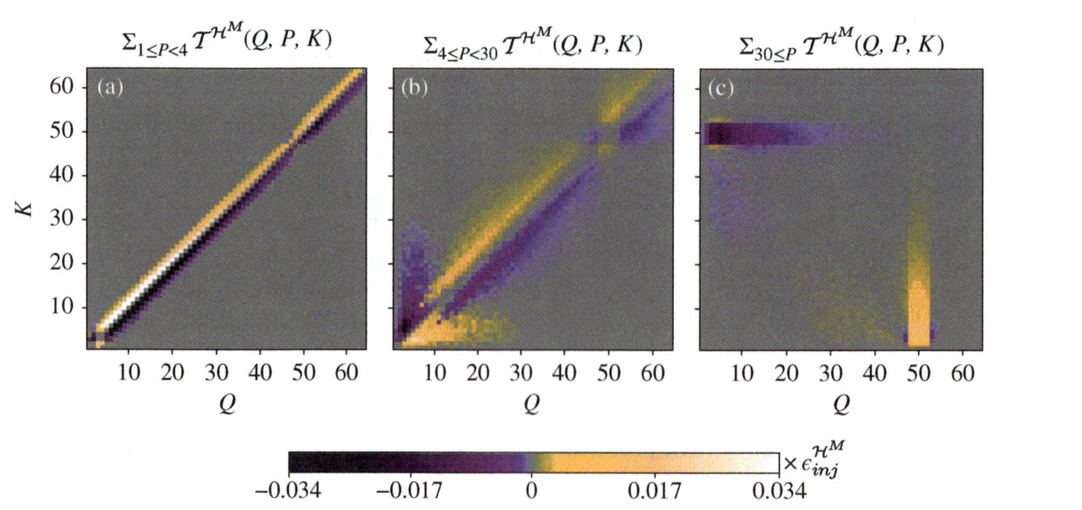

Figure 14.7 Filtering the transfer rates displayed in Fig. 14.6 according to the velocity field shells P. Transfers mediated by: (a) the large-scale velocity field, (b) the intermediate scales, and (c) the small-scale velocity field.

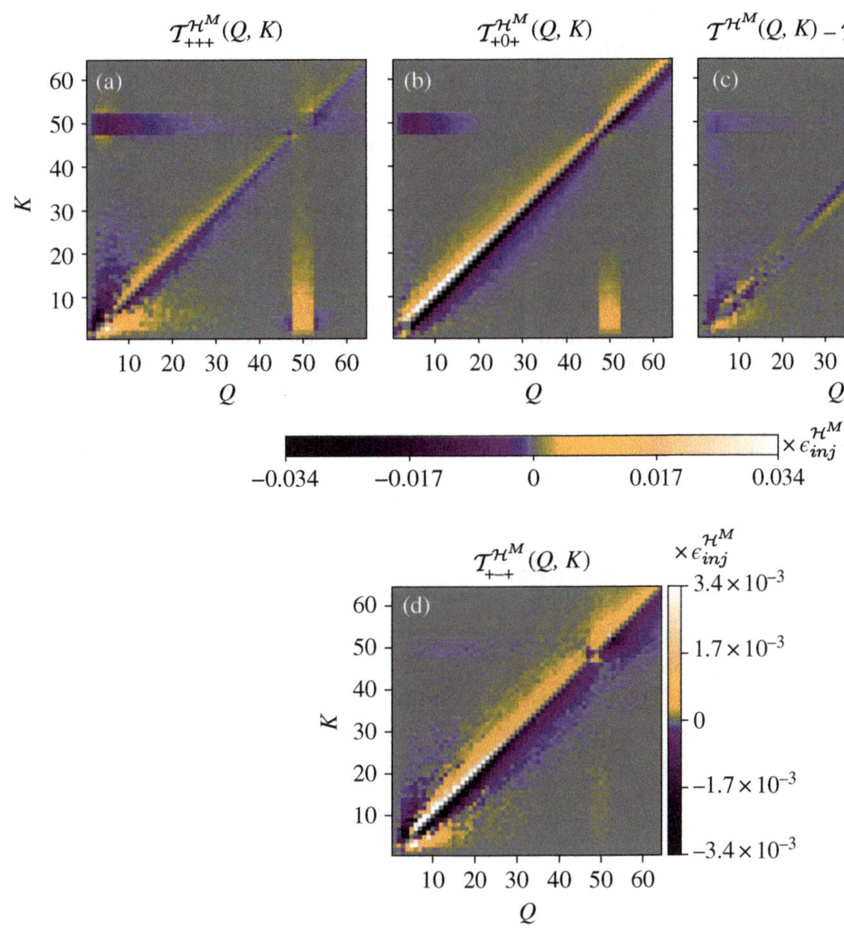

Figure 14.8 Dominant helical transfer terms (equation (14.14)) from Fig. 14.6: (a) mediated by the like-signed helical velocity field; (b) mediated by the compressive part of the velocity field. The remaining transfer terms (c) are relatively small, such as that mediated by the oppositely signed helical velocity field (d). Note the order of magnitude difference between the color bars.

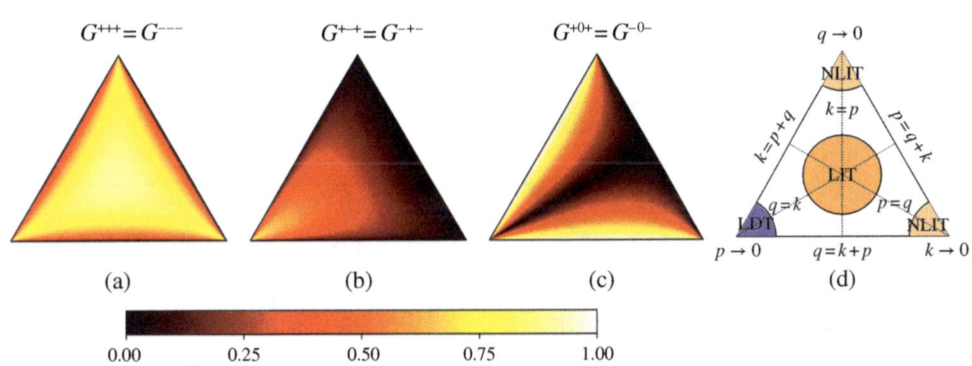

Figure 14.9 (a, b, c): Moduli of the geometric helical factor (equations (14.10)–(14.11)) for the homochiral terms $s_K = s_Q$. (d) Key to read the figures: the triangles are delimited by the lines $k = p + q$, $p = q + k$, and $q = k + p$, where the transfer is from magnetic mode $\boldsymbol{q}$ to magnetic mode $\boldsymbol{k}$ mediated by velocity mode $\boldsymbol{p}$. The region $p \ll q \approx k$ corresponds to the large-scale velocity field (the LDT), the region $p \approx k \approx q$ to the intermediate scale local transfers (the LIT), and the regions $p \approx k \gg q$ and $p \approx q \gg k$ to the small-scale velocity field (the NLIT).

helicity transfer since $\mathcal{T}^{\mathcal{H}^M}(K,K) = 0$. Both solenoidal contributions are on the same order of magnitude for the LDT since $G^{+++} \approx G^{+-+}$ for $p \ll q \approx k$.

These results confirm and extend those from incompressible MHD. Through a comparison of G^{+++} and G^{+-+}, it has been predicted that the transfers mediated by the like-signed helical velocity field should be stronger and more nonlocal than those mediated by the oppositely signed helical velocity field (Linkmann et al., 2017).

A more detailed analysis of the other simulations with higher and lower Mach numbers (Teissier and Müller, 2021b) M01s, M11s, and M1c confirms the results reviewed here.

14.6. SUMMARY AND CONCLUSION

The main results are summarized visually in Fig. 14.10:

1. The magnetic helicity spectra tend to become flatter with higher compressibility. The departure from the incompressible scaling depends more on the driving type than the turbulent Mach number. This suggests that for astrophysical systems of interest, compressible effects are important at already relatively low Mach numbers, provided the turbulence drivers are compressive enough.

2. The Alfvénic helicity (the helicity of the Alfvén velocity $\boldsymbol{b}/\sqrt{\rho}$) presents a scaling law close to the incompressible one, independently of the driving mechanism with a wide range of Mach numbers.

3. The dynamic balance between shear and twisting (Alfvénic balance) can be extended to supersonic MHD when considering the solenoidal part of the specific kinetic energy spectra and the Alfvén velocity in place of the magnetic field.

4. Three phenomena occur in the global picture of the ITMH:

- A local direct transfer (LDT) mediated by the large-scale velocity field. Its compressive part is geometrically favored. The like-signed and oppositely signed helical parts take similar roles.
- A local inverse transfer (LIT) mediated by the intermediate-scale velocity field. The like-signed helical part plays the main role, and the oppositely signed helical part a subdominant role. The importance of the compressive velocity field for this transfer is negligible.
- A nonlocal inverse transfer (NLIT) mediated by the small-scale velocity field. The like-signed helical part

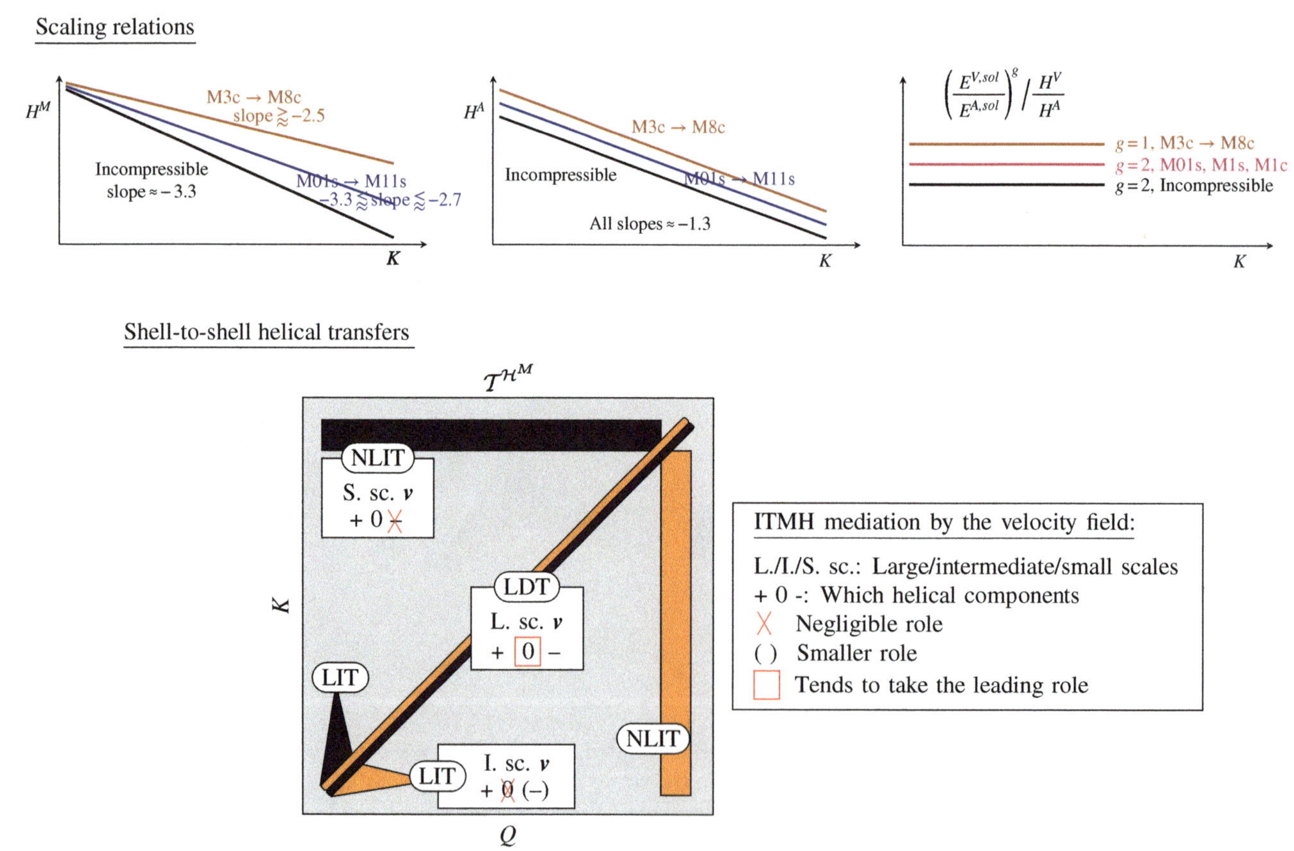

Figure 14.10 Graphical visualisation of the main results.

and the compressive part play the dominant roles, whereas the oppositely signed helical part plays a negligible role.

These results have been gained from direct numerical simulations of the ideal supersonic isothermal MHD equations, subject to both a large-scale mechanical forcing (either purely solenoidal or purely compressive) and a small-scale helical electromotive driving (Teissier and Müller, 2021a,b). A fourth-order numerical scheme allows good accuracy even at the moderate 512^3 resolution. The turbulent Mach numbers span from subsonic to about 10. Even though they are obtained apart from a statistically stationary state, so that transient effects cannot be excluded, confidence in the main results is relatively high because (1) they have been confirmed at higher resolution (1024^3) and (2) their domain of validity spans a wide range of turbulent Mach numbers.

In the presented setting, the magnetic helicity of one sign dominates the system at all scales. However, in geophysical and astrophysical systems of interest, such as dynamos, both signs are expected to be present. This could lead to other scaling laws and to the greater importance of the heterochiral transfer terms, which are more difficult to interpret (see Teissier (2020), sections 7.1 and C.2). Nevertheless, these first results in this simpler setting should be helpful to interpret future investigations.

REFERENCES

Alexakis, A. (2017). Helically decomposed turbulence. *Journal of Fluid Mechanics*, *812*, 752–770.

Alexakis, A., & Biferale, L. (2018). Cascades and transitions in turbulent flows. *Physics Reports*, *767–769*, 1–101.

Alexakis, A., Mininni, P. D., & Pouquet, A. (2006). On the inverse cascade of magnetic helicity. *The Astrophysical Journal*, *640*, 335–343.

Aluie, H. (2011). Compressible turbulence: the cascade and its locality. *Physical Review Letters*, *106*, 174502.

Aluie, H. (2013). Scale decomposition in compressible turbulence. *Physica D*, *247*, 54–65.

Aluie, H. (2017). Coarse-grained incompressible magnetohydrodynamics: analyzing the turbulent cascades. *New Journal of Physics*, *19*, 025008.

Aluie, H., Li, S., & Li, H. (2012). Conservative cascade of kinetic energy in compressible turbulence. *The Astrophysical Journal*, *751*, L29.

Balsara, D., & Pouquet, A. (1999). The formation of large-scale structures in supersonic magnetohydrodynamic flows. *Physics of Plasmas*, *6*(1), 89–99.

Balsara, D. S. (2012). Self-adjusting, positivity preserving high order schemes for hydrodynamics and magnetohydrodynamics. *Journal of Computational Physics*, *231*, 7504–7517.

Balsara, D. S., Garain, S., & Shu, C.-W. (2016). An efficient class of WENO schemes with adaptive order. *Journal of Computational Physics*, *326*, 780–804.

Berger, M. A. (1997). Magnetic helicity in a periodic domain. *Journal of Geophysical Research*, *102*(A2), 2637–2644.

Berger, M. A. (1999). Introduction to magnetic helicity. *Plasma Physics and Controlled Fusion*, *41*, B167–B175.

Berger, M. A., & Field, G. B. (1984). The topological properties of magnetic helicity. *Journal of Fluid Mechanics*, *147*, 133–148.

Biferale, L., & Titi, E. S. (2013). On the global regularity of a helical-decimated version of the 3D Navier-Stokes equations. *Journal of Statistical Physics*, *151*, 1089–1098.

Biferale, L., Musacchio, S., & Toschi, F. (2012). Inverse energy cascade in three-dimensional isotropic turbulence. *Physical Review Letters*, *108*, 164501.

Blackman, E. G. (2015). Magnetic helicity and large scale magnetic fields: a primer. *Space Science Reviews*, *188*, 59–91.

Brandenburg, A. (2001). The inverse cascade and nonlinear alpha-effect in simulations of isotropic helical hydromagnetic turbulence. *The Astrophysical Journal*, *550*, 824–840.

Brandenburg, A., & Subramanian, K. (2005). Astrophysical magnetic fields and nonlinear dynamo theory. *Physics Reports*, *417*, 1–209.

Buchmüller, P., & Helzel, C. (2014). Improved accuracy of high-order WENO finite volume methods on Cartesian grids. *Journal of Scientific Computing*, *61*, 343–368.

Canuto, C., Hussaini, M. Y., Quarteroni, A., & Zang, T. A. (1988). *Spectral methods in fluid dynamics*. Springer-Verlag.

Caroll, B. W., & Ostlie, D. A. (2007). *An introduction to modern astrophysics* (2nd ed.). Pearson Education.

Christensson, M., Hindmarsh, M., & Brandenburg, A. (2001). Inverse cascade in decaying three-dimensional magnetohydrodynamic turbulence. *Physical Review E*, *64*, 056405.

Constantin, P., & Majda, A. (1988). The Beltrami spectrum for incompressible fluid flows. *Communications in Mathematical Physics*, *115*, 435–456.

Elmegreen, B. G., & Scalo, J. (2004). Interstellar turbulence I: observations. *Annual Review of Astronomy and Astrophysics*, *42*, 211–273.

Elsässer, W. M. (1956). Hydromagnetic dynamo theory. *Reviews of Modern Physics*, *28*(2), 135–163.

Federrath, C. (2013). On the universality of supersonic turbulence. *Monthly Notices of the Royal Astronomical Society*, *436*, 1245–1257.

Federrath, C., Roman-Duval, J., Klessen, R. S., Schmidt, W., & Mac Low, M.-M. (2010). Comparing the statistics of interstellar turbulence in simulations and observations, solenoidal versus compressive turbulence forcing. *Astronomy and Astrophysics*, *512*, A81.

Ferrand, R., Galtier, S., Sahraoui, F., & Federrath, C. (2020). Compressible turbulence in the interstellar medium: new insights from a high-resolution supersonic turbulence simulation. *The Astrophysical Journal*, *904*, 160.

Finn, J. M., & Antonsen, Jr., T. M. (1985). Magnetic helicity: what is it and what is it good for? *Comments on Plasma Physics and Controlled Fusion*, *9*(3), 111–126.

Frisch, U., Pouquet, A., Léorat, J., & Mazure, A. (1975). Possibility of an inverse cascade of magnetic helicity in magnetohydrodynamic turbulence. *Journal of Fluid Mechanics*, *68* (Part 4), 769–778.

Galtier, S., & Banerjee, S. (2011). Exact relation for correlation functions in compressible isothermal turbulence. *Physical Review Letters*, *107*, 134501.

Graham, J. P., Mininni, P. D., & Pouquet, A. (2011). High Reynolds number magnetohydrodynamic turbulence using a Lagrangian model. *Physical Review E*, *84*, 016314.

Harten, A., Engquist, B., Osher, S., & Chakravarthy, S. R. (1987). Uniformly high order accurate essentially non-oscillatory schemes, III. *Journal of Computational Physics*, *71*, 231–303.

Jiang, G.-S., & Shu, C.-W. (1996). Efficient implementation of weighted ENO schemes. *Journal of Computational Physics*, *126*, 202–228.

Ketcheson, D. I. (2008). Highly efficient strong stability-preserving Runge-Kutta methods with low-storage implementations. *Society for Industrial and Applied Mathematics Journal on Scientific Computing*, *30*(4), 2113–2136.

Kivelson, M. G. (1995). Physics of space plasmas. In M. G. Kivelson & C. T. Russell (Eds.), *Introduction to space physics* (chapter 2). Cambridge University Press.

Kritsuk, A. G., Wagner, R., & Norman, M. L. (2013). *Energy cascade and scaling in supersonic isothermal turbulence*. *Journal of Fluid Mechanics*, *729*, R1.

Kumar, A., & Rust, D. M. (1996). *Interplanetary magnetic clouds, helicity conservation, and current-core flux-ropes*. *Journal of Geophysical Research*, *101*, 667–684.

Lessinnes, T., Plunian, F., & Carati, D. (2009). Helical shell models for MHD. *Theoretical and Computational Fluid Dynamics*, *23*, 439–450.

Linkmann, M., & Dallas, V. (2016). Large-scale dynamics of magnetic helicity. *Physical Review E*, *94*, 053209.

Linkmann, M., & Dallas, V. (2017). Triad interactions and the bidirectional turbulent cascade of magnetic helicity. *Physical Review Fluids*, *2*, 054605.

Linkmann, M., Berera, A., McKay, M., & Jäger, J. (2016). Helical mode interactions and spectral transfer processes in magnetohydrodynamic turbulence. *Journal of Fluid Mechanics*, *791*, 61–96.

Linkmann, M., Sahoo, G., McKay, M., Berera, A., & Biferale, L. (2017). Effects of magnetic and kinetic helicities on the growth of magnetic fields in laminar and turbulent flows by helical Fourier decomposition. *The Astrophysical Journal*, *836*(1), 26.

Liu, X.-D., Osher, S., & Chan, T. (1994). Weighted essentially non-oscillatory schemes. *Journal of Computational Physics*, *115*, 200–212.

Low, B. C. (1994). Magnetohydrodynamic processes in the solar corona: Flares, coronal mass ejections, and magnetic helicity. *Physics of Plasmas*, *1*, 1684–1690.

McCorquodale, P., & Colella, P. (2011). A high-order finite-volume method for conservation laws on locally refined grids. *Communications in Applied Mathematics and Computational Science*, *6*(1), 1–25.

Meneguzzi, M., Frisch, U., & Pouquet, A. (1981). Helical and nonhelical turbulent dynamos. *Physical Review Letters*, *47*(15), 1060–1064.

Mininni, P. D., & Pouquet, A. (2009). Finite dissipation and intermittency in magnetohydrodynamics. *Physical Review E*, *80*, 025401.

Moffatt, H. K. (1969). The degree of knottedness of tangled vortex lines. *Journal of Fluid Mechanics*, *35*, 117–129.

Müller, W.-C., & Malapaka, S. K. (2010). Understanding nonlinear cascades in magnetohydrodynamic turbulence by statistical closure theory. *ASP Conference Series*, *429*.

Müller, W.-C., Malapaka, S. K., & Busse, A. (2012). Inverse cascade of magnetic helicity in magnetohydrodynamic turbulence. *Physical Review E*, *85*, 015302.

Plunian, F., Stepanov, R., & Verma, M. K. (2019). On uniqueness of transfer rates in magnetohydrodynamic turbulence. *Journal of Plasma Physics*, *85*.

Politano, H., Gomez, T., & Pouquet, A. (2003). Von Kármán-Howarth relationship for helical magnetohydrodynamic flows. *Physical Review E*, *68*, 026315.

Pouquet, A., & Patterson, G. S. (1978). Numerical simulation of helical magnetohydrodynamic turbulence. *Journal of Fluid Mechanics*, *85*(2), 305–323.

Pouquet, A., Frisch, U., & Léorat, J. (1976). Strong MHD helical turbulence and the nonlinear dynamo effect. *Journal of Fluid Mechanics*, *77*(2), 321–354.

Shu, C.-W. (2009). High order weighted essentially nonoscillatory schemes for convection-dominated problems. *SIAM Review*, *51*(1), 82–126.

Stepanov, R., Frick, P., & Mizeva, I. (2015). Joint inverse cascade of magnetic energy and magnetic energy in MHD turbulence. *The Astrophysical Journal Letters*, *798*.

Teissier, J.-M. (2020). *Magnetic helicity inverse transfer in isothermal supersonic magnetohydrodynamic turbulence* (PhD thesis). Technische Universität Berlin. http://dx.doi.org/10.14279/depositonce-9439

Teissier, J.-M., & Müller, W.-C. (2021a). Inverse transfer of magnetic helicity in direct numerical simulations of compressible isothermal turbulence: scaling laws. *Journal of Fluid Mechanics*, *915*, A23.

Teissier, J.-M., & Müller, W.-C. (2021b). Inverse transfer of magnetic helicity in direct numerical simulations of compressible isothermal turbulence: helical transfers. *Journal of Fluid Mechanics*, *921*, A7. Note: The normalization factors of $1/(2\sqrt{2})$ or $1/2$ in equations 5.5, 5.6, A5, A6, and fig. 8 are missing.

Verma, P. S., Teissier, J.-M., Henze, O., & Müller, W.-C. (2019). Fourth-order accurate finite-volume CWENO scheme for astrophysical MHD problems. *Monthly Notices of the Royal Astronomical Society*, *482*, 416–437.

Vishniac, E. T., & Cho, J. (2001). Magnetic helicity conservation and astrophysical dynamos. *The Astrophysical Journal*, *550*, 752–760.

Waleffe, F. (1992). The nature of triad interactions in homogeneous turbulence. *Physics of Fluids A: Fluid Dynamics*, *4*, 350–363.

Woltjer, L. (1958). A theorem on force-free magnetic fields. *Proceedings of the National Academy of Sciences*, *44*(6), 489–491.

Wu, K., & Shu, C.-W. (2019). Provably positive high-order schemes for ideal magnetohydrodynamics: analysis on general meshes. *Numerische Mathematik*, *142*, 995–1047.

15

Nonlinear Mean-Field Dynamos With Magnetic Helicity Transport and Solar Activity: Sunspot Number and Tilt

Nathan Kleeorin[1,2], Kirill Kuzanyan[2,3], Igor Rogachevskii[1,4], and Nikolai Safiullin[2,5]

ABSTRACT

In this chapter, we discuss a mean field solar dynamo model with algebraic and dynamic nonlinearities, various mechanisms of sunspot formation, and prediction of solar activity. The algebraic nonlinearity describes the quenching of the alpha effect, turbulent magnetic diffusion, and the effective pumping velocity due to feedback from the growing large-scale magnetic field on the fluid motion. The dynamic nonlinearity is due to the evolution of the magnetic helicity of the small-scale magnetic field during the nonlinear stage of the dynamo; it is derived from conservation of the total (large-scale plus small-scale) magnetic helicity for very small microscopic magnetic diffusivity. We discuss observations of magnetic helicity in the Sun and their connection with the nonlinear mean field dynamo. We derive a budget equation for sunspot numbers taking into account sunspot formation mechanism due to the negative effective magnetic pressure instability. To predict solar activity, we use dynamo simulations as input to an artificial neural network that learns sunspot dynamics from available observations. Finally, we analyze the contribution of magnetic helicity transport to the formation of tilt in sunspot bipolar regions and compare the results with available observational data over the last 10 solar cycles (15–24).

15.1. INTRODUCTION

Observations of large-scale magnetic fields in the Sun demonstrate their relation to the global properties of solar cyclicity: for example, to the 11- or 22-year periodicity of sunspot formation. The theory of magnetic field generation by turbulent motion of a conducting fluid (dynamo theory) allows us to understand the origin of solar magnetism (Moffatt, 1978; Parker, 1979; Krause & Rädler, 1980; Zeldovich et al., 1983; Rüdiger et al., 2013; Moffatt & Dormy, 2019; Gershberg et al., 2020; Rogachevskii, 2021). The generation (self-excitation) of a magnetic field by a dynamo implies an increase in both magnetic energy and magnetic flux. Not all problems related to the dynamics of magnetic fields relate to dynamo theory. For example, the formation of strongly inhomogeneous magnetic structures, such as sunspots and active regions (ARs), cannot be directly described by dynamo theory because these processes are accompanied not by an increase in magnetic flux but rather by its redistribution in space (Parker, 1955a). However, these effects are extremely important for applying dynamo theory to the Sun because they determine the local magnetic fields on the solar surface.

[1]Department of Mechanical Engineering, Ben-Gurion University of the Negev, Beer-Sheva, Israel
[2]Department of Turbulent Flows, Institute of Continuous Media Mechanics, Russian Academy of Sciences, Perm, Russia
[3]Department of Solar Physics, Pushkov Institute of Terrestrial Magnetism, Ionosphere and Radiowave Propagation (IZMIRAN), Russian Academy of Sciences, Moscow, Russia
[4]Department of Astrophysics, Nordic Institute for Theoretical Physics, KTH Royal Institute of Technology and Stockholm University, Stockholm, Sweden
[5]Department of Information Security, Ural Federal University, Ekaterinburg, Russia

Helicities in Geophysics, Astrophysics, and Beyond, Geophysical Monograph 283, First Edition.
Edited by Kirill Kuzanyan, Nobumitsu Yokoi, Manolis K. Georgoulis, and Rodion Stepanov.

DOI:10.1002/9781119841715.ch15

The complete solar dynamo theory is based on the induction equation, the Navier-Stokes equation, the continuity equation, the entropy transport equation, the equation of state, the Saha equation for ionization, the radiation transfer equation, and the equation for the Rosseland mean absorption coefficient for electromagnetic radiation of solar plasma. The solution to a system of these equations is a very complicated problem with many unknown parameters, so we can only study particular cases.

Historically, the first successful dynamo model of the axisymmetric large-scale solar magnetic field was developed by Parker (1955b). According to this model, the toroidal magnetic field results from stretching the poloidal magnetic field by the nonuniform (differential) rotation. On the other hand, the poloidal magnetic field is produced by helical turbulent convective-cyclonic plasma motions due to the combined effect of the Coriolis force and inhomogeneity of convective turbulence (or density-stratified convection). This dynamo model is referred to as the $\alpha\Omega$ dynamo. The first quantitative theory based on the quasilinear approach was formulated by Steenbeck et al. (1966). This theory allows us to calculate the turbulent electromotive force, including (1) the α effect caused by helical turbulent motions of plasma, (2) the turbulent diffusion of the mean magnetic field, and (3) the effective pumping velocity caused by inhomogeneous turbulence.

Magnetic fields are generated by the helical turbulent motion of solar plasma in the presence of small yet finite magnetic diffusion due to the electrical conductivity of the plasma. In the solar convective zone under the Sun's surface, there is fully developed convective turbulence where rotation is strongly nonuniform (referred to as differential rotation). The solar dynamo mechanism produces large-scale magnetic fields, typically at scales on the order of the solar radius. These large-scale fields govern solar activity, which is observed as sunspots and ARs.

A small-scale dynamo (Kazantsev, 1968; Zeldovich et al., 1990; Childress & Gilbert, 1995) is not caused by rotation and helical motions. Magnetic fields generated by a small-scale dynamo are random and at much smaller scales than the integral scale of turbulence. To generate a small-scale dynamo, it is enough to overcome a small generation threshold in the magnetic Reynolds number, which depends on the magnetic Prandtl number, degree of compressibility of plasma motions, etc. (Batchelor, G. K., 1950; Kazantsev, 1968; Moffatt, 1978; Kulsrud & Anderson, 1992; Rogachevskii & Kleeorin, 1997). The magnetic Reynolds number in the Sun is much larger than this threshold. Nevertheless, the transverse size of the magnetic flux tubes produced by this mechanism can increase from several tens of kilometers to the integral scale of turbulence when the kinetic helicity does not vanish. In solar conditions, the small-scale dynamo can produce magnetic pores.

In this chapter, we discuss nonlinear mean-field dynamos (MFDs) with magnetic helicity transport and prediction of solar activity. In section 15.2, we outline basic ideas of the nonlinear MFD; in section 15.3, we consider dynamical quenching caused by magnetic helicity transport; and in section 15.4, we discuss various mechanisms related to the formation of sunspots. In section 15.5, we discuss observations of magnetic helicity in the Sun and their connection to nonlinear MFD theories. Section 15.6 is devoted to the nonlinear mean-field solar dynamo with a budget equation for the evolution of sunspot numbers, and section 15.7 outlines the possibility of forecasting solar activity by combining numerical simulations of the nonlinear MFD equations with a neural network. Finally, in section 15.8, we consider various mechanisms of formation of the mean tilt of bipolar sunspot regions and analyze the role of magnetic helicity transport in its latitudinal dependence and evolution.

15.2. MEAN-FIELD DYNAMOS

Let us review some basic ideas about the MFD. In the framework of magnetohydrodynamics (MHD), the generation of the magnetic field $\boldsymbol{B}(t, \boldsymbol{x})$ is determined by the induction equation

$$\frac{\partial \boldsymbol{B}}{\partial t} = \boldsymbol{\nabla} \times (\boldsymbol{U} \times \boldsymbol{B} - \eta \boldsymbol{\nabla} \times \boldsymbol{B}), \qquad (15.1)$$

where $\boldsymbol{U}$ is the fluid velocity and η is the coefficient of magnetic diffusivity caused by the electrical conductivity of plasma. For studies of the generation and long-term evolution of the large-scale magnetic field, one may apply a mean-field approach. In this framework, magnetic and velocity fields, fluid pressure, and density are decomposed into mean and fluctuating parts, where the fluctuations have zero mean values (referred to as Reynolds averaging). In this approach, the magnetic field is $\boldsymbol{B} = \overline{\boldsymbol{B}} + \boldsymbol{b}$, where $\overline{\boldsymbol{B}} = \langle \boldsymbol{B} \rangle$ is the mean magnetic field, $\boldsymbol{b}$ are magnetic fluctuations, and $\langle \boldsymbol{b} \rangle = 0$, where the angular brackets $\langle ... \rangle$ denote ensemble averaging. Similarly, the velocity field is decomposed as $\boldsymbol{U} = \overline{\boldsymbol{U}} + \boldsymbol{u}$, where $\overline{\boldsymbol{U}} = \langle \boldsymbol{U} \rangle$ is the mean fluid velocity associated with differential rotation, $\boldsymbol{u}$ are velocity fluctuations, and $\langle \boldsymbol{u} \rangle = 0$. We assume scale separation in space ($L_B \gg \ell_0$) and time ($t_B \gg \tau_0$), where L_B is the mean magnetic field variation scale and t_B is the mean magnetic field variation time scale, with τ_0 being the turbulent time at the integral scale ℓ_0.

The equation for the evolution of a large-scale magnetic field is obtained from equation (15.1) by averaging over an ensemble of fluctuations:

$$\frac{\partial \overline{\boldsymbol{B}}}{\partial t} = \boldsymbol{\nabla} \times \left(\overline{\boldsymbol{U}} \times \overline{\boldsymbol{B}} + \langle \boldsymbol{u} \times \boldsymbol{b} \rangle \right) + \eta \, \Delta \overline{\boldsymbol{B}}, \tag{15.2}$$

where $\boldsymbol{\mathcal{E}} \equiv \langle \boldsymbol{u} \times \boldsymbol{b} \rangle$ is the turbulent electromotive force (EMF) and η is the coefficient of magnetic diffusion caused by the electrical conductivity of plasma. In the pioneering work by Steenbeck et al. (1966), the expression for the turbulent EMF was derived for isotropic and helical turbulence using the quasilinear approach in a high conductivity limit (i.e., for large magnetic Reynolds number):

$$\boldsymbol{\mathcal{E}} = \alpha \overline{\boldsymbol{B}} + \boldsymbol{V}^{\rm eff} \times \overline{\boldsymbol{B}} - \eta_T (\boldsymbol{\nabla} \times \overline{\boldsymbol{B}}), \tag{15.3}$$

where $\alpha = -(\tau_0/3) \, \langle \boldsymbol{u} \cdot (\boldsymbol{\nabla} \times \boldsymbol{u}) \rangle$ describes the kinetic α effect, $\eta_T = (\tau_0/3) \, \langle \boldsymbol{u}^2 \rangle$ is the turbulent magnetic diffusion coefficient, $\langle \boldsymbol{u} \cdot (\boldsymbol{\nabla} \times \boldsymbol{u}) \rangle$ is the kinetic helicity, $\boldsymbol{V}^{\rm eff}$ is the effective pumping velocity, and $\boldsymbol{V}^{\rm eff} = -\boldsymbol{\nabla}\eta_T/2$. The velocity $\boldsymbol{V}^{\rm eff}$ describes the turbulent diamagnetic effect predicted by Zeldovich (1957). This effect causes the mean magnetic field to be pushed out from the region of intense velocity fluctuations: that is, it is pushed out from regions where there is turbulence. In convective density-stratified turbulence, the effective pumping velocity due to the inhomogeneity of density is proportional to $\propto \eta_T (\boldsymbol{\nabla} \overline{\rho})/\overline{\rho}$ (Rogachevskii & Kleeorin, 2006). This effect is caused by the contribution of the vertical heat flux to the effective pumping velocity; in the solar convective zone, it causes downward pumping of the mean magnetic field.

Equation (15.3) was derived in a kinematic approximation where a one-way coupling was considered (Steenbeck et al., 1966; Krause & Rädler, 1980; Roberts & Stix, 1971). This implies that turbulence affects the magnetic field, but there is no feedback effect of the magnetic field on background turbulence. Equations (15.2)–(15.3) were also derived using various analytical approaches (the dimensional analysis, the quasilinear approach, the tau approaches, the path-integral approach, the functional approach) (Moffatt, 1978; Krause & Rädler, 1980; Zeldovich et al., 1983, 1988; Vainshtein & Kichatinov, 1983; Rüdiger & Kichatinov, 1993; Kitchatinov et al., 1994; Rogachevskii & Kleeorin, 1997; Field et al., 1999; Rogachevskii & Kleeorin, 2000, 2001, 2004, 2006; Rüdiger et al., 2013; Rogachevskii, 2021).

The main reason for the large-scale magnetic field generation is the α effect due to the kinetic helicity, $H_u = \langle \boldsymbol{u} \cdot (\boldsymbol{\nabla} \times \boldsymbol{u}) \rangle$. A nonzero kinetic helicity implies that the number of left-handed rotating turbulent eddies at a given instant does not exactly equal the number of right-handed rotating eddies (i.e., the number of eddies rotating clockwise is not equal to the number rotating counterclockwise). A mechanism of the α effect is as follows. Both the left-handed and right-handed rotating eddies cause deformation of the original magnetic field. Since the number of left-handed rotating turbulent eddies at a given instant does not exactly equal the number of right-handed rotating eddies, the total effect of deformation of the original magnetic field is not zero. This causes the generation of a large-scale magnetic field. In real turbulent systems, kinetic helicity is produced in inhomogeneous rotating convective turbulence (or density-stratified rotating turbulent convection). Downward and upward rotating turbulent eddies cause deformation of the magnetic field lines. Because the turbulence is inhomogeneous (or density-stratified), it breaks the symmetry between the downward and upward eddies, resulting in a nonvanishing residual effect on the mean magnetic field and thus generating the mean electric current from the original mean magnetic field.

Since the formulation of the MFD approach (Steenbeck et al., 1966; Krause & Rädler, 1980; Roberts & Stix, 1971), theories of solar, stellar, and galactic dynamos have continued to improve for more than 55 years (Moffatt, 1978; Parker, 1979; Krause & Rädler, 1980; Rädler, 1980, 1982; Zeldovich et al., 1983; Ruzmaikin et al., 1988; Rüdiger et al., 2013; Moffatt & Dormy, 2019; Gershberg et al., 2020; Rogachevskii, 2021). The well-known mechanism for the generation of stellar and galactic magnetic fields is as follows. Solar and galactic dynamos are caused by a combined effect of differential rotation and turbulent motions with net kinetic helicity. In the solar convective zone, kinetic helicity is produced by a rotating density-stratified convection, while in galactic discs, it is caused by rotating density-stratified turbulence produced by supernova explosions. The latter causes the kinetic α effect. The nonuniform large-scale rotation is due to anisotropic turbulent viscosity (Kippenhahn, 1963; Rüdiger, 1989; Rüdiger et al., 2013) and a modification of the Reynolds stress by the turbulent heat flux in density-stratified convective turbulence (Kleeorin & Rogachevskii, 2006; Rogachevskii & Kleeorin, 2018). The combined effect of helical turbulent motions and differential rotation, referred to as an $\alpha\Omega$ or $\alpha^2\Omega$ dynamo, results in dynamo instability. In the solar dynamo, the instability is oscillatory: it generates the dynamo waves predicted by Parker (1955b). In the galactic dynamo, the instability is non-oscillatory (Ruzmaikin et al., 1988).

Applying different approaches, the α effect was derived for rotating density-stratified turbulence (Krause & Rädler, 1980; Rüdiger & Kichatinov, 1993; Rädler et al., 2003; Kleeorin & Rogachevskii, 2003). It was shown using the quasilinear approach (Steenbeck et al., 1966; Krause & Rädler, 1980) that the α effect for weak rotation is $\alpha \approx -\ell_0^2 \, (\boldsymbol{\Omega} \cdot \boldsymbol{\nabla}) \ln \left(\overline{\rho} \, u_{\rm rms} \right)$, where ℓ_0 is the energy-containing turbulence scale, $\overline{\rho}$ is the mean plasma density, and $u_{\rm rms} = \langle \boldsymbol{u}^2 \rangle^{1/2}$. In the convective zone, the mean density decreases by 10^7 times toward the solar surface, while the turbulent kinetic energy decreases by 10^3 times, and $\overline{\rho}\, u_{\rm rms}^3$ does not change in the radial direction (Spruit, 1974). It was found analytically as well as in numerical simulations of strong convective and forced turbulence that the α effect for a weak rotation is given by $\alpha \approx -\ell_0^2(\boldsymbol{\Omega} \cdot \boldsymbol{\nabla}) \ln(\overline{\rho}^{\zeta}\, u_{\rm rms})$, where $\zeta = 1/2$. However, several simulations are consistent with $\zeta = 1/3$ (Brandenburg et al., 2013a).

Note that in the framework of the mixing-length theory, the integral turbulence scale, ℓ_0, is of the order of the mixing-length scale $\ell_{\rm M}$, assuming that it is equal to the height of a typical convective cell. The mixing-length scale, in turn, is of the order of the pressure-height scale $L_{\rm P}$. For example, in the well-known model of the convective zone (Spruit, 1974), it is assumed that $\ell_{\rm M} = 1.6 L_{\rm P}$. On the other hand, according to the results of laboratory experiments with convective turbulence at large Rayleigh number (Bukai et al., 2009), the energy-containing turbulent scale $\ell_0 = 0.1 L_T$. This fact is taken into account when constructing models of the solar dynamo (Kleeorin et al., 2016).

Nonlinear effects limit the exponential growth of the large-scale magnetic field. The simplest feedback effect is modeled (Iroshnikov, 1971; Roberts & Soward, 1975)

$$\alpha(\overline{\boldsymbol{B}}) = \alpha_0 \left(1 + \frac{\overline{\boldsymbol{B}}^2}{\overline{B}_{\rm eq}^2} \right)^{-1}, \qquad (15.4)$$

where $\overline{B}_{\rm eq} = \left(4\pi\overline{\rho}\,\langle \boldsymbol{u}^2 \rangle\right)^{1/2}$ is the equipartition mean magnetic field. Equation (15.4) describes the "alfvenization" of solar plasma as the mean magnetic field increases. This means in addition to fully developed Kolmogorov turbulence; there is an ensemble of nonlinear Alfvén waves (Moffatt, 1978) when the mean magnetic field increases. For weak mean magnetic fields, the turbulent time is $\tau_0 = \ell_0 / \langle \boldsymbol{u}^2 \rangle^{1/2}$. For strong fields, the typical turbulent time is the Alfvén time $\tau_{\rm A} = \ell_0 / V_{\rm A}$, where $V_{\rm A} = \overline{B}/(4\pi\overline{\rho})^{1/2}$. In the intermediate situation, of course, the faster of the two mechanisms dominates: for example, the inverse squared turbulent time can be determined as a linear combination, $C_1 \tau_{\rm A}^{-2} + \tau_0^{-2}$, where $C_1 \sim 1$ is an empirical constant. Therefore, the amplitude of the mean magnetic field when α quenching becomes essential is estimated as the equipartition field $\overline{B}_{\rm eq}$. This estimate for the galactic dynamos agrees with observations (Ruzmaikin et al., 1988).

The previously mentioned nonlinearity is referred to as algebraic nonlinearity and is due to the feedback of the growing mean magnetic field on the fluid motion. Quantitative theories of the algebraic nonlinearities of turbulent EMF have been developed using the quasilinear method for small Rm and Re (Rüdiger & Kichatinov, 1993; Kitchatinov et al., 1994; Rüdiger et al., 2013) and the τ approach for large Rm and Re (Field et al., 1999; Rogachevskii & Kleeorin, 2000, 2001, 2004, 2006), where Re and Rm are the fluid and magnetic Reynolds numbers.

Compressibility effects on the turbulent EMF have also been investigated in a number of studies by applying various analytical approaches (Krause & Rädler, 1980; Rädler et al., 2011; Rogachevskii et al., 2018; Yokoi, N., 2018a,b). In particular, the compressibility of fluid flow reduces the turbulent magnetic diffusivity at both small (Krause & Rädler, 1980) and large Rm (Rogachevskii et al., 2018). The reduction of turbulent magnetic diffusivity by the compressibility of fluid flow has also been confirmed by numerical simulations for an irrotational homogeneous deterministic flow (Rädler et al., 2011). The compressibility of fluid flow decreases the α effect for arbitrary Rm and reduces the magnetic pumping for small Rm. However, compressibility does not influence the magnetic pumping for large Rm (Rogachevskii et al., 2018).

15.3. DYNAMIC NONLINEARITY: TRANSPORT OF MAGNETIC HELICITY

In addition to algebraic nonlinearity, there is also dynamic quenching due to the dynamics of the magnetic helicity density. Indeed, the total α effect is the sum of the kinetic α effect, $\alpha^{\rm v}$, and the magnetic α effect, $\alpha^{\rm m}$, where $\alpha^{\rm v}$ is determined by the kinetic helicity $H_{\rm u} = \langle \boldsymbol{u} \cdot (\boldsymbol{\nabla} \times \boldsymbol{u}) \rangle$, while $\alpha^{\rm m}$ is determined by the current helicity $H_{\rm c} = \langle \boldsymbol{b} \cdot (\boldsymbol{\nabla} \times \boldsymbol{b}) \rangle$ (Pouquet et al., 1976). The dynamics of the current helicity H_c is described by the evolution of the small-scale magnetic helicity density $H_{\rm m} = \langle \boldsymbol{a} \cdot \boldsymbol{b} \rangle$, where $\boldsymbol{b} = \boldsymbol{\nabla} \times \boldsymbol{a}$. The total magnetic helicity, $H^{\rm tot} = \int (H_{\rm M} + H_{\rm m})\, dr^3$, is conserved for negligible η. Here, $H_{\rm M} = \overline{\boldsymbol{A}} \cdot \overline{\boldsymbol{B}}$ is the magnetic helicity density of the large-scale field $\overline{\boldsymbol{B}} = \boldsymbol{\nabla} \times \overline{\boldsymbol{A}}$.

The governing equation for $H_{\rm m}$ was derived for isotropic turbulence by Kleeorin & Ruzmaikin (1982) and for arbitrary anisotropic turbulence by Kleeorin & Rogachevskii (1999). This equation was used for the analytical study of solar dynamos (Kleeorin et al., 1994,1995) as well as for mean-field numerical modeling of solar and galactic dynamos (Covas et al., 1997, 1998; Kleeorin et al., 2000, 2002, 2003a,b; Blackman and Field, 2002; Brandenburg & Subramanian, 2005).

In the process of the MFD, as $\overline{\boldsymbol{B}}$ is amplified, $H_{\rm M}$ grows due to the source term $2\boldsymbol{\mathcal{E}} \cdot \overline{\boldsymbol{B}}$ according to

$$\frac{\partial H_{\rm M}}{\partial t} + \boldsymbol{\nabla} \cdot \boldsymbol{F}_{\rm M} = 2\boldsymbol{\mathcal{E}} \cdot \overline{\boldsymbol{B}} - 2\eta H_C, \qquad (15.5)$$

where $\boldsymbol{\mathcal{E}} = \langle \boldsymbol{u} \times \boldsymbol{b} \rangle$ is the turbulent EMF and $\boldsymbol{F}_{\rm M}$ is the flux of $H_{\rm M}$. Due to the conservation of $H^{\rm tot}$, the small-scale magnetic helicity density $H_{\rm m}$ varies, and its dynamics is described by (Kleeorin & Ruzmaikin, 1982; Zeldovich et al., 1983; Kleeorin et al., 1994, 1995; Kleeorin & Rogachevskii, 1999):

$$\frac{\partial H_{\rm m}}{\partial t} + \nabla \cdot \boldsymbol{F}_{\rm m} = -2\boldsymbol{\mathcal{E}} \cdot \overline{\boldsymbol{B}} - 2\eta H_c, \qquad (15.6)$$

where $-2\boldsymbol{\mathcal{E}} \cdot \overline{\boldsymbol{B}}$ is the source of $H_{\rm m}$, $\boldsymbol{F}_{\rm m}$ is the flux of $H_{\rm m}$, and $2\eta H_c$ describes decay of $H_{\rm m}$. The source of $H_{\rm m}$ and $H_{\rm M}$ is only located in a turbulent region (see equations (15.5) and (15.6)). The typical decay time of $H_{\rm m}$ is $T_{\rm m} = \tau_0\,{\rm Rm}$, which is much larger than $\tau_0 = \ell_0/u_0$, where ${\rm Rm} = \ell_0 u_0/\eta$. The current helicity $H_{\rm c}$ of the small-scale field is proportional to $H_{\rm m}$ when inhomogeneity of turbulence is weak (Kleeorin & Rogachevskii, 1999).

Using the steady-state solution of equation (15.6) with a zero turbulent magnetic helicity flux $\mathbf{F}_{\rm m} = 0$, Vainshtein & Cattaneo (1992) and Gruzinov & Diamond (1994) have concluded that the critical field, $\overline{B}_{\rm cr}$, at which dynamic α quenching becomes significant is $\overline{B}_{\rm cr} = \overline{B}_{\rm eq}\,{\rm Rm}^{-1/2}$. In astrophysics, such as in galactic discs and the solar convective zone, Rm is very large, implying that the saturated magnetic field is much less than $\overline{B}_{\rm eq}$. This effect is known as catastrophic quenching of the α effect. However, the observed magnetic field strength in spiral galaxies is about $\overline{B}_{\rm eq}$ (Ruzmaikin et al., 1988). This conclusion about catastrophic quenching of the α effect is only valid when the turbulent magnetic helicity flux, $\boldsymbol{F}_{\rm m}$, vanishes.

To avoid the catastrophic quenching of the α effect, a nonzero turbulent magnetic helicity flux $\boldsymbol{F}_{\rm m}$ should be taken into account (Kleeorin et al., 2000). This implies that open boundary conditions, which allow magnetic helicity transport, should be used in numerical simulations (Blackman and Field, 2000). It was demonstrated in numerical simulations of a galactic dynamo, by using equation (15.6) as a main nonlinearity with the open boundary conditions, that the saturated mean magnetic field is about $\overline{B}_{\rm eq}$ (Kleeorin et al., 2000, 2002, 2003a,b; Blackman and Brandenburg, 2002; Brandenburg & Subramanian, 2005; Shukurov et al., 2006; Del Sordo et al., 2013).

Note that various fluxes $\mathbf{F}_{\rm m}$ have been suggested in different studies (Kleeorin & Rogachevskii, 1999; Kleeorin et al., 2000, 2002; Vishniac & Cho, 2001; Subramanian & Brandenburg, 2004; Brandenburg & Subramanian, 2005). Turbulent magnetic helicity fluxes have recently been derived (Kleeorin & Rogachevskii, 2022) using the spectral τ approach and applying the Coulomb gauge in density-stratified turbulence. Turbulent magnetic helicity fluxes contain nongradient and gradient contributions. The nongradient magnetic helicity flux is proportional to a nonlinear effective velocity (which vanishes in the absence of the density stratification) multiplied by small-scale magnetic helicity, while the gradient contribution describes turbulent magnetic diffusion of small-scale magnetic helicity. In addition, turbulent magnetic helicity fluxes include source terms proportional to the kinetic α effect or its gradients and contributions caused by large-scale shear. Note also that turbulent magnetic helicity fluxes due to the kinetic α effect and its radial derivative in combination with the nonlinear magnetic diffusion of small-scale magnetic helicity are dominant in the solar convective zone (Kleeorin & Rogachevskii, 2022). The convective zone of the Sun and solar-like stars, as well as galactic discs, are the sources of the production of turbulent magnetic helicity fluxes. Turbulent fluxes of small-scale magnetic helicity have been measured in numerical simulations (Käpylä et al., 2010; Mitra et al., 2010; Hubbard & Brandenburg, 2010–2012; Del Sordo et al., 2013). Numerical simulations have demonstrated that magnetic helicity transport is of crucial importance in the solar dynamo as well (Kleeorin et al., 2003a, 2016, 2020; Zhang et al., 2006, 2012; Sokoloff et al., 2006; Hubbard & Brandenburg, 2012; Del Sordo et al., 2013; Safiullin et al., 2018).

15.4. FORMATION OF SUNSPOTS

The magnetic fields in the Sun are observed as sunspots and ARs. The magnetic fields in sunspots are more than 10^3 G, with a typical size of 10^4 km (Thomas & Weiss, 2008; Weiss & Proctor, 1966), while the typical scale of the dynamo-generated fields is about 10^6 km (Moffatt, 1978; Parker, 1979; Krause & Rädler, 1980; Zeldovich et al., 1983; Schrijver & Zwaan, 2008; Stix, 2012; Rüdiger et al., 2013; Ossendrijver, 2003; Brandenburg & Subramanian, 2005; Jones et al., 2010). The known formation mechanisms of sunspots are as follows.

- Magnetic buoyancy instability of inhomogeneous magnetic fields (Newcomb, 1961; Parker, 1966; Gilman, 1970; Acheson, D. J., 1978; Priest, 1982; Tobias & Hughes, 2004; Kersale et al., 2007). The energy source for this instability is related to the energy of the gravitational field.
- Buoyancy of discrete magnetic flux tubes (Parker, 1955a; Spruit, 1981; Spruit & Van Ballegooijen, 1982; Schüssler et al., 1994; Dikpati & Gilman, 2006; Spiegel & Weiss, 1980; Tobias et al., 2001).
- Magnetic flux expulsion (Weiss, 2010; Weiss & Proctor, 1966).
- Topological magnetic pumping (Weiss, 2010; Drobyshevski & Yuferev, 1974; Zeldovich et al., 1983).
- Negative effective magnetic pressure instability (NEMPI) (Kleeorin et al., 1989, 1990, 1993, 1996; Kleeorin & Rogachevskii, 1994; Rogachevskii & Kleeorin, 2007).

The condition of magnetic buoyancy instability is $L_{\rm B} < H_\rho$, where the length $L_{\rm B}$ and H_ρ characterize inhomogeneities of the original magnetic field and the mean density of plasma, respectively. Since the plasma density decreases by 10^7 times toward the solar surface in the convective zone, an essentially nonuniform magnetic field should exist for the onset of this instability.

A formation mechanism of magnetic spots is associated with NEMPI. This large-scale instability can be driven in small-scale density-stratified turbulence even if an initial mean magnetic field is uniform (Kleeorin et al., 1989, 1990, 1993, 1996; Kleeorin & Rogachevskii, 1994; Rogachevskii & Kleeorin, 2007). The instability mechanism is associated with a strong reduction of total turbulent pressure by a large-scale magnetic field. Velocity and magnetic fluctuations contribute to the total turbulent pressure. Let us define the effective mean magnetic pressure $P^{\rm eff}$ as the sum of turbulent and nonturbulent contributions. Turbulence causes a negative contribution to $P^{\rm eff}$. The latter effect, which for strong turbulence becomes large, is the main reason for NEMPI. The magnetic flux caused by the MFD is redistributed in space by this instability at the expense of the energy of convective or forced turbulence.

This phenomenon for various conditions has been investigated using various theoretical approaches (Kleeorin et al., 1993, 1996; Kleeorin & Rogachevskii, 1994; Rogachevskii & Kleeorin, 2007), as well as by mean-field numerical simulations (Brandenburg et al., 2010, 2016; Kemel et al., 2012b–2013b; Losada et al., 2012; Kemel et al., 2013a; Jabbari et al., 2013, 2014), large-eddy simulations (LESs) (Brandenburg et al., 2014; Käpylä et al., 2012, 2016), and direct numerical simulations (DNSs) (Brandenburg et al., 2011–2013b; Kemel et al., 2012a; Losada et al., 2013; Mitra et al., 2014; Warnecke et al., 2013, 2016; Jabbari et al., 2015–2017). In particular, DNS demonstrates that $P^{\rm eff}$ can be negative in any kind of turbulence, such as in nonstratified (Brandenburg et al., 2010) and density-stratified (Brandenburg et al., 2012) isothermal turbulence, polytropic stably stratified turbulence (Losada et al., 2014), convective turbulence (Käpylä et al., 2012, 2016), and turbulence with an upper coronal layer (Warnecke et al., 2013, 2016; Losada et al., 2019). However, this instability is driven when

- Turbulence is density-stratified.
- The initial mean magnetic field is smaller than the equipartition field.
- There is enough scale separation: that is, the integral turbulence scale is much less than the scale of variations of the mean magnetic field.

For strong turbulence, this instability is weakly dependent on the level of turbulence, and it causes the formation of magnetic spots (Brandenburg et al., 2013b, 2016) and active bipolar regions (Warnecke et al., 2013, 2016; Losada et al., 2019).

The production of sunspots and ARs is related to observed solar activity. A key parameter used to characterize solar activity over more than 250 years is the relative sunspot number, $W = 10g_{\rm w} + f_{\rm w}$ (also known as the International Sunspot Number), originally introduced by Rudolf Wolf (*Wolf number*). It is determined by the number of sunspots, $f_{\rm w}$, observable in the Sun's visible surface and the number of sunspot groups, $g_{\rm w}$. In section 15.6, we describe a nonlinear MFD model containing a budget equation for the evolution of sunspot number density (Kleeorin et al., 2016, 2020; Safiullin et al., 2018).

15.5. OBSERVATIONS OF PROXIES OF MAGNETIC HELICITY IN THE SUN

In this section, we discuss observations of the Sun, the revealed proxies of magnetic helicity, and their connection with nonlinear MFD theories.

15.5.1. Observation of Solar Vector Magnetic Fields

Vector magnetic fields over the solar photosphere can be measured from vector magnetographs, such as the Vector Spectro-Magnetograph (VSM) of the Synoptic Optical Long-term Investigations of the Sun (SOLIS), which is a scanned spectropolarimeter operating at the National Solar Observatory. Other notable vector magnetographs that contributed in the past and present are the Mees Observatory in Hawaii (USA), Udaipur Solar Observatory in India, Photospheric Vector Magnetograph at Okayama Observatory, and Solar Flare Telescope of the National Astronomical Observatory in Japan at Mitaka. The most systematic studies of solar vector magnetic fields over the longest period (since 1988) have been carried out overground by the Solar Magnetic Field Telescope (SMFT) at Huairou Solar Observing Station (HSOS), National Astronomical Observatories of China. The SMFT is a filtergraph with a birefringent filter for wavelength selection and KDP (potassium dihydrogen phosphate) crystals for modulation of polarization signals. The vector magnetogram uses the Fe I 5324.19 Å spectral line and provides four narrow-band (0.125 Å) filtergrams of Stokes I, V and Q, U parameters (for details, see http://sun.bao.ac.cn).

Nowadays, the most precise solar observations have moved into space. The first systematic space-based solar vector magnetograph is the Solar Optical Telescope Spectro-Polarimeter (SOT-SP) on board the joint Japanese/US/UK space mission Hinode (Kosugi et al., 2007); see also Lites et al. (2013) (the data is available at http://csac.hao.ucar.edu/sp_data.php). Hinode/SOT-SP operates at the Fe I 6302 Å line, providing measurements of all four Stokes parameters. Only limited field-of-view vector magnetograms are available. To obtain a two-dimensional map of a certain region, SOT-SP has to scan a set of slits consecutively; it can record only a one-dimensional slit of the solar surface. Hinode has operated since 2006.

The most intensively productive vector magnetic field instrument is the space-based Helioseismic and Magnetic Imager on board the NASA Solar Dynamics Observatory (SDO/HMI) (Scherrer et al., 2012; Schou et al., 2012). SDO/HMI is a filtergraph operating at the Fe I 6173 Å photospheric spectral line. It observes the full solar disc with a spatial resolution of about 1 arcsec, with 0.5 arcsec pixel size (Liu et al., 2012). SDO/HMI has produced continuous data since May 2010 with a cadence of 720 s (the magnetic field data is available at http://jsoc.stanford.edu). The differences between the results obtained from various instruments are discussed by (Xu et al., 2016; Kutsenko et al., 2018).

For reconstruction of the vector magnetic field components on a magnetogram at SFMT of HSOS, one can use, for example, the weak-field approximation. Then one can obtain the values of the magnetic field components from the four Stokes parameters, I, V and Q, U, as follows:

$$B_L = C_L\left(\frac{V}{I}\right), \quad B_T = C_T\left[\left(\frac{Q}{I}\right)^2 + \left(\frac{V}{I}\right)^2\right]^{1/4},$$
$$\phi = \frac{1}{2}\arctan\left(\frac{U}{Q}\right), \qquad (15.7)$$

where C_L and C_T are the calibration coefficients for the longitudinal B_L and transverse B_T magnetic fields, respectively, and ϕ is the azimuth angle. (For further details of SMFT calibration, see, e.g., Wang et al., 1996, Su & Zhang, 2004, Bai et al., 2014, and references therein). For inversion of the spectropolarimetry data from other instruments, one may use more advanced methods.

The data is coming 180° ambiguous with respect to the azimuthal angle ϕ. For low-resolution instruments, one has often used the potential field method and other methods, such as the minimum energy method. At the SDO HMI, the 180° ambiguity for the magnetic field vector is resolved in several ways, such as using a minimum energy method (Metcalf, 1994; Leka et al., 2009). Recently, another very efficient disambiguation code was developed by Rudenko & Anfinogentov (2014).

15.5.2. Computation of Current Helicity

Using the observed full-vector magnetic field measurements in the image plane at a single height in the photosphere, one can calculate a single contribution to current helicity related to the z-direction:

$$H_{Cz} = B_z \cdot J_z, \qquad (15.8)$$

where J_z is the z-component of electric current density $J_z = [\nabla \times \mathbf{B}]_z$. Throughout the community, one often refers to H_{Cz} as current helicity, H_C. The issue of local homogeneity and isotropy is required for such interpretation, as discussed by Xu et al. (2015). Following Abramenko et al. (1996) and using Stokes' theorem, one can rewrite the electric current in the following form:

$$J_z = \frac{1}{A}\oint_L \mathbf{B}_T \cdot d\mathbf{l}, \qquad (15.9)$$

where $\mathbf{B}_T$ is the transverse part of the magnetic field vector. The integration proceeds over the contour L enclosing area A around the central point where the z-component of the electric current J_z is evaluated. When electric currents J_z obtained using equation (15.9) are compared with those obtained by a traditional differential formula with subsequent smoothing, they are in good agreement (Abramenko et al., 1996).

The magnetic helicity in the Sun cannot be observed directly; however, one can use proxy measurements such as electric current helicity density, $H_C = \mathbf{B} \cdot \mathrm{curl}\mathbf{B}$, for interpretation in theoretical modeling. In the high conductivity limit, one can show (see, e.g., Berger & Field, 1984, Seehafer, 1994) that the signs of the current and magnetic helicities are the same and both increase almost simultaneously. Seehafer (1990) was the first to systematically involve signatures of current helicity in observational studies. He used the data on fibrils in the vicinity of 16 ARs and analyzed the curvature of those fibrils and data on the line-of-sight component of the magnetic field to determine the sign of the force-free parameter $\alpha_{\rm ff}$. This was used to describe a force-free field approximation for magnetic field (so-called *Beltrami* field):

$$\mathrm{curl}\,\mathbf{B} = \alpha_{\rm ff}\mathbf{B}. \qquad (15.10)$$

Here we note that the force-free parameter $\alpha_{\rm ff}$ characterizes the magnetic twist. In that pioneering work, Seehafer (1990) established the so-called hemispheric sign rule for current helicity, claiming that the proxies of current helicity (and twist) are predominantly positive/negative in the southern/northern hemispheres of the Sun, thus possibly reflecting the operation of the solar large-scale dynamo based on helical turbulence. Further extensive studies confirmed these results over the available larger statistical samples (see Pevtsov et al., 1994, 1995, Berger & Field, 1984, Abramenko et al., 1996, Hagino & Sakurai, 2004). A number of works reported some violation of the hemispheric sign rule at some solar latitudes and over some phases of the cycle (Bao et al., 2000; Hagino & Sakurai, 2005; Pevtsov et al., 1994; Zhang et al., 2010; Otsuji et al., 2015). Nevertheless, a greater percentage of solar ARs obey the hemispheric sign rule (Liu et al., 2014).

In most of these observational studies, the current helicity is regarded as integral characteristics of a solar AR. They are concerned with the sign of current helicity (Seehafer, 1990), averaged over an AR (Kuzanyan et al., 2000; Zhang et al., 2010), its imbalance (Abramenko et al., 1996). Some other works analyzed the spatial distribution of current helicity over ARs (Pevtsov et al., 1994; Zhang et al., 2002; Zhang, 2014; Hao & Zhang, 2011; Otsuji et al., 2015). Thus, Pevtsov et al. (1994) found characteristic scales of the current helicity pattern of mesogranulation scales of the order of 10 Mm and lifetimes of the order of one day.

This review of the observational methods of measuring proxies of magnetic helicity in the Sun is limited: it may represent some methods only partially and is not exceptional for other more advanced methods. There are other methods of analyzing magnetic helicity injection to mention. The most developed is magnetic helicity injection computation. This method uses the photospheric vector magnetic field data and high cadence observations for foot-point motion detection on the solar surface.

From the implementation of the idea of helicity generation by solar differential rotation (see, e.g., Berger & Ruzmaikin, 2000, DeVore, 2000, Berger, 1988, and references therein), and following observational computations by Chae (2001), the research direction of magnetic helicity injection was developed by a number of studies (Nindos et al., 2003; Démoulin & Berger, 2003; Démoulin & Pariat, 2009; Georgoulis et al., 2009; Yang et al., 2013).

The proxy of relative or mutual helicity has also been computed and thoroughly studied in the solar atmosphere and the corona (Thalmann et al., 2011). For a detailed review of the subject, see the fundamental reviews by Valori et al. (2016) and Thalmann et al. (2021). As shown by Guo et al. (2017), the proxies of helicity computed based on the helicity injection method from the foot-point motion generally agree with computations of the current helicity.

The other method to infer large-scale properties of helicity was developed using the data of synoptic maps of the solar magnetic fields by Carrington rotation, under the assumption that the global structure of the magnetic field does not change much during each rotation. The rotation of the Sun enables us to infer the 3D structure of the global magnetic fields. The idea was originally suggested by Grigoriev, Pescherov, and Latushko and published by Pevtsov & Latushko (2000). The application of this method was used to see the evolution of helicity proxies by Pevtsov et al. (2001) and Weiss (2010) at various times, and thus phases of the solar cycle. Furthermore, the proxies have been used to construct an azimuthal component of the total magnetic helicity (Pipin & Pevtsov, 2014), and the proxies have been obtained for solar cycles 23 and 24 (Pipin et al., 2019). This interesting issue is limited by the resolution and uncertainty of each synoptic map; however, it is very useful, giving an additional constraint to theoretical ideas on global solar helicity.

Recently, spatial spectra of the magnetic and current helicity distribution have been computed from observational data (Zhang, 2014). Under the assumption of isotropy, one can also compute the magnetic helicity and energy spectra from photospheric vector magnetograms of solar ARs. The results show the opposite signs of the magnetic helicity in smaller and larger scales throughout the spectra; see Zhang et al. (2016).

A two-scale formalism has also been developed for computing magnetic helicity spectra, given that the hemispheric rule is obeyed: that is, magnetic helicity has opposite signs in the solar hemispheres across the equator (Brandenburg et al., 2017). These studies use synoptic maps of the solar magnetic fields produced by each Carrington rotation. It enables one to recover bihelical spectra at different phases of the solar cycle. While the magnetic helicity in most of the data is consistent with the hemispheric sign rule, a small portion of the data may reverse (Singh et al., 2018). This method is very promising and developing quickly (Prabhu et al., 2021). The data on current helicity has also been used to study the intermittency spectra (Kutsenko et al., 2018). These effects may be important for understanding the mechanisms and operational prediction of the eruption of coronal mass ejections (CME) and energetic flares.

There are works that study the magnetic helicity distribution over scales; the main finding supports the theoretical idea of separation in the sign of magnetic helicity between small- and large-scale magnetic fields. In real observations, these scales are all greater than the integral turbulent scale. The separation in the helicity sign may also be associated with the twist of magnetic flux tubes connecting opposite polarities. In real observations, however, such sign separation may be effectively detected at larger and, therefore, better resolvable scales of magnetic fields.

From the idea of the EB decomposition of linear polarization signals into parity-even E and parity-odd B polarizations, which are not affected by the azimuthal π ambiguity concerning the line of sight, one can study the correlations of spatial Fourier spectra of B and parity-even quantities such as E (Brandenburg, 2019). These studies have found that the correlation of spatial Fourier spectra of parity-odd and parity-even quantities B and E can be proxies for magnetic helicity in the Sun, and it is indeed independent of the π-ambiguity (Prabhu et al., 2020). Future analysis of available observational data may provide new ways to infer proxies of magnetic helicity.

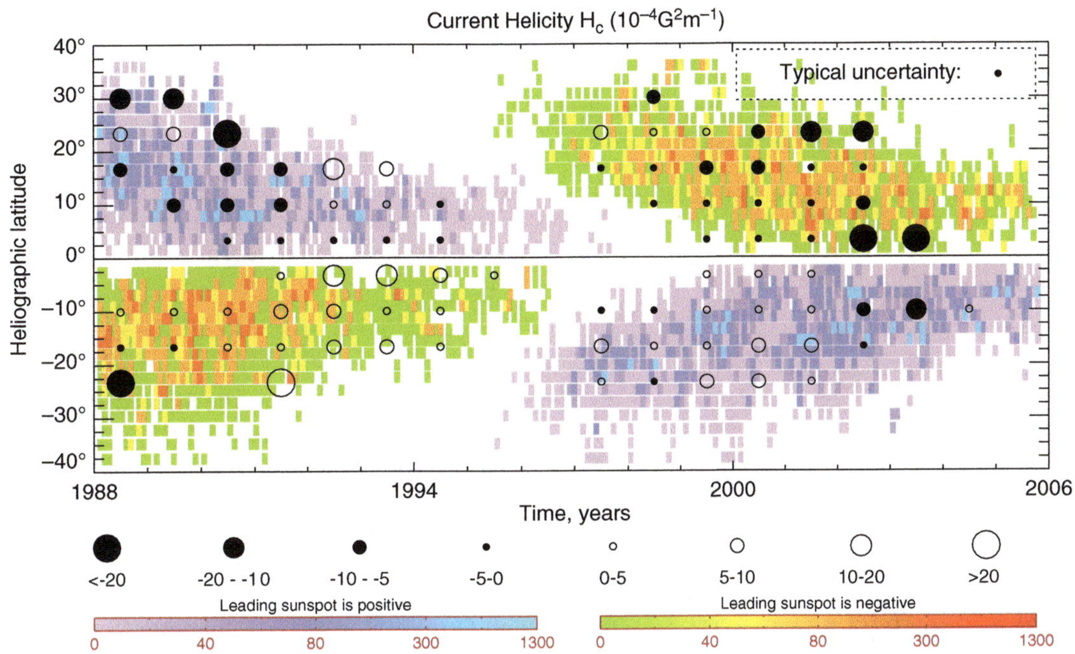

Figure 15.1 Mean observed current helicity for solar active regions (white/black circles for positive/negative values) in solar cycles 22 and 23 averaged over a two-year running window in time and latitudinal bins of 7°, overlaid with the sunspot density (color palette). The circle in the upper-right corner indicates the typical value of observational uncertainty defined by 95% confidence intervals scaled to the same units as the circles. The vertical axis shows the latitude in degrees and the horizontal axis the time in years. Zhang et al., 2012/IOP Publishing.

The major observational result concerning the spatio-temporal distribution of the proxies of current and (apparently also) magnetic helicity in the Sun is *the hemispheric sign rule* (negative/positive sign of helicity in northern/southern hemispheres) – which, however, may vary over the solar cycle. The rule is obeyed for most solar cycle phases (see Fig. 15.1), especially over the phase of solar maximum where most observations of solar ARs are available. This persistence of the sign of helicity is prominent from one solar cycle to the next, even though the sign of the magnetic field vector components changes to the opposite from odd/even cycles. Nevertheless, in some phases of the rise and fall of the solar cycle, one can note the signature of the sign reversal for magnetic helicity proxies (Zhang et al., 2010). This property poses a challenge for solar dynamo theory (Zhang et al., 2012). The departure from simultaneous obeying the hemispheric sign rule, or its nonsimultaneous reversal, may be interpreted as the interplay between the dipole and quadruple parts of the solar global magnetic field.

As we see in this chapter, the relative contribution from the quadruple part of the solar magnetic field increases with recent solar cycles. That may help in understanding some of the irregularities of the hemispheric sign rule for helicity. Note that direct proxies of magnetic helicity of the small-scale magnetic field are not yet available from observations. However, magnetic helicity proxies at the mesoscale (solar ARs) and large-scale (global) magnetic fields are observable. One may hope that the inner part of the twist (which may be associated, e.g., with the last term in equation (15.25) for the tilt of sunspots) may indicate the presence of small-scale magnetic helicity in rising magnetic flux tubes, which can be used as its proxy. We may also interpret the effect of small-scale magnetic helicity on the mean tilt, which may be connected with the sign change of the mean tilt of bipolar sunspot groups from small size to larger scales (Kleeorin et al., 2020).

15.6. NONLINEAR MODEL OF A MEAN-FIELD DYNAMO AND DYNAMICS OF THE SUNSPOT NUMBER

In this section, we outline the nonlinear dynamo model, which contains an equation for the evolution of the sunspot number (Kleeorin et al., 2016, 2020; Safiullin et al., 2018). We use spherical coordinates (r, θ, ϕ) and study an axisymmetric dynamo, where the mean magnetic field is written $\overline{\boldsymbol{B}} = \overline{B}_\phi \boldsymbol{e}_\phi + \boldsymbol{\nabla} \times (\overline{A}\boldsymbol{e}_\phi)$. The following assumptions have been used:

- A thin convective shell is used.

- Strong decrease in density is taken into account by averaging the equations over the radial coordinate (i.e., the no-r approximation is used).
- The curvature of the convective shell is neglected, and the convective shell is replaced by a flat slab.

Thus, the basic equations are (Kleeorin et al., 2003a; Zhang et al., 2006)

$$\frac{\partial \overline{B}_\phi}{\partial t} = GD\sin\theta \frac{\partial \overline{A}}{\partial \theta} + \frac{\partial^2 \overline{B}_\phi}{\partial \theta^2} - \mu^2 \overline{B}_\phi, \qquad (15.11)$$

$$\frac{\partial \overline{A}}{\partial t} = \alpha \overline{B}_\phi + \frac{\partial^2 \overline{A}}{\partial \theta^2} - \mu^2 \overline{A}. \qquad (15.12)$$

In equations (15.11) and (15.12), the turbulent magnetic diffusion along the $r-$ direction is determined by the terms $-\mu^2\overline{B}_\phi$ and $-\mu^2\overline{\mathbf{A}}$; the parameter $G = \partial\Omega/\partial r$, D is the dynamo number defined as $D = R_\alpha R_\omega$, where $R_\alpha = \alpha_0 R_\odot/\eta_T$ and $R_\omega = (\delta\Omega) R_\odot^2/\eta_T$, $\eta_T = \ell_0 u_0/3$; and μ is given by $\int_{2/3}^{1} \left(\partial^2 \overline{B}_\phi/\partial r^2\right) dr = -\mu^2 \overline{B}_\phi/3$. Equations (15.11)–(15.12) are written in nondimensional form, where the length and time are normalized by the solar radius $R_\odot$ and $R_\odot^2/\eta_T$, respectively; the nonuniform rotation $\delta\Omega$ is normalized by the maximum rotational velocity Ω; and α is normalized by the maximum kinetic α effect. The fields $\overline{B}_\phi$ and $\overline{A}$ are normalized by $\overline{B}_{\rm eq}$ and $R_\alpha R_\odot \overline{B}_{\rm eq}$, respectively; $\overline{\rho}$ is normalized by its maximum value $\overline{\rho}_*$; and the turbulent characteristics ℓ_0 and u_0 are normalized by their maximum values. The value $\mu = 3$ corresponds to a convective zone with a thickness of about one-third of the solar radius (Kleeorin et al., 2003a).

The total α effect is given by $\alpha(\theta) = \chi^{\rm v}\phi_{\rm v}(\overline{B}) + \sigma\chi^{\rm c}\phi_{\rm m}(\overline{B})$, where $\alpha^{\rm v} = \chi^{\rm v}\phi_{\rm v}(\overline{B})$ and $\alpha^{\rm m} = \chi^{c}\phi_{\rm m}(\overline{B})$ with $\sigma = \int \left(\overline{\rho}(r)/\overline{\rho}_*\right)^{-1} dr$, $\chi^{\rm v} = -(\tau_0/3) H_{\rm u}$, $\chi^{\rm c} = (\tau_0/12\pi\overline{\rho})H_{\rm c}$, the spatial profile of $\alpha^{\rm v}$ is given by $\alpha(\theta) = \alpha_0 \sin^3\theta \cos\theta$, $\phi_{\rm v}(\overline{B}) = [4\phi_{\rm m}(\overline{B}) + 3L(\overline{B})]/7$, $\phi_{\rm m}(\overline{B}) = 3[1 - \arctan(\sqrt{8}\,\overline{B})/\sqrt{8}\,\overline{B}]/8\overline{B}^2$, and $L(\overline{B}) = 1 - 16\overline{B}^2 + 128\overline{B}^4 \ln[1 + (8\overline{B}^2)^{-1}]$ (Rogachevskii & Kleeorin, 2000, 2001, 2004). Here the squared mean magnetic field is given by $\overline{B}^2 = \xi\{\overline{B}_\phi^2 + R_\alpha^2[\mu^2\overline{A}^2 + (\partial\overline{A}/\partial\theta)^2]\}$ with $\xi = 2(\ell_0/R_\odot)^2$.

The dynamic quenching is characterized by $\chi^{\rm c}(\overline{\boldsymbol{B}})$ and written in nondimensional form as

$$\frac{\partial \chi^{\rm c}}{\partial t} + \left(T^{-1} + \kappa_T \mu^2\right)\chi^{\rm c} = 2\left(\frac{\partial \overline{A}}{\partial \theta}\frac{\partial \overline{B}_\phi}{\partial \theta} + \mu^2 \overline{A}\,\overline{B}_\phi\right) - \frac{\alpha}{\xi}\overline{B}^2 - \frac{\partial}{\partial \theta}\left(\overline{B}_\phi \frac{\partial \overline{A}}{\partial \theta} - \kappa_T \frac{\partial \chi^c}{\partial \theta}\right), \qquad (15.13)$$

where $\boldsymbol{F}_{\rm m} = -\kappa_T \boldsymbol{\nabla}\chi^c$, with κ_T being the turbulent diffusion coefficient, and T is the magnetic helicity relaxation time given by

$$T^{-1} = H^{-1}\int \tilde{T}^{-1}(r)\,dr \sim \frac{\Lambda_\ell R_\odot^2 \eta}{H \ell_0^2 \eta_T}. \qquad (15.14)$$

Here H is the thickness of the convective region, Λ_ℓ is the variation scale of ℓ_0, and $\tilde{T}(r) = (\eta_T/R_\odot^2)(\ell_0^2/\eta)$.

This nonlinear model for the MFD also includes the following equation (Kleeorin et al., 2016; Safiullin et al., 2018):

$$\frac{\partial \tilde{W}}{\partial t} = I(t,\theta) - \frac{\tilde{W}}{\tau_s(\overline{B})}, \qquad (15.15)$$

which takes into account the formation of sunspots associated with NEMPI. Here $\tilde{W}(t,\theta)$ is the surface density of the sunspot number, $I(t,\theta)$ is the production rate of $\tilde{W}(t,\theta)$, $I(t,\theta) = |\gamma_{\rm inst}|\,|\overline{B} - \overline{B}_{\rm cr}|\,\Theta(\overline{B} - \overline{B}_{\rm cr})/\Phi_s$, $\tilde{W}/\tau_s(\overline{B})$ is the decay rate of $\tilde{W}(t,\theta)$, $\tau_s(\overline{B})$ is the decay time of sunspots, Φ_s is the magnetic flux inside a magnetic spot, and the function $\Theta(y)$ is defined as $\Theta(y) = 1$ for $y > 0$ and $\Theta(y) = 0$ for $y \le 0$. Equation (15.15) is coupled with equations (15.11)–(15.12) via the magnetic field $\overline{B}$.

The growth rate, $\gamma_{\rm inst}$, of NEMPI is (Rogachevskii & Kleeorin, 2007; Losada et al., 2012; Brandenburg et al., 2016)

$$\gamma_{\rm inst} \approx \left(\frac{2V_{\rm A}^2 k_x^2}{H_\rho^2 k^2}\left|\frac{dP_{\rm eff}}{d\beta^2}\right| - \frac{4(\boldsymbol{\Omega}\cdot\boldsymbol{k})^2}{\boldsymbol{k}^2}\right)^{1/2} - \eta_T\left(k^2 + \frac{1}{(2H_\rho)^2}\right), \qquad (15.16)$$

where $V_{\rm A} = \overline{B}/\sqrt{4\pi\overline{\rho}}$, $P_{\rm eff}(\beta)$ is the effective magnetic pressure, $\beta = \overline{B}/\overline{B}_{\rm eq}$, and H_ρ is the density stratification height. The necessary condition for the instability is $\overline{B} > \overline{B}_{\rm cr}$, where

$$\frac{\overline{B}_{\rm cr}}{\overline{B}_{\rm eq}} = \frac{\ell_0}{50H_\rho}\left[1 + \left(\frac{10\,{\rm Co}\,H_\rho^2}{\ell_0^2}\right)^2\right]^{1/2}. \qquad (15.17)$$

This is why the rate of production $I(t,\theta) \propto \Theta(\overline{B} - \overline{B}_{\rm cr})$. The typical timescale of $\tilde{W}(t,\theta)$ is associated with $\gamma_{\rm inst}^{-1}$. When the instability is not excited ($\gamma_{\rm inst} < 0$), the source function $I(t,\theta)$ vanishes. Therefore, the function $I(t,\theta) \propto |\gamma_{\rm inst}|\,\Theta(\overline{B} - \overline{B}_{\rm cr})$.

During the nonlinear stage of NEMPI, the decay of sunspots is modeled as $-\tilde{W}/\tau_s(\overline{B})$. The decay time $\tau_s(\overline{B})$ (about a month) is much smaller than the period of the solar cycle (about 11 years). Therefore, a solution for equation (15.15) reads $\tilde{W} = \tau_s(\overline{B})\,I(t,\theta)$. The sunspot number is given by $W = R_\odot^2 \int \tilde{W}(t,\theta)\sin\theta\,d\theta\,d\phi = 2\pi R_\odot^2 \int \tau_s(\overline{B})\,I(t,\theta)\sin\theta\,d\theta$. When the number of sunspots grows during an increase in solar activity, their lifetime also increases, and vice versa. Thus, we assume

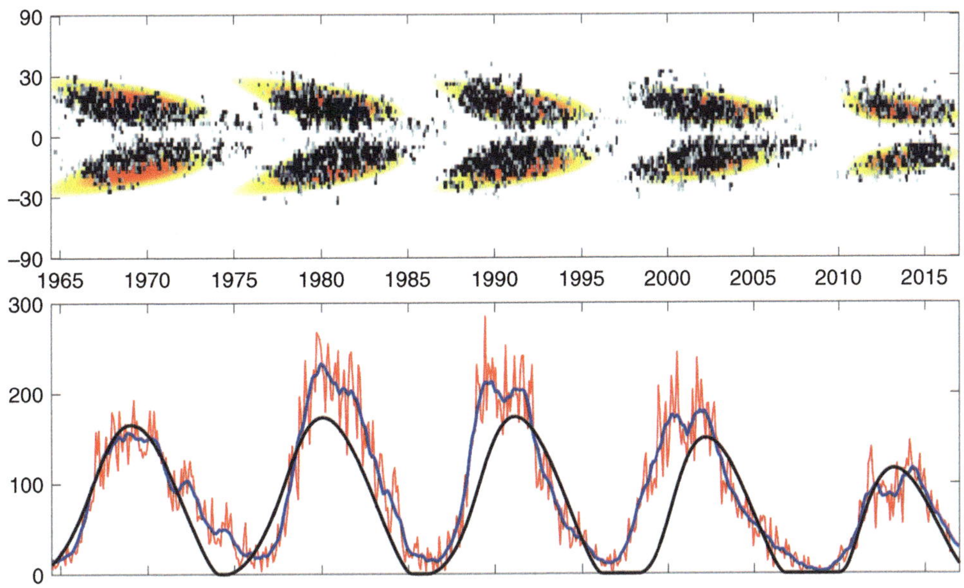

Figure 15.2 Comparison of numerical results with observations. Upper panel: butterfly diagram of the sunspot number variation rate $2\pi \sin\theta\, I(t,\theta)$, from the mean-field numerical simulations (color) and observational data (black). Bottom panel: sunspot numbers obtained from the mean-field numerical simulations (black), observational data (red), and observational data averaged over 13 months (blue). Safiullin et al., 2018/Cambridge University Press.

that $\tau_s(\overline{B}) = \tau_* \exp\left(C_s\, \partial\overline{B}/\partial t\right)$, where $C_s = 2 \times 10^{-3}$ and $\tau_*\, \gamma_{\rm inst} \sim 10$; and $\partial\overline{B}/\partial t$ is normalized by $\xi \overline{B}_{\rm eq}/t_{\rm td}$, with $t_{\rm td}$ being the turbulent diffusion time. The choice of the function $\tau_s(\overline{B})$ does not affect the evolution of the sunspot numbers.

Equations (15.11), (15.12), (15.13), and (15.15), with the initial conditions $\overline{B}_\phi(t=0,\theta) = S_1 \sin\theta + S_2 \sin(2\theta)$ and $\overline{A}(t=0,\theta) = 0$, have been solved numerically for the following parameters: $D = -8450$, $G = 1$, $\sigma = 3$, $\mu = 3$, $\xi = 0.1$, $\kappa_T = 0.1$, $R_\alpha = 2$, $T = 6.3$, $S_1 = 0.051$, $S_2 = 0.95$. For illustration, in Fig. 15.2, we show the results of the mean-field numerical simulations compared with observations, where data for the observed sunspot numbers from the World Data Center SILSO (Royal Observatory of Belgium, Brussels) are used. The dynamo model reproduces the latitude distribution of the ARs well enough. In particular, the butterfly diagram of the sunspot number variation rate $2\pi \sin\theta\, I(t,\theta)$ (Fig. 15.2, upper panel) and the evolution of the sunspot numbers (Fig. 15.2, bottom panel) based on the mean-field numerical simulations and observations are in a good agreement.

Figure 15.3 shows the dynamics of solar activity, including the events analogous to the Dalton minimum and the Maunder minimum. The mean-field numerical simulations include 150 years (14 cycles) of the transient period (not shown here) and the next 50+ years (see Fig. 15.2 as well as the long-term dynamics (see Fig. 15.3). The propagation of sunspot activity to the pole is seen in Fig. 15.3 (upper panel) from $t = 2075$ to $t = 2125$. We use a simplified dynamo model. Depending on $\alpha^{\rm m}$, one can observe either equatorward or poleward propagations of the sunspot belt.

In the framework of these simulations, a parameter scan with 10^3 runs using various sets of parameters was performed to find a high correlation between the numerical simulation results and observations. The dynamo number D and initial magnetic field $B_{\rm init}^{\rm dip}$ for the dipole mode (characterized by the parameter S_2) are two key parameters affecting the magnetic field evolution. For instance, the proper choice of the parameter S_2 allows us to get a short transient behavior and obtain chaotic behavior. The maximum correlation between the simulations and observations is reached for $D = -8450$ and $S_2 = 0.95$ (Kleeorin et al., 2016). The threshold value, $|D_{\rm cr}|$, for the dynamo instability depends on μ. The parameter $\kappa_T \approx 0.1$ characterizes the flux of the magnetic helicity. It cannot be much less than 0.1 to avoid the catastrophic quenching of the α effect. The other parameters are not crucial in the simulations (Kleeorin et al., 2016).

15.7. THE EFFECT OF MAGNETIC HELICITY TRANSPORT AND PREDICTION OF SOLAR ACTIVITY

Different methods of forecasting solar activity have been discussed in the literature (Dikpati et al., 2006; Choudhuri et al., 2007; Kane, 2007; Bushby & Tobias,

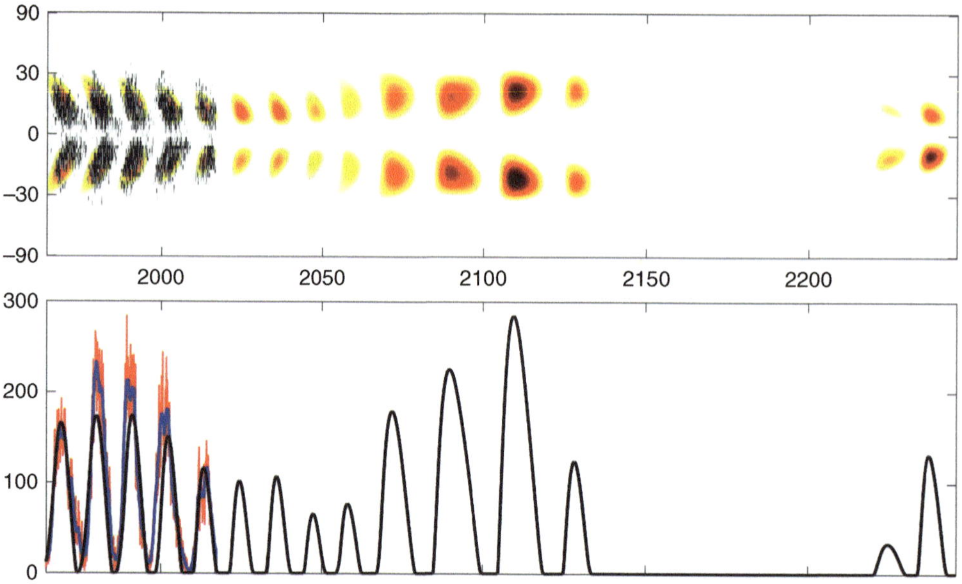

Figure 15.3 The long-term evolution of solar activity. Top panel: butterfly diagram of the sunspot number variation rate $2\pi \sin\theta\, I(t,\theta)$ from the mean-field numerical simulations (color) and observational data (black). Bottom panel: sunspot numbers from the mean-field numerical simulations (black), observational data (red), and observational data averaged over 13 months (blue). Safiullin et al., 2018/Cambridge University Press.

2007; Obridko & Shelting, 2008; De Jager & Duhau, 2009; Tlatov, 2009; Kitiashvili & Kosovichev, 2011; Pesnell, 2012); see also Dmitrieva et al. (2000). Improving the solar activity forecast is a subject of active research. A method based on a combination of numerical simulations and neural network was adopted (Safiullin et al., 2018) for a one-month forecast of solar activity.

In this section, we outline this approach. Our numerical simulations of the sunspot numbers W_i^{model}, applying the model discussed in section 15.6, are used as the basis for the forecast and as input to the neural network. The observational data W_i^{obs} is an additional input to the neural network. For the forecast W_i^{forecast} for the next six months or longer, the learning procedure is applied as $W_i^{\text{forecast}} = f_{\text{out}}\,(\boldsymbol{K}\,\boldsymbol{w} + \boldsymbol{c})$. The latter corresponds to a one-layer artificial neural network. Here $\boldsymbol{w}$ is the input vector, which includes W_i^{obs} and W_i^{model}, $f_{\text{out}}(x)$ is a linear function of an outer layer of neurons, $\boldsymbol{K}$ is the weight matrix of neurons, and $\boldsymbol{c}$ is the vector of biases. To determine $\boldsymbol{K}$, one needs to decrease the error between the deviations of the forecast data W_i^{forecast} from the observational data W_i^{obs}. For a better forecast, it is important to use at least a two-layer neural network:

$$W_i^{\text{forecast}} = f_{\text{out}}\left[\boldsymbol{K}_2 f_{\text{hidden}}\left(\boldsymbol{K}_1\boldsymbol{w} + \boldsymbol{c}_1\right) + \boldsymbol{c}_2\right]. \quad (15.18)$$

The two layers of the network allow us to obtain a complex, stable, adaptable, and adjustable forecast compared to that based on a one-layer artificial neural network. Here $f_{\text{hidden}}(x) = [1 + \exp(-x)]^{-1}$, $\boldsymbol{K}_1$ is the weight matrix 24×8, $\boldsymbol{K}_2$ is the weight matrix 1×24 of outer layer neurons, $\boldsymbol{c}_1$ and $\boldsymbol{c}_2$ are the corresponding bias vectors, and $\boldsymbol{w}$ is the input vector 8×1 consisting of four prior observations $W_{i-1}^{\text{obs}}, \ldots, W_{i-4}^{\text{obs}}$ and four corresponding model estimations $W_{i-1}^{\text{model}}, \ldots, W_{i-4}^{\text{model}}$.

By Bayesian regularization back-propagation, the learning procedure was based on epignosis using the data of the sunspot numbers from cycles 19 to 21, while cycles 22 and 23 were used for the validation procedure. The real forecast started in September 2017 (two years before the end of cycle 24). The input data contains the observational data and the numerical simulation data at the same instant. The one-month forecast of the sunspot number is the output of this neural network.

We minimize the error between the predicted values and available observations over the whole cycle of the learning procedure. Notably, the model output allows us to obtain an initial forecast on which the final forecast is based. The artificial neural network using observable values and the model outputs plays a role in correcting the forecast of the simulated sunspot number. Figures 15.4 and 15.5 compare the results for the one-month forecast of solar activity by the described method with the observational data, and they are in good agreement. Without the MFD simulations, the neural network may agree with observations for at most five years, but the combination of numerical simulations with the neural network can yield high quality forecasts for several solar cycles.

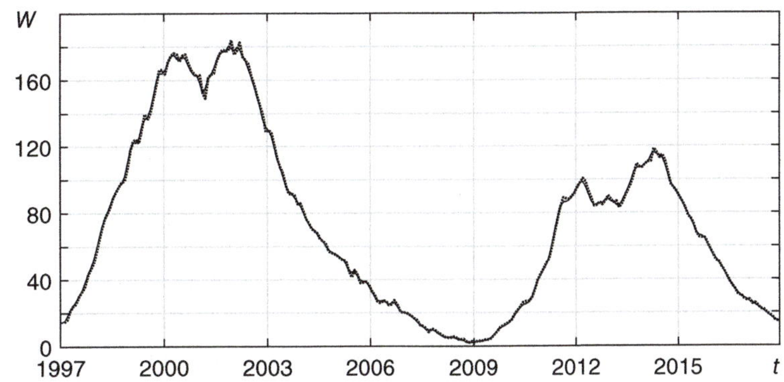

Figure 15.4 Comparison of the results of the one-month preliminary forecast of the sunspot numbers (filled circles) with the observed 13-month running average sunspot numbers (solid line). Safiullin et al., 2018/Cambridge University Press.

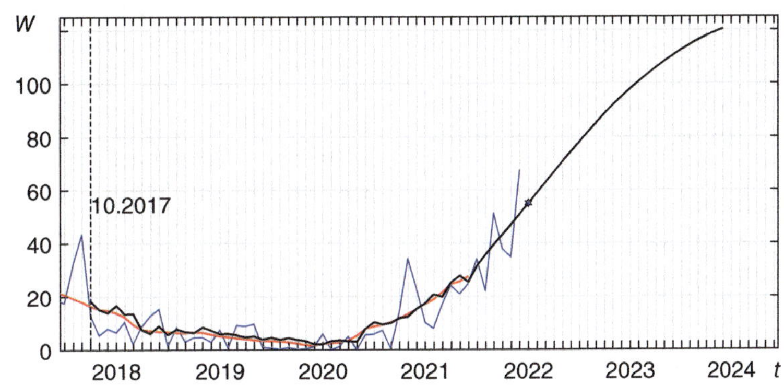

Figure 15.5 Comparison of the one-month forecast of the monthly sunspot number (black line) with the observed 13-month running average sunspot numbers (red line). The blue line corresponds to the observed sunspot number (without averaging).

15.8. THE MEAN TILT OF SUNSPOT BIPOLAR REGIONS

The mean tilt is a key parameter characterizing the sunspot bipolar region. The tilt is the angle between the solar equator and a line that passes through the leading and following sunspots. It has been investigated in numerous theoretical studies and numerical simulations as well as in observations (D'Silva & Choudhuri, 1993; Kosovichev & Stenflo, 2008; Dasi-Espuig et al., 2010; McClintock & Norton, 2013; Illarionov et al., 2015; Tlatov, A. G., 2015; Tlatova et al., 2018; Kleeorin et al., 2020). Let us summarize recent results related to the observations of the averaged tilts of the sunspot bipolar regions (Tlatova et al., 2018):

- The mean tilt varies with the solar latitude ϕ mainly as an antisymmetric odd function that deviates from $\sin \phi$.
- Near the solar equator, the mean tilt does not vanish; it is mainly positive for even cycles and negative for odd cycles.
- The dependence of the mean tilt on the latitude is not universal, varying in different solar cycles.

A recent theory for the formation of the mean tilt by Kleeorin et al. (2020) takes into account three major contributions:

- The effect of the mean magnetic field at low latitudes. In particular, since the Coriolis force is proportional to $\sin \phi$, it decreases in low latitudes so that at some small but finite latitude ϕ, the Coriolis force is balanced by the magnetic tension. Quantitatively, this effect is described by equations (A3)–(A8) in Kleeorin et al. (2020).
- Due to the latitudinal differential rotation in the solar convective zone, the mean tilt deviates from the sine-like Joy's law at higher latitudes.
- The effect of magnetic helicity is that the mean tilt does not vanish near the equator, where it changes its sign in the range $\phi \approx \pm 5^\circ$.

Mean-field simulations using the dynamo model described in section 15.6 have been performed by Kleeorin et al. (2020) to justify the theory. The comparison of the theoretical and numerical results (Kleeorin et al., 2020) with the observed dependence of the mean tilt on the latitude during the last nine solar cycles (Tlatova et al., 2018) shows good agreement between theoretical, numerical, and observational results.

Here we will outline this theory. Let us define the tilt of the sunspot bipolar regions as vector $\boldsymbol{\delta}^{\rm tw}$, characterized by the perturbations of vorticity, $\tilde{\boldsymbol{w}} \equiv \partial \boldsymbol{\delta}^{\rm tw}/\partial t$. The twist of the magnetic field lines connecting sunspots of the opposite magnetic polarity in the bipolar region is described by $|\boldsymbol{\delta}^{\rm tw}| \approx |\tilde{\boldsymbol{w}}|\, \delta t$. Perturbations of the vorticity $\tilde{\boldsymbol{w}}$ during twist time δt do not vanish. The vectors $\boldsymbol{\delta}^{\rm tw}$ and $\tilde{\boldsymbol{w}}$ have the same direction, and they are perpendicular to the twist plane. At the boundary between the convective zone and the photosphere, the radial component of the vector $\boldsymbol{\delta}^{\rm tw}$ describes the tilt of the sunspot bipolar regions, and there the magnetic field inside the sunspots is radial. By averaging the scalar product $\boldsymbol{\delta}^{\rm tw} \cdot \boldsymbol{e}_B$ over a time that is much larger than the Alfvén time $\tau_{\rm A}$, we obtain the mean tilt value $\gamma \equiv \langle \boldsymbol{\delta}^{\rm tw} \cdot \boldsymbol{e}_B \rangle_{\rm time}$. Here $\boldsymbol{e}_B = \boldsymbol{B}^{\rm ar}/B^{\rm ar}$ is the unit vector along the magnetic field $\boldsymbol{B}^{\rm ar}$ of bipolar ARs.

Using the equation of plasma motion and the induction and entropy equations at the boundary between the solar convective zone and the photosphere, as well as adopting the anelastic approximation, we obtain the expression for the mean tilt of the sunspot bipolar regions (Kleeorin et al., 2020):

$$\gamma = -\delta_0 \left[\sin \phi - \delta_M \cos \phi\right], \qquad (15.19)$$

where $\delta_0 = \tau_{\rm A}^2\, \Omega/(2\pi\, \tau_c)$, ϕ is the latitude, $\tau_{\rm A} = L_B/U_{\rm A}$ is the Alfvén time, L_B is the length of the magnetic field line of the sunspot bipolar region, $U_{\rm A} = B^{\rm ar}/(4\pi\overline{\rho})^{1/2}$ is the Alfvén speed, $\overline{\rho}$ is the mean density of plasma, $\tau_c = H_\rho/U^{(c)}$ is the typical time of the super-granular convection with convective velocity $U^{(c)}$, $H_\rho = |\boldsymbol{\nabla} \ln \overline{\rho}|^{-1}$ is the density stratification height, and δ_M is given by

$$\delta_M = C_M \left(\frac{\ell_{\rm top}}{R_\odot}\right)^2 \left(\frac{\overline{\rho}_{\rm bot}}{\overline{\rho}_{\rm top}}\right) \left(\frac{\eta_T^{\rm bot}}{\eta_T^{\rm top}}\right) \left(\frac{\tau_c}{\tau_{\rm bot}}\right) \times \left(\frac{\partial^3}{\partial \phi^3} \frac{\overline{\boldsymbol{B}}^2}{\overline{B}_{\rm eq}^2}\right)_{\rm bot}. \qquad (15.20)$$

Here $R_\odot$ is the solar radius, $\ell_{\rm top}$ is the integral scale of turbulent motions in the upper part of the convective zone; $\overline{\rho}_{\rm bot}$ and $\overline{\rho}_{\rm top}$ are the plasma densities in the bottom and upper parts of the convective zone, respectively; $\eta_T^{\rm bot}$ and $\eta_T^{\rm top}$ are the turbulent magnetic diffusion coefficients in the bottom and upper parts of the convective zone, respectively; $\tau_{\rm bot} = \ell_{\rm bot}/u_{\rm bot}$ is the turbulent time at the bottom of the convective zone; $u_{\rm bot}$ is the turbulent velocity in the bottom part of the convective zone; $\overline{B}_{\rm eq} = u\sqrt{4\pi\overline{\rho}}$ is the equipartition magnetic field; and $C_M \approx 10$ is the numerical coefficient that depends on a turbulent Prandtl number. In equation (15.19), the effect of the solar differential rotation on the mean tilt was neglected.

The physics of the magnetic contribution δ_M to the mean tilt is as follows. The twist of sunspots in the photosphere is caused by the Coriolis force balanced by the magnetic tension at low latitudes. This means that in the vicinity of the equator, δ_M is dominant because the contribution caused by the Coriolis force $\propto \sin\phi$ is small. Notably, the combination of the dipole and quadruple dynamo modes contributes to δ_M near the equator.

To estimate the mean tilt, we take the values of parameters from the solar convective zone model (Spruit, 1974). Here, for illustration, we estimate the mean tilt at different depths in the convective zone. At the bottom of the convective zone – that is, at depth $H \sim 2 \times 10^{10}$ cm – the values of the governing parameters are $u_{\rm bot} \sim 2 \times 10^3$ cm s^{-1}, $\ell_{\rm bot} = 0.8 \times 10^{10}$ cm, $\overline{\rho}_{\rm bot} = 0.2$ g cm^{-3}, $\eta_T^{\rm bot} = 5 \times 10^{12}$ cm^{2}s^{-1}, $H_\rho^{\rm bot} = \overline{\rho}/|\nabla\overline{\rho}| = 6.5 \times 10^9$ cm and $\overline{B}_{\rm eq}^{\rm bot} = 3000$ G. In the upper part of the convective zone, say at $H \sim 2 \times 10^7$ cm, the values of the governing parameters are $u_{\rm top} = 9.5 \times 10^4$ cm s^{-1}, $\ell_{\rm top} = 2.7 \times 10^7$ cm, $\overline{\rho}_{\rm top} = 4.6 \times 10^{-7}$ g cm^{-3}, $\eta_T^{\rm top} = 0.83 \times 10^{12}$ cm^2 s^{-1}, $H_\rho^{\rm top} = 3.7 \times 10^7$ cm, and $\overline{B}_{\rm eq}^{\rm top} = 225$ G.

Using these parameters, we estimate δ_0 and δ_M, which describe the mean value of the tilt of the bipolar region. Taking $U_{\rm A} = 5 \times 10^4$ cm s^{-1}, the length $L_B = 4H_\rho = 4 \times 10^9$ cm, we estimate the Alfvén time $\tau_{\rm A} = L_B/U_{\rm A} = 10^5$ s; similarly, taking $U^{(c)} = (3-5) \times 10^4$ cm s^{-1}, we obtain the convective time $\tau_c = (2-3) \times 10^4$ s. Thus, using these parameter values, we obtain $\delta_0 = 0.3 - 0.5$ (in radians) and $\delta_M = 0.05 - 0.2$. Therefore, the magnetic contribution δ_M to the mean tilt is important at lower latitudes.

Now we will discuss the effect of latitudinally nonuniform solar differential rotation on the mean tilt of bipolar regions. At the surface of the Sun, the solar differential rotation may be determined as $\Omega = \Omega_0(1 - C_2\sin^2\phi - C_4\sin^4\phi)$ (LaBonte & Howard, 1982), where $\Omega_0 = 2.83 \times 10^{-6}$ s^{-1}, $C_2 = 0.121$, and $C_4 = 0.173$. Substituting this expression into equation (15.19), one can arrive at the following expression for the mean tilt of bipolar regions (Kleeorin et al., 2020):

$$\gamma = -\tilde{\delta}_0 \left[\sin\phi + \delta_3 \sin 3\phi - \delta_5 \sin 5\phi - \tilde{\delta}_M \times \left(\cos\phi + \tilde{\delta}_3 \cos 3\phi - \tilde{\delta}_5 \cos 5\phi\right)\right], \qquad (15.21)$$

where $\tilde{\delta}_0 = C_D\,\delta_0$, $\tilde{\delta}_M = \delta_M \tilde{C}_D/16C_D$, $C_D = 1 - (3C_2 + 5C_4)/4 \approx 0.693$, $\tilde{C}_D = 1 - 4C_2 - 2C_4 \approx 0.17$ and $\delta_3 = (4C_2 + 5C_4)/16C_D \approx 0.122$, $\delta_5 = C_4/16C_D \approx 1.56 \times 10^{-2}$, $\tilde{\delta}_3 = (4C_2 + 3C_4)/\tilde{C}_D \approx 4.48$, and $\tilde{\delta}_5 = C_4/\tilde{C}_D \approx 1.02$.

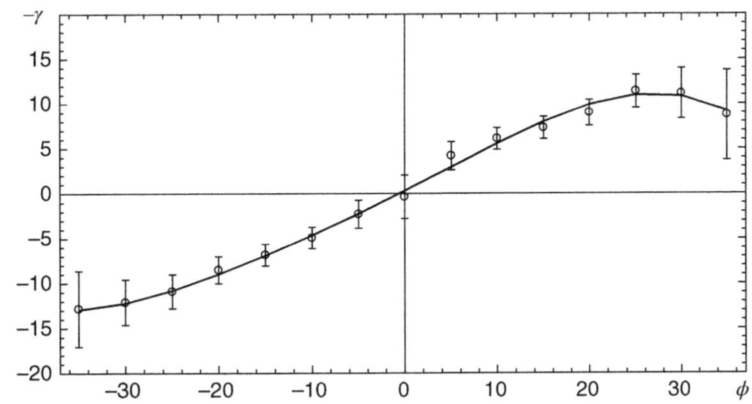

Figure 15.6 The mean tilt denoted as $-\gamma$ (in degrees) as a function of the latitude ϕ (in degrees): equation (15.21) with magnetic contributions and $\tilde{\delta}_0 = 0.31$, $\delta_3 = 1.26$, $\delta_5 = 0.22$, $\tilde{\delta}_3 = 1.87$, $\tilde{\delta}_5 = 0.78$ and $\tilde{\delta}_M = 0.2$ (solid line), and the data from observations (circles) for all solar cycles (see Fig. 3 of Tlatova et al. (2018)). Adapted from Kleeorin et al. (2020).

To derive equation (15.21), we assume the following conditions: $\Omega/H_\rho \gg |\partial\Omega/\partial r|$ and $\Omega/H_\rho \gg r^{-1}|\partial\Omega/\partial\theta|$.

The mean tilt $-\gamma$ is shown in Fig. 15.6 against the latitude ϕ given by equation (15.21), where $\tilde{\delta}_M = 0.2$, $\tilde{\delta}_0 = 0.31$, $\delta_3 = 1.26$, $\delta_5 = 0.22$, $\tilde{\delta}_3 = 1.87$, $\tilde{\delta}_5 = 0.78$, and γ and ϕ are in degrees. To compare the theoretical results (Kleeorin et al., 2020) with the observations (Tlatova et al., 2018), we show in Fig. 15.6 the observational data for all solar cycles (averaged over all available bipolar regions of all sizes; see Tables 1 and 2 by Tlatova et al. (2018)). Inspection of Fig. 15.6 shows that the theoretical results for the mean tilt agree very well with the observations.

Using the nonlinear dynamo model described in section 15.6, we perform mean-field numerical simulations. This allow us to obtain a butterfly diagram (Fig. 15.7a) of the normalized mean tilt $-\gamma/\tilde{\delta}_0$ determined by equation (15.21), where the magnetic contribution is given by

$$\delta_M\left(\overline{\boldsymbol{B}}^2\right) = C_*\left(\frac{\partial^3}{\partial\phi^3}\frac{\overline{\boldsymbol{B}}^2}{B_{\rm eq}^2}\right)_{\rm bot}, \qquad (15.22)$$

with $C_* = 0.8$. As follows from Fig. 15.7a, over the three recent cycles 22–24, the mean tilt increases at the lower range of latitudes. This can be explained as follows. Our mean-field numerical simulations for the last three solar cycles show that the fraction of the quadruple dynamo mode to the overall magnetic field increases, while that of the dipole dynamo mode decreases. This conclusion agrees with the observations (Livshits & Obridko, 2006). The mean-field numerical simulations demonstrate that during a transition from high to low solar cycles, the mean tilt decreases, which agrees with available observations (Dasi-Espuig et al., 2010).

Figure 15.7b shows the butterfly diagram of $\delta^*_M = -\tilde{\delta}_M(\cos\phi + \tilde{\delta}_3\cos 3\phi - \tilde{\delta}_5\cos 5\phi)$, where we take into account the latitudinal differential rotation. It shows that the mean tilt in each hemisphere does not vary much with the phase of the solar cycle, except at low latitudes, where it reverses. In Fig. 15.7c, we show for comparison the butterfly diagram for the density of the sunspot numbers, which varies strongly with the phase of the solar cycle.

Around the solar maximum, the contributions to the mean tilt due to the large-scale magnetic field, δ^*_M, and the Coriolis force have the same sign at middle latitudes (Fig. 15.7a). At lower latitudes (below 10°), the magnetic contribution to the mean tilt δ^*_M is positive/negative in the southern/northern hemispheres (Fig. 15.7b). This effect is stronger for solar cycles 23 and 24, with lower activity, while the magnetic activity zone migrates to the lower latitudes (Fig. 15.7c).

In Fig. 15.8, we plot the mean tilt $-\gamma$ against latitude determined by equation (15.21). Here we use the mean-field numerical simulation to determine δ_M, considering the latitudinally nonuniform differential rotation's effect on the mean tilt. The parameters of the simulations are the following: $C_* = 0.8$, $\delta_0 = 0.29$, $\delta_3 = 0.122$, $\delta_5 = 1.56\times10^{-2}$, $\tilde{\delta}_3 = 4.48$, and $\tilde{\delta}_5 = 1.02$. We compare the observational data for the mean tilt averaged over individual solar cycles (for cycles 15–24) with the numerical simulations for $-\gamma$. We find that the theoretical numerical data of the northern and southern hemispheres are asymmetric (Fig. 15.8), which is in agreement with the observations.

In this section, we consider two effects contributing to the mean tilt: (1) the latitudinal differential rotation in the solar convective zone and (2) the effect of a large-scale magnetic field. In Fig. 15.8, we used the value of the parameter δ_M obtained from the

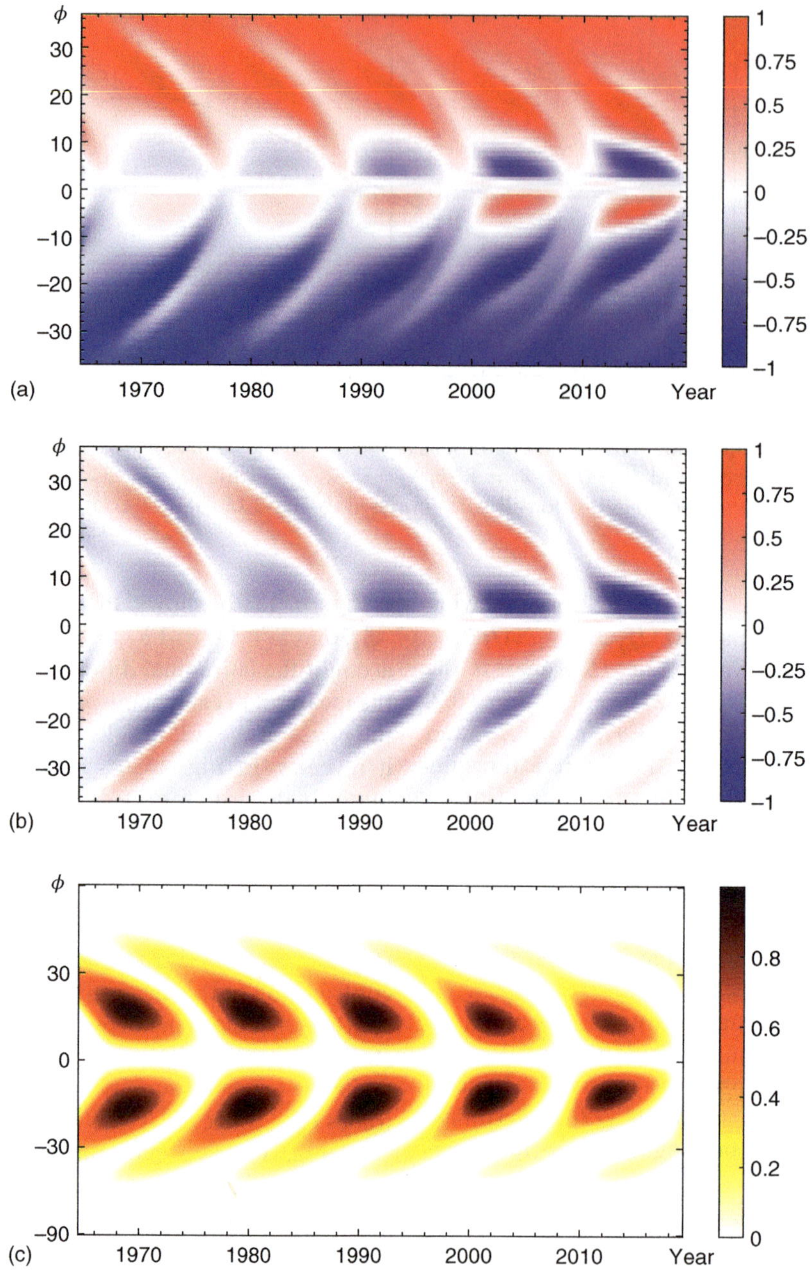

Figure 15.7 Butterfly diagrams of (a) the normalized mean tilt $-\gamma/\tilde{\delta}_0$ (see equation (15.21)); (b) the magnetic contribution δ^*_M to the mean tilt; (c) the surface density of the sunspot numbers, where the color bars are normalized by their maximum values. Adapted from Kleeorin et al. (2020).

mean-field numerical simulations. On the other hand, in Fig. 15.6, we used the value of the parameter δ_M obtained from a theoretical model that gives an estimate for the third-order derivative of the normalized magnetic energy over the latitude $(\partial^3/\partial\phi^3)(\overline{\boldsymbol{B}}^2/B^2_{\rm eq})_{\rm bot}$. In this case, the solution is considered as a combination of dipole and quadrupole modes. The first set of parameters of δ_i (given below equation (15.21) and used to produce Fig. 15.6) is given just for illustration. The second set of parameters of δ_i used to produce Fig. 15.8 takes into account the following effect. The latitudinal differential rotation in the solar convective zone is characterized by the parameters C_2 and C_4, known at the solar surface, and varies with the depth in the convective zone. The variations of the parameters C_2 and C_4 with the depth in the convective zone cause variations of the parameters of δ_i, which are used to produce Fig. 15.8.

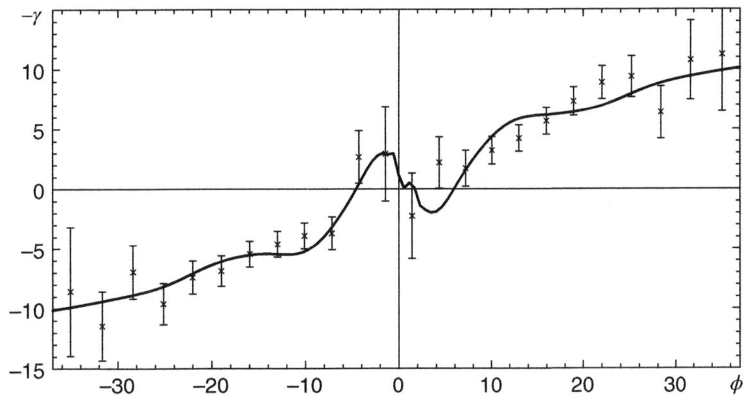

Figure 15.8 The mean tilt is determined as $-\gamma$ (in degrees) against the latitude ϕ (in degrees): numerical simulations (solid line) and observations of sunspot bipolar regions averaged over individual cycles 15–24 (crosses). Adapted from Kleeorin et al. (2020).

However, near the equator, Figs. 15.6 and 15.8 are very different. Indeed, the mean tilt in Fig. 15.6 averaged over ARs of all sizes tends to zero near the equator. On the other hand, the mean tilt in Fig. 15.8 averaged only over larger bipolar regions does not vanish near the equator. It changes its sign at $\phi \approx \pm 5^\circ$. This difference is caused by the fact that the current helicity, $H_c^{ar} = \langle \boldsymbol{B}^{ar} \cdot \mathbf{curl}\, \boldsymbol{B}^{ar} \rangle$, of the ARs contributes to the mean tilt. Here H_c^{ar} characterizes the twist of the magnetic field, $\boldsymbol{B}^{ar}$ is the magnetic field in the AR, and the averaging is performed over the surface within the entire AR. The contribution of the current helicity of ARs to the total mean tilt is $\gamma_H = L_{ar} H_c^{ar} / \langle (\boldsymbol{B}^{ar})^2 \rangle$, where L_{ar} is the AR size. The current helicity H_c^{ar} depends on the magnetic helicity $H_m^{ar} = \langle \boldsymbol{A}^{ar} \cdot \boldsymbol{B}^{ar} \rangle$ of the AR as $H_c^{ar} \approx L_{ar}^{-2} H_m^{ar} + O(L_{ar}^2 / R_\odot^2)$ (Zhang et al., 2012), where $R_\odot$ stands for the radius of the Sun. Now we substitute this expression into γ_H, so

$$\gamma_H = \frac{H_m^{ar}}{L_{ar} \langle (\boldsymbol{B}^{ar})^2 \rangle}. \tag{15.23}$$

The total magnetic helicity of large-scale and small-scale magnetic fields, H_{total}, is conserved since magnetic diffusion in the solar convective zone is negligibly small. A fraction of the total magnetic helicity is transported from the AR to the solar chromosphere and corona by a nonzero magnetic helicity flux. Thus, the total magnetic helicity, $H_{total} \equiv H_m^{ar}\, V$, is the sum of the transported H_{transp} and residual H_{observ} fractions of the magnetic helicity, $H_{total} = H_{transp} + H_{observ}$. The residual magnetic helicity, H_{observ}, is the observable magnetic helicity. Here V is the AR volume. We assume that the transported magnetic helicity, H_{transp}, is the sum of the original magnetic helicity of the AR, $C_w H_m^{ar}\, V$ (with $C_w < 1$), and the magnetic helicity $C_\Omega \langle (\boldsymbol{B}^{ar})^2 \rangle \gamma L_{ar} V$ produced by the twisting of magnetic flux tubes due to the Coriolis force, $H_{transp} = \left(C_w H_m^{ar} + C_\Omega \langle (\boldsymbol{B}^{ar})^2 \rangle \gamma L_{ar} \right) V$, where $-\gamma$ is the mean tilt. The observed mean tilt is defined as

$$\gamma_{observ} = \frac{H_{observ}}{\langle (\boldsymbol{B}^{ar})^2 \rangle L_{ar} V}. \tag{15.24}$$

The total magnetic helicity, $H_{total} \equiv \gamma_H L_{ar} \langle (\boldsymbol{B}^{ar})^2 \rangle V$, is conserved. Substituting equations for the transported magnetic helicity, H_{transp}, into the equation for $H_{total} = H_{transp} + H_{observ}$, and using equations (15.23) and (15.24), we obtain $\gamma_{observ} = (1 - C_w)\gamma_H - C_\Omega \gamma$. It follows from $\gamma_H \propto L_{ar}^{-1}$ and $\gamma \propto L_{ar}^2$ that for large ARs, $|\gamma_H| \ll |\gamma|$, where $\delta_0 \propto L_B^2 \geq L_{ar}^2$. Therefore, $C_\Omega = -1$, since $\gamma_{observ} \approx \gamma$. For small ARs, $|\gamma_H| \gg |\gamma|$. Since the observable tilt for different sizes of ARs is $\gamma_{observ} = \gamma + (1 - C_w) H_m^{ar} / \left[L_{ar} \langle (\boldsymbol{B}^{ar})^2 \rangle \right]$, we can find the size of the AR where both contributions to the mean tilt are of the same order. This result agrees with that suggested by Illarionov et al. (2015). Since $H_m^{ar} = -\overline{\boldsymbol{A}} \cdot \overline{\boldsymbol{B}}$, we find $\gamma_{observ} = \gamma - (1 - C_w) \overline{\boldsymbol{A}} \cdot \overline{\boldsymbol{B}} / \left[L_{ar} \langle (\boldsymbol{B}^{ar})^2 \rangle \right]$. The latter expression and equation (15.21) yield the final equation for the observable tilt of bipolar regions:

$$\begin{aligned} \gamma_{observ} = &-\tilde{\delta}_0 \left[\sin\phi + \delta_3 \sin 3\phi - \delta_5 \sin 5\phi - \tilde{\delta}_M \right. \\ &\left. \times \left(\cos\phi + \tilde{\delta}_3 \cos 3\phi - \tilde{\delta}_5 \cos 5\phi \right) \right] \\ &-(1 - C_w) \frac{\overline{\boldsymbol{A}} \cdot \overline{\boldsymbol{B}}}{L_{ar} \langle (\boldsymbol{B}^{ar})^2 \rangle}. \end{aligned} \tag{15.25}$$

Since $\overline{\boldsymbol{A}} \cdot \overline{\boldsymbol{B}}$ is negative in the northern hemisphere, the sign of the last term on the right-hand side of equation (15.23) due to the magnetic helicity of the AR is opposite to that caused by the Coriolis force. Therefore, the signs of the mean tilts of small and large ARs are opposites. Thus, we can explain the different behaviors near the solar equator observed in Figs. 15.6 and 15.8. Since the mean tilt in Fig. 15.8 is averaged over only large bipolar regions, it does not vanish near the equator. On the other hand, the mean tilt averaged over all ARs in Fig. 15.6 vanishes near the equator because the contributions of the large and small ARs compensate for each other. Notably, since the effect of the Coriolis force on the mean tilt tends to zero near the equator, the compensation effect for large and small bipolar regions is stronger near the equator.

15.9. CONCLUDING REMARKS

In this chapter, we discussed the mean-field solar dynamo with algebraic and dynamic nonlinearities and various mechanisms of sunspot formation. Algebraic nonlinearity describes the quenching of the alpha effect, turbulent magnetic diffusion, and the effective pumping velocity by the back reaction of a large-scale magnetic field on fluid motion. Dynamic nonlinearity is due to the evolution of small-scale magnetic helicity during the nonlinear stage of the dynamo, and it is derived from the conservation law for total magnetic helicity.

We analyzed the results of nonlinear MFD with magnetic helicity transport. This nonlinear dynamo model also contains a budget equation for the dynamics of the sunspot number. In addition, we analyzed the possibility of predicting solar activity by combining the mean-field solar dynamo model with an artificial neural network. We also discussed observations of magnetic helicity in the Sun and their relation to the nonlinear MFD models. Finally, we analyzed a mechanism of formation of the mean tilt of sunspot bipolar regions and the contribution of magnetic helicity transport. We compared the results of the numerical simulations of the mean tilt with the latitudinal distribution of the averaged tilts of bipolar sunspots observed over a century.

Here, we briefly highlight recent developments and current challenges in investigations of the nonlinear mean-field dynamo with magnetic helicity transport and solar activity. In recent decades, essential progress has been achieved in the following directions:

(1) Development of a mean-field theory of turbulent magnetic helicity fluxes as an important ingredient in the equation describing dynamical quenching.

(2) Numerical and theoretical modeling of the formation of magnetic spots and bipoles in developed strongly stratified turbulence and turbulent convection based on the negative effective magnetic pressure instability (NEMPI).

(3) Development of a method for a short-term forecast of solar activity based on a combination of mean-field numerical simulations of the nonlinear dynamo model (which includes the dynamic nonlinearity of small-scale magnetic helicity evolution and an equation for the evolution of the sunspot number) and a neural network that uses observational data (Wolf numbers).

(4) Theoretical and numerical development of the mechanism for the mean tilt of solar bipolar regions.

(5) Development of an improved theory of solar differential rotation, which considers the effect of turbulent heat flux on the formation of strong shear near the solar surface.

(6) Direct numerical simulations of the formation of solar and stellar differential rotation.

(7) Diagnostics of the formation of solar active regions based on new methods in helioseismology.

Finally, we briefly highlight future directions in investigations of nonlinear mean-field dynamos with magnetic helicity transport and solar activity:

(1) Development of the theory of two-year oscillations of solar magnetic activity.

(2) Theoretical, numerical, and observational investigations of stellar magnetic activity.

(3) Development of a method for a long-term solar activity forecast based on a combination of the mean-field numerical simulations of the nonlinear dynamo model and the neural network, which uses observational data (Wolf numbers).

(4) Development of an improved theory of sunspot formation based on a combination of NEMPI and thermomagnetic instability, including the effect of differential rotation on NEMPI.

ACKNOWLEDGMENTS

We would like to thank the Online Advanced Study Program in 2020–2021 on Helicities in Astrophysics and Beyond (helicity2020.izmiran.ru), which stimulated and consolidated our discussions and ideas. The work of NK, KK, and NS on formulation, review, and further generalization of the results on magnetic helicity transport and development of solar dynamo models was supported by Russian Science Foundation grant No. 21–72–20067. The authors would like to thank the anonymous reviewers for their comments and suggestions that helped improve the manuscript.

REFERENCES

Abramenko, V. I., Wang, T., & Yurchishin, V. B. (1996). Analysis of electric current helicity in active regions on the basis of vector magnetograms. *Solar Physics*, *168*(1), 75–89.

Acheson, D. J. (1978). On the instability of toroidal magnetic fields and differential rotation in stars. *Philosophical Transactions of the Royal Society of London. Series A, Mathematical and Physical Sciences*, *289*, 459–500.

Bai, X. Y., Deng, Y. Y., Teng, F., Su, J. T., Mao, X. J., & Wang, G. P. (2014). Improved magnetogram calibration of Solar

Magnetic Field Telescope and its comparison with the Helioseismic and Magnetic Imager. *Monthly Notices of the Royal Astronomical Society*, *445*(1), 49–55.

Bao, S. D., Ai, G. X., & Zhang, H. Q. (2000). The hemispheric sign rule of current helicity during the rising phase of cycle 23. *Journal of Astrophysics and Astronomy*, *21*, 303.

Batchelor, G. K. (1950). On the spontaneous magnetic field in a conducting liquid in turbulent motion. *Proceedings of the Royal Society of London. Series A, Mathematical and Physical Sciences*, *201*, 405–416.

Berger, M. A. (1988). An energy formula for nonlinear force-free magnetic fields. *Astronomy and Astrophysics*, *201*(2), 355–361.

Berger, M. A. & Field, G. B. (1984). The topological properties of magnetic helicity. *Journal of Fluid Mechanics*, *147*, 133–148.

Berger, M. A., & Ruzmaikin, A. (2000). Rate of helicity production by solar rotation. *Journal of Geophysical Research*, *105*(A5), 10481–10490.

Blackman, E. G., & Brandenburg, A. (2002). Dynamic nonlinearity in large-scale dynamos with shear. *The Astrophysical Journal*, *579*(1), 359.

Blackman, E. G., & Field, G. B. (2000). Constraints on the magnitude of alpha in dynamo theory. *The Astrophysical Journal*, *534*(2), 984.

Blackman, E. G., & Field, G. B. (2002). New dynamical mean-field dynamo theory and closure approach. *Physical Review Letters*, *89*(26), 265007.

Brandenburg, A. (2019). A global two-scale helicity proxy from pi-ambiguous solar magnetic fields. *The Astrophysical Journal*, *883*(2), 119.

Brandenburg, A. (2021). Chirality in astrophysics. arXiv:2110.08117

Brandenburg, A., Gressel, O., Jabbari, S., Kleeorin, N., & Rogachevskii, I. (2014). Mean-field and direct numerical simulations of magnetic flux concentrations from vertical field. *Astronomy and Astrophysics*, *562*, A53.

Brandenburg, A., Gressel, O., Käpylä, P. J., Kleeorin, N., Mantere, M. J., & Rogachevskii, I. (2013a). New scaling for the alpha effect in slowly rotating turbulence. *The Astrophysical Journal*, *762*(2), 127.

Brandenburg, A., Kemel, K., Kleeorin, N., & Rogachevskii, I. (2012). The negative effective magnetic pressure in stratified forced turbulence. *The Astrophysical Journal*, *749*(2), 179.

Brandenburg, A., Kemel, K., Kleeorin, N., Mitra, Dh., & Rogachevskii, I. (2011). Detection of negative effective magnetic pressure instability in turbulence simulations. *Astrophysical Journal Letters*, *740*(2), L50.

Brandenburg, A., Kleeorin, N., & Rogachevskii, I. (2010). Large-scale magnetic flux concentrations from turbulent stresses. *Astronomische Nachrichten*, *331*(1), 5–13.

Brandenburg, A., Kleeorin, N., & Rogachevskii, I. (2013b). Self-assembly of shallow magnetic spots through strongly stratified turbulence. *Astrophysical Journal Letters*, *776*(2), L23.

Brandenburg, A., Petrie, G. J. D., & Singh, N. K. (2017). Two-scale analysis of solar magnetic helicity. *The Astrophysical Journal*, *836*(1), 21.

Brandenburg, A., Rogachevskii, I., & Kleeorin, N. (2016). Magnetic concentrations in stratified turbulence: the negative effective magnetic pressure instability. *New Journal of Physics*, *18*(12), 125011.

Brandenburg, A., & Subramanian, K. (2005). Astrophysical magnetic fields and non-linear dynamo theory. *Physics Reports*, *417*, 1–209.

Bukai, M., Eidelman, A., Elperin, T., Kleeorin, N., Rogachevskii, I., & Sapir Katiraie, I. (2009). Effect of large-scale coherent structures on turbulent convection. *Physical Review E*, *79*(6), 066302.

Bushby, P. J., & Tobias, S. M. (2007). On predicting the solar cycle using mean-field models. *The Astrophysical Journal*, *661*(2), 1289.

Chae, J. (2001). Observational determination of the rate of magnetic helicity transport through the solar surface via the horizontal motion of field line footpoints. *Astrophysical Journal Letters*, *560*(1), L95–L98.

Childress, S., & Gilbert, A. D. (1995). *Stretch, twist, fold: the fast dynamo* (Vol. 37). Berlin: Springer.

Choudhuri, A. R., Chatterjee, P., & Jiang, J. (2007). Predicting solar cycle 24 with a solar dynamo model. *Physical Review Letters*, *98*(13), 131103.

Covas, E., Tavakol, R., Tworkowski, A., & Brandenburg, A. (1998). Axisymmetric mean field dynamos with dynamic and algebraic alpha–quenchings. *Astronomy and Astrophysics*, *329*, 350–360.

Covas, E., Tworkowski, A., Tavakol, R., & Brandenburg, A. (1997). Robustness of truncated alpha omega dynamos with a dynamic alpha. *Solar Physics*, *172*(1–2), 3–9.

Dasi-Espuig, M., Solanki, S. K., Krivova, N. A., Cameron, R., & Penuela, T. (2010). *Sunspot group tilt angles and the strength of the solar cycle. Astronomy and Astrophysics*, *518*, A7.

De Jager, C., & Duhau, S. (2009). Forecasting the parameters of sunspot cycle 24 and beyond. *Journal of Atmospheric and Solar-Terrestrial Physics*, *71*(2), 239–245.

Del Sordo, F., Guerrero, G., & Brandenburg, A. (2013). Turbulent dynamos with advective magnetic helicity flux. *Monthly Notices of the Royal Astronomical Society*, *429*(2), 1686–1694.

Démoulin, P., & Berger, M. A. (2003). Magnetic energy and helicity fluxes at the photospheric level. *Solar Physics*, *215*(2), 203–215.

Démoulin, P., & Pariat, E. (2009). Modelling and observations of photospheric magnetic helicity. *Advances in Space Research*, *43*(7), 1013–1031.

DeVore, C. R. (2000). Magnetic helicity generation by solar differential rotation. *The Astrophysical Journal*, *539*(2), 944–953.

Dikpati, M., & Gilman, P. A. (2006). Simulating and predicting solar cycles using a flux-transport dynamo. *The Astrophysical Journal*, *649*(1), 498.

Dikpati, M., De Toma, G., & Gilman, P. A. (2006). Predicting the strength of solar cycle 24 using a flux-transport dynamo-based tool. *Geophysical Research Letters*, *33*(5).

Dmitrieva, I. V., Kuzanyan, K. M., & Obridko, V. N. (2000). Amplitude and period of the dynamo wave and prediction of the solar cycle. *Solar Physics*, *195*(1), 209–218.

Drobyshevski, E. M., & Yuferev, V. S. (1974). Topological pumping of magnetic flux by three-dimensional convection. *Journal of Fluid Mechanics*, *65*(1), 33–44.

D'Silva, S., & Choudhuri, A. R. (1993). A theoretical model for tilts of bipolar magnetic regions. *Astronomy and Astrophysics*, *272*, 621.

Field, G. B., Blackman, E. G., & Chou, H. (1999). Nonlinear alpha-effect in dynamo theory. *The Astrophysical Journal*, *513*(2), 638.

Georgoulis, M. K., Rust, D. M., Pevtsov, A. A., Bernasconi, P. N., & Kuzanyan, K. M. (2009). Solar magnetic helicity injected into the heliosphere: magnitude, balance, and periodicities over solar cycle 23. *Astrophysical Journal Letters*, *705*(1), L48–L52.

Gershberg, R. E., Kleeorin, N. I., Pustilnik, L. A., & Shlyapnikov, A. A. (2020). Physics of stars of intermediate and low masses with solar type activity. Moscow: PHYSMATLIT.

Gilman, P. A. (1970). Instability of magnetohydrostatic stellar interiors from magnetic buoyancy. *The Astrophysical Journal*, *162*, 1019.

Gruzinov, A. V., & Diamond, P. H. (1994). Self-consistent theory of mean-field electrodynamics. *Physical Review Letters*, *72*(11), 1651.

Guo, Y., Pariat, E., Valori, G., Anfinogentov, S., Chen, F., Georgoulis, M. K., et al. (2017). Magnetic helicity estimations in models and observations of the solar magnetic field. III. Twist number method. *The Astrophysical Journal*, *840*(1), 40.

Hagino, M., & Sakurai, T. (2004). Latitude variation of helicity in solar active regions. *Publications of the Astronomical Society of Japan*, *56*, 831–843.

Hagino, M., & Sakurai, T. (2005). Solar-cycle variation of magnetic helicity in active regions. *Publications of the Astronomical Society of Japan*, *57*, 481–485.

Hao, J., & Zhang, M. (2011). Hemispheric helicity trend for solar cycle 24. *Astrophysical Journal Letters*, *733*(2), L27.

Hubbard, A., & Brandenburg, A. (2010). Magnetic helicity fluxes in an alpha^2 dynamo embedded in a halo. *Geophysical and Astrophysical Fluid Dynamics*, *104*(5–6), 577–590.

Hubbard, A., & Brandenburg, A. (2011). Magnetic helicity flux in the presence of shear. *The Astrophysical Journal*, *727*(1), 11.

Hubbard, A., & Brandenburg, A. (2012). Catastrophic quenching in alpha omega dynamos revisited. *The Astrophysical Journal*, *748*(1), 51.

Illarionov, E., Tlatov, A., & Sokoloff, D. (2015). The properties of the tilts of bipolar solar regions. *Solar Physics*, *290*(2), 351–361.

Iroshnikov, R. S. (1971). Viscous torsional oscillations and the cycles of solar activity. *Soviet Astronomy*, *15*, 432.

Jabbari, S., Brandenburg, A., Kleeorin, N., Mitra, Dh., & Rogachevskii, I. (2013). Surface flux concentrations in a spherical alpha^2 dynamo. *Astronomy and Astrophysics*, *556*, A106.

Jabbari, S., Brandenburg, A., Kleeorin, N., Mitra, Dh., & Rogachevskii, I. (2015). Bipolar magnetic spots from dynamos in stratified spherical shell turbulence. *The Astrophysical Journal*, *805*(2), 166.

Jabbari, S., Brandenburg, A., Kleeorin, N., & Rogachevskii, I. (2017). Sharp magnetic structures from dynamos with density stratification. *Monthly Notices of the Royal Astronomical Society*, *467*(3), 2753–2765.

Jabbari, S., Brandenburg, A., Losada, I. R., Kleeorin, N., & Rogachevskii, I. (2014). Magnetic flux concentrations from dynamo-generated fields. *Astronomy and Astrophysics*, *568*, A112.

Jabbari, S., Brandenburg, A., Mitra, Dh., Kleeorin, N., & Rogachevskii, I. (2016). Turbulent reconnection of magnetic bipoles in stratified turbulence. *Monthly Notices of the Royal Astronomical Society*, *459*(4), 4046–4056.

Jones, C. A., Thompson, M. J., & Tobias, S. M. (2010). The solar dynamo. *Space Science Reviews*, *152*, 591–616.

Kane, R. P. (2007). Solar cycle predictions based on solar activity at different solar latitudes. *Solar Physics*, *246*(2), 471–485.

Käpylä, P. J., Brandenburg, A., Kleeorin, N., Käpylä, M. J., & Rogachevskii, I. (2016). Magnetic flux concentrations from turbulent stratified convection. *Astronomy and Astrophysics*, *588*, A150.

Käpylä, P. J., Brandenburg, A., Kleeorin, N., Mantere, M. J., & Rogachevskii, I. (2012). Negative effective magnetic pressure in turbulent convection. *Monthly Notices of the Royal Astronomical Society*, *422*, 2465–2473.

Käpylä, P. J., Korpi, M. J., & Brandenburg, A. (2010). Open and closed boundaries in large-scale convective dynamos. *Astronomy and Astrophysics*, *518*, A22.

Kazantsev, A. P. (1968). Enhancement of a magnetic field by a conducting fluid. *Soviet Journal of Experimental and Theoretical Physics*, *26*(5), 1031–1034.

Kemel, K., Brandenburg, A., Kleeorin, N., & Rogachevskii, I. (2012b). Properties of the negative effective magnetic pressure instability. *Astronomische Nachrichten*, *333*(2), 95–100.

Kemel, K., Brandenburg, A., Kleeorin, N., & Rogachevskii, I. (2013b). Non-uniformity effects in the negative effective magnetic pressure instability. *Physica Scripta*, *2013*(T155), 014027.

Kemel, K., Brandenburg, A., Kleeorin, N., Mitra, Dh., & Rogachevskii, I. (2012a). Spontaneous formation of magnetic flux concentrations in stratified turbulence. *Solar Physics*, *280*(2), 321–333.

Kemel, K., Brandenburg, A., Kleeorin, N., Mitra, Dh., & Rogachevskii, I. (2013a). Active region formation through the negative effective magnetic pressure instability. *Solar Physics*, *287*(1–2), 293–313.

Kersale, E., Hughes, D. W., & Tobias, S. M. (2007). The nonlinear evolution of instabilities driven by magnetic buoyancy: A new mechanism for the formation of coherent magnetic structures. *Astrophysical Journal Letters*, *663*(2), L113.

Kippenhahn, R. (1963). Differential rotation in stars with convective envelopes. *The Astrophysical Journal*, *137*, 664.

Kitchatinov, L. L., Pipin, V. V., & Rüdiger, G. (1994). Turbulent viscosity, magnetic diffusivity, and heat conductivity under the influence of rotation and magnetic field. *Astronomische Nachrichten*, *315*(2), 157–170.

Kitiashvili, I. N., & Kosovichev, A. G. (2011). Modeling and prediction of solar cycles using data assimilation methods. In *The Pulsations of the Sun and the Stars*, 121–137. Springer.

Kleeorin, N., Kuzanyan, K., Moss, D., Rogachevskii, I., Sokoloff, D., & Zhang, H. (2003a). Magnetic helicity evolution during the solar activity cycle: observations and dynamo theory. *Astronomy and Astrophysics*, *409*(3), 1097–1105.

Kleeorin, N., Mond, M., & Rogachevskii, I. (1993). Magnetohydrodynamic instabilities in developed small-scale turbulence. *Physics of Fluids B: Plasma Physics*, *5*(11), 4128–4134.

Kleeorin, N., Mond, M., & Rogachevskii, I. (1996). Magnetohydrodynamic turbulence in the solar convective zone as a source of oscillations and sunspots formation. *Astronomy and Astrophysics*, *307*, 293.

Kleeorin, N., Moss, D., Rogachevskii, I., & Sokoloff, D. (2000). Helicity balance and steady-state strength of the dynamo generated galactic magnetic field. *Astronomy and Astrophysics*, *361*, L5–L8.

Kleeorin, N., Moss, D., Rogachevskii, I., & Sokoloff, D. (2002). The role of magnetic helicity transport in nonlinear galactic dynamos. *Astronomy and Astrophysics*, *387*(2), 453–462.

Kleeorin, N., Moss, D., Rogachevskii, I., & Sokoloff, D. (2003b). Nonlinear magnetic diffusion and magnetic helicity transport in galactic dynamos. *Astronomy and Astrophysics*, *400*(1), 9–18.

Kleeorin, N., & Rogachevskii, I. (1994). Effective ampère force in developed magnetohydrodynamic turbulence. *Physical Review E*, *50*(4), 2716.

Kleeorin, N., & Rogachevskii, I. (1999). Magnetic helicity tensor for an anisotropic turbulence. *Physical Review E*, *59*(6), 6724.

Kleeorin, N., & Rogachevskii, I. (2003). Effect of rotation on a developed turbulent stratified convection: The hydrodynamic helicity, the alpha effect, and the effective drift velocity. *Physical Review E*, *67*(2), 026321.

Kleeorin, N., & Rogachevskii, I. (2006). Effect of heat flux on differential rotation in turbulent convection. *Physical Review E*, *73*(4), 046303.

Kleeorin, N., & Rogachevskii, I. (2022). Turbulent magnetic helicity fluxes in solar convective zone. *Monthly Notices of the Royal Astronomical Society*, *515*(4), 5437–5448.

Kleeorin, N., Rogachevskii, I., & Ruzmaikin, A. (1994). Solar irradiance variations and nonlinear mean field dynamo. *Solar Physics*, *155*(2), 223–234.

Kleeorin, N., Rogachevskii, I., & Ruzmaikin, A. (1995). Magnitude of the dynamo generated magnetic field in solar-type convective zones. *Astronomy and Astrophysics*, *297*, 159–167.

Kleeorin, N., Rogachevskii, I., & Ruzmaikin, A. A. (1989). Negative magnetic pressure as a trigger of large-scale magnetic instability in the solar convective zone. *Soviet Astronomy Letters*, *15*, 274–277.

Kleeorin, N., Rogachevskii, I., & Ruzmaikin, A. A. (1990). Magnetic force reversal and instability in a plasma with advanced magnetohydrodynamic turbulence. *Soviet Journal of Experimental and Theoretical Physics*, *70*, 878–883.

Kleeorin, N., & Ruzmaikin, A. (1982). Dynamics of the average turbulent helicity in a magnetic field. *Magnetohydrodynamics*, *18*(3), 116–122.

Kleeorin, N., Safiullin, N., Kuzanyan, K., Rogachevskii, I., Tlatov, A., & Porshnev, S. (2020). The mean tilt of sunspot bipolar regions: theory, simulations and comparison with observations. *Monthly Notices of the Royal Astronomical Society*, *495*(1), 238–248.

Kleeorin, Ya., Safiullin, N., Kleeorin, N., Porshnev, S., Rogachevskii, I., & Sokoloff, D. (2016). The dynamics of wolf numbers based on nonlinear dynamos with magnetic helicity: comparisons with observations. *Monthly Notices of the Royal Astronomical Society*, *460*(4), 3960–3967.

Kosovichev, A. G., & Stenflo, J. O. (2008). Tilt of emerging bipolar magnetic regions on the sun. *Astrophysical Journal Letters*, *688*(2), L115.

Kosugi, T., Matsuzaki, K., Sakao, T., Shimizu, T., Sone, Y., Tachikawa, S., et al. (2007). The Hinode (Solar-B) mission: an overview. *Solar Physics*, *243*(1), 3–17.

Krause, F., & Rädler, K.-H. (1980). Mean-field magnetohydrodynamics and dynamo theory. Oxford: Pergamon Press.

Kulsrud, R. M., & Anderson, S. W. (1992). The spectrum of random magnetic fields in the mean field dynamo theory of the galactic magnetic field. *The Astrophysical Journal*, *396*, 606–630.

Kutsenko, A. S., Abramenko, V. I., Kuzanyan, K. M., Xu, H., & Zhang, H. (2018). Intermittency spectra of current helicity in solar active regions. *Monthly Notices of the Royal Astronomical Society*, *480*(3), 3780–3787.

Kuzanyan, K., Bao, S., & Zhang, H. (2000). Probing signatures of the alpha-effect in the solar convection zone. *Solar Physics*, *191*(2), 231–246.

LaBonte, B. J., & Howard, R. (1982). Torsional waves on the sun and the activity cycle. *Solar Physics*, *75*(1), 161–178.

Leka, K. D., Barnes, G., Crouch, A. D., Metcalf, T. R., Gary, G., et al. (2009). Resolving the 1800 ambiguity in solar vector magnetic field data: evaluating the effects of noise, spatial resolution, and method assumptions. *Solar Physics*, *260*(1), 83–108.

Lites, B. W., Akin, D. L., Card, G., Cruz, T., Duncan, D. W., Edwards, C. G., et al. (2013). The Hinode spectro-polarimeter. *Solar Physics*, *283*(2), 579–599.

Liu, Y., Hoeksema, J. T., Scherrer, P. H., Schou, J., Couvidat, S., Bush, R. I., et al. (2012). Comparison of Line-of-sight magnetograms taken by the Solar Dynamics Observatory/Helioseismic and Magnetic Imager and Solar and Heliospheric Observatory/Michelson Doppler Imager. *Solar Physics*, *279*(1), 295–316.

Liu, Y., Hoeksema, J. T., & Sun, X. (2014). Test of the hemispheric rule of magnetic helicity in the sun using the Helioseismic and Magnetic Imager (HMI) data. *Astrophysical Journal Letters*, *783*(1), L1.

Livshits, I. M., & Obridko, V. N. (2006). Variations of the dipole magnetic moment of the sun during the solar activity cycle. *Astronomy Reports*, *50*(11), 926–935.

Losada, I. R., Brandenburg, A., Kleeorin, N., Mitra, Dh., & Rogachevskii, I. (2012). Rotational effects on the negative magnetic pressure instability. *Astronomy and Astrophysics*, *548*(A49).

Losada, I. R., Brandenburg, A., Kleeorin, N., & Rogachevskii, I. (2013). Competition of rotation and stratification in flux concentrations. *Astronomy and Astrophysics*, *556*(A83).

Losada, I. R., Brandenburg, A., Kleeorin, N., & Rogachevskii, I. (2014). Magnetic flux concentrations in a polytropic atmosphere. *Astronomy and Astrophysics*, *564*(A2).

Losada, I. R., Warnecke, J., Brandenburg, A., Kleeorin, N., & Rogachevskii, I. (2019). Magnetic bipoles in rotating turbulence with coronal envelope. *Astronomy and Astrophysics*, *621*(A61).

McClintock, B. H., & Norton, A. A. (2013). Recovering Joy's Law as a function of solar cycle, hemisphere, and longitude.

In *Solar Dynamics and Magnetism from the Interior to the Atmosphere* (pp. 215–227). Springer.

Metcalf, T. R. (1994). Resolving the 180-degree ambiguity in vector magnetic field measurements: The "minimum" energy solution. *Solar Physics*, *155*(2), 235–242.

Mitra, Dh., Brandenburg, A., Kleeorin, N., & Rogachevskii, I. (2014). Intense bipolar structures from stratified helical dynamos. *Monthly Notices of the Royal Astronomical Society*, *445*(1), 761–769.

Mitra, Dh., Candelaresi, S., Chatterjee, P., Tavakol, R., & Brandenburg, A. (2010). Equatorial magnetic helicity flux in simulations with different gauges. *Astronomische Nachrichten*, *331*(1), 130–135.

Moffatt, H. K. (1978). *Magnetic field generation in electrically conducting fluids*. Cambridge: Cambridge University Press.

Moffatt, H. K., & Dormy, E. (2019). *Self-exciting fluid dynamos* (Vol. 59). Cambridge: Cambridge University Press.

Newcomb, W. A. (1961). Convective instability induced by gravity in a plasma with a frozen-in magnetic field. *Physics of Fluids*, *4*(4), 391–396.

Nindos, A., Zhang, J., & Zhang, H. (2003). The magnetic helicity budget of solar active regions and coronal mass ejections. *The Astrophysical Journal*, *594*(2), 1033–1048.

Obridko, V. N., & Shelting, B. D. (2008). On prediction of the strength of the 11-year solar cycle no. 24. *Solar Physics*, *248*(1), 191–202.

Ossendrijver, M. (2003). The solar dynamo. *The Astronomy and Astrophysics Review*, *11*(4), 287–367.

Otsuji, K., Sakurai, T., & Kuzanyan, K. (2015). A statistical analysis of current helicity and twist in solar active regions over the phases of the solar cycle using the spectro-polarimeter data of Hinode. *Publications of the Astronomical Society of Japan*, *67*(1), 6.

Parker, E. N. (1955a). The formation of sunspots from the solar toroidal field. *The Astrophysical Journal*, *121*, 491.

Parker, E. N. (1955b). Hydromagnetic dynamo models. *The Astrophysical Journal*, *122*, 293.

Parker, E. N. (1966). The dynamical state of the interstellar gas and field. *The Astrophysical Journal*, *145*, 811.

Parker, E. N. (1979). *Cosmical magnetic fields: their origin and their activity*. Oxford: Clarendon Press.

Pesnell, W. D. (2012) Solar cycle predictions (invited review). *Solar Physics*, *281*(1), 507–532.

Pevtsov, A. A., & Latushko, S. M. (2000). Current helicity of the large-scale photospheric magnetic field. *The Astrophysical Journal*, *528*(2), 999–1003.

Pevtsov, A. A., Canfield, R. C., & Latushko, S. M. (2001). Hemispheric helicity trend for solar cycle 23. *Astrophysical Journal Letters*, *549*(2), L261–L263.

Pevtsov, A. A., Canfield, R. C., & Metcalf, T. R. (1994). Patterns of helicity in solar active regions. *Astrophysical Journal Letters*, *425*(L117).

Pevtsov, A. A., Canfield, R. C., & Metcalf, T. R. (1995). Latitudinal variation of helicity of photospheric magnetic fields. *Astrophysical Journal Letters*, *440*(L109).

Pipin, V. V., & Pevtsov, A. A. (2014). Magnetic helicity of the global field in solar cycles 23 and 24. *The Astrophysical Journal*, *789*(1), 21.

Pipin, V. V., Pevtsov, A. A., Liu, Y., & Kosovichev, A. G. (2019). Evolution of magnetic helicity in solar cycle 24. *Astrophysical Journal Letters*, *877*(2), L36.

Pouquet, A., Frisch, U., & Léorat, J. (1976). Strong MHD helical turbulence and the nonlinear dynamo effect. *Journal of Fluid Mechanics*, *77*(2), 321–354.

Prabhu, A., Brandenburg, A., Käpylä, M. J., & Lagg, A. (2020). Helicity proxies from linear polarisation of solar active regions. *Astronomy and Astrophysics*, *641*, A46.

Prabhu, A. P., Singh, N. K., Käpylä, M. J., & Lagg, A. (2021). Inferring magnetic helicity spectrum in spherical domains: method and example applications. *Astronomy and Astrophysics*, *654*, A3.

Priest, E. R. (1982). *Solar magnetohydrodynamics*. Dordrecht: Reidel Publ. Co.

Rädler, K.-H. (1980). Mean-field approach to spherical dynamo models. *Astronomische Nachrichten*, *301*(3), 101–129.

Rädler, K.-H. (1982). On dynamo action in the high-conductivity limit. *Geophysical and Astrophysical Fluid Dynamics*, *20*(3–4), 191–211.

Rädler, K.-H., Brandenburg, A., Del Sordo, F., & Rheinhardt, M. (2011). Mean-field diffusivities in passive scalar and magnetic transport in irrotational flows. *Physical Review E*, *84*(4), 046321.

Rädler, K.-H., Kleeorin, N., & Rogachevskii, I. (2003). The mean electromotive force for MHD turbulence: the case of a weak mean magnetic field and slow rotation. *Geophysical and Astrophysical Fluid Dynamics*, *97*, 249–274.

Roberts, P. H., & Soward, A. M. (1975). A unified approach to mean field electrodynamics. *Astronomische Nachrichten*, *296*(2), 49–64.

Roberts, P. H., & Stix, M. (1971). *The turbulent dynamo*. Boulder: NCAR.

Rogachevskii, I. (2021). *Introduction to turbulent transport of particles, temperature and magnetic fields*. Cambridge: Cambridge University Press.

Rogachevskii, I., & Kleeorin, N. (1997). Intermittency and anomalous scaling for magnetic fluctuations. *Physical Review E*, *56*(1), 417.

Rogachevskii, I., & Kleeorin, N. (2000). Electromotive force for an anisotropic turbulence: intermediate nonlinearity. *Physical Review E*, *61*(5), 5202.

Rogachevskii, I., & Kleeorin, N. (2001). Nonlinear turbulent magnetic diffusion and mean-field dynamo. *Physical Review E*, *64*(5), 056307.

Rogachevskii, I., & Kleeorin, N. (2004). Nonlinear theory of a "shear-current" effect and mean-field magnetic dynamos. *Physical Review E*, *70*(4),046310.

Rogachevskii, I., & Kleeorin, N. (2006). Small-scale magnetic buoyancy and magnetic pumping effects in a turbulent convection. *Geophysical and Astrophysical Fluid Dynamics*, *100*(3), 243–263.

Rogachevskii, I., & Kleeorin, N. (2007). Magnetic fluctuations and formation of large-scale inhomogeneous magnetic structures in a turbulent convection. *Physical Review E*, *76*(5), 056307.

Rogachevskii, I., Kleeorin, N., & Brandenburg, A. (2018). Compressibility in turbulent MHD and passive scalar transport: mean-field theory. *Journal of Plasma Physics*, *84*, 735840502.

Rudenko, G. V., & Anfinogentov, S. A. (2014). Very fast and accurate azimuth disambiguation of vector magnetograms. *Solar Physics*, *289*(5), 1499–1516.

Rüdiger, G. (1989). *Differential rotation and stellar convection: Sun and solar-type stars*. New York: Gordon and Breach.

Rüdiger, G., Hollerbach, R., & Kitchatinov, L. L. (2013). *Magnetic processes in astrophysics: theory, simulations, experiments*. Weinheim: John Wiley & Sons.

Rüdiger, G., & Kichatinov, L. L. (1993). Alpha-effect and alpha-quenching. *Astronomy and Astrophysics*, *269*, 581–588.

Ruzmaikin, A., Shukurov, A. M., & Sokoloff, D. D. (1988). *Magnetic fields of galaxies*. Dordrecht: Kluwer Academic.

Safiullin, N., Kleeorin, N., Porshnev, S., Rogachevskii, I., & Ruzmaikin, A. (2018). Nonlinear mean-field dynamo and prediction of solar activity. *Journal of Plasma Physics*, *84*, 735840306.

Schüssler, M., Caligari, P., Ferriz-Mas, A., & Moreno-Insertis, F. (1994). Instability and eruption of magnetic flux tubes in the solar convection zone. *Astronomy and Astrophysics*, *281*, L69–L72.

Scherrer, P. H., Schou, J., Bush, R. I., et al. (2012). The Helioseismic and Magnetic Imager (HMI) investigation for the Solar Dynamics Observatory (SDO). *Solar Physics*, *275*(1–2), 207–227.

Schou, J., Scherrer, P. H., Bush, R. I., et al. (2012). Design and ground calibration of the Helioseismic and Magnetic Imager (HMI) Instrument on the Solar Dynamics Observatory (SDO). *Solar Physics*, *275*(1–2), 229–259.

Schrijver, C. J., & Zwaan, C. (2008). *Solar and stellar magnetic activity, volume 34*. Cambridge: Cambridge University Press.

Seehafer, N. (1990). Electric current helicity in the solar atmosphere. *Solar Physics*, *125*(2), 219–232.

Seehafer, N. (1994). Alpha effect in the solar atmosphere. *Astronomy and Astrophysics*, *284*, 593–598.

Shukurov, A., Sokoloff, D., Subramanian, K., & Brandenburg, A. (2006). Galactic dynamo and helicity losses through fountain flow. *Astronomy and Astrophysics*, *448*(2), L33–L36.

Singh, N. K., Käpylä, M. J., Brandenburg, A., et al. (2018). Bihelical spectrum of solar magnetic helicity and its evolution. *The Astrophysical Journal*, *863*(2), 182.

Sokoloff, D., Bao, S. D., Kleeorin, N., et al. (2006). The distribution of current helicity at the solar surface at the beginning of the solar cycle. *Astronomische Nachrichten*, *327*(9), 876–883.

Spiegel, E. A., & Weiss, N. O. (1980). Magnetic activity and variations in solar luminosity. *Nature*, *287*(5783), 616–617.

Spruit, H. C. (1974). A model of the solar convection zone. *Solar Physics*, *34*(2), 277–290.

Spruit, H. C. (1981). Motion of magnetic flux tubes in the solar convection zone and chromosphere. *Astronomy and Astrophysics*, *98*, 155–160.

Spruit, H. C., & Van Ballegooijen, A. A. (1982). Stability of toroidal flux tubes in stars. *Astronomy and Astrophysics*, *106*, 58–66.

Steenbeck, M., Krause, F., & Rädler, K.-H. (1966). Berechnung der mittleren lorentzfeldstärke für ein elektrisch leitendes medium in turbulenter, durch corioliskräfte beeinflußter bewegung. *Zeitschrift für Naturforschung A*, *21*(4), 369–376.

Stix, M. (2012). *The Sun: an introduction*. Berlin: Springer.

Su, J.-T., & Zhang, H. (2004). Calibration of vector magnetogram with the non-linear least-squares fitting technique. *Chinese Journal of Astronomy and Astrophysics*, *4*, 365–376.

Subramanian, K., & Brandenburg, A. (2004). Nonlinear current helicity fluxes in turbulent dynamos and alpha quenching. *Physical Review Letters*, *93*(20), 205001.

Thalmann, J. K., Georgoulis, M. K., Liu, Y., et al. (2021). Magnetic helicity estimations in models and observations of the solar magnetic field. IV. Application to solar observations. *The Astrophysical Journal*, *922*(1), 41.

Thalmann, J. K., Inhester, B., & Wiegelmann, T. (2011). Estimating the relative helicity of coronal magnetic fields. *Solar Physics*, *272*(2), 243–255.

Thomas, J. H., & Weiss, N. O. (2008). *Sunspots and star spots*. Cambridge: Cambridge University Press.

Tlatov, A. G. (2009). The minimum activity epoch as a precursor of the solar activity. *Solar Physics*, *260*(2), 465–477.

Tlatov, A. G. (2015). The change of the solar cyclicity mode. *Advances in Space Research*, *55*(3), 851–856.

Tlatova, A., Tlatov, A., Pevtsov, A. A., et al. (2018). Tilt of sunspot bipoles in solar cycles 15 to 24. *Solar Physics*, *293*(8), 1–12.

Tobias, S. M., & Hughes, D. W. (2004). The influence of velocity shear on magnetic buoyancy instability in the solar tachocline. *The Astrophysical Journal*, *603*(2), 785.

Tobias, S. M., Brummell, N. H., Clune, T. L., & Toomre, J. (2001). Transport and storage of magnetic field by overshooting turbulent compressible convection. *The Astrophysical Journal*, *549*(2), 1183.

Vainshtein, S. I., & Cattaneo, F. (1992). Nonlinear restrictions on dynamo action. *The Astrophysical Journal*, *393*, 165–171.

Vainshtein, S. I., & Kichatinov, L. L. (1983). The macroscopic magnetohydrodynamics of inhomogeneously turbulent cosmic plasmas. *Geophysical and Astrophysical Fluid Dynamics*, *24*(4), 273–298.

Valori, G., Pariat, E., Anfinogentov, S., et al. (2016). Magnetic helicity estimations in models and observations of the solar magnetic field. Part I: Finite volume methods. *Space Science Reviews*, *201*(1–4), 147–200.

Vishniac, E. T., & Cho, J. (2001). Magnetic helicity conservation and astrophysical dynamos. *The Astrophysical Journal*, *550*(2), 752.

Wang, Ch., & Zhang, M. (2010). A hemispheric helicity sign rule indicated by large-scale photospheric magnetic fields at three phases of solar cycle 23. *The Astrophysical Journal*, *720*(1), 632–638.

Wang, T., Ai, G., & Deng, Y. (1996). Calibration of nine-channel solar magnetic field telescope. II. The results of the observational calibration. *Astrophysics Reports Publications of the Beijing Astronomical Observatory*, *28*, 41–48.

Warnecke, J., Losada, I. R., Brandenburg, A., Kleeorin, N., & Rogachevskii, I. (2013). Bipolar magnetic structures driven by stratified turbulence with a coronal envelope. *Astrophysical Journal Letters*, *777*(2), L37.

Warnecke, J., Losada, I. R., Brandenburg, A., Kleeorin, N., & Rogachevskii, I. (2016). Bipolar region formation in stratified two-layer turbulence. *Astronomy and Astrophysics*, *589*, A125.

Weiss, N. O. (1966). The expulsion of magnetic flux by eddies. *Proceedings of the Royal Society of London. Series A, Mathematical and Physical Sciences*, *293*(1434), 310–328.

Weiss, N. O., & Proctor, M. R. E. (2014). *Magnetoconvection*. New York: Cambridge University Press.

Xu, H., Stepanov, R., Kuzanyan, K., Sokoloff, D., Zhang, H., & Gao, Y. (2015). Current helicity and magnetic field anisotropy in solar active regions. *Monthly Notices of the Royal Astronomical Society*, *454*(2), 1921–1930.

Xu, H., Zhang, H., Kuzanyan, K., & Sakurai, T. (2016). On the origin of differences in helicity parameters derived from data of two solar magnetographs. *Solar Physics*, *291*(8), 2253–2267.

Yang, S., Büchner, J., Santos, J. C., & Zhang, H. (2013). Evolution of relative magnetic helicity: method of computation and its application to a simulated solar corona above an active region. *Solar Physics*, *283*(2), 369–382.

Yokoi, N. (2018a). Mass and internal-energy transports in strongly compressible magnetohydrodynamic turbulence. *Journal of Plasma Physics*, *84*(6).

Yokoi, N. (2018b). Electromotive force in strongly compressible magnetohydrodynamic turbulence. *Journal of Plasma Physics*, *84*(5).

Zeldovich, Ya. B. (1957). The magnetic field in the two-dimensional motion of a conducting turbulent fluid. *Soviet Journal of Experimental and Theoretical Physics*, *4*, 460–462.

Zeldovich, Ya. B., Molchanov, S. A., Ruzmaikin, A. A., & Sokolov, D. D. (1988). Intermittency, diffusion and generation in a nonstationary random medium. *Soviet Scientific Reviews. Section C, Mathematical Physics*, *7*, 1–110.

Zeldovich, Ya. B., Ruzmaikin, A. A., & Sokoloff, D. D. (1983). *Magnetic fields in astrophysics*. New York: Gordon and Breach.

Zeldovich, Ya. B., Ruzmaikin, A. A., & Sokoloff, D. D. (1990). *The almighty chance*. Singapore: World Scientific.

Zhang, H., Bao, S., & Kuzanyan, K. M. (2002). Twist of magnetic fields in solar active regions. *Astronomy Reports*, *46*(5), 424–434.

Zhang, H., Brandenburg, A., & Sokoloff, D. D. (2014). Magnetic helicity and energy spectra of a solar active region. *Astrophysical Journal Letters*, *784*(2), L45.

Zhang, H., Brandenburg, A., & Sokoloff, D. D. (2016). Evolution of magnetic helicity and energy spectra of solar active regions. *The Astrophysical Journal*, *819*(2), 146.

Zhang, H., Moss, D., Kleeorin, N., Kuzanyan, K., Rogachevskii, I., Sokoloff, D., et al. (2012). Current helicity of active regions as a tracer of large-scale solar magnetic helicity. *The Astrophysical Journal*, *751*(1), 47.

Zhang, H., Sakurai, T., Pevtsov, A., Gao, Y., Xu, H., Sokoloff, D. D., & Kuzanyan, K. (2010). A new dynamo pattern revealed by solar helical magnetic fields. *Monthly Notices of the Royal Astronomical Society*, *402*(1), L30–L33.

Zhang, H., Sokoloff, D., Rogachevskii, I., Moss, D., Lamburt, V., Kuzanyan, K., & Kleeorin, N. (2006). The radial distribution of magnetic helicity in the solar convective zone: observations and dynamo theory. *Monthly Notices of the Royal Astronomical Society*, *365*(1), 276–286.

16

The Spatial Segregation of Kinetic Helicity in Geodynamo Simulations

Avishek Ranjan[1] and Peter Davidson[2]

ABSTRACT

Kinetic helicity and its spatial distribution are thought to be important for the maintenance of a dipolar magnetic field in the geodynamo through the α-effect. Outside the inner core tangent cylinder, spatial segregation of kinetic helicity has been observed in multiple simulations; it is negative in the north and positive in the south. However, the origin of this pattern in these simulations is not well understood. In this chapter, we discuss the various sources of kinetic helicity, including the buoyancy force, magnetic force, and helical waves, such as inertial waves. These waves are known to carry kinetic helicity such that those traveling parallel (antiparallel) to the rotation axis carry negative (positive) helicity. Recent numerical simulations show a strong spatial correlation between the segregation pattern of kinetic helicity and the helicity source due to the buoyancy force. We propose that this spatial correlation is due to helical wave propagation. For turbulent dynamo simulations, this effect is statistical and manifests most clearly in averaged quantities.

16.1. INTRODUCTION

In most three-dimensional turbulent flows, such as those in the Sun and the Earth's core, the vortex lines can be twisted, linked, and entangled. Kinetic helicity measures the degree of *writhe* and *twist* in the flow (Moffatt, 1969; 2016) and is defined as the inner product of velocity, $\boldsymbol{u}$, and vorticity, $\boldsymbol{\omega}$, integrated over a material volume V that has vorticity everywhere parallel to the surface:

$$H_k = \int h\, dV \quad \text{where } (\boldsymbol{\omega} \cdot d\boldsymbol{S} = 0).$$

The quantity $h = \boldsymbol{u} \cdot \boldsymbol{\omega}$ is usually referred to as *kinetic helicity density*, which we will refer to as *helicity* in this chapter. H_k is an invariant of the inviscid flow as the vortex lines remain frozen into the fluid, thus preserving the flow topology (Thomson, 1867; Helmholtz, 1867). Moreover, helicity can be positive (or negative) depending on whether the nature of the linkage between vortex lines is right- or left-handed (Davidson, 2017).

[1] *Department of Mechanical Engineering, Indian Institute of Technology Bombay, Mumbai, India*
[2] *Department of Engineering, University of Cambridge, Cambridge, UK*

16.2. IMPORTANCE OF KINETIC HELICITY IN DYNAMO ACTION

The geomagnetic field is known to be strongly dipolar and is thought to be generated by the fluid motion of electrically conducting liquid iron in the rapidly rotating outer core. Exactly how such an organized field can arise from a chaotic flow is a curiosity that has spawned several decades of research. It is known that the presence of nonzero helicity in a conducting fluid *can* lead to the generation of the magnetic field through a "lift and twist" mechanism, also known as dynamo action, through the α-effect (Parker, 1955; Krause & Radler, 1980; Moffatt & Dormy, 2019). In the presence of large-scale helicity, a seed magnetic field can grow exponentially until the back reaction due to the Lorentz

Helicities in Geophysics, Astrophysics, and Beyond, Geophysical Monograph 283, First Edition.
Edited by Kirill Kuzanyan, Nobumitsu Yokoi, Manolis K. Georgoulis, and Rodion Stepanov.

DOI:10.1002/9781119841715.ch16

force on the flow establishes a statistical equilibrium. Moreover, the α-effect can act twice: once to create a poloidal field, $\boldsymbol{B}_\mathrm{P}$, from an initially azimuthal field, B_ϕ, and again to create B_ϕ from $\boldsymbol{B}_\mathrm{P}$ so that the magnetic field can be maintained against the natural Ohmic dissipation (Fig. 16.1a). Since helicity is needed for the lift (denoted by vertical velocity) and the twist (by vorticity), indicated in Fig. 16.1b, its sign or handedness decides the relative orientation of the magnetic field and current lines. Therefore, a negative (positive) sign of helicity in the north (south) is crucial for this dynamo to operate. Remarkably, the α^2-dynamo is very well supported by numerical simulations. This is illustrated in Fig. 16.1c,d, where the color (lines) denote azimuthally averaged J_ϕ ($\boldsymbol{B}_\mathrm{P}$) and B_ϕ ($\boldsymbol{J}_\mathrm{P}$), respectively. (Here and elsewhere, an angled bracket indicates an azimuthal average.) The flow in the rapidly rotating simulations is strongly columnar (Fig. 16.1e), and the helicity is negative (positive) in the north (south) in most of geodynamo simulations (e.g., Fig. 16.2). However, the reason such a helicity

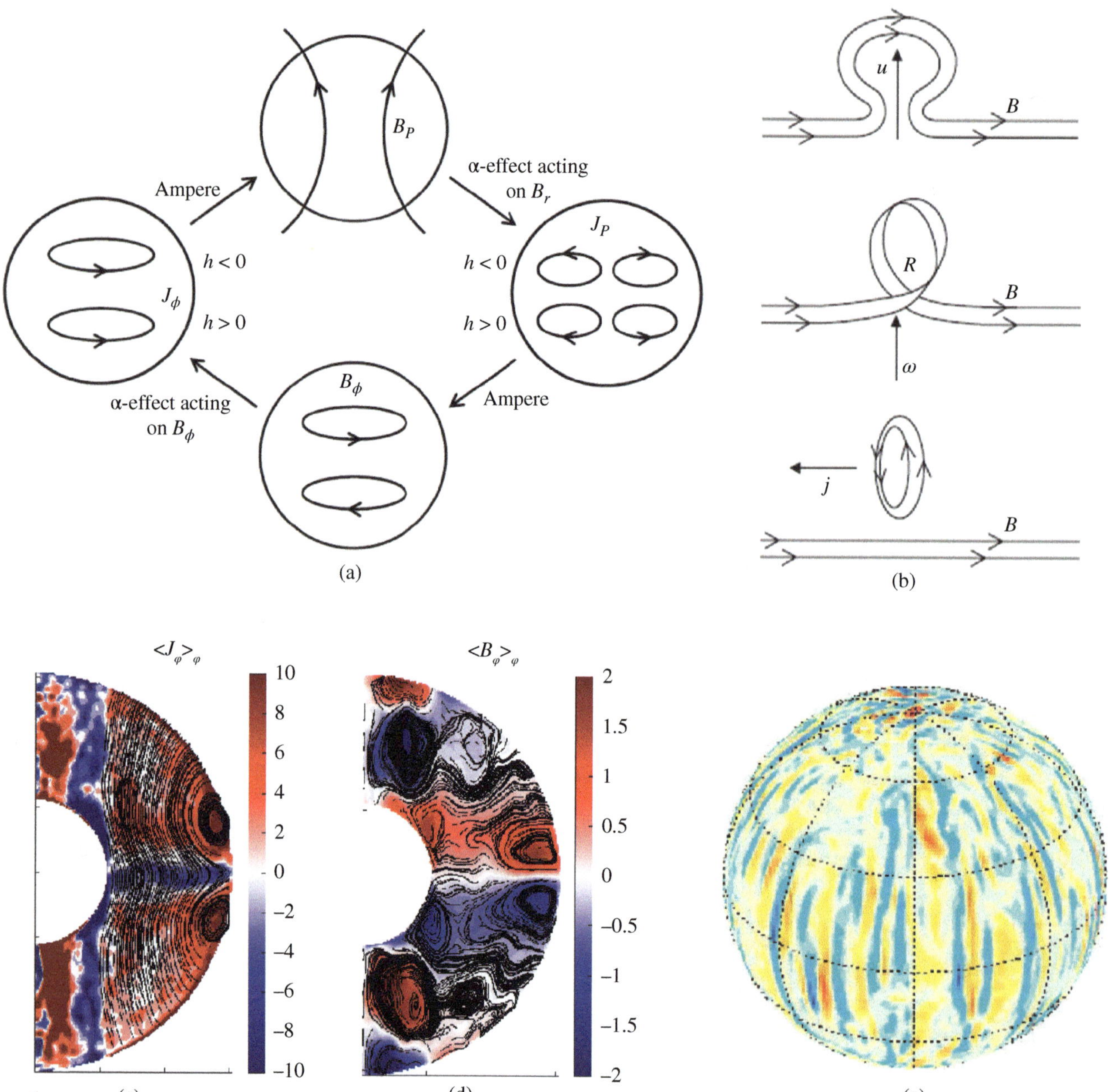

Figure 16.1 (a) The α^2-dynamo. Davidson (2014). (b) Parker's lift-and-twist mechanism. Roberts & King, 2013/IOP Publishing. For a moderately large *Ra* dynamo simulation, S1, (c) $\langle J_\varphi \rangle_\varphi$ with poloidal magnetic field lines; (d) azimuthally averaged, $\langle B_\varphi \rangle_\varphi$, with poloidal current density lines; (e) radial velocity close to the outer radius.

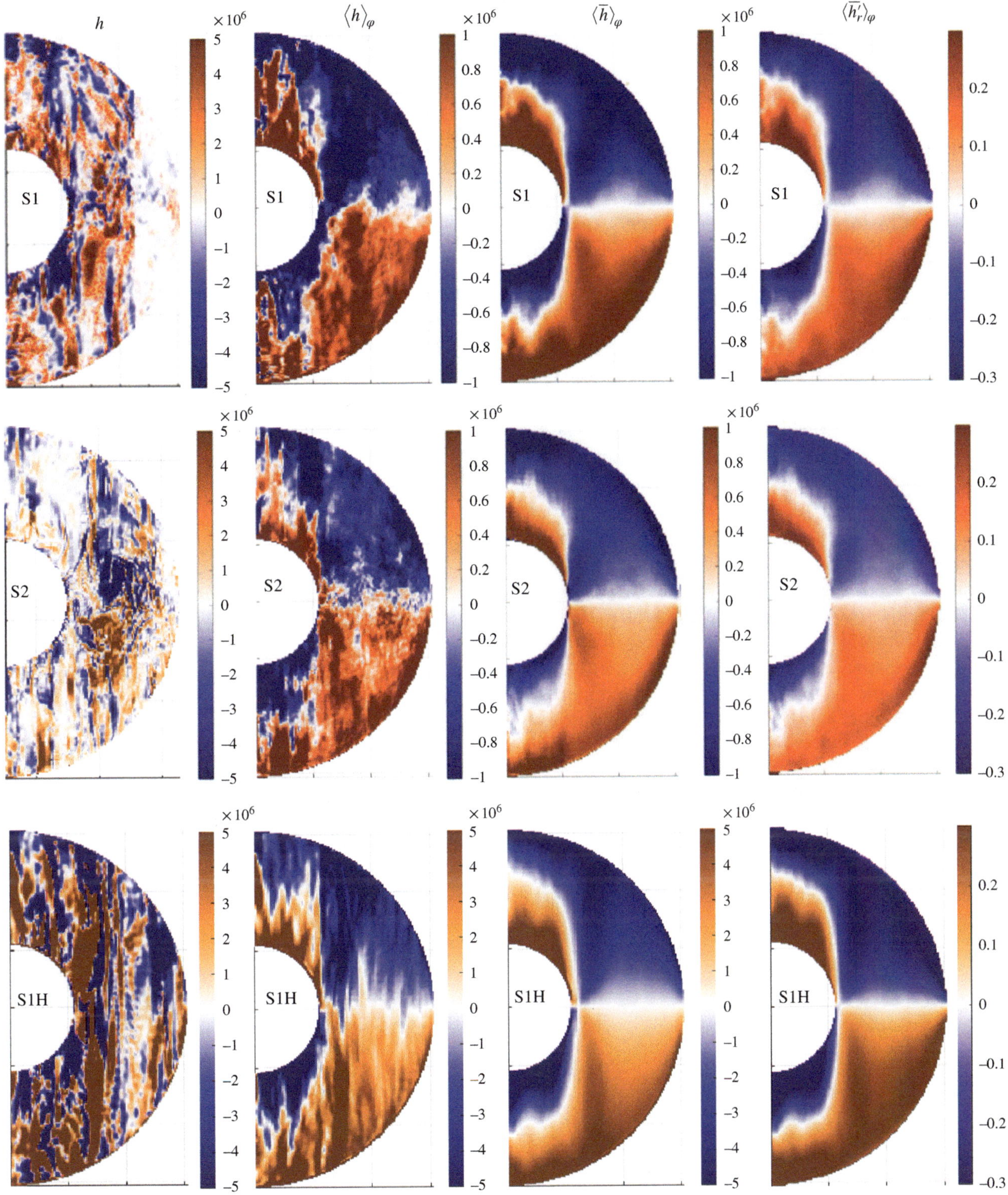

Figure 16.2 Helicity in two dynamo simulations, S1 (top row) and S2 (middle row), from Ranjan et al. (2020), and hydrodynamic simulation S1H (bottom row). Instantaneous h at a fixed longitude $\varphi = \pi$, φ-averaged helicity $\langle h \rangle_\varphi$, φ-averaged and time-averaged helicity $\langle \overline{h} \rangle_\varphi$, and relative helicity $\langle \overline{h'_r} \rangle_\varphi$ (left to right).

distribution (which we call *helicity segregation*) exists remains unknown.

16.2.1. Helicity in Dynamo Simulations

In most dynamo simulations, outside the tangent cylinder (TC), the helicity is negative in the north and positive in the south (Glatzmaier & Roberts, 1995; Kageyama & Sato, 1997; Olson et al., 1999; Ishihara & Kida, 2002; Sreenivasan & Jones, 2011; Sreenivasan et al., 2014; Schaeffer et al., 2017, Ranjan et al., 2020). In Fig. 16.2, for two dynamo simulations taken from Ranjan et al. (2020) and a hydrodynamic simulation of rotating convection, we show the instantaneous h plots at a fixed azimuth ($\varphi = \pi$), azimuthally averaged helicity at the same time, $\langle h \rangle_\varphi$, and azimuthally averaged and time-averaged helicity, $\langle \overline{h} \rangle_\varphi$ (more details about these simulations are given later). We also show, in the final column, the fluctuating relative helicity, defined as $\langle \overline{h'_r} \rangle_\varphi = \langle \overline{\boldsymbol{u}' \cdot \boldsymbol{\omega}' / |\boldsymbol{u}'||\boldsymbol{\omega}'|} \rangle_\varphi$, where $\boldsymbol{u}'(\boldsymbol{x}, t) = \boldsymbol{u}(\boldsymbol{x}, t) - \overline{\boldsymbol{u}}(\boldsymbol{x})$ and $\boldsymbol{\omega}' = \nabla \times \boldsymbol{u}'$. $\langle \overline{h'_r} \rangle_\varphi$ denotes the additional helicity that arises due to the unsteady flow. Clearly, the helicity segregation becomes evident only on spatial and/or temporal averaging; instantaneous plots reveal a complex and dynamic flow. This is also clear in Fig. 16.3, which shows $\langle \overline{h'_r} \rangle_\varphi$ in the dynamo simulations carried out at the most extreme parameter regimes computationally feasible (Schaeffer et al. (2017)). These plots indicate that the observed segregation outside the TC is a dynamic feature of the flow. The Ekman number ($E = \nu/\Omega d^2$, the ratio of the viscous forces to Coriolis force) is the smallest, and the Rayleigh number ($Ra = \alpha g d^3 \Delta T / \nu\kappa$, a measure of the strength of convection) is the largest for run S2, as compared to S1 and S0. (Here, ΔT is the temperature difference, ν is the kinematic viscosity, κ the thermal diffusivity, α is the thermal expansion coefficient, d is the width of the spherical shell, and Ω is the rotation rate.)

Apart from the clear segregation in the sign, the following points can be inferred from Fig. 16.3:

(a) Outside the TC, the helicity peaks are further from the boundaries as simulations become more strongly driven.
(b) Helicity segregation is clearly visible even in the hydrodynamic simulation, S1*, even though the helicity distribution (e.g., location of peaks) is different from S1.
(c) As simulations become more turbulent (S2), nearly all the regions of strong helicity are outside the TC.

Inference (a) indicates that as the flow is more strongly driven, helicity generation is more likely to be due to internal sources in the flow. From (b), it is evident that the origin of helicity segregation is likely to be nonmagnetic, even though magnetic forces may alter the location of the peak helicity. Finally, from (c), it is clear that the mechanism for helicity generation outside the TC differs significantly from that inside (where a boundary-driven mechanism based on stability theory, suggested by Veronis (1959) and observed in rotating Rayleigh-Benard investigations (Stellmach & Hansen, 2004; Giesecke et al., 2005; Grooms et al., 2010) can be invoked to explain the observed pattern).

16.2.2. Lack of a Satisfactory Explanation for Helicity Segregation outside the TC

Before the famous dynamo simulations that showed dipolar fields and reversals (Glatzmaier & Roberts, 1995)

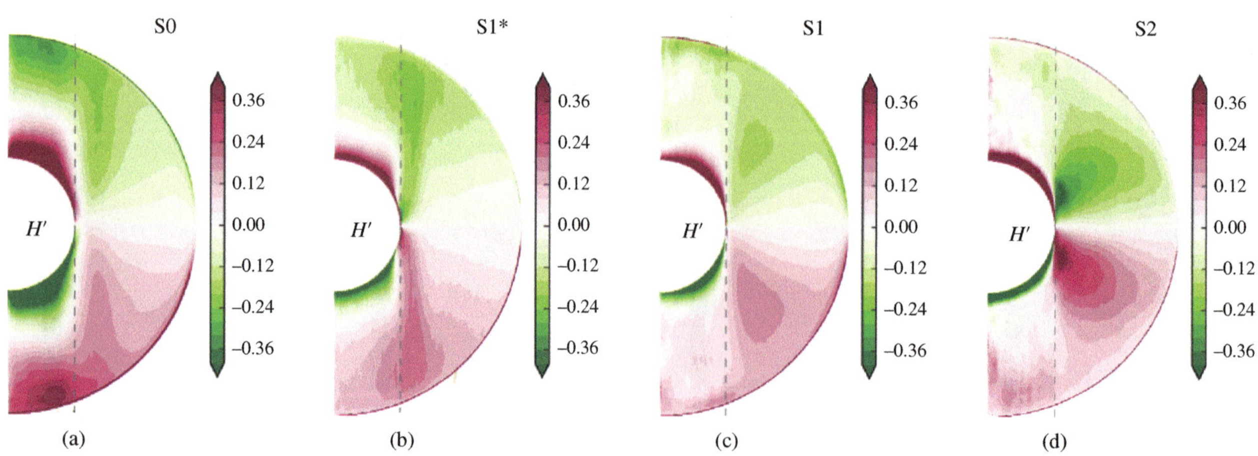

Figure 16.3 Fluctuating relative helicity in four simulations. Simulation parameters for S0: $Ra = 6.34 \times 10^9$, $E = 10^{-5}$; S1: $Ra = 1.27 \times 10^{11}$, $E = 10^{-6}$; S2: $Ra = 2.54 \times 10^{12}$, $E = 10^{-7}$ (Table 2 in the paper has other details). S1* is a hydrodynamic simulation with all other parameters kept the same as in S1. The boundary of the TC is shown by grey lines. Schaeffer et al., 2017/Oxford University Press.

resembling the Earth's, Busse (1975, 1976) suggested the following three sources of helicity generation in his analytical studies: (1) Ekman pumping, (2) spatial variations of the Lorentz force, and (3) the spherical boundary. Of these, Ekman pumping can also lead to the same expected pattern of helicity segregation: negative in the north, positive in the south (Kono et al., 2000; Roberts & King, 2013). This was likely to be the case in most early simulations that were weakly driven and highly viscous. However, as we approach the core regimes ($E \sim 10^{-15}$), it is clear from Figs. 16.2 and 16.3 that the helicity must come from the interior. Moreover, the same helicity segregation pattern exists in dynamo simulations with free-slip boundary conditions (Kageyama et al., 1993; Yadav et al., 2013), thus ruling out Ekman pumping as a major helicity source in the core.

Due to the spherical boundary, the changing height of convective columns moving radially outward or inward in a spherical shell (also called the β-effect) can be a source of additional helicity (Busse, 2002; Reshetnyak, 2006). This may be important where the vertical height of the outer shell boundary changes rapidly with a change in cylindrical radius, but such a mechanism, based on conservation of angular momentum, can explain only a local intensification of helicity and not the origin of the segregation pattern. Similarly, although the Lorentz force may influence local helicity (Sreenivasan & Jones, 2011), it is unlikely to create helicity segregation since the same helicity segregation pattern has also been observed in hydrodynamic rotating convection simulations (Kitauchi et al., 1997; Glatzmaier & Olson, 1993; Olson et al., 1999; Schaeffer et al., 2017; Ranjan et al., 2020).

Hide (1975) speculated about helicity generation due to the interplay of buoyancy and Coriolis forces. In their seminal work, Olson et al. (1999) point out that spatial variation in the buoyancy force along the convection columns can also lead to negative (positive) helicity in the north (south). They suggested that this occurs by a transfer of vorticity from equatorial regions, where temperature perturbations are larger, to higher latitudes through a secondary flow along the column axes, similar to the Ekman suction mechanism but internally driven. In some (relatively viscous) simulations with larger values of E, it is not obvious whether the helicity arises due to boundary-driven Ekman pumping or internal buoyancy force. For instance, Olson et al. (1999), in their simulations performed at $E = 3.0 \times 10^{-4}$ and $Ra = 3.1 \times 10^5$ (which they called "case 1"), suggested that Ekman pumping was unlikely to be important for helicity generation as the peak h was far from the boundaries. However, Schaeffer et al. (2017), in their simulation S0 performed at a lower $E = 10^{-5}$ and larger $Ra = 6.34 \times 10^9$, noted that most helicity was produced at the boundaries, in contrast to their simulation S2 (with $E = 10^{-7}$, $Ra = 2.54 \times 10^{12}$). Note that the product $Ra.E^2$ for Olson's case 1 is 0.028, whereas $Ra.E^2$ for Schaeffer's S0 is 0.634 and for S2 is 0.025. Assuming $\nu = \kappa$, we have $Ra.E^2 = \alpha g \Delta T / \Omega^2 d$, which indicates that it may be the relative strengths of the buoyancy and the Coriolis forces that influence the helicity distribution: that is, whether the helicity is close to the boundary. Further investigations are required to verify this hypothesis.

Having discussed various sources of helicity, an obvious question arises: "What is the real mechanism for the helicity segregation?" According to Davidson (2014), this segregation could be achieved through the propagation of helical waves, such as inertial waves, preferentially excited in and around the equatorial regions where temperature perturbations are statistically more dominant. These waves arise due to the restoring action of the Coriolis force and can exist in a rapidly rotating system such as the Earth's core. It has long been known that inertial wave packets traveling parallel (antiparallel) to the rotation axis carry negative (positive) helicity (Moffatt, 1970). As we will see later, the helical wave mechanism proposed by Davidson (2014) and extended in Davidson and Ranjan (2018a) is a promising candidate to explain the helicity segregation observed in dynamo simulations.

16.3. SOURCES OF HELICITY

16.3.1. Evolution Equation

For an electrically conducting fluid at a very small Rossby number ($Ro = u/2\Omega l$), the evolution equation for momentum in a rotating reference frame under the Boussinesq approximation is

$$\frac{\partial \boldsymbol{u}}{\partial t} = -\frac{\nabla p^*}{\rho} + 2(\boldsymbol{\Omega} \cdot \nabla)\boldsymbol{a} - \alpha T \mathbf{g} + \frac{\boldsymbol{J} \times \boldsymbol{B}}{\rho} + \nu \nabla^2 \boldsymbol{u}, \tag{16.1}$$

with the mass conservation

$$\nabla \cdot \boldsymbol{u} = 0. \tag{16.2}$$

Here, $\boldsymbol{a}$ is the solenoidal vector potential for $\boldsymbol{u}$ (defined by $\boldsymbol{u} = \nabla \times \boldsymbol{a}$, $\nabla \cdot \boldsymbol{a} = 0$), so the Coriolis force can be written $2(\boldsymbol{\Omega} \cdot \nabla)\boldsymbol{a}$ (Davidson, 2016; Davidson & Ranjan, 2018a); T is the temperature perturbation; p^* is the modified pressure that includes the centrifugal force and a contribution from the Coriolis force, $p^* = p - (\rho/2)(\boldsymbol{\Omega} \times \boldsymbol{x})^2 + 2\rho(\boldsymbol{a} \cdot \boldsymbol{\Omega})$ (note that p^* can be considered a residual pressure that remains after the geostrophic part is removed); $\boldsymbol{J} = (\nabla \times \boldsymbol{B})/\mu$ is the current density; and μ is the permeability of free space. The evolution of the magnetic field $\boldsymbol{B}$ is governed by the induction equation

$$\frac{\partial \boldsymbol{B}}{\partial t} + (\boldsymbol{u} \cdot \nabla)\boldsymbol{B} = (\boldsymbol{B} \cdot \nabla)\boldsymbol{u} + \lambda \nabla^2 \boldsymbol{B}, \qquad \nabla \cdot \boldsymbol{B} = 0, \tag{16.3}$$

where $\lambda = 1/\mu\sigma$ is the magnetic diffusivity and σ is the electrical conductivity. Then corresponding to equation (16.1), the equation for vorticity, $\boldsymbol{\omega} = \nabla \times \boldsymbol{u}$, is

$$\frac{\partial \boldsymbol{\omega}}{\partial t} = 2(\boldsymbol{\Omega} \cdot \nabla)\boldsymbol{u} - \alpha \nabla T \times \mathbf{g} + \frac{\nabla \times (\boldsymbol{J} \times \boldsymbol{B})}{\rho} + \nu \nabla^2 \boldsymbol{\omega}. \quad (16.4)$$

The evolution equation for helicity can thus be derived (Davidson & Ranjan, 2018a):

$$\frac{\partial h}{\partial t} + \nabla \cdot \boldsymbol{F} = H_T + H_B + H_\nu, \quad (16.5)$$

where

$$\boldsymbol{F} = -2(\boldsymbol{u}^2)\boldsymbol{\Omega} - 2\boldsymbol{u} \times (\boldsymbol{\Omega} \cdot \nabla \boldsymbol{a}) + \frac{p^* \boldsymbol{\omega}}{\rho} \quad (16.6)$$

$$H_T = -\alpha[\boldsymbol{u} \cdot (\nabla T \times \mathbf{g}) + \boldsymbol{\omega} \cdot (T\mathbf{g})], \quad (16.7a)$$

$$H_B = [\boldsymbol{\omega} \cdot (\boldsymbol{J} \times \boldsymbol{B}) + \boldsymbol{u} \cdot (\nabla \times (\boldsymbol{J} \times \boldsymbol{B})] / \rho, \quad (16.7b)$$

$$H_\nu = \nu (\boldsymbol{\omega} \cdot \nabla^2 \boldsymbol{u} + \boldsymbol{u} \cdot \nabla^2 \boldsymbol{\omega}). \quad (16.7c)$$

Clearly, $\boldsymbol{F}$ represents a flux of helicity, and H_T, H_ν, and H_B can be considered *sources or production terms* of helicity due to the buoyancy, viscous, and Lorentz forces, respectively. Note that interpreting H_T, H_B and H_ν as sources comes with a caveat: as they depend on $\boldsymbol{u}$, they are not independent of the flow and, therefore, its helicity.

16.3.2. Body Forces and Flux of Helicity

Thermal Source

The first term in the thermal source of helicity, H_T, comes in part from thermal source of z-vorticity $(\nabla T \times \mathbf{g})_z$ multiplied by the vertical velocity, whereas the second term depends on temperature perturbations, T. In most dynamo simulations, outside the TC, T is localized near the equatorial regions outside which the latter term may be considered relatively less important. Nevertheless, it is useful to rearrange H_T as $H_T = H_{T1} + H_{T2}$, where

$$H_{T1} = \nabla \cdot (\boldsymbol{u} \times (-\alpha T\mathbf{g})), \quad (16.8a)$$

$$H_{T2} = -2\alpha\, \boldsymbol{u} \cdot (\nabla T \times \mathbf{g}). \quad (16.8b)$$

The component H_{T1} will integrate to zero over the spherical shell, but H_{T2} will not (see Davidson and Ranjan (2018a)). The thermal source for helicity H_T has been found to be spatially correlated with helicity, h (compare Fig. 16.2 with the first column of Fig. 16.4), showing the same signs north and south of the equator (Ranjan et al., 2020). This is an intriguing result to which we will return in section 16.4.

Magnetic Source

In the magnetic source, H_B, one of the terms is the dot product of vorticity with the Lorentz force, $H_{B1} = \boldsymbol{\omega} \cdot (\boldsymbol{J} \times \boldsymbol{B})/\rho$. For successful dynamo action, $\boldsymbol{u} \cdot (\boldsymbol{J} \times \boldsymbol{B})$ must be negative on average, so the magnetic energy grows and the kinetic energy falls. If the helicity is negative (positive), we expect H_{B1} to acquire a positive (negative) sign. Indeed, we observe this in the $\langle \overline{H}_B \rangle_\varphi$ computed in our dynamo simulations (Fig. 16.4).

Flux of Helicity

The flux of helicity, given by equation (16.6), comprises three terms: $-2(\boldsymbol{u}^2)\boldsymbol{\Omega}$, $-2\boldsymbol{u} \times (\boldsymbol{\Omega} \cdot \nabla \boldsymbol{a})$, and $p^* \boldsymbol{\omega}/\rho$. In the absence of the Lorentz force, the modified pressure is governed by

$$\nabla^2 p^* = -\rho \nabla \cdot (\alpha T \mathbf{g}), \quad (16.9)$$

which may be inverted to find p^*. In regions away from buoyancy sources, equation (16.9) suggests that the contribution from the last term in the helicity flux equation (16.6) is likely relatively small (Davidson & Ranjan, 2015). In a strongly columnar, quasisteady flow, $(\boldsymbol{\Omega} \cdot \nabla)\boldsymbol{u} = 0$ from the Taylor Proudman theorem (Davidson, 2013), so we can also expect $(\boldsymbol{\Omega} \cdot \nabla)\boldsymbol{a} \approx 0$. This suggests we can write $\boldsymbol{F} \approx -(2\boldsymbol{u}^2)\boldsymbol{\Omega} = \boldsymbol{F}_\Omega$ at locations far from the buoyancy source and the boundaries. We will use $\boldsymbol{F}_\Omega$ to denote this estimate of the helicity flux. In Davidson and Ranjan (2018a), $\boldsymbol{F}_\Omega$ is used to denote the helicity flux by inertial wave packets propagating away from a localized layer of buoyant anomalies.

16.3.3. Helical Waves

Helical waves such as inertial and magnetic-Coriolis waves have been known to propagate helicity and lead to dynamo action through the α-effect (Moffatt, 1970; Olson, 1981). It is also known that low-frequency inertial wave packets triggered from a rapidly rotating buoyant blob (Fig. 16.5) travel nearly parallel to the rotation axis and form columnar structures that consist of an alternate cyclone-anticyclone pair (Fig. 16.5) with negative (positive) helicity above (below) the blob (Davidson, 2014; Davidson & Ranjan, 2018a). As is evident from the bottom row of Fig. 16.5, this columnar flow structure and the helicity pattern is present even with a transverse mean magnetic field, $\boldsymbol{B}_0$, where the structures are stretched along the field (Ranjan, 2015; Davidson & Ranjan, 2015). Note that the extent of elongation of the structures along the rotation axis versus that along the field depends on the relative strengths of the Coriolis force and the Lorentz force. The helical columnar structure was evident in the analytical predictions by Loper and colleagues in a series of papers (e.g., see Shimizu & Loper (2000), Loper et al. (2003), and Chulliat et al. (2003)). However, the helicity in these columns was attributed to inertial wave packets much later by Davidson (2014).

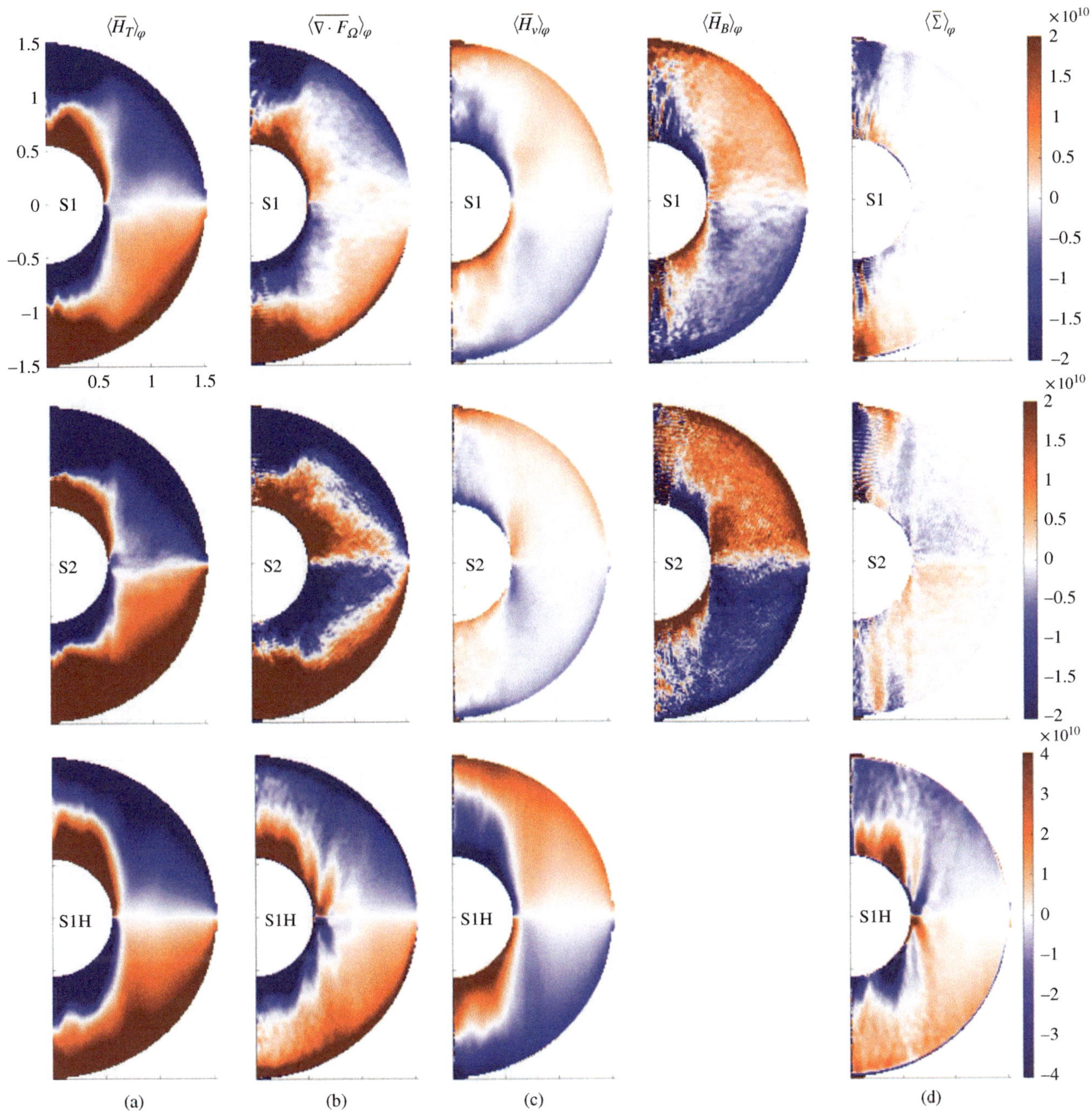

Figure 16.4 Helicity sources (left to right): $\langle \overline{H}_T \rangle_\varphi$, $\langle \overline{\nabla \cdot \boldsymbol{F}_\Omega} \rangle_\varphi$, $\langle \overline{H}_v \rangle_\varphi$, $\langle \overline{H}_B \rangle_\varphi$, and $\langle \overline{\Sigma} \rangle_\varphi$for S1 (first row), S2 (second row), and S1H (last row), from Ranjan et al. (2020). S1H is nonmagnetic.

To derive the dispersion relation for helical waves, let us consider the linearized diffusionless equations of fluctuating magnetic field and vorticity (Davidson, 2013):

$$\frac{\partial \boldsymbol{b}}{\partial t} = (\boldsymbol{B}_0 \cdot \nabla)\boldsymbol{u}; \qquad (16.10a)$$

$$\frac{\partial \boldsymbol{\omega}}{\partial t} = (\boldsymbol{B}_0 \cdot \nabla)\boldsymbol{j} + 2(\boldsymbol{\Omega} \cdot \nabla)\boldsymbol{u}, \qquad (16.10b)$$

where $\boldsymbol{j} = \nabla \times \boldsymbol{b}$ and $\boldsymbol{B}$ is in Alfven units. This yields the wave equation

$$\left[\frac{\partial^2}{\partial t^2} - (\boldsymbol{B}_0 \cdot \nabla)^2\right]^2 \nabla^2 \boldsymbol{u} + 4(\boldsymbol{\Omega} \cdot \nabla)^2 \frac{\partial^2 \boldsymbol{u}}{\partial t^2} = \boldsymbol{0}. \qquad (16.11)$$

This equation supports plane waves of the form $\boldsymbol{u} = \hat{\boldsymbol{u}} \exp(\boldsymbol{k} \cdot \boldsymbol{x} - \varpi t)$ with the dispersion relation given by

$$\varpi^2 \mp \varpi_I \varpi - \varpi_B^2 = \boldsymbol{0}, \qquad (16.12)$$

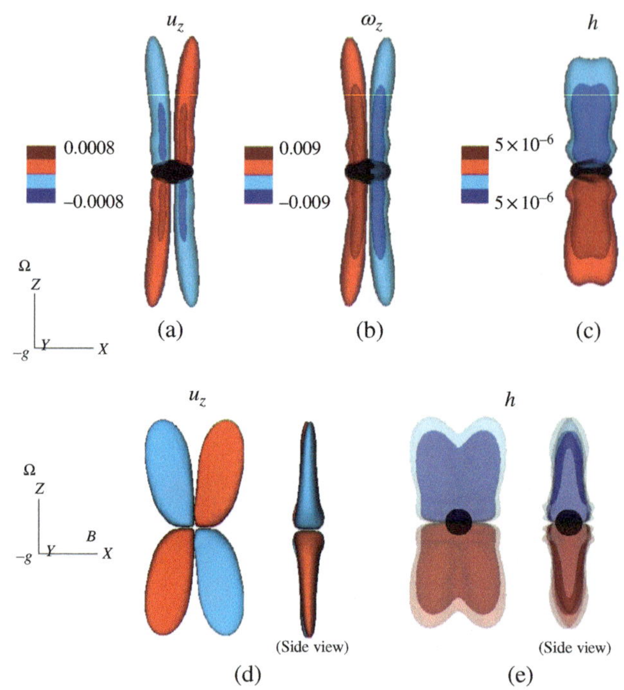

Figure 16.5 Gaussian blob under rotation: (a) isosurfaces of u_z, (b) ω_z, (c) h. Ranjan & Davidson, 2018a/Cambridge University Press. The Gaussian blob is stretched under the presence of a magnetic field: (d) u_z, (e) h; red (blue) colors denote positive (negative) values. Ranjan 2015/University of Cambridge.

where $\varpi_B = \boldsymbol{k}\cdot\boldsymbol{B}_0$ and $\varpi_I = 2(\boldsymbol{k}\cdot\boldsymbol{\Omega})/k$ are the Alfvén frequency and the inertial wave frequency, respectively. The waves are maximally helical as $\hat{\boldsymbol{\omega}} = \mp k\,\hat{\boldsymbol{u}}$: that is, the vorticity and velocity amplitudes are aligned for a monochromatic wave. Therefore, we can write the wave helicity as $h = \mp k\,\boldsymbol{u}^2$, indicating that wave packets that travel parallel (antiparallel) to the rotation axis have negative (positive) helicity (Moffatt, 1970). The group velocity of these waves, defined by $\boldsymbol{C}_g = \partial\varpi/\partial\boldsymbol{k}$, is given by the equation

$$\left[1 + \left(\frac{\varpi_B}{\varpi}\right)^2\right]\boldsymbol{C}_g = \frac{2\varpi_B}{\varpi}\boldsymbol{B}_0 + \boldsymbol{C}_{gI}, \qquad (16.13)$$

where $\boldsymbol{C}_{gI} = \pm 2\boldsymbol{k}\times(\boldsymbol{\Omega}\times\boldsymbol{k})/k^3$ is the group velocity of inertial waves, so $\boldsymbol{\Omega}\cdot\boldsymbol{C}_{gI} > 0$ for inertial wave packets traveling parallel, and $\boldsymbol{\Omega}\cdot\boldsymbol{C}_{gI} < 0$ for wave packets traveling antiparallel to the rotation vector. For inertial waves with low frequency, $\boldsymbol{\Omega}\cdot\boldsymbol{k}\approx 0$ and $\boldsymbol{C}_{gI}\approx \pm 2\boldsymbol{\Omega}/k$, so energy is dispersed along the rotation axis (Davidson et al., 2006). Among the different classes of helical waves described by equation (16.12), there are weakly modified inertial waves ($\varpi \approx \varpi_I \gg \varpi_B$), inertial-Alfvén waves ($\varpi \approx \pm\varpi_B$; $\varpi_I \approx 0$), and intermediate magnetic-Coriolis (MC) waves ($|\varpi_B| \approx |\varpi_I|$) (Bardsley & Davidson, 2016, 2017, McDermott & Davidson, 2020). The helicity segregation characteristic of inertial waves applies to all these helical waves and to wave packets comprising linearly superimposed monochromatic waves (Ranjan, 2017). Bardsley and Davidson (2017) conjecture that columnar flow in Earth's core is likely to be maintained by a combination of weakly modified inertial waves and intermediate MC waves, which are, once again, likely to retain the helicity segregation characteristics of inertial waves. Another class of helical waves are MAC waves arising from the interplay of magnetic, Archimedes (buoyancy), and Coriolis forces in the stably stratified layer near the core-mantle boundary (Buffett et al., 2016). If rotation is much stronger than stratification ($N/2\Omega \ll 1$), the inertia-gravity waves retain the helicity segregation characteristics of inertial waves (Ranjan, 2017).

The α^2-dynamo model based on helical waves proposed by Davidson (2014) predicted a relationship between the induced e.m.f. $\langle \boldsymbol{u}\times\boldsymbol{b}\rangle$ and h at low magnetic Reynolds number

$$\langle \boldsymbol{u}\times\boldsymbol{b}\rangle = -\frac{\delta'^2}{2\lambda}\langle h\rangle\langle \boldsymbol{B}_0\rangle, \qquad (16.14)$$

where δ' is a constant having dimensions of length that depends on the sign of helicity. Note that positive helicity implies $\langle \boldsymbol{u}\times\boldsymbol{b}\rangle.\ \langle \boldsymbol{B}_0\rangle < 0$, while negative helicity leads to $\langle \boldsymbol{u}\times\boldsymbol{b}\rangle.\ \langle \boldsymbol{B}_0\rangle > 0$, in line with the α-effect. This model was found to work well for a rapidly rotating localized layer of buoyancy (Davidson & Ranjan, 2015) where the helicity is close to maximal and also under a mean magnetic field (McDermott & Davidson, 2020). Furthermore, the helical wave model leads to scaling laws consistent with dynamo simulations (Davidson, 2016; Davidson & Ranjan, 2018b).

16.4. RECENT RESULTS FROM NUMERICAL SIMULATIONS

16.4.1. Model Problems: Rapidly Rotating, Localized Buoyant Anomalies

In numerical dynamo simulations, the dynamics is complicated partly by the variety of forces and the spherical boundary (e.g., Fig. 16.2a,d). In such a case, it is useful to consider simpler model problems such as a localized buoyant blob (Fig. 16.5) under rapid rotation. For a Gaussian blob, the formation of an alternate cyclone-anticyclone column pair was explained by Davidson (2014) as follows. We will consider $\mathbf{g} = -g\hat{\mathbf{e}}_y$ and $\boldsymbol{\Omega} = \Omega\hat{\mathbf{e}}_z$. The Rossby number for the blob is small, given by $Ro_\delta = cg/2\Omega^2\delta = 0.01$, and obtained by taking $u \sim cg/\Omega$, where $c = \rho'/\rho$ is the normalized density perturbation and δ is now the blob size. In the absence of a magnetic field, and assuming inviscid, low-frequency waves, the vorticity from equation (16.4) near the blob is

$$2(\mathbf{\Omega} \cdot \nabla)\boldsymbol{u} \approx -\nabla c \times \mathbf{g}, \tag{16.15}$$

where c and T are related by $c = -\alpha T$. Taking the curl of equation (16.15) leads to

$$2(\mathbf{\Omega} \cdot \nabla)\boldsymbol{\omega} \approx \mathbf{g}\nabla^2 c - (\mathbf{g} \cdot \nabla)\nabla c. \tag{16.16}$$

Using these equations, we can obtain vertical jump conditions across the buoyant blob after the passage of inertial waves. Since the waves within the tangent cylinder region that circumscribes the blob are of low frequency, the time dependence is significant only near the advancing fronts of the wave packets. Integrating vertically through the blob yields $\Delta u_x \approx \Delta u_y \approx 0$, $\Delta\omega_z \approx 0$, and

$$\Delta u_z \approx \frac{g}{2\Omega}\int (\partial c/\partial x)\, dz,$$

where Δ denotes a vertical jump across the blob. The condition $\Delta\omega_z \approx 0$ predicts that the cyclone with $\omega_z > 0$ (anticyclone, $\omega_z < 0$) above the blob corresponds to a cyclone (anticyclone) below the blob. Moreover, for the Gaussian blob considered here, we have $\Delta u_z > 0$ for $x > 0$ and $\Delta u_z < 0$ for $x < 0$. For the vortex pair above and below the blob, u_z diverges away from the blob at $x > 0$ (in the anticyclone) and converges toward the blob at $x < 0$ (in the cyclone), as observed in Fig. 16.5. Since the upward (downward) propagating inertial waves carry negative (positive) helicity, the dispersion pattern within the tangent cylinder that circumscribes the blob consists of a cyclone-anticyclone pair of columnar vortices above the blob matched to a cyclone-anticyclone pair below. If there are several buoyant blobs of varying sizes that may or may not overlap, is the dispersion pattern similar? As we will now see, this is exactly the case.

Inspired by the equatorial regions of the core, where rotation is perpendicular to gravity and buoyant anomalies are preferentially distributed (Fig. 16.6), Davidson and Ranjan (2018a) considered a rapidly rotating, localized layer of buoyant anomalies as a model problem (Figs. 16.6 and 16.7). To create this layer, 2,000 Gaussian blobs of size $\delta/2 \le \delta_i \le 2\delta$ are randomly distributed in the vertical center of a triply periodic box of size $50\delta \times 50\delta \times 150\delta$ with the density distribution given by $c_i = c_0 \exp\left(-\left(\boldsymbol{x}^2 - \boldsymbol{x}_i^2\right)/\delta_i^2\right)$. The blob centers, $\boldsymbol{x}_i$, are located in the region defined by $-2.8\delta < z_i < 2.8\delta$, filling the box in the horizontal direction. A pseudospectral direct numerical simulation (DNS) code described in Yeung and Zhou (1998) and Ranjan and Davidson (2014) was used to solve the (nonmagnetic) momentum equation in the rotating reference frame along with an advection-diffusion equation for c with $\kappa=\nu$ (note that

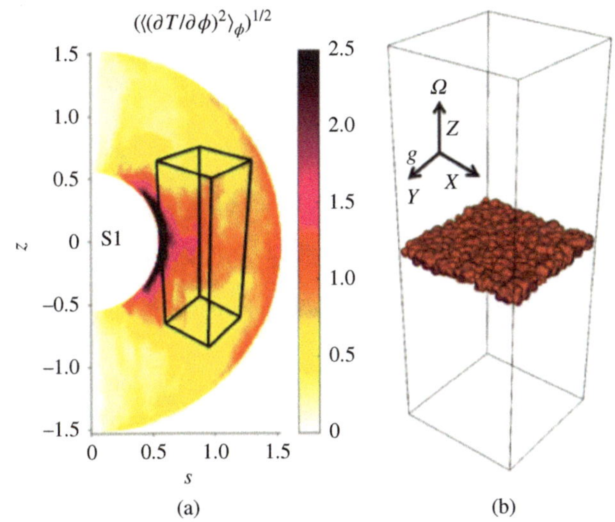

Figure 16.6 (a) Azimuthal root-mean-square of the thermal source of vorticity. Adapted from Ranjan et al., 2018. (b) A layer of buoyant anomalies in a long box inspired by the equatorial regions in (a).

the same DNS code was used for the single-blob computations in Fig. 16.5). Keeping the Ekman number $E_\delta = \nu/\Omega^2\delta$ fixed, the simulations were carried out for two initial Rossby numbers: $Ro_\delta = c_0 g/2\Omega^2\delta = 0.01$ and $Ro_\delta = 1$, which drops to 0.3 by $\Omega t = 12$. The simulations were stopped at $\Omega t = 12$ before the waves could reach the box boundary.

Figure 16.7 shows snapshots of c, h, and u_z in the xz plane at $y = 0$ for the $Ro_\delta = 0.01$ case. Much like the single blob (Fig. 16.5), we see inertial wave packets, in the form of alternate cyclone–anticyclone pairs, emerging from the buoyant layer, carrying negative (positive) helicity upward (downward). Such pairs have been observed in dynamo simulations as well, particularly near the equator, where $\partial u_z/\partial z > 0$ in the cyclone and <0 in the anticyclone (Olson et al., 1999; Ranjan et al., 2018). Similar observations were found to hold for the $Ro_\delta = 1$ case (not shown here), except that the columns tilt with respect to the rotation axis. This happens because horizontal advection in the buoyant anomalies is now significant, leading to a tilt in the wave packets. Moreover, the buoyant region is elongated vertically by ≈ 1.8 times, so depending on the Peclet number, the larger advection also leads to heat transport along the columns (Ranjan & Davidson, 2021).

In Fig. 16.8, we show the y-averaged thermal source of helicity, $\langle H_T \rangle_y$, and its constituents $\langle H_{T1} \rangle_y$ and $\langle H_{T2} \rangle_y$ for the $Ro_\delta = 0.01$, 1 cases. Near the top and bottom edges of the layer, H_T acquires the same sign as the helicity, h, primarily due to the pattern observed in H_{T2}. In our model problems, since $c = -\alpha T$, H_{T2} depends on the sign of the product $u_z g \partial c/\partial x$. For a single Gaussian blob, $\partial c/\partial x < 0$ for $x > 0$ (right half) and $\partial c/\partial x > 0$ for $x < 0$

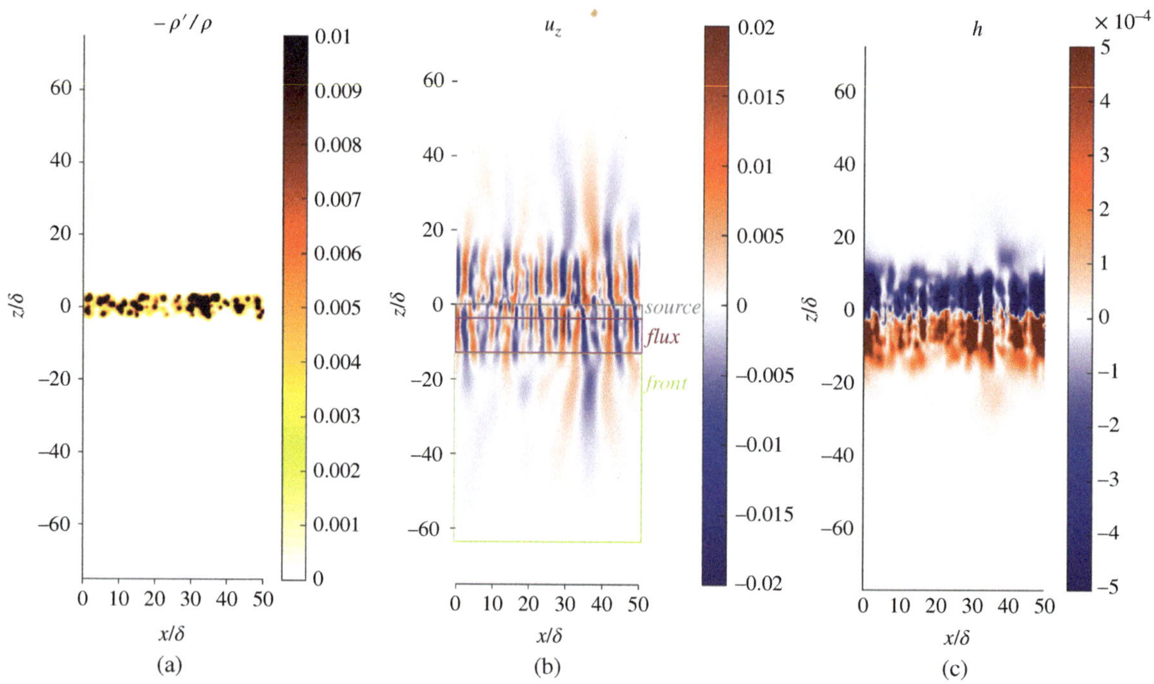

Figure 16.7 Snapshot in the mid xz plane of (a) $-c$, (b) u_z, (c) h at $\Omega t = 12$. The source, flux and front regions for $z < 0$ are marked in (b). Ranjan & Davidson, 2018a/Cambridge University Press.

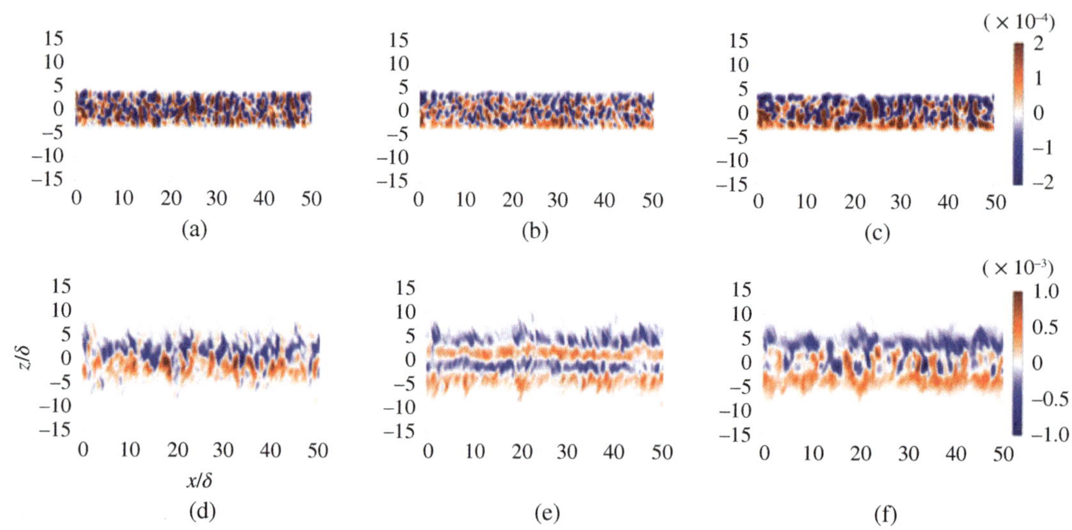

Figure 16.8 Thermal helicity source $\langle H_{T1}\rangle_y$, $\langle H_{T2}\rangle_y$, and $\langle H_T\rangle_y$ for $Ro_\delta = 0.01$ (a–c), $Ro_\delta = 1$ (d–f), at $\Omega t = 12$. Ranjan & Davidson, 2018a/Cambridge University Press.

(left half), leading to the negative (positive) sign of H_{T2} above (below) the blob. The same applies to the edges of the buoyant layer, but not in the interior where the blobs overlap. The pattern seen in H_{T1} is also interesting, although harder to interpret. It differs significantly for the two Ro_δ cases and matches h for the higher Ro_δ case. Note also that just as with the volume integral of total helicity, the volume integral of H_T and its constituents is zero. The spatial correlation between helicity and its thermal source was also observed for the dynamo simulations (Fig. 16.4). In section 16.5, we will return to this observation with an explanation based on dispersion of inertial waves.

McDermott and Davidson (2020) investigated a rapidly rotating layer of buoyancy with a transverse mean magnetic field in a similar triply periodic box (Fig. 16.6).

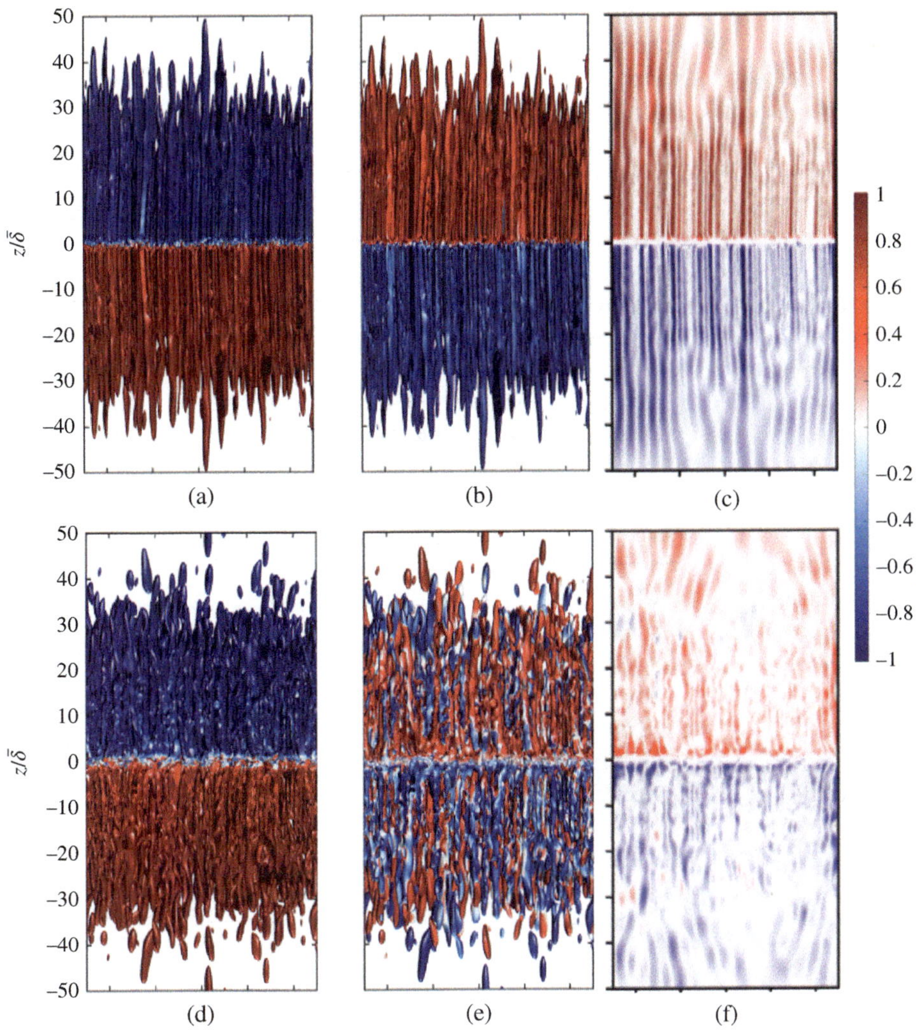

Figure 16.9 Isosurfaces of u_z colored by (a) h, (b) e.m.f, (c) y-averaged e.m.f. at $\Omega t = 30$ (McDermott & Davidson, 2020) for two simulations $Le = 0.01$ (a–c), 0.1 (d–f).

We briefly show here (in Fig. 16.9) the results from two of their simulations at similar Ro but different magnetic field strengths, denoted by the Lehnert number, $Le = (B_0/\sqrt{\rho\mu})/2\Omega\delta = 0.01$ and 0.1 (in the same range as the core with $Le \sim 0.001$–0.1). In Figs. 16.9a,b, the isosurfaces of vertical velocity colored by relative helicity and normalized induced e.m.f., respectively, are shown, along with the y-averaged e.m.f. in Fig. 16.9c. The helicity segregation property of magnetic Coriolis waves is retained with increased field strength. Note, however, the off-axis radiation of waves at larger Le. The induced e.m.f. gradually becomes spatially uncorrelated with helicity as Le increases. Nevertheless, the fact that the α-effect induced by the waves remains strong and coherent for realistic values of Le lends support to the α^2-dynamo model from equation (16.14) of Davidson (2014).

16.4.2. Numerical Geodynamo Simulations

Several dynamo simulations have been conducted to study the force balance (Christensen & Aubert, 2006; Sreenivasan et al., 2014; Yadav et al., 2016; Schaeffer et al., 2017; Schwaiger et al., 2019). However, a systematic study of helicity budgets was carried out recently by Ranjan et al. (2020), where both instantaneous and time-averaged helicity sources and flux were computed for three dynamo simulations. Here we present the time-averaged results for two dynamo simulations, S1 and S2, and the hydrodynamic simulation S1H for which the helicity was shown in Fig. 16.2 and the helicity sources are shown in Fig. 16.4. Several of the observations hold for instantaneous plots as well (see Ranjan et al. [2020] for details).

These simulations are carried out with the dynamo code MagIC (Wicht, 2002), where the governing equations of mass, momentum, energy, and magnetic induction are solved for a conducting fluid in a spherical shell of width d. The dimensionless governing equations are converted to spectral space using spherical harmonics in the azimuthal and latitudinal directions and Chebyshev polynomials in the radial direction (Christensen & Wicht, 2015). Simulations S1 and S2 were initially run to achieve a statistically stationary dynamo and then run again to collect 100 or more instantaneous snapshots at short time intervals from which the time average was computed. For the latter, they were integrated for ≈ 12 and ≈ 17 eddy turnover times, respectively, for S1 and S2. A nonmagnetic solution S1H was obtained, corresponding to case S1 by switching off the magnetic field and integrating in time until the kinetic energy reached a statistically stationary state (at a value approximately four times higher than that in S1). Case S1H was integrated for another ≈ 21 eddy turnover times to take 100 or more instantaneous snapshots. Along with the control parameters, Table 16.1 shows the relevant diagnostic parameters averaged over the run time: the magnetic Reynolds number ($Rm = ud/\lambda$) and the Rossby number ($Ro = u/\Omega d$), where u is the r.m.s. velocity. The simulation output is transformed to cylindrical coordinates for ease in computation (see the appendix of Ranjan et al. [2018] for details).

Helicity sources and the flux in equation (16.5) are first computed from the instantaneous variables and then time-averaged. The sources, $\langle \overline{H_T} \rangle_\varphi$, $\langle \overline{H_B} \rangle_\varphi$, and $\langle \overline{H_\nu} \rangle_\varphi$, and the divergence of helicity flux, $\langle \nabla \cdot \overline{\boldsymbol{F}_\Omega} \rangle_\varphi$, are shown in Fig. 16.4 for runs S1 and S2 (for the nonmagnetic case S1H, all except $\langle \overline{H_B} \rangle_\varphi$ are shown). As mentioned earlier, an interesting feature in Fig. 16.4 is the segregation in $\langle \overline{H_T} \rangle_\varphi$, with a negative (positive) sign on the north (south). This correlates very well with the segregation observed in $\langle \overline{h} \rangle_\varphi$ (Figs. 16.2 and 16.3) both outside and inside the TC. As this is observed even in S1H, the correlation is likely to be independent of the presence of the magnetic field.

It is also evident in Fig. 16.4 that $\langle H_\nu \rangle_\varphi \ll \langle H_T \rangle_\varphi$, which is not unexpected as the Ekman number is small in all the cases. The observation that $\langle H_\nu \rangle_\varphi$ shows a segregation pattern approximately the opposite of $\langle H_T \rangle_\varphi$ and $\langle \overline{h} \rangle_\varphi$ may be understood in the following way. If we write the vorticity of a strongly helical flow field as $\boldsymbol{\omega} \sim \pm k\boldsymbol{u}$, where $1/k$ is a length scale, then we can write

$$H_\nu = \nu(\boldsymbol{\omega} \cdot \nabla^2 \boldsymbol{u} + \boldsymbol{u} \cdot \nabla^2 \boldsymbol{\omega}) \sim -2\nu k^2\, h\,. \qquad (16.17)$$

Therefore, on average, H_ν is likely to be oppositely correlated to h if the flow has maximal helicity. This is the case in the dynamo simulations.

The contributions from $\langle \nabla \cdot \overline{\boldsymbol{F}_\Omega} \rangle_\varphi (= -2\Omega \langle \overline{\partial \boldsymbol{u}^2/\partial z} \rangle_\varphi)$ and $\langle \overline{H_B} \rangle_\varphi$ are almost comparable to that of $\langle \overline{H_T} \rangle_\varphi$. For S1 and S2, near the inner boundary, $\langle \nabla \cdot \overline{\boldsymbol{F}_\Omega} \rangle_\varphi$ is mostly positive (negative) in the north (south), but it has the opposite sign near the outer boundary. (Also note the difference in $\langle \nabla \cdot \overline{\boldsymbol{F}_\Omega} \rangle_\varphi$ between S1 and S2 near the inner core boundary.) In the quasisteady limit, we can write $\langle \nabla \cdot \overline{\boldsymbol{F}_\Omega} \rangle_\varphi \approx \langle \overline{H_T} \rangle_\varphi + \langle \overline{H_B} \rangle_\varphi$ since $\langle H_\nu \rangle_\varphi \ll \langle \overline{H_T} \rangle_\varphi, \langle \overline{H_B} \rangle_\varphi$. Then we can infer that $\langle \nabla \cdot \overline{\boldsymbol{F}_\Omega} \rangle_\varphi$ must take the appropriate sign to approximately balance $\langle \overline{H_T} \rangle_\varphi + \langle \overline{H_B} \rangle_\varphi$. For instance, in the southern hemisphere, $\langle \nabla \cdot \overline{\boldsymbol{F}_\Omega} \rangle_\varphi$ takes a negative (positive) sign near the inner (outer) boundary to balance $\langle \overline{H_B} \rangle_\varphi$, which is negative (positive) on average. In the south and near the inner boundary, $\langle \overline{H_B} \rangle_\varphi$ and $\langle \overline{H_T} \rangle_\varphi$ act as sink and source, respectively, of positive helicity; therefore, $\langle \overline{H_B} \rangle_\varphi > \langle \overline{H_T} \rangle_\varphi$ for a negative $\langle \nabla \cdot \overline{\boldsymbol{F}_\Omega} \rangle_\varphi$. On the other hand, near the outer boundary, $\langle \overline{H_T} \rangle_\varphi > \langle \overline{H_B} \rangle_\varphi$, so that there is an overall source of positive helicity; this explains the positive sign of $\langle \nabla \cdot \overline{\boldsymbol{F}_\Omega} \rangle_\varphi$. Similar inferences can be drawn for the sources and sinks of negative helicity in the northern hemisphere. Obviously, the regions where $\langle \nabla \cdot \overline{\boldsymbol{F}_\Omega} \rangle_\varphi \approx 0$ are those where there is a local balance between the source and the sink and the helicity is transported without being deposited.

The distribution of $\langle \overline{H_B} \rangle_\varphi$ is the opposite that of $\langle \overline{H_T} \rangle_\varphi$ and, therefore, of $\langle \overline{h} \rangle_\varphi$, but it is similar to that of $\langle \overline{H_\nu} \rangle_\varphi$. Through the work done by the Lorentz force, the magnetic energy grows at the expense of kinetic energy until it saturates. This can be understood by considering the magnetic energy equation (Davidson, 2017)

$$\frac{d}{dt} \int \frac{\boldsymbol{B}^2}{2\mu} dV = -\int \boldsymbol{u} \cdot (\boldsymbol{J} \times \boldsymbol{B}) dV - \int \frac{\boldsymbol{J}^2}{\sigma} dV. \quad (16.18)$$

Therefore, $\langle \overline{\boldsymbol{u} \cdot (\boldsymbol{J} \times \boldsymbol{B})} \rangle_\varphi$ must be negative on average for a successful dynamo. This requires a positive (negative) $\langle \overline{H_{B1}} \rangle_\varphi = \langle \overline{\boldsymbol{\omega} \cdot (\boldsymbol{J} \times \boldsymbol{B})/\rho} \rangle_\varphi$ in the north (south) for a

Table 16.1 Parameters in the dynamo simulations. Here, $\nu = \kappa = 2.5\lambda$

Case	Resolution $(N_r, N_\theta, N_\varphi)$	Ra $(\times 10^8)$	Ra/Ra_c	E $(\times 10^{-5})$	Rm	Re	Ro $(\times 10^{-2})$
S1	(81,192,384)	1.2	42.4	3	921	368	1.1
S1H	(81,192,384)	1.2	42.4	3	–	759	2.2
S2	(73,256,512)	3	28.4	1	962	385	0.38

strongly helical flow, $\boldsymbol{\omega} \sim \pm \delta \boldsymbol{u}$, which is indeed observed (Fig. 16.4). A similar explanation can be provided for the other term in $\langle \overline{H}_B \rangle_\varphi$: *that is,* $\langle \overline{\boldsymbol{u} \cdot \nabla \times (\boldsymbol{J} \times \boldsymbol{B})/\rho} \rangle_\varphi$, which has nearly the same segregation pattern as $\langle \overline{H}_{B1} \rangle_\varphi$ (not shown here). Finally, in the last column of Fig. 16.4, we show the quantity $\langle \overline{\Sigma} \rangle_\varphi$

$$\langle \overline{\Sigma} \rangle_\varphi = \langle -\overline{\nabla \cdot \boldsymbol{F}_\Omega} + \overline{H}_T + \overline{H}_B + \overline{H}_\nu \rangle_\varphi, \qquad (16.19)$$

which is the sum of the previously mentioned source and flux terms in equation (16.5) with $\boldsymbol{F}$ replaced by $\boldsymbol{F}_\Omega$. In the quasisteady limit, we can say that $\langle \overline{\Sigma} \rangle_\varphi$ represents the sum of all the terms ignored in equation (16.6). For both S1 and S2, $\langle \overline{\Sigma} \rangle_\varphi$ is much weaker in magnitude than $\langle \overline{H}_T \rangle_\varphi$, $\langle \overline{\nabla \cdot \boldsymbol{F}_\Omega} \rangle_\varphi$, and $\langle \overline{H}_B \rangle_\varphi$, suggesting that $\boldsymbol{F} \approx \boldsymbol{F}_\Omega$ in the numerical dynamos. However, this may not be true for the hydrodynamic case S1H, where $\langle \overline{\Sigma} \rangle_\varphi$ is relatively larger than the corresponding plots in S1. This suggests that some other terms may need to be included in the helicity budget for S1H.

Let us now return to segregation in the helicity and the spatial similarity between $\langle \overline{H}_T \rangle_\varphi$ and $\langle \overline{h} \rangle_\varphi$. At a glance, this spatial similarity can suggest an in situ or local generation of helicity by its thermal source. However, in that scenario, the helicity equation (16.5) suggests a spatial similarity of H_T with $\partial h/\partial t$ but not with h. In other words, the net local production of negative (positive) helicity in the north (south) in a given time should be more than its destruction if it is a case of local generation. However, the plots of $\langle \overline{\Delta h/\Delta t} \rangle_\varphi$ in Ranjan et al. (2020), not shown here, indicate that there is no obvious segregation pattern in $\partial h/\partial t$ or any spatial correlation with H_T.

Let us first identify the part of $\langle \overline{H}_T \rangle_\varphi$ that correlates best with $\langle \overline{h} \rangle_\varphi$. Of $\langle \overline{H}_{T1} \rangle_\varphi$ and $\langle \overline{H}_{T2} \rangle_\varphi$, Ranjan et al. (2020) found that most of the contribution to $\langle \overline{H}_T \rangle_\varphi$ outside the TC comes from $\langle \overline{H}_{T2} \rangle_\varphi$; this, in turn, depends primarily on the term $\langle (2\alpha g_0/r_0)(\overline{-u_z \partial T/\partial \varphi}) \rangle_\varphi (= \langle \overline{H}_{T\varphi} \rangle_\varphi$ say), where we have assumed a linearly varying gravity $\mathbf{g} = -g_0\ r/r_0$, with g_0 the reference gravity at radius r_0. Figure 16.10 shows that $\langle \overline{H}_{T\varphi} \rangle_\varphi$ is well correlated with $\langle \overline{u_z \omega_z} \rangle_\varphi$ for all three simulations S1, S2, and S1H. This spatial correlation reminds us of our results in section 16.3.1, where we saw that when inertial wave packets are launched from a localized buoyant blob or the layer, the source, $-u_z \partial T/\partial x$, at the edges takes the same sign as the local helicity in the inertial waves (Fig. 16.8). This occurs because ω_z and $-\partial T/\partial x$ (or $\partial c/\partial x$ in Fig. 16.8) are positively correlated when inertial wave packets are launched from a buoyant anomaly. Ranjan et al. (2020) found a very good statistical correlation between ω_z and $-\partial T/\partial \varphi$ in dynamo simulations, indicating that the conjecture based on the model problems may indeed be valid, albeit in a statistical sense, as our simulations are turbulent rather than laminar.

16.5. A CONJECTURE FOR HELICITY SEGREGATION IN DYNAMO SIMULATIONS

Returning to our model problem, let us now discuss in more detail how $\langle H_T \rangle$ acquires the same sign as $\langle h \rangle$ in the buoyant layer (Figs. 16.7 and 16.8). We will find that the velocity field in the layer automatically adjusts the local sign of H_T so that, on average, the upper (lower) regions develop a negative (positive) sign that occurs due to the dispersion of inertial waves (Davidson & Ranjan, 2018a). Let us consider only the half-space below the mid-horizontal plane (i.e., $z < 0$) where $h > 0$. Taking our inviscid, nonmagnetic, helicity equation (16.5),

$$\frac{\partial h}{\partial t} + \nabla \cdot \boldsymbol{F} = H_T, \qquad (16.20)$$

let us divide the domain into three regions: V_{source}, in the lower part of the buoyant layer; V_{front}, which includes the downward advancing wave front; and V_{flux}, between them (Fig. 16.7b). Recall that we are dealing with low-frequency waves, which are responsible for column formation, so the time dependence is significant only in V_{front}.

In this model problem, the helicity flux $\boldsymbol{F}$ outside the source is only due to the propagation of inertial wave packets. Therefore, the helicity equation in these three regions takes the form

$$V_{\text{front}}:\ \partial h/\partial t \approx -\nabla \cdot \boldsymbol{F}; \qquad (16.21)$$

$$V_{\text{flux}}:\ \nabla \cdot \boldsymbol{F} \approx 0; \qquad (16.22)$$

$$V_{\text{source}}:\ \nabla \cdot \boldsymbol{F} \approx H_T \qquad (16.23)$$

Equation (16.21) says that the downward advancing front, which has $h > 0$, needs a flux of positive helicity entering this region, so $\partial h/\partial t > 0$ (see Fig. 16.7b,c). From equation (16.22), this helicity flux must arise from the source region above to supply the front, so in V_{flux}, we have $\nabla \cdot \boldsymbol{F} \approx 0$. Finally, equation (16.23) tells us that this downward flux of positive helicity must come from the integral of H_T over the lower half of the buoyant layer. In short, the integral of H_T over $z < 0$ must be positive to supply the positive helicity (here, the flux leaves the source) to the downward advancing front below. Of course, a similar argument applied to $z > 0$ tells us that the integral of H_T over the upper half of the blob must be negative. This is exactly what we observe (see Fig. 16.8). Figure 16.11 shows the evolution of horizontally averaged h, $\partial h/\partial t$, and H_T over a range of times, $\Omega t = 4 \rightarrow 20$. Note that the magnitude of the helicity above and below the layer eventually saturates, leading to $\nabla \cdot \boldsymbol{F} \approx 0$. Finally, within the buoyant layer, while h correlates well with H_T, there is no such correlation between $\partial h/\partial t$ and H_T at large times. Most of these results apply to both the Ro_δ cases, except that the buoyancy field declines with time in the higher Ro_δ case. It is interesting to note concerning the V_{source} in equation (16.23) that $\partial h/\partial z$ and H_T change

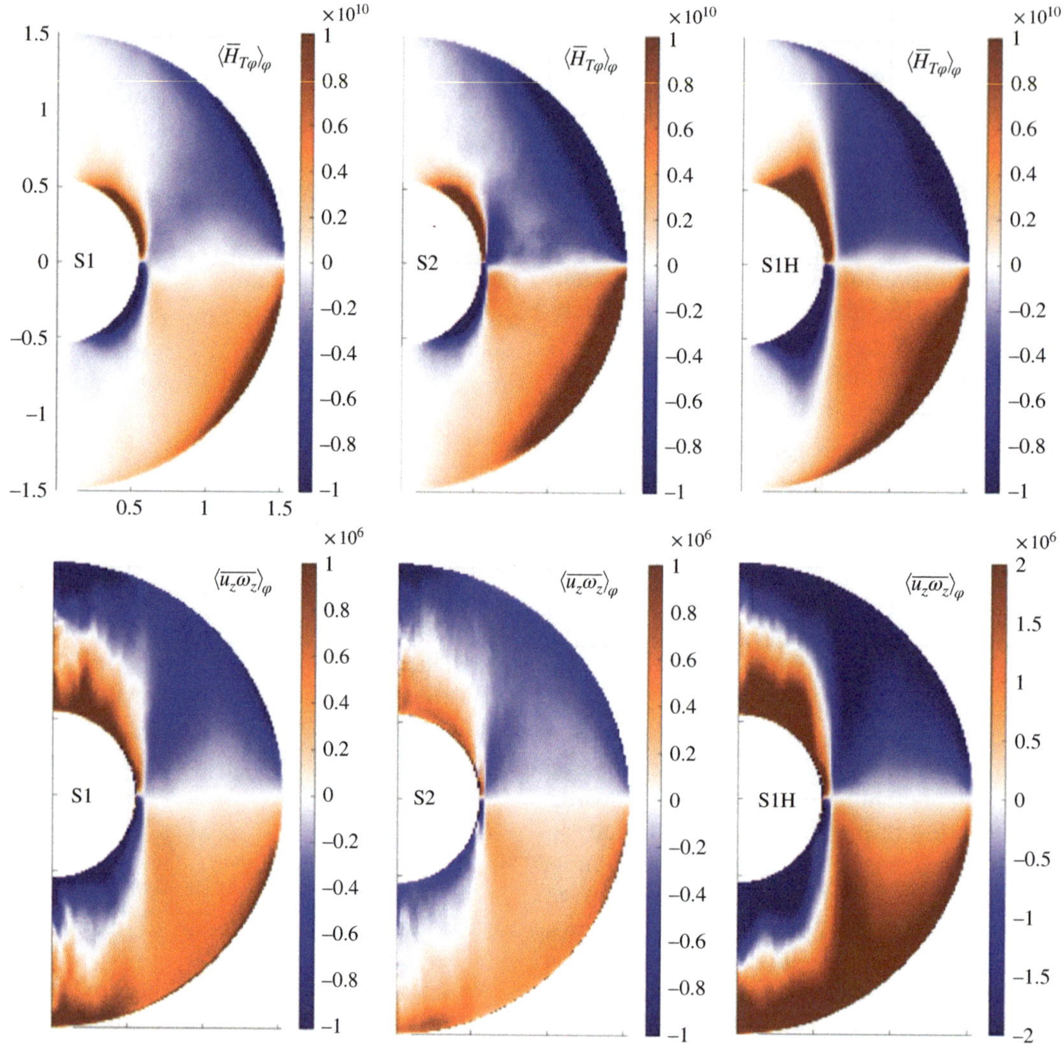

Figure 16.10 $\langle \overline{H}_{T\varphi} \rangle_\varphi$ (top row) and $\langle \overline{u_z \omega_z} \rangle_\varphi$ (bottom row) for the simulations S1, S2, and S1H (Ranjan et al., 2020).

significantly due to the changing buoyancy field, but H_T retains the same sign.

In dynamo simulations, fluctuations in density are present everywhere, even though we observe a statistical bias in the concentration of buoyant anomalies toward the equatorial regions (Fig. 16.6). Therefore, there will be cancellation in helicity, so the helicity segregation mechanism proposed here is effective only if the buoyancy sources are statistically more localized near the equatorial regions (Davidson & Ranjan, 2023). Moreover, by integrating the helicity equation (16.20) in a region above and below the equator, it becomes evident that relatively large fluctuating kinetic energy in the equatorial regions favors the helicity segregation shown in Figs. 16.2 and 16.3 (Davidson & Ranjan, 2018a). If there is more wave activity near the equator than the mantle, waves will tend to disperse away from the equatorial regions, carrying with them helicity of the observed signs. Moreover, the mechanism described here relies on continual dispersion of helical waves from the equatorial regions and assumes that the waves dissipate before they can reach the mantle through Ohmic dissipation.

16.6. CONCLUSIONS AND FUTURE SCOPE

In this chapter, we have discussed an important observation in numerical dynamo simulations that has remained unexplained. On average, the helicity in most simulations is negative (positive) in the north (south) in regions outside the tangent cylinder (Figs. 16.2 and 16.3). We obtained the various helicity sources in the helicity evolution equation arising from the buoyancy (H_T), Lorentz (H_B), and viscous (H_ν) forces and the divergence of helicity flux ($\nabla \cdot \boldsymbol{F}$) in the numerical dynamos. Another

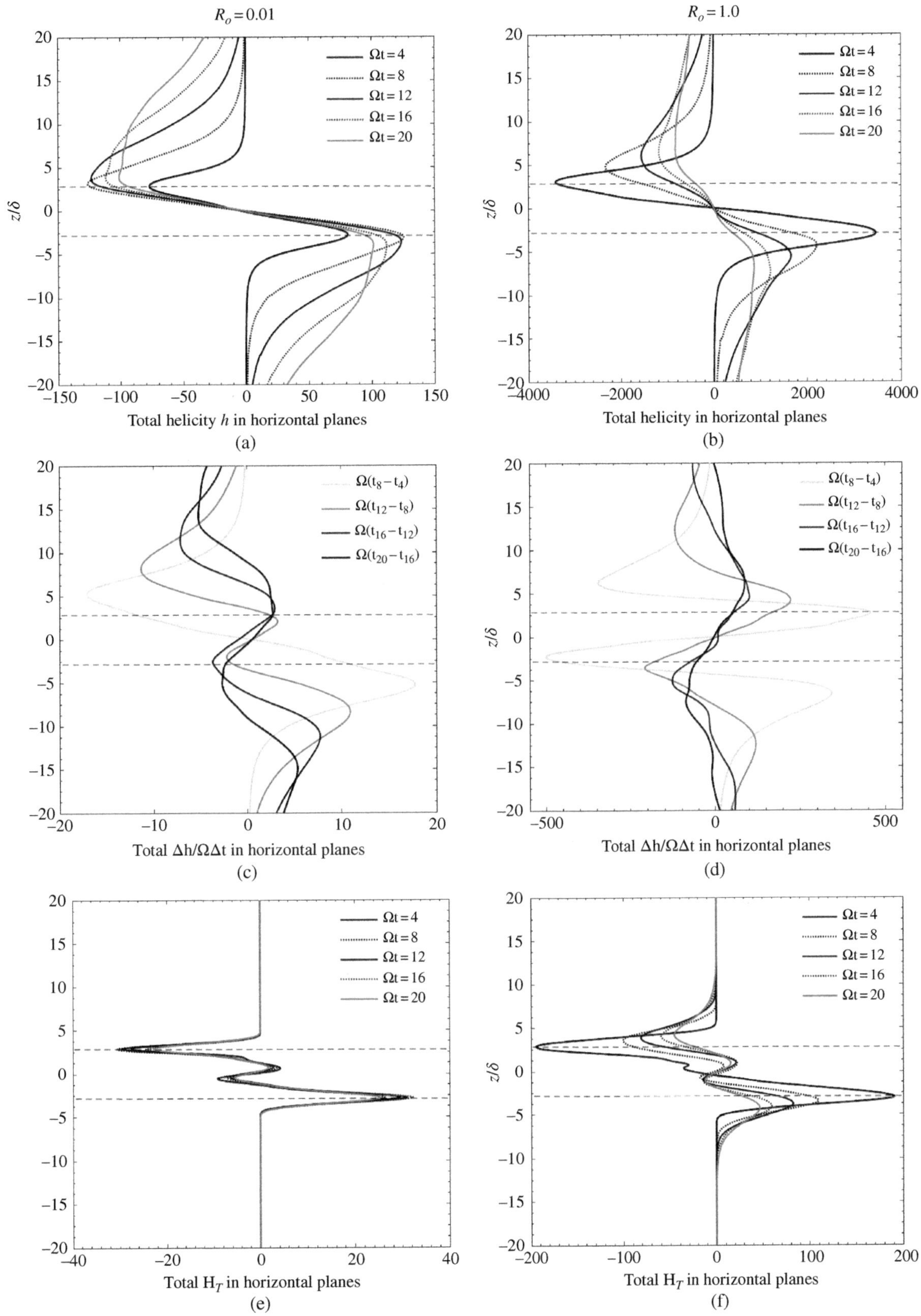

Figure 16.11 The evolution of the horizontally averaged values of (a, b) h, (c,d) $\partial h/\partial t$, (e, f) H_T for (a, c, e) 0.01, (b, d, f) $Ro_\delta = 1$ during $\Omega t = 4 \rightarrow 20$. Dashed horizontal lines show the initial location of the layer. Ranjan & Davidson, 2018a/Cambridge University Press.

important observation in the dynamo simulations is the spatial similarity between the helicity, $\overline{h}$, and its thermal source, $\overline{H_T}$, which also shows the same segregation pattern north and south of the equator (Fig. 16.4). We find that the major contribution to $\overline{H_T}$ comes from $\overline{-u_z \partial T/\partial\varphi}$, which is correlated with $\overline{u_z\omega_z}$. The segregation pattern in $\overline{-u_z \partial T/\partial\varphi}$ is thus a result of the fact that $-\partial T/\partial\varphi$ is positively correlated with ω_z. Davidson and Ranjan (2018a) have shown that such a spatial correlation exists when inertial waves are launched from localized buoyant anomalies so that H_T takes the same sign as the local helicity (Davidson & Ranjan, 2018a). Earlier, inertial waves were identified in the same set of dynamo simulations (Ranjan et al., 2018). It is known that helical waves, such as weakly modified inertial waves or magnetic-Coriolis waves, are essential in maintaining columnar flow. We conjecture, therefore, that the helicity segregation in the dynamo simulations arises from the presence of helical waves in the interior regions, particularly at higher *Ra* and smaller *E*.

The dynamics in the numerical dynamos is much more complex than the simpler configuration in the model problems considered by Davidson and Ranjan (2018a) due to the presence of spherical boundaries and magnetic fields. For instance, the helicity flux, $\boldsymbol{F}$, in the model problems is clearly due to wave propagation where $\nabla \cdot \boldsymbol{F}$ and H_T have the same sign inside the source. This is not as clear in the dynamo simulations, where close to the inner boundary, $\langle\overline{\nabla \cdot \boldsymbol{F}}\rangle_\varphi$ has the opposite sign of $\langle\overline{H_T}\rangle_\varphi$ (but the same sign as $\langle\overline{H_B}\rangle_\varphi$) and $\langle\overline{\nabla \cdot \boldsymbol{F}}\rangle_\varphi$ changes sign as we move toward the outer boundary (Fig. 16.4). Further studies are needed to understand helicity flux in dynamos and the complex interplay of helical waves with both buoyancy and magnetic fields.

Because dynamo simulations are turbulent, they contain a wide range of length scales, and it is not well understood at which scale helicity generation occurs. In their simulations, Sreenivasan and Kar (2018) found that the magnetic field leads to helicity generation in larger length scales and depletion in smaller scales relative to nonmagnetic convection. Earlier, Sreenivasan and Jones (2011) predicted that a strong magnetic field should lead to additional negative (positive) helicity in the north (south). However, in our results, $\overline{H_B}$ has a clear sign opposite that of $\overline{h}$, indicating that $\overline{H_B}$ counteracts further generation of helicity (observed even in instantaneous results; see Ranjan et al. [2020]). The precise role of magnetic field in helicity generation at various scales and parameter regimes requires further investigations.

In most older dynamo simulations, with modest parameter regimes, helicity generation was due to Ekman pumping and/or buoyancy, probably depending on the value of $Ra.E^2$. Ekman pumping is still the primary mechanism near the boundaries even in some present-day simulations and is likely to remain so in the near future due to the prohibitive computational cost that rises exponentially for a small increase in *Ra* and decrease in *E*. It is therefore useful to understand how helicity is produced in the Ekman layers (Deusebio & Lindborg, 2014; Ojha et al., 2022) and influenced by the presence of helical waves, buoyancy, and magnetic fields. Model problems such as those used in the present studies can be designed to help isolate specific phenomena and understand the effects on helicity generation and segregation. The flow near the core–mantle boundary can be computed using the observed geomagnetic field data and assuming frozen magnetic flux (Amit & Olson, 2004; Holme et al., 2015). Since only the horizontal flow can be obtained, it is not possible to compute the helicity or evaluate its segregation pattern without any assumptions, but this may become possible in the future.

ACKNOWLEDGMENTS

The authors thank Prof. U. R. Christensen and Dr. Johannes Wicht (MPI, Gottingen) for sharing the MagIC code simulation data set, Prof. P. K. Yeung (Georgia Tech.) for sharing the DNS code, and the Leverhulme Trust for sponsoring the research.

REFERENCES

Amit, H., & Olson, P. (2004). Helical core flow from geomagnetic secular variation. *Physics of the Earth and Planetary Interiors*, *147*, 1–25.

Bardsley, O. P., & Davidson, P. A. (2016). Inertial Alfvén waves as columnar helices in planetary cores. *Journal of Fluid Mechanics Rapids*, *805*, R2.

Bardsley, O. P., & Davidson, P. A. (2017). The dispersion of magnetic-Coriolis waves in planetary cores. *Geophysical Journal International*, *210*, 18–26.

Buffett, B., Knezek, N., & Holme, R. (2016). Evidence for MAC waves at the top of Earth's core and implications for variations in length of day. *Geophysical Journal International*, *204*(3), 1789–1800.

Busse, F. (1975). A model of the geodynamo. *Geophysical Journal Royal Astronomical Society*, *42*, 437–459.

Busse, F. (1976). Generation of planetary magnetism by convection. *Physics of the Earth and Planetary Interiors*, *12*, 350–358.

Busse, F. (2002). Convective flows in rapidly rotating spheres and their dynamo action. *Physics of Fluids*, *14*, 1301–1314.

Christensen, U. R., & Aubert, J. (2006). Scaling properties of convection-driven dynamos in rotating spherical shells and application to planetary magnetic fields. *Geophysical Journal International*, *166*(1), 97–114.

Christensen, U. R., & Wicht, J. (2015). Numerical dynamo simulations. In G. Schubert (Ed.), *Treatise on Geophysics* (2nd ed., Vol. *8*, pp. 245–277). Elsevier.

Chulliat, A., Shimizu, H., & Loper, D. E. (2003). Buoyancy-driven perturbations in a rapidly rotating, electrically conducting fluid: part II – dynamo action. *Geophysical & Astrophysical Fluid Dynamics*, *97*(6), 471–487.

Davidson, P. A., Staplehurst, P. J., & Dalziel, S. B. (2006). On the evolution of eddies in a rapidly rotating system. *Journal of Fluid Mechanics*, *557*, 135–144.

Davidson, P. A. (2013). *Turbulence in rotating, stratified and electrically conducting fluids*. Cambridge University Press.

Davidson, P. A. (2014). The dynamics and scaling laws of planetary dynamos driven by inertial waves. *Geophysical Journal International*, *198*(3), 1832–1847.

Davidson, P. A., & Ranjan, A. (2015). Planetary dynamos driven by helical waves – II. *Geophysical Journal International*, *202*, 1646–1662.

Davidson, P. A. (2016). Dynamos driven by helical waves: scaling laws for numerical dynamos and for the planets. *Geophysical Journal International*, *207*(2), 680–690.

Davidson, P. A. (2017). *Introduction to Magnetohydrodynamics* (2nd ed.). Cambridge University Press.

Davidson, P. A., & Ranjan, A. (2018a). On the spatial segregation of helicity by inertial waves in dynamo simulations and planetary cores. *Journal of Fluid Mechanics*, *851*, 268–287.

Davidson, P. A., & Ranjan, A. (2018b). Are planetary dynamos driven by helical waves? *Journal of Plasma Physics*, *84*, 735840304.

Davidson, P. A., & Ranjan, A. (2023). The connection between the equatorial temperature bias and north–south helicity segregation in numerical simulations of the geodynamo *Geophysical Journal International*, *233*, 2253–2267.

Deusebio, E., & Lindborg, 2014. Helicity in the Ekman boundary layer. *Journal of Fluid Mechanics*, *755*, 654–671.

Duarte, L. D., Wicht, J., Browning, M. K., & Gastine, T. (2015). Helicity inversion in spherical convection as a means for equatorward dynamo wave propagation. *Monthly Notices of the Royal Astronomical Society*, *456*(2), 1708–1722.

Giesecke, A., Ziegler, U., & Rüdiger, G. (2005). Geodynamo α-effect derived from box simulations of rotating magnetoconvection. *Physics of the Earth and Planetary Interiors*, *152*(1–2), 90–102.

Glatzmaier, G. A., & Olson, P. (1993). Highly supercritical thermal convection in a rotating spherical shell: centrifugal vs. radial gravity. *Geophysical & Astrophysical Fluid Dynamics*, *70*(1–4), 113–136.

Glatzmaier, G. A., & Roberts, P. H. (1995). A three-dimensional convective dynamo solution with rotating and finitely conducting inner core and mantle. *Physics of the Earth and Planetary Interiors*, *91*(1–3), 63–75.

Grooms, I., Julien, K., Weiss, J. B., & Knobloch, E. (2010). Model of convective Taylor columns in rotating Rayleigh-Bénard convection. *Physical Review Letters*, *104*(22), 224501.

Holme, R., Olson, P., & Schubert, G. (2015). Large-scale flow in the core. In G. Schubert (Ed.), *Treatise on Geophysics* (2nd ed., Vol. *8*, pp. 107–130). Elsevier.

Ishihara, N., & Kida, S. (2002). Dynamo mechanism in a rotating spherical shell: competition between magnetic field and convection vortices. *Journal of Fluid Mechanics*, *465*, 1–32.

Kageyama, A., Watanabe, K., & Sato, T. (1993). Simulation study of a magnetohydrodynamic dynamo: convection in a rotating spherical shell. *Physics of Fluids B*, *5*(8), 2793–2805.

Kageyama, A., & Sato, T. (1997). Generation mechanism of a dipole field by a magnetohydrodynamic dynamo. *Physical Review E*, *55*(4), 4617–4626.

Kitauchi, H., Araki, K., & Kida, S. (1997). Flow structure of thermal convection in a rotating spherical shell. *Nonlinearity*, *10*(4), 885–904.

Kono, M., Sakuraba, A., & Ishida, M. (2000). Dynamo simulation and palaeosecular variation models. *Philosophical Transactions of the Royal Society of London A*, *358*, 1123–1139.

Krause, F., & Rädler, K. H. (1980). *Mean-field electrodynamics and dynamo theory*. Berlin, Germany: Akademie Verlag.

Loper, D. E., Chulliat, A., & Shimizu, H. (2003). Buoyancy-driven perturbations in a rapidly rotating, electrically conducting fluid: part I – flow and magnetic field. *Geophysical & Astrophysical Fluid Dynamics*, *97*(6), 429–469.

Helmholtz, H. V. (1867). On Integrals of the hydrodynamical equations, which express vortex-motion. *Philosophical Magazine*, *33*, 485–512.

Hide, R. (1975). A short note on helicity. *Geophysical and Fluid Dynamics*, *7*(1), 157–161.

McDermott, B., & Davidson, P. A. (2020). On the helicity characteristics and induced emf of magnetic-Coriolis wave packets. *Geophysical Journal International*, *223*, 1398–1411.

Moffatt, H. K. (1969). The degree of knottedness of tangled vortex lines. *Journal of Fluid Mechanics*, *35*(1), 117–129.

Moffatt, H. K. (1970). Dynamo action associated with random inertial waves in a rotating conducting fluid. *Journal of Fluid Mechanics*, *44*(4), 705–719.

Moffatt, H. K. (1978). *Magnetic field generation in electrically conducting fluids*. Cambridge University Press.

Moffatt, H. K., & Dormy, E. (2019). *Self-exciting fluid dynamos*. Cambridge University Press.

Moffatt, H. K. (2016). Helicity and celestial magnetism. *Proceedings of the Royal Society of London, Series A*, *472*, 20160183.

Olson, P. (1981). A simple physical model for the terrestrial dynamo. *Journal of Geophysical Research: Solid Earth*, *86*(B11), 10875–10882.

Olson, P., Christensen, U. R., & Glatzmaier, G. A. (1999). Numerical modelling of the geodynamo: mechanisms of field generation and equilibration. *Journal of Geophysical Research*, *104*(B5), 10383–10404.

Ojha, A., Anas, M., Ranjan, A., Joshi, P., & Verma, M. K. (2022). Helicity segregation by Ekman pumping in laminar rotating flows with gravity orthogonal to rotation. *Physical Review Fluids*, *7*, 034801.

Parker, E. N. (1955). Hydromagnetic dynamo models. *The Astrophysical Journal*, *122*, 293–314.

Ranjan, A., & Davidson, P. A. (2014). Evolution of a turbulent cloud under rotation. *Journal of Fluid Mechanics*, *756*, 488–509.

Ranjan, A. (2015). Inertial waves in rotating turbulence (PhD thesis). University of Cambridge.

Ranjan, A. (2017). Segregation of helicity in inertial wave packets. *Physical Review Fluids*, *2*(3), 033801.

Ranjan, A., Davidson, P. A., Christensen, U. R., & Wicht, J. (2018). Internally-driven inertial waves in geodynamo simulations. *Geophysical Journal International*, *213*, 1281–1295.

Ranjan, A., Davidson, P. A., Christensen, U. R., & Wicht, J. (2020). On the generation and segregation of helicity in geodynamo simulations. *Geophysical Journal International*, *221*, 741–757.

Ranjan, A., & Davidson, P. A. (2021). Columnar heat transport via advection induced by inertial waves. *International Journal of Heat and Fluid Flow*, *87*.

Roberts, P. H., & King, E. M. (2013). On the genesis of the Earth's magnetism. *Reports on Progress in Physics*, *76*(9), 096801.

Reshetnyak, M. Y. (2006). Hydrodynamic helicity in Boussinesq-type models of the geodynamo. *Izvestiya, Physics of the Solid Earth*, *42*, 449.

Shimizu, H., & Loper, D. E. (2000). Small-scale helicity and the α-effect in the Earth's core. *Physics of the Earth and Planetary Interiors*, *121*, 139–155.

Schaeffer, N., Jault, D., Nataf, H.-C., & Fournier, A. (2017). Turbulent geodynamo simulations: a leap towards Earth's core. *Geophysical Journal International*, *211*, 1–29.

Schwaiger, T., Gastine, T., & Aubert, J. (2019). Force balance in numerical geodynamo simulations: a systematic study. *Geophysical Journal International*, *219*, S101–S114.

Sreenivasan, B., & Jones, C. A. (2011). Helicity generation and subcritical behaviour in rapidly rotating dynamos. *Journal of Fluid Mechanics*, *688*, 5–30.

Sreenivasan, B., Sahoo, S., & Dhama, G. (2014). The role of buoyancy in polarity reversals of the geodynamo. *Geophysical Journal International*, *199*, 1698–1708.

Sreenivasan, B., & Kar, S. (2018). Scale dependence of kinetic helicity and selection of the axial dipole in rapidly rotating dynamos. *Physical Review Fluids*, *3*, 093801.

Stellmach, S., & Hansen, U. (2004). Cartesian convection driven dynamos at low Ekman number. *Physical Review E*, *70*(5), 056312.

Thomson, W. (Lord Kelvin). (1867). On vortex atoms. *Philosophical Magazine*, *34*, 15–24.

Veronis, G. (1959). Cellular convection with finite amplitude in a rotating fluid. *Journal of Fluid Mechanics*, *5*(3), 401–435.

Wicht, J. (2002). Inner-core conductivity in numerical dynamo simulations. *Physics of the Earth and Planetary Interiors*, *132*(4), 281–302.

Yadav, R. K., Gastine, T., & Christensen, U. R. (2013). Scaling laws in spherical shell dynamos with free-slip boundaries. *Icarus*, *225*, 184–193.

Yadav, R. K., Gastine, T., Christensen, U. R., Wolk, S. J., & Poppenhaeger, K. (2016). Approaching a realistic force balance in geodynamo simulations. *Proceedings of the National Academy of Sciences*, *113*(43), 12065–12070.

Yeung, P. K., & Zhou, Y. (1998). Numerical study of rotating turbulence with external forcing. *Physics of Fluids*, *10*, 2895.

INDEX

Helicities in Geophysics, Astrophysics, and Beyond, Geophysical Monograph 283, First Edition.
Edited by Kirill Kuzanyan, Nobumitsu Yokoi, Manolis K. Georgoulis, and Rodion Stepanov.

DOI:10.1002/9781119841715.bindex

Printed in the USA/Agawam, MA
September 16, 2026

910661.211